Advances in Intelligent Systems and Computing

Volume 712

Series editor

Janusz Kacprzyk, Polish Academy of Sciences, Warsaw, Poland
e-mail: kacprzyk@ibspan.waw.pl

The series "Advances in Intelligent Systems and Computing" contains publications on theory, applications, and design methods of Intelligent Systems and Intelligent Computing. Virtually all disciplines such as engineering, natural sciences, computer and information science, ICT, economics, business, e-commerce, environment, healthcare, life science are covered. The list of topics spans all the areas of modern intelligent systems and computing such as: computational intelligence, soft computing including neural networks, fuzzy systems, evolutionary computing and the fusion of these paradigms, social intelligence, ambient intelligence, computational neuroscience, artificial life, virtual worlds and society, cognitive science and systems, Perception and Vision, DNA and immune based systems, self-organizing and adaptive systems, e-Learning and teaching, human-centered and human-centric computing, recommender systems, intelligent control, robotics and mechatronics including human-machine teaming, knowledge-based paradigms, learning paradigms, machine ethics, intelligent data analysis, knowledge management, intelligent agents, intelligent decision making and support, intelligent network security, trust management, interactive entertainment, Web intelligence and multimedia.

The publications within "Advances in Intelligent Systems and Computing" are primarily proceedings of important conferences, symposia and congresses. They cover significant recent developments in the field, both of a foundational and applicable character. An important characteristic feature of the series is the short publication time and world-wide distribution. This permits a rapid and broad dissemination of research results.

More information about this series at http://www.springer.com/series/11156

Vikrant Bhateja · João Manuel R. S. Tavares
B. Padmaja Rani · V. Kamakshi Prasad
K. Srujan Raju
Editors

Proceedings of the Second International Conference on Computational Intelligence and Informatics

ICCII-2017

 Springer

Editors
Vikrant Bhateja
Department of Electronics
and Communication Engineering
Shri Ramswaroop Memorial Group
of Professional Colleges
Lucknow, Uttar Pradesh
India

V. Kamakshi Prasad
Department of Computer Science
and Engineering
JNTUH College of Engineering Hyderabad
(Autonomous)
Hyderabad, Telangana
India

João Manuel R. S. Tavares
Departamento de Engenharia Mecânica
Universidade do Porto
Porto
Portugal

K. Srujan Raju
CMR Technical Campus
Hyderabad, Telangana
India

B. Padmaja Rani
Department of Computer Science
and Engineering
JNTUH College of Engineering Hyderabad
(Autonomous)
Hyderabad, Telangana
India

ISSN 2194-5357   ISSN 2194-5365   (electronic)
Advances in Intelligent Systems and Computing
ISBN 978-981-10-8227-6   ISBN 978-981-10-8228-3   (eBook)
https://doi.org/10.1007/978-981-10-8228-3

Library of Congress Control Number: 2018931430

Printed on acid-free paper

This Springer imprint is published by the registered company Springer Nature Singapore Pte Ltd.
The registered company address is: 152 Beach Road, #21-01/04 Gateway East, Singapore 189721, Singapore

Preface

The Second International Conference on Computational Intelligence and Informatics (ICCII-2017) was hosted by the Department of Computer Science and Engineering, JNTUHCEH, Hyderabad, in association with CSI during September 25–27, 2017. It provided a great platform for researchers from across the world to report, deliberate, and review the latest progress in the cutting-edge research pertaining to computational intelligence and its applications to various engineering fields.

The response to ICCII-2017 was overwhelming with a good number of submissions from different areas relating to computational intelligence and its applications in main tracks. After a rigorous peer review process with the help of program committee members and external reviewers, 66 papers were accepted for publication in this volume of AISC series of Springer.

ICCII-2017 was organized in the honor of Prof. K. Venkateswara Rao's retirement, commemorating his services rendered in the JNTUHCEH University in general and the Department of Computer Science and Engineering, JNTUHCEH, Hyderabad, in particular.

For the benefit of authors and delegates of the conference, a preconference workshop was held on emerging technologies in the field of computer science on Information Security, Machine Learning, and Internet of Things. Professor Suresh Chandra Satapathy, PVPSIT, Vijayawada, Prof. L. Pratap Reddy, JNTUHCEH, Mr. Kiran Chandra, FSMI, Prof. C. Raghavendra Rao, HCU, Prof. R. Padmavathi, NIT Warangal, have delivered lectures that were very informative.

Dr. Hock Ann Goh from MMU, Malaysia, and Prof. Ravi from IDRBT have delivered keynote speeches on September 26–27, 2017.

We take this opportunity to thank all the speakers and session chairs for their excellent support in making ICCII-2017 a grand success. The quality of a refereed volume depends mainly on the expertise and dedication of the reviewers. We are indebted to the program committee members and external reviewers who not only produced excellent reviews but also reviewed in the stipulated period of time without any delay. We would also like to thank CSI, Hyderabad, for coming forward to support us in organizing this mega-convention. Our thanks are due to

Dr. Ramakrishna Murthy, ANITS, Vizag, for his valuable support in reviews. Also, we thank special session chairs Dr. Tanupriya and Dr. Praveen of Amity University, Noida.

We express our heartfelt thanks to our Chief Patrons, Prof. A. Venugopal Reddy, Vice Chancellor, JNTUHCEH, Prof. N. V. Ramana Rao, Rector, JNTUHCEH, Prof. N. Yadaiah, Registrar, JNTUHCEH, Prof. A. Govardhan, Principal, JNTUHCEH, faculty and administrative staff of JNTUHCEH for their continuous support during the course of the convention.

We would also like to thank the authors and participants of this convention, who have considered the convention above all hardships. Finally, we would like to thank all the volunteers who spent tireless efforts in meeting the deadlines and arranging every detail to make sure that the convention ran smoothly. All the efforts are worth and would please us all, if the readers of this proceedings and participants of this convention found the papers and event inspiring and enjoyable. Our sincere thanks to the press, print, and electronic media for their excellent coverage of this convention.

Lucknow, India   Vikrant Bhateja
Porto, Portugal   João Manuel R. S. Tavares
Hyderabad, India   B. Padmaja Rani
Hyderabad, India   V. Kamakshi Prasad
Hyderabad, India   K. Srujan Raju

Organizing Committee

Conference Chair

Dr. K. Venkateswara Rao, JNTUHCEH, Telangana, India

Organizing Chairs

Prof. B. Padmaja Rani, JNTUHCEH, Telangana, India
Prof. V. Kamakshi Prasad, JNTUHCEH, Telangana, India

Technical Chairs

Prof. L. Pratap Reddy, JNTUHCEH, Telangana, India
Prof. G. Vijaya Kumari, JNTUHCEH, Telangana, India
Prof. O. B. V. Ramanaiah, JNTUHCEH, Telangana, India

Technical Co-Chairs

Prof. R. Sridevi, JNTUHCEH, Telangana, India
Prof. M. Chandra Mohan, JNTUHCEH, Telangana, India
Prof. D. Vasumathi, JNTUHCEH, Telangana, India
Prof. S. Viswanadha Raju, CSE, JNTUHCEJ, Telangana, India
Prof. B. Vishnu Vardhan, CSE, JNTUHCEJ, Telangana, India

International Advisory Committee

Dr. Kun-lin Hsieh, NTU, Taiwan
Dr. Ahamad J. Rusumdar, KIT, Germany
Dr. V. R. Chirumamilla, EUT, Netherland
Dr. Halis Altun, MU, Turkey
Dr. P. N. Suganthan, Singapore
Dr. Boka Kumsa, Ethiopia

Advisory Committee

Dr. B. L. Deekshatulu, IDRBT, Telangana, India
Prof. Y. N. Narasimhulu, VC, Rayalseema University, Andhra Pradesh, India
Prof. A. Damodaram, VC, SVU, Uttar Pradesh, India
Prof. L. S. S. Reddy, VC, KLU, Andhra Pradesh, India
Prof. Allam Appa Rao, CRR Institute of Technology, Delhi, India
Prof. Hema A. Murthy, IIT Madras, Chennai, Tamil Nadu, India
Dr. P. Sateesh Kumar, IIT Roorkee, Uttarakhand, India
Prof. N. Somayajulu, NIT Warangal, Telangana, India
Prof. Arun Agarwal, HCU, Telangana, India
Prof. Atul Negi, HCU, Telangana, India
Prof. C. Raghavendra Rao, HCU, Telangana, India
Prof. V. N. Sastry, IDRBT, Hyderabad, Telangana, India
Mr. Gautam Mahapatra, DRDO, New Delhi, India
Prof. A. Ananda Rao, JNTUA, Andhra Pradesh, India
Prof. S. Satyanarayana, JNTUK, Andhra Pradesh, India
Dr. A. Ansari, JMIU, Delhi, India
Prof. M. Srinivasa Rao, SIT, JNTUHCEH, Telangana, India
Prof. S. Durga Bhavani, SIT, JNTUHCEH, Telangana, India
Prof. S. V. L. Narasimham, SIT, JNTUHCEH, Telangana, India
Prof. M. Madhavi Latha, JNTUHCEH, Telangana, India
Prof. D. Srinivasa Rao, JNTUHCEH, Telangana, India
Prof. M. Asha Rani, JNTUHCEH, Telangana, India
Prof. B. N. Bhandari, JNTUHCEH, Telangana, India

Publication Chair

Dr. S. C. Satapathy, PVPSIT, Andhra Pradesh, India

Editorial Board

Prof. V. Kamakshi Prasad, JNTUHCEH, Telangana, India
Prof. B. Padmaja Rani, JNTUHCEH, Telangana, India
Dr. João Manuel R. S. Tavares, FEUP, Portugal
Prof. Vikrant Bhateja, SRMGPC, Lucknow, Uttar Pradesh, India
Prof. K. Srujan Raju, CMR Technical Campus, Telangana, India

Organizing Committee

Mr. B. Rama Mohan, JNTUHCEH, Telangana, India
Ms. K. Neeraja, JNTUHCEH, Telangana, India
Dr. M. Nagarathna, JNTUHCEH, Telangana, India
Ms. J. Ujwala Rekha, JNTUHCEH, Telangana, India
Dr. K. P. Supreethi, JNTUHCEH, Telangana, India
Ms. Kavitha, JNTUHCEH, Telangana, India
Ms. Hemalatha, JNTUHCEH, Telangana, India
Ms. I. Lakshmi Manikyamba, JNTUHCEH, Telangana, India

Technical Committee

Prof. P. V. S. Srinivas Rao, JBIET, Telangana, India
Prof. T. V. Rajinikanth, SNIST, Telangana, India
Prof. Khaleel Ur Rahman Khan, ACE, Telangana, India
Prof. C. R. K. Reddy, CBIT, Telangana, India
Prof. A. Nagesh, MGIT, Telangana, India
Prof. A. Jagan, BVRIT, Telangana, India
Dr. G. Narasimha, JNTUHCEJ, Telangana, India
Dr. P. Sammulal, JNTUHCEJ, Telangana, India
Dr. K. Shahu Chatrapathi, JNTUHCEM, Telangana, India
Dr. Kranthi Kiran, JNTUHCEJ, Telangana, India
Dr. T. Venu Gopal, JNTUHCES, Telangana, India
Dr. K. Rama Krishna, CMRCET, Telangana, India

Web and Publicity Committee

Ms. B. Kezia Rani, JNTUHCEH, Telangana, India
Ms. P. Subhashini, JNTUHCEH, Telangana, India
Mr. R. Sarath Babu, JNTUHCEH, Telangana, India

Contents

About the Editors

Prof. Vikrant Bhateja is Associate Professor, Department of Electronics and Communication Engineering, Shri Ramswaroop Memorial Group of Professional Colleges (SRMGPC), Lucknow, and also the Head (Academics and Quality Control) in the same college. His areas of research include digital image and video processing, computer vision, medical imaging, machine learning, pattern analysis and recognition, neural networks, soft computing, and bio-inspired computing techniques. He has more than 90 quality publications in various international journals and conference proceedings. He has been on TPC and chaired various sessions from the above domains in international conferences of IEEE and Springer. He has been the track chair and served in the core technical/editorial teams for international conferences: FICTA 2014, CSI 2014, and INDIA 2015 under Springer AISC series and INDIACom-2015, ICACCI-2015 under IEEE. He is Associate Editor in *International Journal of Convergence Computing (IJConvC)* and also serving in the editorial board of *International Journal of Image Mining (IJIM)* under Inderscience Publishers. At present, he is Guest Editor for two special issues floated in *International Journal of Rough Sets and Data Analysis (IJRSDA)* and *International Journal of System Dynamics Applications (IJSDA)* under IGI Global Publications.

João Manuel R. S. Tavares graduated in Mechanical Engineering from the University of Porto, Portugal (1992); M.Sc. in Electrical and Computer Engineering, in the field of Industrial Informatics, University of Porto (1995); Ph.D. in Electrical and Computer Engineering, University of Porto (2001). From 1995 to 2000, he was a researcher at the Institute of Biomedical Engineering (INEB). He is co-author of more than 350 scientific papers in national and international journals and conferences, co-editor of 18 international books and guest editor of several special issues of international journals. In addition, he is Editor-in-Chief of the *Computer Methods in Biomechanics and Biomedical Engineering: Imaging & Visualization (CMBBE: Imaging & Visualization)*; Editor-in-Chief of the *International Journal of Biometrics and Bioinformatics (IJBB)*; Co-Editor-in-Chief of the *International Journal for Computational Vision and Biomechanics (IJCV & B)*;

Co-Editor of the Lecture Notes in *Computational Vision and Biomechanics (LNCV & B)*; Associate Editor of the EURASIP *Journal on Advances in Signal Processing (JASP)*, *Journal of Engineering, ISRN Machine Vision, Advances in Biomechanics & Applications*, and of the *Journal of Computer Science (INFOCOMP)*, and reviewer of several international scientific journals. Since 2001, he has been Supervisor and Co-Supervisor of several M.Sc. and Ph.D. theses and involved in several research projects, both as researcher and as scientific coordinator. Additionally, he is co-author of three international patents and two national patents. His main research areas include Computational Vision, Medical Imaging, Computational Mechanics, Scientific Visualization, Human-Computer Interaction and New Product Development.

Dr. B. Padmaja Rani is a Professor in Computer Science and Engineering Department at JNTUH College of Engineering, Hyderabad. Her interest area is information retrieval embedded systems. She has published more than 25 papers in reputed journals and conferences in the areas of agile modeling, Web services and mining, etc. She was the former Head of Department of CSE, JNTUHCEH. She is a professional member of CSI.

Dr. V. Kamakshi Prasad is a Professor of Computer Science and Engineering Department at JNTUH College of Engineering, Hyderabad. He completed his Ph.D. in speech recognition from IIT Madras, India. He did his M.Tech. from Andhra University and B.Tech. from K. L. College of Engineering. He has completed over 12 years in JNTU on various positions. He has 21 years of teaching and 11 years of research experience. He has been teaching subjects like speech processing, pattern recognition, computer networks, digital image processing, artificial neural, artificial intelligence and expert systems, computer graphics, object-oriented analysis and design through UML, and soft computing. He has supervised 12 Ph.D. and 2 MS students. His research areas are speech recognition and processing, image processing, neural networks, data mining, and ad hoc networks. He has authored two books published by Lambert Academic Publishing and over 50 papers in national- and international-level journals.

Dr. K. Srujan Raju is the Professor and Head, Department of CSE, CMR Technical Campus, Hyderabad, India. He earned his Ph.D. in the field of network security, and his current research includes computer networks, information security, data mining, image processing, intrusion detection, and cognitive radio networks. He has published several papers in refereed international conferences and peer-reviewed journals and also he was in the editorial board of CSI 2014 Springer AISC series; 337 and 338 volumes. In addition to this, he has served as reviewer for many indexed journals. He is also awarded with Significant Contributor, Active Member Awards by Computer Society of India (CSI).

Efficient Video Indexing and Retrieval Using Hierarchical Clustering Technique

D. Saravanan

"We take the full responsibility of the ethical issue of the work. We have taken permission to use the information in our work. In case of any issue we are only responsible."

Abstract Technology has brought the use of image-based information on the World Wide Web. Many professionals, academicians, researchers, and many other users use images for their work. An image retrieval system based on the image content brings the image based on the semantic similarity between the input and stored image. For this, it uses many image properties such as color, shape, and texture of the image. In such situations, user needs extra care to bring the preferred output in large and complex database. The process is based on ranking according to the similarity measure which is computed from the low-level image. The proposed technique is based on hierarchical clustering algorithm, and it consists of two steps. In the first step, images are trained and stored in the database. In step two, input query image properties are extracted using this property and preferred images are extracted based on image property matching technique. This method is less time-consuming and is more user-friendly than text-based technique.

Keywords Image retrieval · Hierarchical cluster · Content retrieval
Data mining · Image comparison · Image feature

1 Introduction

Today, most of the users spent their time for browsing, searching, and retrieving information from the large database. Traditional technique is supported for key-based information retrieval but technology made this process simple. Extracting the preferred output is difficult for the user [1], as retrieving the required information is comparitively more tough than storing images or videos in the web. Retrieving the preferred output required by users is a challenge due to the increased collection of image database on web. For this, efficient methods are needed to

D. Saravanan (✉)
Faculty of Operations & IT, The ICFAI Foundation for Higher Education
(Deemed-to-be-University), IBS Hyderabad, Hyderabad, Telangana, India
e-mail: sa_roin@yahoo.com

© Springer Nature Singapore Pte Ltd. 2018
V. Bhateja et al. (eds.), *Proceedings of the Second International Conference on Computational Intelligence and Informatics*, Advances in Intelligent Systems and Computing 712, https://doi.org/10.1007/978-981-10-8228-3_1

1

retrieve the needed information from this collection. For searching the image, user needs to specify query in terms of text, key frame image, or feed; any image information for this system will return similar type of image information contents. This similarity is used for searching the image content using the image low-level feature such as color, image attributes, etc. [2]. Due to increasing demand, retrieve the preferred images that are hot spot for many researchers [3]. This content-based searching is used in many areas such as artificial intelligence, in the field of medicine, security, and education training. Most of this searching technique is based on text-based today; user enters the query by a set of keywords, and system finds the match based on keyword available in the query [4]. This method has some drawbacks such as keyword search; considering the huge collection of image, it is not reasonable to automatically interpret them and images are not defined by keyword; this drawback brings the new technology in the information retrieval.

1.1 Information Retrieval Based on the Content

Technology brings digital images based on the visual content [5]. This technique is based on low-level image features such as an image color, texture, shape, pixel value, motion, etc.; and these features help to find the similarity between two images. Input image is classified based on this property and helps the user for increasing image retrieval in the complex database that increases the efficiency when it performs the image retrieval operation. In the proposed technique, hierarchical clustering technique is used based on the hierarchical decomposition of image database. Most clustering technique used is distance measure for finding cluster similarity [6]. This technique is applied for various video data files and identify the best clustering technique.

2 Existing System

- Traditionally, image retrieval is based on text-based query and images are stored in unordered way; it brings the searching more difficult.
- No proper image indexing techniques with this collection of images increases the searching time.
- Retrieval of the needed image from this complex database increases user burden.
- Performance of the existing system is very slow and time-consuming.

3 Proposed System

The proposed system works as follows: the input video is converted into frames, using frame extract action process; the unwanted frames, i.e., duplicate frames, are removed. Using hierarchical clustering, technique frames are clustered. Finally, user retrieves relevant frames using image query input; the proposed model is shown in the figure.

3.1 Advantage of Proposed Technique

- Proposed technique is based on hierarchical clustering algorithms.
- Cluster formation is based on the input illustration of pixels value.
- It is based on clustering to identify a needed image from small relevant group that increases the operation efficiency.
- Information retrieval then can be stored for later user reference.
- Proposed technique can be easily automated.

3.2 Proposed Architecture

See Fig. 1.

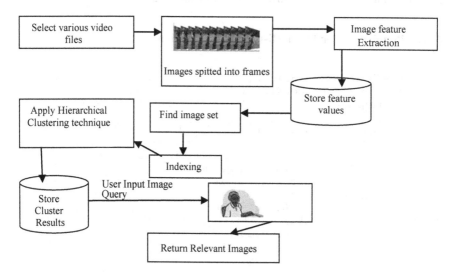

Fig. 1 Proposed architecture

3.3 Proposed Algorithm

Step 1   Get the various video files.
Step 2   Input the image and perform hierarchical clustering.
Step 3   Identify pixel as its own cluster.
Step 4   Find similar values and group of cluster.
Step 5   Find similarity between groups of cluster and merge most similar groups into one cluster.
Step 6   Process gets continued until all pixels are combined into one group.
Step 7   Apply proposed technique and find number of cluster k.
Step 8   Each cluster group identifies which point center it is closest to.
Step 9   Generate centric points with the help of Step 8.
Step 10   Repeat steps 7–10 up to process get terminated

4 Experimental Setup

4.1 Image Elements Extraction

Compared to the traditional text-based technique, the proposed technique works differently. Proposed technique works based on stored images from a huge complex image collection by comparing image features [7]. The common features used here are color and pixel value. In the proposed technique, find the histogram value of the given input image. These values are used in future for image processing. Image extraction process finds the equivalent values which are stored in the database; corresponding images are returned as output of the input.

4.1.1 Pseudocode for Color Histogram Calculation

```
double A = Image.getHeight();
  double B = Iimage.getWidth();
ColorHistogram        =        ColorStuctureExtraction((int)A,
(int)B1);
       ColorHistogram = reQuantization(ColorHistogram);
if(Image descriptor of Colorformationexecution)
       {               Colorformationexecution        =
(Colorformationexecution) image descriptor;
          if(Colorformationexecution.quantizationLevels
== quantizationLevels)
           {f    =    0.0F;   for(int   i   =   0;   i   <
ColorHistogram.length;   i++)f+=Math.abs(ColorHistogram[i]
- Colorformationexecution.ColorHistogram[i]);
```

4.2 Indexing of Images

Video is an effective tool to exchange the information, instead of typing big text-based information, and it brings easy communication between the user groups. For almost all digital image processing, the RGB color space is utilized in normal way for color monitors. Image indexing reduces the user searching time and also improves the performance of image retrieval system. Getting a better result avoids numerous measurement of images. Proposed technique uses image histogram value. Experimental output verifies that proposed techniques work well for different videos types.

4.3 Image Retrieval Using Clustering Technique

Clustering algorithms can follow a hierarchical or a partitioning model. Both use some specific points within the cluster to represent it in order to find out the similarities [8]. Clustering not only faces the time complexity and quality but also frequency estimation, order statistics, and compute data stream [9].

4.3.1 Pseucdocode for Cluster Calculation

```
int l = 0;     double A = Cluster(closest, lk, f, r);
          while(d3 > EPSGLOB)
          { Centroids(closest, lk, f, c);
              B = Cluster(closest, lk, i, r);
              if(A > 0.0D) C = (A - B) / A; else     C =
0.0D;
              A = B;     if(l == 0 || C < pixelspliet&&
m_CurrSize < m_MaxSize)
                 { Split(closest, af, f, c);
                    B = Cluster(closest, lk, f, r);
                    C= 1.0D; }for(j = 0; j < m_CurrSize; j++)
{                    for(int k2 = 0; k2 < j; k2++)}}
```

5 Experimental Outcomes

See Figs. 2, 3, 4, 5, and 6 and Table 1.

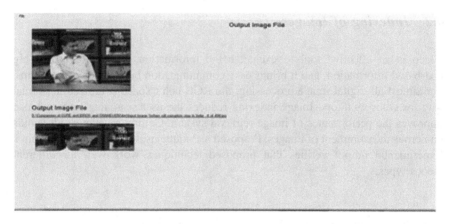

Fig. 2 Based on user image input query 1 output

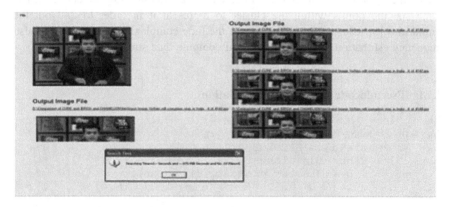

Fig. 3 Based on user image input query 3 output

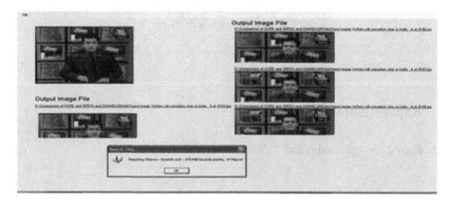

Fig. 4 Based on user image input query 4 output

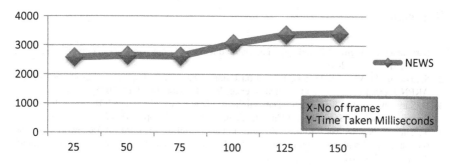

Fig. 5 Performance graph of news video file

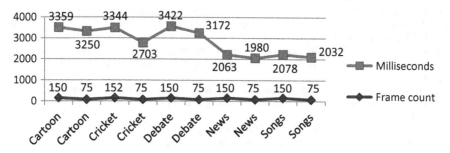

Fig. 6 Comparison performance graph of various video files

Table 1 Frame count versus time for different video files

Frame count	Milliseconds	Category of video file
150	3359	Cartoon
75	3250	Cartoon
152	3344	Cricket
75	2703	Cricket
150	3422	Debate
75	3172	Debate

6 Conclusion

In this work, an efficient framework for image retrieval from the given video is proposed. Hierarchical clustering technique for image retrieval is presented; images are initially clustered into groups based on image color; experimental output verified the proposed technique which brings more image outputs for given input image query, and it also works well in all types of videos. From the observations made, it is found that the existing techniques can be used only for limited sets of videos. Proposed technique fetches accurate image results for all types of video files and the outcomes have verified this. The technique may be further combined with other clustering algorithms to design other cross-algorithms.

References

1. Saravanan. D, Segment Based Indexing Technique for Data file. Procedia of computer Science, 87(2016), 12–17. (2016).
2. Saravanan. D, Video Substance Extraction Using image Future Population based Techniques. ARPN Journal of Engg., and applied science, Vol. 11, No(11), 7041–7045, (2016).
3. D. Saravanan and S. Srinivasan, "Data mining framework for video data," Recent Advances in Space Technology Services and Climate Change (RSTSCC), pp. 167–170, (2010).
4. Y. Yang, F. Nie, D. Xu, J. Luo, Y. Zhuang, Y. Pan. A multimedia retrieval framework based on semisupervised ranking and relevance feedback, IEEE Trans. Pattern Anal. Mach, 2012, Intell. 34: 723–742, (2012).
5. Martin Ester, Hans-Peter Kriegel, and Xiaowei Xu, A database interface for clustering in large spatial databases In Int'l Conference on Knowledge Discovery in Databases and Data Mining (KDD-95), Montreal, Canada, (1995).
6. D. Saravanan, "Various Assorted Cluster Performance Examination using Vide Data Mining Technique", Global journal of pure and applied Mathematics, Volume 11, No. 6 (2015), ISSN 0973-1768, Pages 4457–4467, (2015).
7. D. Saravanan, "Video Substance Extraction Using image Future Population based Techniques", ARPN Journal of Engg and applied science, Vol. 11, No(11), Pages 7041–7045, (2016).
8. D. Saravanan, "Design and implementation of feature matching procedure for video frame retrieval", International journal of control theory and applications, 9(7), Pages 3283–3293, (2016).
9. Tian Zhang, Raghu Ramakrishnan, and Miron Livny, (1996) Birch: An efficient data clustering method for very large databases In Proceedings of the ACM SIGMOD Conference on Management of Data, pages 103–114, Montreal, Canada, (1996).

A Data Perturbation Method to Preserve Privacy Using Fuzzy Rules

Thanveer Jahan, K. Pavani, G. Narsimha and C. V. Guru Rao

Abstract Data mining methods analyze the patterns found in data, irrespective of the confidential information of an individual. It has led to raise privacy concerns about confidential data. Different methods are inhibited in data mining to protect these data. Privacy preserving data mining plays a major role in protecting confidential data. The paper focuses on data perturbation method to preserve confidential data present in the real-world datasets. These identified confidential data are perturbed using fuzzy membership function (FMF) and obtains fuzzy data. The mining utility such as classification and clustering methods are used. The accuracy is determined and compared between an original data and fuzzy data. The results shown in the paper proves the proposed method is efficient in preserving confidential data.

Keywords Privacy · Fuzzy logic · Fuzzy membership function
Data mining

1 Introduction

Privacy has become a major concern in data mining applications, cloud computing, and service computing. Due to increase in era of internet, data is shared and is available publicly. The valuable information is extracted from large amounts of

T. Jahan (✉) · K. Pavani
Vaagdevi College of Engineering, Warangal, India
e-mail: thanvijahan@gmail.com

K. Pavani
e-mail: bandaripavani@gmail.com

G. Narsimha
JNTU Jagityal, Karimnagar, Telangana, India
e-mail: narsimha06@gmail.com

C. V. Guru Rao
Department of CSE, S R Engineering College, Warangal, Telangana, India
e-mail: guru_cv_rao@hotmail.com

© Springer Nature Singapore Pte Ltd. 2018
V. Bhateja et al. (eds.), *Proceedings of the Second International Conference on Computational Intelligence and Informatics*, Advances in Intelligent Systems and Computing 712, https://doi.org/10.1007/978-981-10-8228-3_2

data, for business analysis and scientific computing using data mining methods. A major problem occurred while unauthorized users have access on private data. The private data is well preserved along with accuracy in data mining through privacy preserving data mining. PPDM protects sensitive information present in individuals before releasing. In the field of research, privacy preserving data mining has become novel [1]. The problem is challenging, primarily focusing to protect crucial data and secondary on data usage, to meet specific requirements of an organization. Violation occurred when confidential data is known from the patterns extracted using data mining. Thus, developing efficient data mining methods is needed to preserve privacy, in order to making data available publicly. Only by ensuring sensitive private data has been protected. Data perturbation is well known to preserve privacy in data mining techniques. Sensitive information of an individual is not efficiently protected by data mining, this can lead violation. Data mining extracts information hidden without disclosure of private data of individuals. The major challenge of privacy preserving is to concentrate on factors to achieve privacy guarantee and data utility.

1.1 Privacy Preserving Data Mining

Privacy is extensively used in two areas, they are centralized and distributed environments. In centralized environment, the data is located at a single place. Privacy preserving data mining algorithms are used to protect the sensitive information in data. In distributed database environment, data is distributed on various different sites either by horizontal or vertical partition. In this environment, the privacy preserving techniques are applied to protect individual information by integrated data from multiple sites. Privacy preserving is a technique to study mining patterns which mask the private information and preserves data. Privacy preserving algorithms are used to protect the sensitive information in data. It is classified into reconstruction methods, heuristics methods, and cryptographic methods (secure multiparty computations). Reconstruction is classified into data perturbation, data swapping and data swapping and randomization. The main research aspects of PPDM are applicability, privacy metrics, mining accuracy, and computation [2]. The disadvantages of the existing methods in PPDM methods are mostly concerned about protection metric and accuracy of mining. Data perturbation is the popular model for privacy preserving data mining.

1.2 Data Perturbation in PPDM

Data perturbation methods are used to modify data or add noise to data, data mining techniques have proved that original and perturbed data are relatively same and accuracy is measured by different classifiers. Data perturbation is the popular model for privacy preserving data mining. Data perturbation is mainly categorized as probability distribution and value distortion approaches. The value distortion approaches perturb confidential data using noise. Recent methods on perturbation focus on random noise applied to the datasets, but not considering various privacy requirements of the different users. Mainly, there are two types of data perturbation methods used on continuous data: additive and matrix multiplicative methods such as, SVD, SSVD [3], fuzzy logic, and other detailed multiplicative data perturbation techniques. Data perturbation methods are used by the data owners before data is outsourced. Perturbation means changing of an attribute value by a new value. The perturbed data meet the two conditions: first an attacker cannot discover the original data from the perturbed or distorted data and second, the distorted data maintains the statistical properties of original data and can be derived. Data perturbation is extensively used in two areas, they are centralized and distributed environments. In centralized environment, the data is located at a single place. Distributed database environment relies on the data that is located/distributed on different sites or places. In this environment, the privacy preserving techniques are applied to protect individual information by integrated data from multiple sites.

The paper is mainly focused on the centralized environment, where data is placed under an authorized user (Data owner). Data owner needs to respect these privacy concerns when sharing, publishing, or otherwise releasing the data. The data owner mainly primarily aims to protect data before publishing it to out-source. Data perturbation methods are applied on the original set of data. These perturbed data are then analyzed to check the accuracy along with the original data. Since, there is no trade-off difference between perturbed and original data, the perturbed data can therefore be released. Second, data owner should also assume the role of attackers and develop techniques for breaching by estimating the original data from the perturbed data and any available additional prior knowledge. The attackers work can offers into vulnerabilities on different type's data perturbation methods.

2 Background and Related Work

Huge volumes of data are collected during data publishing as there is rapid growth of internet technology [4]. Many applications such as defense, medical care, transactions based on finance rely on issues of data privacy. Data is collected and gathered during the process of data mining. There are many data mining techniques to analyze raw data and is prone to threat for individuals privacy [5]. Many PPDM techniques are seen in literature such as data swapping, aggregation, Fourier and signal transformation, data anonymization including generalization and suppression. Among all these methods, data perturbation plays a major role in PPDM. In data perturbation, original dimension of data matrix was reduced by transforming using feature selection and SSVD for analysis purpose. Small distorted values are discarded as features for classification purpose [6, 7]. A fuzzy membership function (FMF) was used on original data to provide increased data privacy and decreased

number of passes to perform clustering. The advantages of fuzzy sets are impre-
cision and uncertainty that rely on real-world knowledge [8, 9]. Different fuzzy
membership functions are used to modify or distort data having confidential attributes [10]. These confidential attributes are distorted and analyzed using data
mining techniques such as classification that has decreased complexity and processing time [11]. In privacy preserving data mining, different methods are used in
the recent years, based on noise addition methods [12]. Rotation perturbation is
used as well as condensation-based perturbation [13]. Projection-based perturbation
and geometric based perturbation as a perturbation and large body of literatures on
k-anonymity model [9]. We mainly focus on perturbation methods.

3   Data Perturbation Using Fuzzy Logic

Fuzzy logic is used to solve imprecise problem with better solutions. To decrease
complexity of a system, we need to understand a system well. Fuzzy sets are the
extensions of generic set theory. The fuzzy sets are crisp set and progressive
transition. It deals with a fuzzy set of a pair (D, μ_d) where D is a set and μ_d:
$D \rightarrow [0, 1]$. For all $x \in D$, $\mu_d(x)$ is named as the grade of membership of x.
Different linguistic variables are defined by shape such as triangular, S-shaped,
Z-shaped fuzzy membership functions [14, 15]. Different FMF are defined such as
S-shaped membership function is represented as

$$f(x; a, b) = \begin{cases} 0, & x \le a \\ 2\left(\frac{x-a}{b-a}\right)^2, & a \le x \le \frac{a+b}{2} \\ 1-2\left(\frac{x-b}{b-a}\right)^2, & \frac{a+b}{2} \le x \le b \\ 1, & x \ge b \end{cases}, \tag{1}$$

where x is the confidential attribute or column of the dataset, a is calculated as maximum value of the attribute, and b is calculated as the minimum value of the attribute
for a dataset. The FMF such as Z-shaped membership function is represented as:

$$f(x; a, b) = \begin{cases} 1, & x \le a \\ 1-2\left(\frac{x-a}{b-a}\right)^2, & a \le x \le \frac{a+b}{2} \\ 2\left(\frac{x-b}{b-a}\right)^2, & \frac{a+b}{2} \le x \le b \\ 0, & x \ge b \end{cases} \tag{2}$$

3.1   Fuzzy Rule Based Systems

Fuzzy rules are based on linguistic that constructs If-Then in the form of "IF X
THEN Y" where X and Y are the linguistic variables. X is named as premise
whereas Y is named as consequence of the rule. Fuzzy classification is developed

for p training tuples $A_m = (A_{m1}, A_{m2}, \ldots A_{mq})$, m = 1, 2, ... p from p different classes where A_m is an q-dimensional vector of attributes in which A_{mi} is the i-th attribute value of the m-th training pattern (i = 1, 2, ... q). the fuzzy if-then rules is shown below:

$$\text{Rule } F_r: \text{if } A_1 \text{ is } X_{r1} \text{ and } \ldots \text{ and } A_r \text{ is } X_{rq} \text{ then class } C_r \text{ with } CF_r, \quad (3)$$

where F_r is the r-th fuzzy rule, A = (A1, ... Aq) is the q-dimensional vector of a pattern, X_{r1} is the antecedent fuzzy set, C_r ias a class label, and CF_r is the weight of the r-th fuzzy rule [16].

4 Proposed Work

It presents a new data perturbation technique for privacy preserving publishing or outsourced data mining. A data analysis system is represented and it consists of two parts, namely, the data manipulation and data analyst as illustrated in Fig. 1. The original data D is perturbed or manipulated by the authorized user or data owner using data distortion process. The distorted data or perturbed data D″ is collected by analysts to perform all actions such as classification and clustering, etc. The protected data maintains privacy as analysts are unknown with actual values of the original data. Data perturbation or data distortion methods are based on random value data perturbation methods [17].

The data perturbation methods proposed in the work are implemented and executed on MATLAB (version 7, R2012a). The data mining tool used is Tanagra data mining tool. The tool is used to mine the original and perturbed datasets. The accuracy of both perturbed and original are calculated. The datasets are downloaded from UCI machine Learning Depository. A real-world dataset is used such as iris dataset, water treatment plant dataset, and stone flakes dataset. Iris dataset contains 50 instances and 3 classes where each class is a type of iris plant. Water treatment plant contains 527 instances and 38 attributes where the domain is stated as ill-structured. Stone flake dataset contains 79 instances and 8 attributes having history of mankind. It contains information about stone tool. The original dataset has sensitive information of the above datasets are perturbed with S-based FMF and Z-based fuzzy membership function. For experimental purpose, Tanagra data mining tool and performance is evaluated using MATLAB package.

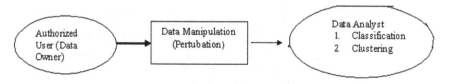

Fig. 1 Data analysis system

4.1 Proposed Algorithm

Input: Original dataset D
  Output: Fuzzy data D'
  Step 1: An authorized user owns an original dataset.
  Step 2: An original dataset is identified with confidential attributes.
  Step 3: The identified confidential attributes are perturbed using S-based FMF and Z-based FMF (fuzzy data).
  Step 4: The fuzzy data is published by a user to an analyst for analysis.
  Step 5: Analyst receives the fuzzy data and identifies fuzzy rules to classify and cluster using mining techniques.

The above algorithm is applied on original datasets such as iris, water treatment plant, and stone flakes. The fuzzy data obtained is analyzed using to determine fuzzy rules and thus classified. The results obtained after classification and clustering are discussed in the next section.

5 Results

The datasets of iris, water treatment plant, and stone flakes are downloaded from UCI Machine Learning Repository. These datasets are given as input to the proposed algorithm to produce fuzzy datasets. The fuzzy datasets obtained are further used to generate fuzzy rules. The experimental results on the fuzzy data and original data are analyzed using fuzzy rule based classification method. The classification accuracy is calculated between original and fuzzy data using Eq. 3.

The above graph in Fig. 2 represents the classification accuracy on original dataset and perturbed datasets. The accuracy is better on perturbed dataset than original dataset. Similarly, the clustering utility is used on original dataset and perturbed dataset. The different clustering techniques are compared on original and distorted datasets. The clustering methods are k-means, nearest neighbor and DB-scan. A F1 measure is used to compare the different datasets. The values obtained on the datasets are tabulated in Table 1.

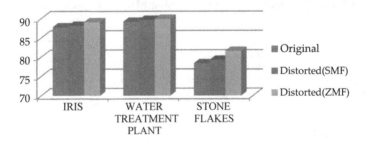

Fig. 2 Resultant accuracy on datasets

Table 1 Comparison of F-measures for the three different clustering algorithms

Clustering method	Datasets	F-measure
K-means	Iris	0.67
	Water treatment plant	0.75
	Stone flakes	0.64
Nearest neighbor	Iris	0.75
	Water treatment plant	0.70
	Stone flakes	0.71
DB-scan	Iris	0.79
	Water treatment plant	0.80
	Stone flakes	0.85

Precision P_{ij} equals: $\frac{|C_i \cap C_j'|}{|C_j'|}$ and recall R_{ij} equals: $\frac{|C_i \cap C_j'|}{|C_i|}$. The F-measure of a cluster C_i is given by $F_i = \max_j F_{ij}$. Finally the overall F-measure is:

$$F = \sum_{i=0}^{n} \frac{|C_i|}{N} F_i, \text{ where N is the total number of records (Eq. 2).}$$

6 Conclusion

Data perturbation techniques on multivariate datasets are used to perturb the original data. The techniques proposed are well classified by different classifier and clustering methods. The fuzzy transformation methods are efficient to perturb the original dataset. The confidential attributes present in the original dataset are only perturbed using the fuzzy logic. The classifiers are showing the better accurate results to the original and fuzzy data. Fuzzy-based transformations have proved its efficiency both in univariate and multivariate datasets. Further, the proposed method is made resilient by incorporating the attack analysis. The proposed privacy preserving methods are efficient to perturb the original data and can publish the perturbed data to the data analyst.

References

1. R. Agrawal and R. srikant "Privacy–Preserving data mining", In proceedings of the 2000 ACM SIGMOD International Conference on management of data, pp 86–97, San Diego, CA, 2003.
2. V. Estvill–Castro, L. Brankovic, D.L. Dowe, "Privacy in datamining", Australian Computer Society NSW Branch, Australia, Available at www.acs.org.au/nsw/articles/199082.html.
3. J. Wang, W. Zhong, S. Xu, J. Zhang. "Selective Data Distortion via Structural Partition and SSVD for privacy Preseervation", In Proceedings of the 2006 International Conference on Information & knowledge Engineering, pp: 114–120, CSREA Press, Las Vegas.

4. R. Agrawal, A. Evfimieski and R. Srikanth. "Information sharing across private databases". In proceedings' of the 2003 ACM SIGMOD International Conference on management of data, pp. 86–97, San Diego, CA, 2003.
5. S. Xu, J. Zhang, D. Han, J. Wang. Data distortion for privacy protection in a terrorist analysis system. in: Proceedings of the 2005 IEEE International Conference on Intelligence and Security Informatics, 2005, 459–464.
6. Thanveer, G. Narasimha and C.V. Guru Rao "Data Perturbation and Feature selection in Preserving Privacy" In Proceeding of the 2012 Ninth International conference in wireless and optical communication networks (WOCN), Indore, Madhya Pradesh, India. IEEE catalog number: CFP12604-CDR, ISBN: 978-1-4673-1989-8/12.
7. T. Jahan, G. Narsimha, C.V Guru Rao, Privacy preserving clustering on distorted data in International Organization of Scientific Research. J. Comput. Eng. ISSN: 2278–0661, ISBN: 2278–8727 5(2), 25–29 (2012).
8. Thanveer Jahan, Dr. G. Narsimha, Dr. C.V. Guru Rao "A Hybrid Data Perturbation Approach To Preserve Privacy" International Journal of Scientific & Engineering Research, Volume 6, Issue 6, June-2015 1528 ISSN 2229-5518.
9. B. Gillburd, A. Schuster and R. Wolff. "K-TTP: A new Privacy model for large–scale distributed environments". In Proceedings of the 10th ACM SIGKDD International Conference on Knowledge Discovery and Data Mining (kdd'04), Seattle, WA, USA, 2004.
10. B. Karthikeyan, G. Manikandan, V. Vaithiyanathan, A fuzzy based approach for privacy preserving clustering. J. Theor. Appl. Inf. Technol. 32(2), 118–122 (2011).
11. V. Vallikumari, S. Srinivasa Rao, KVSVN Raju, KV Ramana, BVS Avadhani "Fuzzy based approach for privacy preserving publication of data", IJCSNS, Vol. 8 No. 1, January 2008.
12. P. Kamakshi, A. Vinaya Babu, "A Novel Framework to Improve the Quality of Additive Perturbation Technique", In proceeding of International Journal of Computer Applications, Volume 30, No. 6, September 2011.
13. Aggarwal, C. C. and Yu, P. S. (2004), A condensation approach to privacy preserving data mining, in 'Proceedings of International Conference on Extending Database Technology (EDBT)', Vol. 2992, Springer, Heraklion, Crete, Greece, pp. 183–199.
14. Zadeh L "Fuzzy sets", Inf. Control. Vol. 8, PP, 338–353, 1965.
15. Thanveer Jahan, G. Narasimha, and C.V. Guru Rao "A Comparative Study of Data Perturbation Using Fuzzy Logic to Preserve Privacy" Networks and Communications (NetCom2013), Lecture Notes in Electrical Engineering 284, https://doi.org/10.1007/978-3-31903692-2_13, © Springer International Publishing Switzerland 2014.
16. Ishibuchi H., Yamamoto T.: Comparison of heuristic criteria for fuzzy rule selection in classification problems. Fuzzy Optimization and Decision Making, Vol 3, No. 2, 119–139, (2004).
17. K. Chen, and L. Liu, A Random Rotation Perturbation Approach to Privacy Data Classification, In Proc of IEEE Intl. Conf. on Data Mining (ICDM), pp. 589–592, 2005.

An Enhanced Unsupervised Learning Approach for Sentiment Analysis Using Extraction of Tri-Co-Occurrence Words Phrases

Midde. Venkateswarlu Naik, D. Vasumathi and A. P. Siva Kumar

Abstract This article reveals an unsupervised learning approach for determining the polarity of unstructured text in big data environment. The key inspiration for sentiment analysis research is essential for end users or e-commerce firms with local and global languages who expressed views about certain entities or subjects in social media or blogs or web resources. In proposed approach, applied an unsupervised learning approach with the help of idiom pattern extraction in determining favorable or unfavorable opinions or sentiments. Prior methods have achieved precision of sentiment classification accuracy on English language text up to 81.33% on a movie dataset with two co-occurrences of sentiment words phrases. This approach addressed the enhancement of sentiment classification accuracy in unstructured text in a big data environment with the help of extracting phrase patterns with tri-co-occurrences sentiment words. Proposed approach used two datasets such as cornel movie review and university selection datasets that are publicly available. Lastly, a review document is classified after comprehensive computation of semantic orientation of the phrases into positive or negative.

Keywords Unstructured text · Sentiment computation · Phrase pattern extraction with tri-co-occurrence · Semantic orientation (SO) · Unsupervised machine learning approach

Midde. Venkateswarlu Naik (✉)
Department of CSE, JNTUHCEH, KMIT, Hyderabad, India
e-mail: vvnaikcse@gmail.com

D. Vasumathi
Department of CSE, JNTUHCEH, Hyderabad 500085, Telangana, India
e-mail: vasukumar_devara@jntuh.ac.in

A. P. Siva Kumar
Department of CSE, JNTUA, Anantapur, India
e-mail: sivakumar.cse@jntua.ac.in

© Springer Nature Singapore Pte Ltd. 2018
V. Bhateja et al. (eds.), *Proceedings of the Second International Conference on Computational Intelligence and Informatics*, Advances in Intelligent Systems and Computing 712, https://doi.org/10.1007/978-981-10-8228-3_3

17

1   Introduction

In the current era, fast escalation of social media platforms like Twitter, Facebook, LinkedIn, Google plus, Yelp, Flickr, blogging and microblogging, forums, news reports, click streams lead to mine, transform, load, analyze, heterogeneous, unstructured, huge volume cross-domain and multilingual data [1]. This kind of data is called as big data. Unstructured type of data to be analyzed, with a various granularity levels, traditional approaches or techniques such as statistical or data mining, Information retrieval does not outperform well. In this context, it is a very tedious task to analyze or determine sentiment regarding customer opinion or reviews posed by numerous people about various entities over social media platforms. Numerous individuals are influenced by the opinionated information available over the social media applications. Actually, it is absolutely right for any entity reviews to influence buying activities. Furthermore, the information provided by the individual customer is more reliable than the feedback given by the vendor. To derive certain accurate decisions by the firm, considering likes and dislikes posed by the individuals would assist a lot in many factors. Usually, classification of sentiment is essential for plenty of applications. However, sentiment is uttered differently in various domains. Sentiment analysis (SA) is a computational process to extract opinion as positive or negative from text about an entity. Discover polarity in unstructured text needs polar words such as nice, thrilled, best, worst, bad, etc., these words are rooted pointers for designing machine intelligent learning models for sentiment classification. Lexicon-based methods compute the semantic orientation values for sentiment words to compute the overall opinion of the document. An individual polar words are not able to determine the actual sentiment of the text. Sentiment words are context sensitive or domain based such as TV is curved is positive sentiment but in mobile domain it is negative sentiment [2]. As a result, syntactical and contextual information is vital for sentiment classification. In this situation, phrases patterns play an optimal role to accommodate syntactical and contextual information. The proposed article focuses on the approach for retrieving pattern phrases that are essential for sentiment analysis to enhance classifier accuracy.

Lexicon-based approaches use dataset with predefined polarity or polarity score to conclude the opinion or sentiment of giving text. The vital merit of these approaches is that it does not require any training data. The author [3] proposed few phrase patterns such as two-word co-occurrences from reviews. He achieved sentiment classification accuracy with 77.83% on the movie review dataset. The authors [4] proposed new phrases for sentiment classification to enhance accuracy than earlier authors. He achieved accuracy with proposed phrases and including Turney rules such as 79.50%, with dependency phrases 80.50%, and combined all phrases 81.33%.

In this article, sect. 2 deals about the associated work. Subsequently third section deals about proposed model with detailed discussion. Consequently fourth section discussed about results and discussions. Last section discussed regarding conclusion of this research work.

Following are the key contributions obtained by these authors in this article. (1) Innovative POS-based tri-co-occurrence phrase patterns are recognized for sentiment classification. (2) We used Stanford NLP parser for extracting relations among sentiment words syntactically and contextually with the help of phrase patterns. (3) Obtained enhanced accuracy of the sentiment classifier on movie dataset and educational dataset than existing accuracies.

2 Associated Work

Sentiment Analysis approaches are broadly classified into three categories such as machine learning, intelligent techniques [5–7]. This article deals with sentiment classification with the help of unsupervised approach phrase extraction with various possibilities. The author [3] discovered a two-word co-occurrence of phrases for determining the polarity of the document and subsequently employed on the movie review dataset. First, this method retrieves phrases with two-word co-occurrences with the help of predefined POS based prototype. Subsequently, SO is computed for retrieving phrases with the help of PMI. Lastly, the overall sentiment of the document is determined by aggregating the SO of all the retrieving phrases. The authors [4] proposed few two-word co-occurrence phrase patterns other than the [3] phrase patterns to overcome the accuracy of existing methods. This author has employed different combinations such as POS-based phrase patterns Turney rules, Turney [4] rules, [4] combining all rules, as a result [4] author obtained 81% of accuracy on movie dataset. This author uses the Stanford dependency parser tree for providing syntactical relations between words in the statement. These dependency relations are also useful to determine sentiment analysis. The authors [8] designed a model for automatic recognition of phrase patterns with the help of feature selection methods such as document frequency and information gain to determine the overall polarity of the document. The authors [9] proposed a model for POS tagging with the help of semi-supervised associative classification method for various languages such as Telugu, Hindi, Tamil, Bengali, and English. This method outperformed average accuracy 82.5% for English, 84.3% of Telugu language dataset, etc. The authors [10] have employed genetic algorithm for recognition named entities with the ensemble classifier method. This author obtained precision with 81.38% for Telugu language. The authors in [11] discovered few phrase patterns using adjectives, adverbs, verbs, conjunctions, noun, and prepositions.

3 Proposed Model

In the proposed model, sentiment classification is determined with the help of three major phases. Phase 1 performs data preprocessing. Phase 2 performs a sentiment assessment (SA) from supplied preprocessed data. Phase 3 performs an accuracy computation for employed technique. Figure 1 demonstrates the proposed model to enhance accuracy of existing methods for sentiment classification.

3.1 Steps for Preprocessing Dataset

In *Sentence Segmentation* module, the raw text of the review document is split into sentences with the help of sentence segmenter. In *Tokenized Sentence module*, output returned by sentence segmenter will become input and further each sentence is subdivided into tokenized sentences. *Filter Stop Words module* filters English stop words from a document by eliminating words which are equal to the stop words from the built-in stop word list. *Stemming module* stems English language words using a porter stemming algorithm. It works an iterative, rule-based substitute word suffixes to obtain meaningful word, for example, playing is stemmed as play and studying is stemmed as study.

Parts of Speech Tagging does the tagging each word in the text with associated class of word according to the English grammar. POS tagging is also known as word group disambiguation. Chunking process determines types of sentiment or polarity based on predefined phrase pattern.

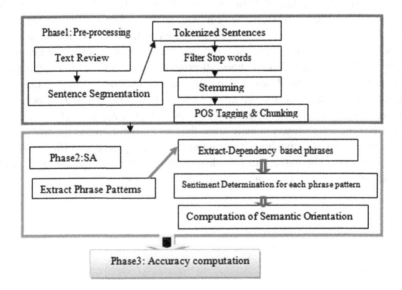

Fig. 1 Proposed model to enhance accuracy of existing methods for sentiment classification

3.2 Extract Phrase Patterns

In sentiment analysis, phrases are very crucial for retrieving syntactic and contextual information in a given text. For instance, one-word polar word "good" may give positive sentiment but a two-word co-occurrence an adverb followed by adjective would enhance the sentiment intensity like "very good". Although, as per the current world, users review comments in social media, and intensity of sentiment expressed is very high such as more than two-word phrases like three-word phrases like a nail (NN) biting (VBG) movie (NN). Apart from this phrase pattern extraction, gathering contextual information such as negation-related statements like "not good", "erratic behavior", etc. In this article, we proposed tri-word co-occurrence phrases that contain syntactic and contextual information that would be useful for sentiment analysis. Lastly, with the help of the Stanford dependency parser tree, dependency relations based phrases are also identified. In [3], the author discovered the phrase patterns, in which one member of two-word co-occurrence phrases is adjective or adverb or noun. In [4], the author recognized new phrase patterns, in which new phrase patterns with two-word co-occurrence and their employed dependency parser to parse syntactic and semantic information about the occurrence of the sentiment words. These phrase patterns achieved higher level of accuracies than author [3]. In dependency relation phrase pattern extraction, more useful phrases are extracted than POS based phrases. For instance, "drama is very inspiring and effectual", this sentence is tagged as per POS tagging as follows "drama_NN is_VBZ very_RB inspiring_VBG and _CC effectual_JJ". The phrase "very inspiring" will be retrieved with the help of POS-based phrase pattern. Following are phrases extracted with the help of dependency relations [4], parse tree (ROOT (S (NP (NN drama)) (VP (VBZ is) (ADJP (RB very) (JJ inspiring) (CC and) (JJ effectual)))))). Below are the possible dependency relations both uncollapsed and enhanced. Uncollapsed dependencies: root (ROOT-0, inspiring-4), nsubj (inspiring-4, drama-1), cop (inspiring-4, is-2), advmod (inspiring-4, very-3), cc (inspiring-4, and-5), conj (inspiring-4, effectual-6).

Fig. 2 Dependency mapping between words in basic and enhanced

Basic Dependencies:

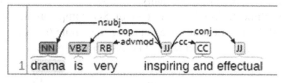

Enhanced Dependencies:

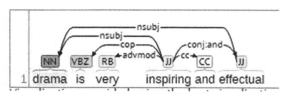

Table 1 Discovered innovative phrase patterns

Si. no	1st Word occurrence	2nd Word occurrence	3rd Word occurrence	Polarity/sentiment score
(i)	RB	RB	JJ	Ex: extremely very sorry Polarity: neg, pos. 0.2, neg. 0.8
(ii)	JJ*	NN	VBN	Ex: spectacular (JJ) majestic (JJ) grandeur (NN) personified (VBN) Polarity: pos. 0.7 neg. 0.3
(iii)	RB	JJ	NN	Ex: very (RB) good (JJ) faculty (NN) Polarity: pos: 0.8, neg.0.2 Ex: good (JJ) faculty (NN) Polarity: pos: 0.6, neg: 0.4
(iv)	NN	VBG/VBD	NN	Ex: nail (NN) biting (VBG) movie (NN) Polarity: positive, pos. 0.5, neg. 0.5 Ex: lovely (NN) love (VBD) story (NN) Ex: half baked drama Polarity: neg
(v)	JJ	NN	NN	Ex: likable/beautiful love story Polarity: pos. 0.7, neg. 0.3
(vi)	VBP/VBD	RB	JJ	Ex: are not right/were hardly right Polarity: neg

Enhanced dependencies: except nsubj (effectual-6, drama-1), the rest of the phrases are same as uncollapsed phrases. Figure 2 shows a pictorial representation of dependencies among the words.

Table 1 lists innovative tri-co-occurrence phrase patterns discovered as our contribution to enhance the accuracy level of existing approaches.

3.3   Computation of Semantic Orientation

To compute semantic orientation for each phrase, extraction of phrase patterns should be done before. A text document is said to be a positive sentiment if it has a more positive phrase occurrence, subsequently semantic orientation of that document is also proportionally high, else given text document is said to be negative sentiment. Although, as part of SO computation, Pointwise Mutual Information (PMI) is essential for computing the strength among phrases in terms of positive and negative statements [11]. PMI is computed as follows:

$$PMI\,(x, positive) = \log\frac{p(x, positive)}{p(x)\,p(positive)} \tag{1}$$

$$PMI(x, negative) = \log\frac{p(x, negative)}{p(x)\,p(negative)}, \tag{2}$$

where P (x, positive) is the probability of a phrase that occurs in positive documents. It is the ratio of frequency phrase pattern in positive documents divided by total number of positive documents. P (x, negative) is the ratio of frequency phrase pattern in negative documents divided by total number of negative documents. As a result, a phrase polarity value is computed from their PMI value difference [3].

$$SO(x) = PMI(x, positive) - PMI(x, negative) \tag{3}$$

$$SO(x) = \log \frac{p(x, positive)/p(positive)}{p(x, negative)/p(negative)} \tag{4}$$

$$SO(x) = \log \frac{p(x, positive)}{p(x, negative)} \tag{5}$$

3.4 Pseudocode Procedure for Implementation

Function (D,m)
Intend: Computation of Sentiment classification
Input: movie review text document d_t, maximum threshold co-occurrence
       pattern phrases m
Output: obtain sentiment classification accuracy
    1. Pre-process the review document D_t
    2. Scan all the review documents in D and extract phrase patterns
       $p_1, p_2, \ldots p_n$ from Segmented tagged sentences $s_1, s_2, \ldots s_n$ along with
       respective Polarities from all supplied review documents such
       that sentiment words $p_j \subseteq s_{ij}$, j=1,2...n
       for each pattern
       {
           if (polarity = = pos)
               *fp1 ← p;
               pcount++;
           else
               *fp2 ← p;
               ncount++;
       } //end of for statement
    3. Computation of point wise mutual information for each pattern
       phrase stored in *fp1, *fp2
       For each phrase
          {
              $PMI_{POS} = PMI(x, pos)$;
              $PMI_{NEG} = PMI(x, neg)$;
          }
    4. Computation of semantic orientation with the help of step3 PMI
       values.
    5. Computation of sentiment classifier accuracy

4   Results and Discussions

In order to estimate the performance of our tri-word co-occurrence method for sentiment analysis, current method utilized freely available cornel review movie dataset [12]. It possesses 2 K reviews, 1 K positive reviews and 1 K negative review. In training dataset, arbitrarily selected 700 and 700 positive, negative documents. The rest of the documents 300 positive, 300 negative are utilized for testing the new model. Although, this research work used university selection review documents with equal number of training and testing documents like a movie review dataset. Evaluation of the proposed approach is measured with the help of accuracy metric as below.

$$\text{Accuracy} = \frac{\text{Total No. of Correctly classified documents}}{\text{Total No. of Testing documents}} \qquad (6)$$

Table 2 List of accuracies obtained using various phrase patterns on movie and university selection review dataset

Dataset	Phrase pattern	True predicted positively	True predicted negatively	Total no. of true classified	Accuracy in %
Cornel movie reviews	Turney	250	217	467	77.83
	Basant	254	234	488	81.33
	Tri-word-occurrence + existing phrases	284	260	544	**90.66**
University selection review (own)	Tri-word-occurrence + existing phrases	260	243	470	83.83

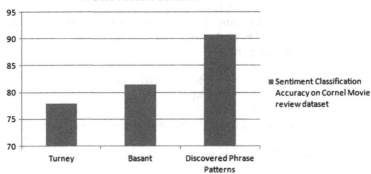

Fig. 3 Sentiment classification accuracies using various phrase patterns

Before computing accuracy of an approach, POS tagging is done with the help of Stanford tagger on respective movie review documents. Further, according to fixed predefined phrase pattern specified in [3, 4], Table 2. Extracted respectively. Subsequently, with help of dependency parser tree, dependency relation based phrases are extracted. Lastly, SO for each phrase is computed with the help of equation given in (5). Table 2 lists the accuracies of applied phrase patterns.

Figure 3 reveals the accuracy of sentiment classification for new pattern phrases and reference [3, 4]. Proposed method has got 90.66% of the sentiment classification accuracy than past methods as shown in the graph.

5   Conclusion and Future Scope

This research article utilized unsupervised learning method for sentiment classification. Proposed method used phrase patterns for tri-word co-occurrence sentiment words extraction. Apart from proposed phrase patterns, the proposed method utilized earlier designed phrases to enhance performance for sentiment classifier accuracy. The proposed method is outperformed by 90.66% than previous applied methods accuracies. Therefore, proposed phrase patterns are essential and helpful to enhance overall performance of the sentiment classifier.

Future work includes discovering improved phrase patterns which might match more sentences, various possibilities of expressing sentiments or opinion, applying the current proposed method for multilingual reviews for sentiment determination, and applying the proposed method for cross-domain sentiment analysis.

Acknowledgements First and foremost, I would like to thank my supervisor Dr. D. Vasumathi, Professor, JNTUHCEH and co-supervisor Dr. A. P. Siva Kumar, Professor, JNTUA for continuous support for my research contributions. Although, I must be thankful to the Keshav Memorial Institute of Technology for their resource, moral, and financial support professionally.

References

1. Kumar Ravi., Vadlamani Ravi.: Survey on opinion mining and sentiment analysis: tasks, approaches and applications. ELSEVIER Knowledge-Based Systems, vol. 89, pp. 14–46 (2015).
2. Danushka Bollegala., David Weir., John Carroll.: Cross—Domain Sentiment Classification using a Sentiment Sensitive Thesaurus, IEEE Transactions on Knowledge and Data Engineering, vol. 25, pp. 1719–1731 (2013).
3. Turney P.D.: Thumbs Up or Thumbs Down? Semantic Orientation Applied to Unsupervised Classification of Reviews, ACL, pp. 417–424, (2002).
4. Basant Agarwal., Vijay Kumar Sharma., and Namita Mittal.: Sentiment Classification of Review Documents using Phrase Patterns, IEEE International Conference on Advances in Computing, Communications and Informatics (ICACCI), pp. 1577–1580 (2013).

5. B. Pang Lee., S. Vaithyanathan.: Thumbs down? Sentiment classification using machine learning techniques, Proceedings of the ACL-02, Conference on Empirical methods in Natural Language Processing, vol. 10, pp. 79–86 (2002).
6. Sonia Saini., Shruti Kohli.: Machine Learning techniques for effective text analysis of social network E-health data, IEEE International Conference on Computing for Sustainable Global Development (INDIACom), pp. 3783–3788 (2016).
7. Ebru Aydogan., M. Ali Akcayol.: A Comprehensive Survey for Sentiment Analysis tasks using Machine Learning techniques, IEEE International Symposium on Innovations in Intelligent systems and Applications (INISTA), https://doi.org/10.1109/inista.2016.7571856 (2016).
8. Mukras R., Wiratunga N., Lothian R.: Selecting Bi-Tags for sentiment analysis of text, International Conference of Innovative Techniques and Applications of Artificial Intelligence, pp. 181–194 (2008).
9. Pratibha Rani., Vikram Pudi., Dipti Misra Sharma.: A semi supervised associative classification method for POS tagging, Springer International Journal Data Science Anal, vol. 1, pp. 123–136 (2016).
10. Asif Ekbal., Sriparna Saha.: Classifier Ensemble Selection Using Genetic Algorithm for Named Entity Recognition, Springer Res on Lang and Comput. vol. 8, pp. 73–99 (2010).
11. Kaji N., Kitsurgawa M.: Building Lexicon for Sentiment Analysis from Massive Collection of HTML Documents, ACL, pp. 1075–1083 (2007).
12. Pang B., Lee L.: Sentiment education: sentiment analysis using subjectivity summarization based on minimum cuts, in Proceedings of the Association for Computational Linguistics (ACL), pp. 271–278 (2004).

Land Use/Land Cover Segmentation of Satellite Imagery to Estimate the Utilization of Earth's Surface

D. R. Sowmya, Aditya N. Kulkarni, S. Sandeep, P. Deepa Shenoy and K. R. Venugopal

Abstract Land Use/Land Cover (LULC) mapping plays a major role in land management applications such as to generate community map, proper urban planning, and disaster risk management. Proposed algorithm efficiently segments different land use/land cover classes such as buildings, trees, bare land, and water body. RMS value based multi-thresholding technique is used to segment various land use/land cover classes and consequently using the binning technique to accurately estimate the utilization of earth's surface. The proposed algorithm is tested on two different data sets of Bengaluru city, India. The percentage utilization of surface objects for grid 7 image of dataset I is found to be 96.68% building, 1.05% vegetation, and 0.22% barren land, the area covered in grid 7 of dataset I is identified as overutilized land. Percentage utilization of surface objects for grid 8 of dataset II is found to be 68.95% building, 11.93% vegetation, 0.15% bare land, and 1.14% water body. The area covered in grid 8 of dataset II is identified as overutilized land.

Keywords Binning · Land use/land cover · Segmentation · Superpixel
Utilization

1 Introduction

Remote sensing is the process of acquisition of information about an object, phenomenon, or a geographic location without making physical contact with it. Remote sensing is one of the most efficient techniques available currently for the observation of earth's surface.

Each piece of earth's surface is unique in its features and broadly it is classified into land use and land cover classes. Land use is how human beings utilize land for

D. R. Sowmya (✉) · A. N. Kulkarni · S. Sandeep · P. Deepa Shenoy · K. R. Venugopal
Department of Computer Science and Engineering, University Visvesvaraya
College of Engineering, Bangalore University, Bengaluru, India
e-mail: sowmyadr8@gmail.com

© Springer Nature Singapore Pte Ltd. 2018
V. Bhateja et al. (eds.), *Proceedings of the Second International Conference
on Computational Intelligence and Informatics*, Advances in Intelligent Systems
and Computing 712, https://doi.org/10.1007/978-981-10-8228-3_4

socioeconomic activities and land cover is the geographical state of the earth's surface (physical cover) which covers grass, asphalt, trees, bare land, and water.

Land use/Land cover (LULC) mapping has many applications including military reconnaissance, urban planning, generating community map, and disaster valuation. Early days, the details of land cover are recorded by field survey, today there exist algorithms which automatically segment and classify various land use and land cover classes of earth's surface and the classification of satellite imagery is still an active research topic in remote sensing.

It is well known that the growth of urban landscape is rapidly on the rise due to the increase in human population. Proper urban planning is an absolute necessity so as to prevent various calamities such as hydrological consequences, transportation issues, environmental damage, etc. An important step in urban planning is to determine the utilization of the earth's surface over a geographic area.

Segmentation is the first step in analyzing remote sensing image. Numerous algorithms exist to perform segmentation. Threshold-based segmentation is one such, which efficiently segments image but traditional threshold-based segmentation has many disadvantages, i.e., it considers only the intensity of each pixel and neglects the differences between pixels thus arrives conflict in segmenting noisy pixels. Traditional threshold-based segmentation performs segmentation operation on binary or grayscale image. Multi-thresholding concept based on RMS value of superpixel overcomes these disadvantages.

This paper describes land use/land cover segmentation using RMS value based multi-thresholding technique and the result is used to estimate the utilization of the earth's surface based on binning technique which enables to identify overutilized land.

This paper is organized as follows: Section 2 deals with related work. Section 3 is about the study area. Section 4 is the discussion of datasets and methodology. Section 5 presents the proposed algorithm. Section 6 is the implementation and performance analysis and Sect. 7 presents the conclusions of proposed work.

2 Related Work

Gueguen [1] discussed multi-scale segmentation and hierarchical clustering using min-tree and kd-tree algorithms, respectively. Algorithm is applied on 2 m multi-spectral UC Merced 21-class dataset to retrieve compound objects. The main focus is to improve computational time. The author of paper [2] used graph theory concept to detect buildings in urban area. Key points represent nodes in a graph, spatial distance, and intensity values represent edges. Multiple subgraph matching method is employed to accurately extract building in urban area. This method is tested on ikonos 1 m resolution data set. Two-level-set-evolution algorithm is proposed by Li et al. [3] to extract man-made objects from high-resolution multi-spectral satellite data. Algorithm efficiently classified airport runways, building roofs, and road networks.

Authors of [4] discussed improved random decision tree algorithm and Genetic Rule based Fuzzy Classifier (GRFC) to classify land cover types which is implemented on ALOS images of Longmen city, China and Greek island of Thasos using hyper spectral satellite image and obtained effective thematic map with accuracy higher than maximum likelihood classification method.

Hu et al. [5] developed object-based method for mapping land cover/land use on Google Earth (GE) image and quickbird image of Wuhan city, China. The classification results show that GE has different ghettoizing capacities for specific land cover type because of lower spectral properties. However, there was no such significant difference found between two images.

3 Study Area

Bengaluru is a major city situated in the southeast of Karnataka and it is the fifth most urban agglomeration in India. Bengaluru is positioned at longitude 12.97° N and latitude 77.56° E, covering an area of 1741 km^2 having an average elevation of 920 m or 3020 feet above sea level. Bengaluru has about 40% of the land used for residential purposes and another 23% of the land for transport. Land use/land cover mainly consists of buildings, roads, bare land, water bodies, trees, and parking area. Estimating the land utilization of Bengaluru city is very much essential for proper urban planning.

4 Dataset and Methodology

4.1 Dataset Description

Dataset I: The availability of spaceborne remote sensing provides multispectral satellite imagery with high spatial resolution. IRS-P5 multispectral data with spatial Resolution 2.5 m is used for the study, which is a four-band (Red, Green, NIR, and SWIR) spectral data.

Dataset II: GE Imagery is a mosaicking of images obtained from Satellite Imagery and aerial photography with Geographic Information System (GIS) on the 3D globe. The spatial resolution of GE images varies from 15 to 15 m GE image is used to validate the accuracy of the proposed system. High-quality GE image is acquired through GE explorer 8 and ArcGis 10.2 is used for geo-referencing and to export the study map (Fig. 1).

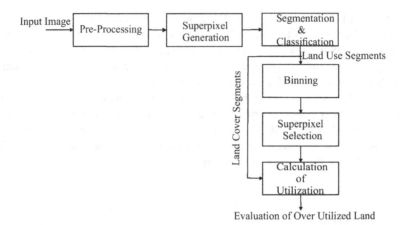

Fig. 1 Proposed architecture

4.2 Methodology

a. Preprocessing

Input image of 2.5 m resolution multispectral satellite image is subjected to pre-processing. Preprocessing includes radiometric corrections, atmospheric corrections, and geometric corrections and these corrections are already done in the obtained data while geo-referencing is performed using ArcMap. Geo-referencing is the process of assigning real-world coordinates to the map. GPS-based control points are mapped on to the image to extract the study map of Bengaluru city.

b. Superpixel Generation

Superpixels form an object by grouping similar kind of pixels. Superpixel generation is based on SLIC algorithm, SLIC is Simple Linear Iterative Clustering and the structure of this algorithm is inherited from K-means clustering algorithm. The image is randomly divided into number of regular grids and center pixel of each grid is named as k, the number of k cluster centers is sampled with step size of b pixels. In order to avoid the localization on the edges of each grid, the sampled k clusters are moved to its lowest gradient position which also avoids overlapping of noisy pixels. Each pixel is assigned to its respective cluster based on distance formula given in Eq. (1).

$$P = \sqrt{q_r^2 + \left(\frac{p_b}{b}\right)^2 a^2} \tag{1}$$

Lab color space gives three-dimensional perceivable colors, where L is the illumination lightness ranging from darkest black to brightest white (00 to 100), a and b are color components. q_r is the Euclidian distance between pixels in Lab color space. p_b is the Euclidean distance between the two pixel locations, b is the maximum spatial distance between two k's and a is a constant and the value of a is in the range of 1–40. Given pixel is grouped into specific cluster based on distance P and in updating step k is recalculated by assigning mean of all values in the given cluster based on Lab color space and its location values. RMS value of each superpixel is also calculated to perform segmentation.

c. Segmentation and Classification

Segmentation is the process of dividing an image into its constituent components or multiple segments. RMS value based adaptive thresholding method is used for segmentation, if the RMS value of a given superpixel is within a predefined threshold for a specific class (land use/land cover) then the superpixel is classified as belonging to that class.

For each land cover type, a superpixel is selected and labeled as reference descriptor. Superpixels are mapped into respective class of land cover based on the RMS values of the saved reference descriptors, RMS value of RD (Reference Descriptor) is compared to that of each superpixel, if the difference between the two is less than the thresholds which are predefined then the superpixels are mapped into its specific class (land use/land cover).

Segmented land use type (building) has to be refined so as to calculate the land utilization more accurately and refining process is done using binning technique. The other land cover classes do not require refining process as it is clearly distinguishable.

d. Binning

Binning is a technique which is used to reduce the noise effects such as minor errors in observation. It is the process of grouping similar or continuous values into a smaller number of bins. Following certainty and postulates are considered to perform refining process using binning technique.

- The extracted land use type (buildings) majorly consists of similar buildings with close pixel RGB values.
- Considered fixed number of bins.

A number of bins are assumed and assigned each and every pixel to a specific bin based on the RGB value of the pixel. For example for $11 \times 11 \times 11$ bin, bin number is calculated using Eq. (2).

$$\text{Bin number} = R + (G \times 11) + (B \times 121) \tag{2}$$

Once all pixels are positioned into their respective bins, the bin with the highest or second highest number of pixels is considered and this selected bin is an input to the next step.

e. Superpixel Selection

Instead of selecting all the pixels in selected bin, pixels in the superpixel of that selected bin are considered which greatly improves the accuracy of utilization.

f. Calculation of Utilization

The percentage utilization of land is calculated by dividing an image into multiple smaller grids of equal size and then the utilization of each grid is calculated using Eq. (3).

$$\text{Utilization \%} = \frac{\text{Total no. of pixels in the grid belonging to the selected class}}{\text{Total no. of pixels in grid}} \tag{3}$$

The utilization estimation of surface objects leads to evaluate the overutilized land.

5 Algorithm

Problem Definition: For a given multispectral satellite image, the main objective is:

(i) To develop segmentation algorithm.
(ii) To estimate utilization of surface objects.

Algorithm: The main objective is to obtain better segmentation from the satellite image and to estimate utilization of surface objects.

Input: m: Multispectral image
Output: a. Land/use mapping.
          b. Utilization of surface objects.

```
Sp = SLIC (I(x, y), No_of_superpixels);
Rd = RMS (Sp[R, G, B]).
for each i in Sp do
        rms = RMS(Sp).
        Diff = rms – Rd
if (diff<threshold)
        Mark that the superpixel is similar to that RD.
else do
        Mark as unsegment.
end if
end for
for each j in Sp do:
        if the superpixel is marked as any non-natural segment do :
Im(x, y) = Sp(x, y)
end if
end for
```

Fig. 2 Proposed algorithm to perform segmentation

Fig. 3 Proposed algorithm to estimate the utilization of surface objects

```
Bin(Im):
for each row and column
    Bin_no = r + (g*11) + (b*121)
    Count[Bin_no]++
end for
Gr = divide( Im )
for each j in Gr do:
    Utilization= util( j )
end for
```

SLIC algorithm shown in Fig. 2 is implemented on multispectral image to extract the superpixel. RMS value is computed for each Sp (SuperPixel) based on its RGB values; if the difference between RMS of Sp and RD is less than T_h (Threshold), then it is grouped into particular class of land use/cover.

Binning is applied on to the image by assigning each pixel into its respective bins based on its RGB. Maximum frequency bin is considered for calculation. Image is divided into grids and percentage utilization is calculated for each grid as shown in Fig. 3.

6 Implementation and Performance Analysis

MATLAB 2013a is used for the implementation. SLIC algorithm is applied on to the input images IRS-P5 and GE Imagery (Fig. 4a, b) of size 587 × 1119 and generates about 7000 superpixels with the new cluster. RMS value of each superpixel is calculated based on median of all the pixels in superpixel. Segmentation algorithm is then applied to the image and it took around 45 s to segment various land use/land cover, shown in Fig. 5a, b. The RMS values of Reference Descriptor for each land objects were selected to perform fine segmentation of land objects.

(a)   **(b)**

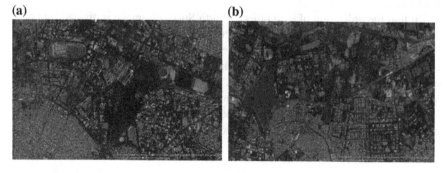

Fig. 4 **a** Input image (IRS-P5). **b** Input image (GE Imagery)

(a)   **(b)**

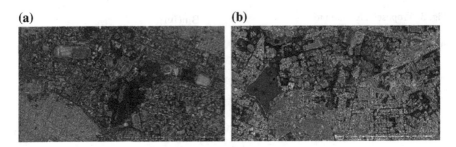

Fig. 5 **a** Segmented image (IRS-P5). **b** Segmented image (GE Imagery)

Table 1 Percentage utilization of surface objects for dataset I and dataset II

Grid	Buildings after binning and sp selection		Vegetation		Bare land		Water body
	DS-I	DS-II	DS-I	DS-II	DS-I	DS-II	DS-II
1	38.5	19.57	21.43	49.97	0.35	2.71	0.59
2	24.69	9.09	51.1	39.63	0.86	5.55	0.13
3	58.7	26.21	15.98	30.62	0.32	3.36	0
4	47.92	23.46	9.65	23.7	1.19	2.02	27.22
5	18.65	18.21	58.87	45.2	1.07	1.91	0.11
6	26.65	30.93	40.94	39.83	5.23	2.11	0.35
7	93.77	47.53	1.05	23.89	0.22	0.3	3.32
8	33.97	64.03	41.44	11.93	0.72	0.15	1.14
9	40.87	40.32	30.65	37.59	2.76	0.41	0.18

Five bins are selected out of which pixels belonging to the second highest frequency bin are selected for superpixel selection to get better accuracy and to eliminate majority of the noise. Certain bare land, roads, and wastelands are not being considered after binning and superpixel selection, thus the data is refined. We obtained accurate result for binning technique with superpixel selection shown in Table 1.

To find the utilization of land objects (Earth Surface), image is divided into nine equal grids (Fig. 6) and utilization is calculated for each object in the grid. This is done by dividing the total number of pixels tagged in objects (buildings, trees and bare land) by the total number of pixels in the grid. Percentage utilization of each object in the grid is tabulated in Table 1. The area covered in grid 7 of dataset I identifies the highest percentage of buildings and less percentage of trees and barren land which signifies overutilization of land. The area covered in grid 8 of dataset II is identified as overutilized land as shown in Fig. 7.

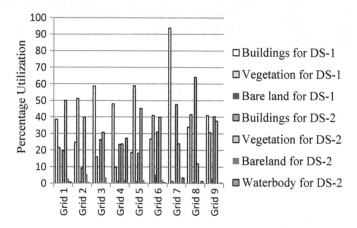

Fig. 6 Grid images of dataset I and dataset II

Fig. 7 Utilization graph for dataset II

7 Conclusions

We have proposed RMS value based multi-thresholding method to perform segmentation operation and by implementing binning technique, we found that the percentage utilization of buildings, trees, and bare land for grid 7 image of dataset I are 96.68%, 1.05%, and 0.22%, respectively, hence it is identified as overutilized land and percentage utilization of buildings, trees, bare land, and water body for grid 8 of dataset II are 68.95%, 11.93%, 0.15%, and 1.14%, respectively.

We have done the accuracy comparison between multispectral satellite imagery and the GE Imagery. Accurate result is found for multispectral satellite imagery whereas GE imagery shows misclassification in classifying water body and green patches (lawn) as the spectral properties of GE Imagery is low. The avenues of future work are to use automation for the selection of reference descriptors and focus more on segmenting various surface objects like roads, soil, rocks, etc.

References

1. L. Gueguen.: Classifying Compound Structures in Satellite Images: A Compressed Representation for Fast Queries, IEEE Transactions on Geoscience and Remote Sensing, vol. 53, no. 4, pp. 1803–1818, (2015).
2. B. Sirmacek and C. Unsalan.: Urban-Area and Building Detection using SIFT Key-points and Graph Theory, IEEE Transactions on Geo-science and Remote Sensing, vol. 47, no. 4, pp. 1156–1167, (2009).
3. Z. Li, W. Shi, Q. Wang, and Z. Miao.: Extracting Man-Made Objects from High Spatial Resolution Remote Sensing Images via Fast Level Set Evolutions, IEEE Transactions on Geo-science and Remote Sensing, vol. 53, no. 2, pp. 883–889, (2015).
4. D. G. Stavrakoudis, G. N. Galidaki, I. Z. Gi-tas, and J. B. Theocharis.: A Genetic Fuzzy-Rule-Based Classifier for Land Cover Classification from Hyperspectral Imagery, IEEE Transactions on Geoscience and Remote Sensing, vol. 50, no. 1, pp. 130–148, (2012).
5. Q. Hu, W. Wu, T. Xia, Q. Yu, P. Yang, Z. Li and Q. Song.: "Exploring the Use of Google Earth Imagery and Object-Based Methods in Land Use/Cover mapping," Remote Sensing, vol. 5, no. 11, pp. 6026–6042, (2013).

Grammar Error Detection Tool for Medical Transcription Using Stop Words Parts-of-Speech Tags Ngram Based Model

B. R. Ganesh, Deepa Gupta and T. Sasikala

Abstract Medical transcription is the conversion of audio files, dictated by medical experts, to electronic data files in a predetermined format. The doctor's thoughts are documented, covering procedures carried out on a patient starting from the time the patient enters the hospital, until the ailment is treated. The transcripts are important to track a patient's medical history and need to be errorless. Most tools are specifically designed to detect wrong grammar in the generic English language. It is important to improve the intelligence of a grammar checker in an unknown domain and to improve the level of accuracy set by the existing tools. These are the driving factors to propose a new approach to an old problem. Using the stop words as the backbone of a sentence and by figuring out the common parts-of-speech tags which surround them, a sentence's grammatical structure can be better understood using statistical methods.

Keywords Grammar checker · Parts-of-speech tagging · Probability model
Language model · Stop words and natural language processing

B. R. Ganesh (✉) · T. Sasikala
Department of Computer Science and Engineering,
Amrita School of Engineering, Amrita Vishwa Vidyapeetham,
Amrita University, Bengaluru, Karnataka, India
e-mail: ganeshbelgur@gmail.com

T. Sasikala
e-mail: jivishnaa@gmail.com

D. Gupta
Department of Mathematics, Amrita School of Engineering,
Amrita Vishwa Vidyapeetham, Amrita University, Bengaluru, Karnataka, India
e-mail: deepagupta.verma@gmail.com

© Springer Nature Singapore Pte Ltd. 2018
V. Bhateja et al. (eds.), *Proceedings of the Second International Conference on Computational Intelligence and Informatics*, Advances in Intelligent Systems and Computing 712, https://doi.org/10.1007/978-981-10-8228-3_5

1 Introduction

Medical transcription is the process of conversion of audio files dictated by medical experts to electronic data files. The doctor's thoughts as well as medical procedures carried out on a patient needs to be documented. These instructions are dictated and stored in an audio file. A medical transcriptionist does the text-to-speech conversion by listening to the audio file and simultaneously typing the content. When these audio files are converted to text files, they contain a vast amount of errors such as grammatical mistakes, semantically incorrect use of words and wrong usage of medical terms. The errors which have occurred might be due to diversity in the dictator's accent, inaudible medical terms, possible audio discrepancies, or grammatical errors in the doctor's dictation. Therefore, these transcripts need to undergo a numerous level of manual proofreading. A proofreader looks for mistakes in the data generated by the medical transcriptionist and makes necessary changes to it. A quality assurance manager scrutinizes the transcript one last time and forwards it to the hospital's archives for future references. The hospital picks up each transcript file from the archives, scrutinizes it one more time, and appends a feedback to the transcript file. This feedback is a copy of the original data of the transcript with the necessary modifications included. The feedback data is the most authentic form of a transcribed audio file. The proposed model addresses the problem of correcting grammatical errors in the medical transcripts but this method is not confined to any particular domain. It can be extended as required if sufficient training data is available.

The amount of audio files that needs to be transcribed is huge. Cerebra Integrated Technologies Limited is a medical transcription company which handles 800 h of medical dictation each month resulting up to 5700 thousand lines of transcribed text, each year [1]. Therefore, medical transcription companies look for ways to automate and aid the proofreaders. To address the issue of grammatical mistakes, companies look to buy commercial grammar checker tools available in the market or use open source tools.

Open source grammar error detection tools are not accurate enough and most of the features are either deprecated or not completely developed. Commercial grammar error detection tools mostly follow a subscription model instead of offering a one-time purchase. This is very expensive. Commercial tools and open source tools could miss errors related to the medical terms since they are mostly trained to detect errors in the generic English language. The feedback data from the hospital serves as a great source of training data to build a grammar error detection model, in-house. This enables the tool to identify any grammatical issues related to medical terms since the training is performed on medical transcripts themselves.

1.1 Prior Work

A grammar checker generally uses techniques such as a parser, a rule-based approach or a statistical approach to isolate the errors. Most of the approaches first separate a corpus of text into a group of sentences before looking for any errors. The individual sentences from the list of sentences are then checked. There are two important things carried out in every checker: detection of sentences and tagging parts-of-speech (POS) tags. Therefore, the total accuracy of the sentence detector and the POS tagger decide the dependent accuracy of the grammar checker itself.

The detectors which find the sentence boundaries have pretty high precision rates (above 95%) for sentences which are from news articles, books, and other well-written sources. There are two methods for detecting grammatical errors which have been popular. There is a method to generate a parse tree of a sentence to detect errors. The sentences are parsed into a tree structure that detects the part of speech for each word. The detector generates parse trees for sentences which are syntactically right. An incorrect sentence structure will either fail during a parse or be parsed into an error tree [2]. Link Grammar is an English grammar parser by The School of Computer Science of Carnegie Mellon University. Link Grammar follows a context-free formula to explain natural language. It consists of a set of terminal symbols which are the words in the language and each has linkage requirement. The words in a sentence and the grammatical dependencies are converted into a graph and analyzed [3]. Relative Position Language Model is an error detection technique which stores relative lexical position relationships between any two given words in a huge corpus. Parse Template Language Model records the possible parse trees to each word in all the sentences in the whole corpus. To avoid the issue of data sparseness, the model does not consider only parse templates but also the sub-trees that constitute them [4]. The first problem with the use of a parser is that it should be able to understand the entire grammar of the languages. The second problem is that parsing some sentences into a single tree may not be possible. Some of the natural ambiguities of a language cannot be handled by a parser.

The other method is to use a checker that is rule based to identify sequences of text that does not seem to be regular. Natural Language Processing (NLP) problems such as POS tagging have been successfully solved using rule-based systems. A collection of governing rules can be revamped from time to time to cover a larger number of errors. Specific errors can be found by tweaking the rules [2]. Rules Combined Character with Instantiation is a method that uses the three important reference groups in the Chinese grammar. These groups are considered to find the most consistent grammatical mistakes made in the China National Matriculation Examination (NME). An exhaustive rule set is generated by systematically analyzing the most common errors made in a 10-year period of the examinations [5]. One other approach is the indexing method of Chinese grammar rules based on corpus. Rules are indexed based on their frequency and their accuracy. To further improve their performance in the real world, an iterative enhancement approach is followed [6]. Using Statistical Machine Translation is also a method where an erroneous sentence is translated to the corrected sentence in the same language [7]. This concept was further clubbed with a classification approach to improve the results of detecting grammatical errors [8]. The common grammatical errors made by the second language learners are a good source to learn and capture the structure of errors where a statistical model is used to classify the associations of the errors based on parts of speech [9]. Apart from these, text mining has been used to uncover the underlying themes or concepts contained in large texts [10]. Healthcare data is already being used to predict trends in the patient conditions and their behaviors [11]. To perform such studies, the data needs to be in the cleanest form possible. Removing grammatical errors before conducting such studies yields better results.

Although, all the abovementioned approaches might be using stop words and POS tags to aid them in finding the errors, they do not extensively explore the associations that can be made by using just the combination of stop words and POS tags alone.

The proposed approach generates a sentence skeleton in which the stop words are rigid structures being linked together by the POS tags of the rest of the words in the sentence. By statistically analyzing the rigid stop words and the surrounding linking local POS tags, a deeper insight into the grammatical structure of the sentence can be gained. This allows for the creation of a more efficient grammar error detection model to spot the errors.

2   Research Method

The design of the *Stop Words POS tags Ngram* model consists of two stages, training and testing stages. The training of the model uses the feedback data which is found appended to all the medical transcripts. The feedback data is the most accurate form of any given transcript. A corpus was formed, for training the model, by concatenating a vast number of feedbacks from many different transcripts. The corpus was preprocessed and a language model was generated by using an open source toolkit called IRSTLM. The IRSTLM tool takes in a huge chunk of text data and splits it into a set of ngram where the maximum level of n can be set as needed [12]. A probability is calculated for each ngram based on its number of occurrence.

During the preprocessing stage, the individual sentences in the corpus were identified and the words in the sentences were transformed into a sequence of processed tokens. The language model holds the probabilities of all the ngrams which are formed out of these processed tokens. This statistical analysis of the occurrences of the processed tokens in the form of ngrams is passed onto the testing phase. The actual grammar checker is a part of the testing phase. The user gives an error-ridden transcript to the system and a similar type of preprocessing is done as explained in the training phase but this time a language model is not created. The preprocessed data is broken up into trigrams and quad grams. The statistical information available from the training phase, in the form of a language model, is

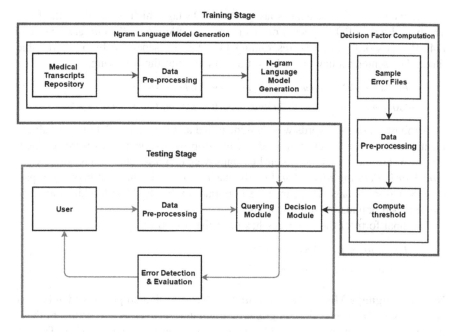

Fig. 1 Dataflow and workflow of the *Stop Words POS tags Ngram* model

used to score these trigrams and the quad grams. In cases where there is no statistical information available on a particular ngram, a feature called back-off weights in the language model is used to predict the probability score. A cutoff had to be set in order to separate out the rogue ngrams. These ngrams were then reported to the user as errors in the system. To set the cutoff, a large number of error-ridden transcripts were processed and the distribution of the probability scores of the ngrams from those files was analyzed. The probability distribution was converted into a box plot to understand and find the outliers in the distribution. The cutoff which was set was based on a combination of *outlier analysis* and *heuristic analysis*. For the sake of simplicity, we have considered only unigrams and bigrams to explain the proposed approach. Figure 1 shows the data flow in the system.

2.1 Training Stage

Data Preprocessing: The sentence boundaries in the corpus need to be identified. A sentence tokenizer in the NLTK toolkit was used for this purpose [13]. Consider the following set of sentences:

I am working in Cerebra. I am in Cerebra right now.

The words in these sentences are given a POS tag which is based on the Penn Treebank English POS tag set. For this purpose, the *Stanford POS tagger* was used [14, 15]. A word and its corresponding POS tag are separated by a ~~ (double tilde). The sentences have now been transformed into the following:

I ~~PRP am ~~VBP working ~~VBG in ~~IN Cerebra ~~NNP. ~~.

I ~~PRP am ~~VBP in ~~IN Cerebra ~~NNP right ~~RB now ~~RB . ~~.

A robust set of stop words was then identified and listed. The POS tags of all the identified stop words were removed but the stop words themselves were retained. For the rest of the words, only their POS tags were retained. This step is done by the *Stop Words POS tag algorithm*. Finally, each sentence is enclosed between a pair of <s> and </s> tags. These tags are important to the IRSTLM toolkit to identify the sentence boundaries in the corpus.

The final form of the sentences has been listed below:

<s> I am working in ~~NNP ~~. </s>

<s> I am in ~~NNP right now ~~. </s>

Ngram Language Model Generation: The new modified corpus was then broken up into a set of non-repeating unigrams and bigrams. The respective counts of these individual ngrams were calculated based on their occurrences in the corpus. Table 1 shows the tokens and the corresponding counts. Using these counts, a language model was generated which specifies the probabilities of sequences of words of varying length, occurring in a given corpus [16]. The language model is stored in the ARPA format. This format is a popular way of representing all the statistical information contained in a language model. According to the format specification, every ngram should have a corresponding probability score and an optional back-off weight. The back-off weights have a role in predicting the probability scores of the ngrams which do not turn out in the language model itself. Further, the logarithm of the probabilities is considered since the probabilities of occurrences of the ngrams are very small.

Table 1 Tokens and the corresponding counts

Unigram tokens	Counts	Bigram tokens	Counts
<s>	2	<s> I	2
I	2	I am	2
am	2	am working	1
working	1	am in	1
in	2	working in	1
~~NNP	2	in ~~NNP	2
~~.	2	~~NNP ~~.	1
</s>	2	~~NNP right	1
right	1	~~. </s>	2
now	1	</s> <s>	2
		right now	1
		now ~~.	1

2.2 Testing Stage

User: The training stage provides the necessary statistical information on a huge corpus of correct data (language model). This information can now be used to analyze error-ridden files which will be passed by a user to the tool.

Preprocessing: The preprocessing step in the testing phase is slightly different from the training phase. Consider the following test sentence:

I am right am working in Cerebra.

After sentence boundary detection, POS tagging and running the *Stop Words POS tag algorithm*, the words in the sentences will be tokenized and will look like the following sentence:

<s> I am right am working in ~~NNP ~~. </s>

The text is then directly broken up into unigrams and bigrams. It can be imagined as if two windows, one token taken at a time and two tokens taken at time, were sliding across the sentences while extracting the contents inside the window at any given point of time. Table 2 shows the statuses of the windows at the various intervals.

Querying Module: The unigram and bigram tokens from the previous step should now be assessed based on their occurrences in the corpus used in the training stage. In order to do that, their probabilities should be queried from the language model. The KenLM toolkit's querying module was used to score the test unigrams and bigrams [17]. The querying module is designed to retrieve the probabilities of recorded sequences of words from the ARPA file (Fig. 2). In cases where the probabilities are missing, a conditional probability is calculated based on the history of occurrences of the known subsequences of words in a given query. The calculations of such non-occurring sequences from the history of their subsequences are governed by the concept of backing-off. One such governing model is the Katz's back-off model [18].

Table 2 Unigram and bigram tokens in the test sentence

Unigram tokens	Bigram tokens
<s>	*<s> I*
I	*I am*
am	*am right*
right	*right am*
am	*am working*
working	*working in*
in	*in ~~NNP*
~~NNP	*~~NNP ~~.*
~~.	*~~. </s>*
</s>	

Sample Queries

	Unigrams			Bigrams	
Words	**log p**	**log b**	**Words**	**log p**	
<s>	-1.26717	-0.477121	<s> I	-0.443263	
I	-1.09108	-0.477121	I am	-0.158832	
am	-1.09108	-0.30103	am working	-0.557478	
working	-1.26717	-0.30103	am in	-0.536793	
right	-1.26717	-0.30103	working in	-0.267172	
...	...	...	...	...	

Query: **am working**
log p (working | am) = -0.557478

Query: **am right**

log p (right)	-1.26717
log backoff (am)	+ -0.30103
log p (right | am)	= -1.5682

Fig. 2 Working of the querying module

Computing the Decision Factor: The probabilities of all the test unigrams and test bigrams have been found. Now, the errors have to be identified, for which a decision parameter has to be set.

A huge sample of error-ridden test files was taken and preprocessed just like the way the sample test file was processed in the previous section. The test unigrams and test bigrams were assigned probability values from the ARPA file. This generated a huge probability distribution, on which an outlier analysis was performed by using a *box plot*. The outlier formula on the minimum side of the box plot was used to compute the cutoff:

$$< Q1 - (1.5 \times IQR). \tag{1}$$

However, in order to verify this cutoff value, the probability scores were further analyzed heuristically.

Decision Module and Detecting Errors: The example that is being followed here shows that the training has been performed only on two lines of transcribed text.

This is no way exhaustive and is only meant for the sake of reader's understanding of the concept. However, it is easy to notice from Table 3 that the probabilities of the bigrams, "am right" and "right am" being the least of the lot. The Decision Module identifies such ngrams from the sequences of tokenized ngrams based on the preset cutoff. These particular bigrams also happen to be the site of the error in the test sentence. The errors are the ones whose probabilities go well below the cutoff. These errors are then reported to the user.

Table 3 Tokens with the assigned probability scores

Unigram tokens	Probability score	Bigram tokens	Probability score
<s>	−1.26717	<s> I	−0.443263
I	−1.09108	I am	−0.158832
Am	−1.09108	**am right**	**−1.5682**
right	−1.26717	**right am**	**−1.39211**
Am	−1.09108	am working	−0.557478
working	−1.26717	working in	−0.267172
In	−1.09108	in ~~NNP	−0.158832
~~NNP	−1.09108	~~NNP ~~.	−0.536793
~~.	−1.09108	~~. </s>	−0.157866
</s>	−1.09108		

2.3 Data Statistics and Parameter Setting

As mentioned earlier, the training has been performed here on just two sentences. This is not exhaustive and is meant only for the sake of reader's understanding. The actual model was trained on a huge corpus of text.

Table 4 summarizes the statistical information about the data which was used for building the actual model. The stop words were taken from various reliable sources on the Internet [19]. The *decision factor* or the *cutoff* was set at −16.3 after performing an outlier analysis (Fig. 3) which was verified heuristically.

In the actual model, trigrams and quad grams were considered in place of the unigrams and bigrams. Trigrams are particularly well suited in identifying errors with respect to tense of the verb in a given sentence. On the other hand, quad grams are used to identify errors relating to subject–verb agreement [2].

Table 4 Statistical specification of data

Item	Quantity
Number of transcripts used for training	8446
Number of words in the transcripts	1494232
Number of sentences in the transcripts	105725
Number of stop words	571
Number of transcripts used to compute decision factor	1100
Number of words for computing decision factor	222276
Number of sentences to compute decision factor	12836

Fig. 3 Probability distribution, box plot, and outlier analysis

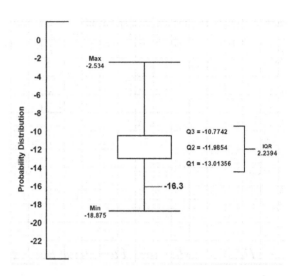

3 Results and Analysis

The evaluation of the grammar checker is done by a corpus of 180 annotated sentences. A chunk (67%) of the corpus is grammatically right and the rest are subjected to few errors. Instead of using a tiny corpus of sentences, using a large-scale evaluation would use a larger variety of error sentences for the evaluation. The same set of sentences was used to evaluate both Microsoft Word 2016 and the *Stop Words POS tags Ngram* model. Table 5 explains the categories into which the sentences were placed. The corpus to test the grammar checker is a collection of many sentences. Let us suppose the total number of sentences is e. About a sentences $(a < e)$ are grammatically wrong where $e = a + b + c + d$. The analysis and evaluation of the model were carried out based on the Precision, Recall, F-measure, and Accuracy parameters. Many sentences in the test corpus were taken from the medical transcripts which were yet to be proofread. A sample of sentences from news articles on different topics was used as the set of correct sentences. The most common types of grammatical errors mentioned on the web were taken to generate the set of incorrect sentences.

An information retrieval system is evaluated by using two standard parameters called *Recall* and *Precision* [2]. By minimizing the value of b, recall can be

Table 5 Sentence categories as confusion matrix

Total population		Probability score	
		Yes	No
True condition	Yes	True positive (a)	False negative (b)
	No	False positive (c)	True negative (d)

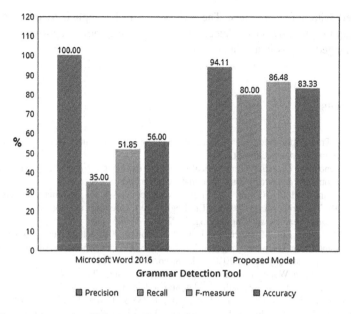

Fig. 4 Comparative study of Microsoft Word 2016 and the proposed model

controlled, i.e., all the actual errors can be found. One of the most important parameter is the accuracy. The verification is not based on the error type detected to match a particular grammatical error. A sentence is only defined as right or wrong. The accuracy parameter combines all the results in Table 5 into one unique percentage score. Accuracy is a ratio of the number of grammatically wrong sentences and grammatically right sentences detected to the total number of sentences in the corpus (Fig. 4).

The Microsoft Word 2016 fails to find many errors since its *recall* is very low. On the other, the *proposed model* can find many complex errors relating to subject–verb agreement and the tense of the verbs. But there are issues relating to some sentences where many stops words form a long continuous chain and in cases where very less POS tags are present.

4 Conclusion

The proposed model is still in its nascent form. The approach explained here should be the key highlight. Further, the following aspects of the tool can be highlighted: The *Stop Words POS tags Ngram* model can help capture the complete sentence structure. A new set of training data can always be patched to the existing model as and when available. This flexibility helps in increasing the robustness of the tool. If the necessary training data is available in the right format, training can be performed within few minutes. The technique can be applied to other domains which require grammar error detection, e.g., blogging, news reporting, etc., provided a tagged corpus is available.

References

1. Medical Transcription at Cerebra Integrated Technologies Limited. Retrieved from http://www.cerebracomputers.com/mt.htm.
2. Manu Konchady. Detecting Grammatical Errors in Text using a Ngram-based Ruleset. http://emustru.sourceforge.net/detecting_grammatical_errors.pdf, 2009.
3. Y. H. Wang and C. H. Lin. An English sentence parser for grammar error detection. TENCON '02. Proceedings. 2002 IEEE Region 10 Conference on Computers, Communications, Control and Power Engineering, 2002; vol. 1: 445–448.
4. C. H. Wu, C. H. Liu, M. Harris and L. C. Yu. Sentence Correction Incorporating Relative Position and Parse Template Language Models in IEEE Transactions on Audio, Speech, and Language Processing, August, 2010; vol. 18, no. 6: 1170–1181.
5. Ying Jiang, Tong Wang, Tao Lin, Fangjie Wang, Wenting Cheng, Xiaofei Liu, Chenghui Wang. A rule based Chinese spelling and grammar detection system utility. 2012 International Conference on System Science and Engineering (ICSSE), Dalian, Liaoning, 2012; 437–440.
6. Ying Jiang, Zechao Lin, Junyue Wang, Miaojuan Dai, Liwei Zhen, Ningran Li, Zhouyang Hu, Shuzhou Chen, Yang Meng. Corpus Based Chinese Grammar Error Detection Rules Evaluation Method and System. Intelligent System Design and Engineering Applications (ISDEA), 2013 Third International Conference on, Hong Kong, 2013; 496–499.
7. N. Ehsan and H. Faili. Statistical Machine Translation as a Grammar Checker for Persian Language. Sixth Int. Multi-Conference Comput. Glob. Inf. Technol., 2011; 20–26.
8. Hendy Raymond Susanto, Peter Phandi, and Tou Hwee, Ng. System combination for grammatical error correction. In Proceedings of the 2014 Conference on Empirical Methods in Natural Language Processing (EMNLP), 951–962.
9. Rozovskaya, A. and Roth, D. Building a State-of-the-Art Grammatical Error Correction System. Transactions of the Association for Computational Linguistics, 2014; 2: 419–434.
10. Naw Naw and Ei Ei Hlaing. Relevant Words Extraction Method for Recommendation System. Bulletin of Electrical Engineering and Informatics, September 2013; vol. 2, no. 3:169–176.
11. Milovic and M. Milovic. Prediction and decision making in health care using data mining. International Journal of Public Health Science, 2012; vol. 1, no. 2: 69–78.
12. M. Federico, N. Bertoldi, M. Cettolo. IRSTLM: An Open Source Toolkit for Handling Large Scale Language Models, Proceedings of Interspeech, Brisbane, Australia, 2008.
13. Bird, Steven Klein, Ewan Loper, Edward Baldridge, Jason. Multidisciplinary instruction with the Natural Language Toolkit. Proceedings of the Third Workshop on Issues in Teaching Computational Linguistics, ACL.
14. Kristina Toutanova, Dan Klein, Christopher Manning, and Yoram Singer. Feature-Rich Part-of-Speech Tagging with a Cyclic Dependency Network. In Proceedings of HLT—NAACL 2003; 252–259.
15. Kristina Toutanova and Christopher D. Manning. Enriching the Knowledge Sources Used in a Maximum Entropy Part-of-Speech Tagger. In Proceedings of the Joint SIGDAT Conference on Empirical Methods in Natural Language Processing and Very Large Corpora (EMNLP/VLC-2000), pp. 63–70.
16. Wikipedia, Retrieved from https://en.wikipedia.org/wiki/Language_model, September 2016.

17. KenLM: Faster and Smaller Language Model Queries Kenneth Heafield. WMT at EMNLP, Edinburgh, Scotland, United Kingdom, 30–31 July, 2011.
18. Katz, S. M. Estimation of probabilities from sparse data for the language model component of a speech recogniser. IEEE Transactions on Acoustics, Speech, and Signal Processing, 1987; 35(3): 400–401.
19. Stop words list, Retrieved from https://pypi.python.org/pypi/stop-words, September 2016.

Churn and Non-churn of Customers in Banking Sector Using Extreme Learning Machine

Ramakanta Mohanty and C. Naga Ratna Sree

Abstract With an invention been made in computational techniques, there has been increasing interest in the field of neural networks as major potential machine learning techniques have come to the front. In the current past, gradient descent-based algorithm is used in feed forward neural network and the parameter utilized, which devours extensively more opportunity for learning and tuned iteratively. The single hidden layer feed forward neural network have the input weights, the hidden layer and biases randomly assigned. The straightforward backward operation is utilized to discover output weight which thus relies on upon handling from hidden layer to output layer. In this paper, we propose to utilize Extreme Learning Machine (ELM) to foresee client churn. The goal of this paper is to propose a novel approach that enhances the precision of churn and non-churn of clients using banking data.

Keywords Extreme learning machine · Artificial neural networks
Customer churn · Single-layer feed forward network

1 Introduction

Customer churn implies loss of clients. Customer churn can be seen in numerous ventures like banking, broadcast communications and insurance organizations. Nowadays, Customer churn has turned into a vital issue in the banking industry, as well as in media transmission industry excessively [1]. To discover the early cautioning signs in lessened exchanges of clients, steps should be taken to retain the customer. The principal target of the customer retention is to cut through the

R. Mohanty (✉) · C. Naga Ratna Sree
Computer Science and Engineering, Keshav Memorial Institute of Technology,
Narayanaguda, Hyderabad 500011, India
e-mail: ramakanta5a@gmail.com

C. Naga Ratna Sree
e-mail: cnratnasree@gmail.com

© Springer Nature Singapore Pte Ltd. 2018
V. Bhateja et al. (eds.), *Proceedings of the Second International Conference
on Computational Intelligence and Informatics*, Advances in Intelligent Systems
and Computing 712, https://doi.org/10.1007/978-981-10-8228-3_6

51

customer's behaviour based on predictions. Because of the mechanical headways, information has been expanded in an awesome space, size and dimensionality step by step. Accordingly, it is essential to investigate some viable machine learning strategies that can be used to analyse the information and haul out the valuable information from the information.

Of late, Extreme Learning Machines (ELMs) has become one of the prominent machine learning methods for predictive analysis. ELMs are a feed forward neural network recognized by the introduction of their hidden layer weights, alongside the training algorithm [2–5]. These models can create great execution and are utilized to find a huge number of times quicker than other types of neural network like probabilistic neural network, decision trees and by use of ensemble method [6, 7].

Since client agitation has turned into a noteworthy issue, to keep the beat, we utilize a machine learning model where it is utilized to discover which sort of clients will probably churn.

This paper is organized as follows: Sect. 2 presents about related work on churn and non-churn different domain. Section 3 examines about the review of an extraordinary extreme learning machine. In Sects. 4 and 5 the outcomes are discussed and the paper is concluded.

2 Literature Survey

Building a functional model for customer churn has now become a decisive topic in recent days. The emergence of e-commerce has increased the information availability and offered for companies to respond efficiently to the clients, hence this gave rise to the notion of the customer churn by Lejeune et al. [8]. Kotler and Keller [9] analysed that customer relationship management is nothing but solving the customer's problem such as customer dissatisfaction, switching cost, service message and customer status and so on. Kincaid [10] defined CRM as the essential usage of information processing, use of technology, and people to manage the client relationship with the governing body for the industry to develop. Burez and Van den Poel [11, 12] proposed to manage the customer churn by two methods, viz. reactive and proactive. According to him, a customer requests the organization to offset the service supplied by the organization but on the other hand, in the proactive approach, organization solved the problem brought up by the customer and produced it as suits for the churn and find out a constructive way to solve the issues. Shaw et al. [13] proposed that genuine client relationship management can be found by accumulating the information revelation handle with the administration and by seeking after the promoting the market plans. Salman et al. [14] examined about logistic regression and time oriented, analytical technique to dissect client agitation. Zhang et al. [15] developed a model which consisted of logistic regression, decision tree and neural network (NN) learning algorithms by using Chinese mobile telecommunication dataset and also used SAS Enterprise Miner for training the models for prediction of customer churn. Coussement and Van sanctum Poel [16] utilized support vector machines and the result acquired by

SVM is that SVM performed superior to other methods. Mutanen [17] employed logistic regression in personal retail banking dataset to predict customer churn. Most recently, Mohanty and Jhansi Rani [18] employed FuzzyArtMap and counter propagation neural network on Indian Telecommunication data to predict churn of customers. From this, we can conclude that neural networks can be used to predict customer churn in different domains such as retails [19], banking [20] and finance [21]. This paper proposes a neural network-based approach, i.e., extreme learning machine to predict the customer churn with respect to the banking domain.

3 Methodology

3.1 Extreme Learning Machine

The architecture of extreme learning machine is depicted in Fig. 1. Extreme Learning Machine (ELM) is a learning algorithm proposed by Huang et al. [2–5]. It consists of a single-layer feed forward neural network, which consists of an input layer, hidden layer and output nodes. The input weight is accessed to the outputs by series of weights.

The main advantages of ELM are that its parameters, hidden nodes, input weights and biases are randomly allocated and need not required to be tuned.

Let D be the training data of M samples, which can be represented as $(X_1 t_1)$, where $\{X_1\} \in R^m$, R^m is m times input vector. $t_1 \in R^n$, where R^n is n times Xl target vector. The ELM is a single-layer network having N nodes and let f(x) be the activation which can be defined as

$$\sum_{i=1}^{\bar{N}} \beta_i f_i(x_j) = \sum_{i=1}^{\bar{N}} \beta_i f(w_i, b_i, x_j) = o_j, 1 \le j \le. \tag{1}$$

where w_i is weight vector of the input hidden nodes.

β_i be the weight vector of both the hidden nodes and the output nodes, whereas b_i is the threshold of the hidden node.

Fig. 1 Architecture of ELM

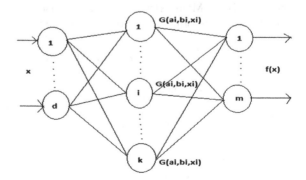

We need to derive the relation between X_i and t_i from β_i, w_i and b_i such that Eq. (2) can be written as

$$\sum_{i=1}^{\bar{N}} \beta_i f(w_i, b_i, x_j) = t_j, 1 \le j \le N. \tag{2}$$

Equation (2) can be written alternatively as

$$H\beta = T. \tag{3}$$

where

$$H = \begin{bmatrix} h(x_1) \\ h(x_N) \end{bmatrix} = \begin{bmatrix} f(w_i, b_i, x_j) & \cdots & f(w_N, b_N, x_1) \\ \vdots & \ddots & \vdots \\ f(w_i, b_i, x_N) & \cdots & f(w_N, b_N, x_N) \end{bmatrix}_{NXn} \tag{4}$$

H is the hidden layer output matrix of the network.

$h(x) = f(w_i, b_i, x) \dots f(w_N, b_N, x)$ is the hidden layer feature mapping. Equation (4) provides the hidden node output with respect to the input vector. If we know the activation function, we can assign randomly the value of input weight vector w_i and hidden layer biases b_i. Therefore, parameters need not required to be tuned without changing the hidden layer output matrix. Accordingly, ELM provides smallest training error and the smallest output weight as well to predict accurately.

$$Min = \|H\beta = T\| \tag{5}$$

If N (the no. of hidden neuron) = training sample, then, the output weight β can be found out by inverting the H. Thus, we can get less error in training samples by using ELM.

Further,

$$\beta = H^+ T, \tag{6}$$

where H^+ is the Moore–Penrose pseudo inverse of matrix H, and it can be computed by using the orthogonal projection method as

$$H^+ = \left(HH^T\right)^{-1}H^T, \tag{7}$$

when HH^T is remarkable, or

$$H^+ = H^T \left(HH^T \right)^{-1}. \tag{8}$$

When HH^T is nonremarkable.

Where T means matrix transposition.

3.2 Algorithm

A dataset N with the number of hidden neurons L and Activation Function $f(x)$ is taken then

Step 1. Randomly assign input weights w_i and biases b_i i.e. hidden layer parameters $(w_i, b_i,)$, i = 1———L randomly.

Step 2. Calculate the hidden layer output matrix H.

Step 3. Calculate the output weight $\beta: \beta = H^+ T$.

4 Results and Discussions

The churn and non-churn of customers is a Portuguese Banking Sector dataset, where it consists of both categorical and numerical values. The dataset consisting of 41,189 samples and having 20 attributes are presented in Table 1.

The dataset is divided into training and testing in the ratio of 80:20 and followed by tenfold cross-validation. We developed the Java code for ELM and the experiment is carried out in MATLAB environment. We use the input function for ELM as follows: training data, testing data, ELM type, number of hidden neurons, and activation function type. The outputs of ELM algorithm are average training time, average testing time, average training, and testing accuracy. We used the following formula for calculating the training and testing accuracies as follows:

Training Accuracy = 1 − *MisclassificationRate_Training/Number of Training data*

and

Testing Accuracy = 1 − *MisclassificationRate_Testing/Number of Testing data.*

In our experiment, we chose four types of activation function, viz. Sigmoid, Radbias, Hardlim and Tribias, respectively. From our experiment, we found that Tribias activation function took less time for testing data to be processed. Accordingly, it also provides best training accuracy of 0.0044 and testing accuracy of 0.0420 by using the Tribias activation function. We got training accuracy of

Table 1 Attributes of Portuguese telemarketing dataset

Sl. no	Attributes
1	Age
2	Job (type of job)
3	Marital (marital status)
4	Education
5	Default (has credit in default)
6	Housing (housing loan)
7	Loan (personal loan)
8	Contact (contact communication)
9	Month (last contact month year)
10	Day_of_week (last contact day of week)
11	Duration (last contact duration)
12	Campaign (no. of contact performed)
13	Pdays (no. of days that passed the client contacted)
14	Previous (no. of contact before campaign)
15	Poutcome (outcome of previous market campaign)
16	Emp.var.rate (employment variation date)
17	Cons.price.idx (consumer price index)
18	Cons.conf.idx (consumer confidence index)
19	Eribur3m (daily indicator)
20	nr.employed (no. of employees)
21	Output variable (yes or no)

0.0431 and testing accuracy of 0.0420 by using Radbias activation function compared to other activation functions, which are presented in Table 2.

Further, we simulate our experiment folds wise, we found that in case of fold 1 the best training accuracy is of 0.0045 and followed by fold 6 value of 0.0500, respectively. But, on the other hand, the best testing accuracy value is on fold 1 of value 0.0024 and followed by fold 5 of value 0.0344, respectively, which is shown in Table 3.

Table 2 The training and testing accuracies and time taken by applying different ELM activation functions

Type of activation function	No. of hidden neurons	Average training time (ms)	Average testing time (ms)	Training accuracy	Testing accuracy
Sigmoid	700	39.6875	0.3281	0.0566	0.0582
Radbias	600	45.1250	0.2031	0.0431	0.0420
Hardlim	500	16.5313	0.2500	0.0541	0.0585
Tribias	400	8.4375	0.1718	0.0044	0.0420

Table 3 The training and testing accuracies fold wise of dataset

Average	Training time	Testing time	Training accuracy	Testing accuracy
Fold 1	7.43	0.14	0.0045	0.0024
Fold 2	11.75	0.28	0.0509	0.0449
Fold 3	16.35	0.28	0.0520	0.0398
Fold 4	11.84	0.28	0.0540	0.0473
Fold 5	13.64	0.20	0.0536	0.0344
Fold 6	12.34	0.20	0.0500	0.0565
Fold 7	14.75	0.31	0.0509	0.0548
Fold 8	14.07	0.31	0.0515	0.0424
Fold 9	16.87	0.31	0.0517	0.0388
Fold 10	15.06	0.21	0.0532	0.0376
Total average	13.410	0.23	0.04714	0.03989

The novelty of this algorithm is that processing time for both training and testing data is in the order of milliseconds compared to other techniques what we observed earlier [22, 23]. In our experiment, for training data, it took an average time of 13.41 ms and testing data of 23 ms out of 41.189 samples.

5 Conclusions

This paper analyses the systematic way to predict customer churn by employing Extreme Learning Machine. The different activation functions used to predict the actual churners relatively well. Bank customers churning average Training time, Average Testing Time and Average Training Accuracy, Average Testing Accuracy is being computed from the given dataset. The ELM model gives more accurate results compared to other machine learning techniques, viz. SVM, Gradient descent, backpropagation neural networks, etc. Time to time, many data mining techniques have been implemented on the banking data to predict the customer churn and non-churn but our results outperformed all other machine learning techniques in terms of accuracies and time taken to process the dataset. This is the significant study in this paper.

References

1. Athanassopoulos, A. D.: Customer satisfaction cues to support market segmentation and explain switching behavior. Journal of Business Research, Warwick Business School, pp. 191–207, (2000).

2. Huang, G. B., Chen, L., Siew, C. K.: Universal Approximation Using Incremental Constructive Feedforward Networks with Random Hidden Nodes. IEEE Transactions on Neural Networks, vol. 17, no. 4, pp. 879–892, (2006).
3. G. B. Huang, G. B., L. Chen., Convex Incremental Extreme Learning Machine: Neurocomputing, vol. 70, pp. 3056–3062, (2007).
4. Huang, G. B., Zhou, H., Ding, H., and Zhang, R.: Extreme Learning Machine for Regression and Multiclass Classification, IEEE Transactions on Systems, Man, and Cybernetics - Part B: Cybernetics, vol. 42, no. 2, pp. 513–529, (2012).
5. Huang, G. B.: An Insight into Extreme Learning Machines: Random Neurons, Random Features and Kernels, Cognitive Computation, vol. 6, pp. 376–390, (2014).
6. Huang, G. B., Chen L.: Enhanced random search based incremental extreme learning machine, Neuro computing 71, 3460–3468, (2008).
7. Huang, G. B., and Ding, X., Zhou, H.: Optimization method based extreme learning machine for classification. Neuro computing 74, pp. 155–163, (2010).
8. Lejeune, M. A.: Measuring the impact of data mining on churn management. Internet research: Electronic Networking Applications and Policy, 11, pp. 375–387 (2001).
9. Kotler, P., Keller, L.: Maketing Management (12th ed.). New Jersey, Pearson prentice Hall (2006).
10. Kincaid, J.: Customer relationship management: Getting it Right NJ: Pretice-Hall PTR (2003).
11. Burez, J., Van den Poel, D.: Handling class imbalance in customer churn prediction, Expert system with Applications, 36, pp. 4626–4636 (2009).
12. Bue, J., Van den Poel, D.: Separating financial from commercial customer churn: Expert Systems with Applications, 35, pp. 497–514 (2008).
13. Shaw, M. Subramaniam, C., Tan, G., Weldge, M.: Knowledge management and data mining for marketing, Decision Support System, 31(1), pp. 127–137 (2001).
14. Salman, S.: Value based Time dimensioned Churn Prediction, Journal of Emerging trends in Computing and Information, vol. 4, pp. 180–183, (2013).
15. Zhang, X., Z. Liu, Z., Yang, X., Shi, W., and Wang, Q.: Predicting Customer Churn by Integrating the Effect of the Customer Contact Network: IEEE International conference on Service Operations, Logistics and Informatics (SOLI), Shandong, China, pp. 392–397, (2010).
16. Coussement, K., Van den Poel: Improving Customer attrition Prediction by integrating emotions from Client/Company interaction emails and evaluating multiple classifiers. Experts Systems with Applications, 36, pp. 6127–6134 (2009).
17. Mutanen, T. Customers churn analysis – a case study, Research Report No. VTT-R01184-06, Dated March 15, Available at: http://www.vtt.fi/inf/julkaisut/muut/2006/customer_churn_ case_study.pdf, Retrieved on 19 August (2008).
18. Mohanty R, Jhansi Rani K.: Application of Computational Intelligence to predict churn and Non churn of customers in Indian Telecommunication. IEEE International Conference on Computational Intelligence and Communication Networks, pp. 598–603, (2015). India.
19. Buckinx, W., Van den Poel, D.: Customer base analysis: partial defection of behaviourally loyal clients in a non-contractual FMCG retail setting. European Journal of Operational Research, 164(1), pp. 252–268, (2005).
20. Van den Poel, D., Larivie're, B.: Customer attrition analysis for financial services using proportional hazard models. European Journal of Operational Research, 157, pp. 196–217, (2004).
21. Chiang, D. A., Wang, Y. F., Lee, S. L., Lin, C. J.: Goal-oriented sequential pattern for network banking churn analysis. Expert Systems with Applications, 25, pp. 293–302, (2003).
22. Mohanty, R., Ravi, V., Patra, M. R.: Hybrid Intelligent Systems for Predicting Software Reliabilit, Elsevier, Applied Soft Computing 13, PP. 180–200, (2013).
23. Mohanty, R., Ravi, V., Patra, M. R.: Web Services Classification using intelligent Techniques, Elsevier, Expert Systems with Applications 37, PP. 5484–5490, (2010).

Identifying the Local Business Trends in Cities Using Data Mining Techniques

B. Pallavi Reddy and Durga Toshniwal

Abstract With the growing popularity of social media, and several users using them, humongous amount of information is being generated. This information may be 140 words like tweets, posts, and images shared on Facebook or reviews written on Yelp. It would be valuable to both consumers and businesses if the current local business trends followed by the people of different cities can be identified. Local business reviews are available, which when mined can be used to find out the local business trends across cities. With this information, the users can watch out for the prevalent local businesses (yoga, beauty and spas, ballet, restaurant, etc.) in each city, and the businesses can chart their b-plans, accordingly. To accomplish this, the present work uses data mining technique of clustering on benchmark dataset.

Keywords Local businesses · Trends · Data mining techniques
Clustering · Yelp

1 Introduction

With the boom in the field of information technology, came the Internet and then came the social media. Social media is of varied types like social networking sites (Facebook, LinkedIn), microblogging like Twitter, news aggregation like Google Reader, photo sharing (Flickr, Picasa, Instagram), video sharing (YouTube), recommendation platforms like Yelp, etc. 2016 statistics say that almost two billion people are currently using social networks and it is expected to grow even more because of the increasing mobile technologies. Due to the tremendous amount of time people spend on these sites, they tend to give their views on the various current

B. Pallavi Reddy (✉) · D. Toshniwal
Department of Computer Science & Engineering,
Indian Institute of Technology Roorkee, Roorkee, India
e-mail: bpallavi.reddy9@gmail.com

D. Toshniwal
e-mail: durgatoshniwal@gmail.com

© Springer Nature Singapore Pte Ltd. 2018
V. Bhateja et al. (eds.), *Proceedings of the Second International Conference
on Computational Intelligence and Informatics*, Advances in Intelligent Systems
and Computing 712, https://doi.org/10.1007/978-981-10-8228-3_7

trends happening in the world. Be it their opinions on the elections of their respective countries, the products they have bought, institutions and organizations they have attended, etc. This generates huge quantities of data on social media, which can be called in other terms as social media data. Collecting such huge information, representing it, analyzing it, and mining the useful contents (like patterns) out of this social media data is basically what is called as social media mining.

Among so many social networking sites, Yelp.com [1, 2] is a popular social media website which provides an online platform in which users can review local businesses. It is basically a crowd-sourced review platform and a social networking website. Not only the users but also the businesses can provide their information like the hours they are open, location, contact information, etc. Figure 1 shows one such local business. Yelpers can thus see which local businesses are prevalent in their city and also recommend the same to other Yelpers.

The aim of this study is to find the trends of cities like which local businesses are prevalent in a city from the reviews on Yelp dataset. Say, for example, a city "A" is known for rafting and kayaking, a city "B" is known for beauty and spas and nightlife, and so on. Work has been done in fields like recommendation (of restaurants, cuisines, and movies among various services), star rating analysis, sentiment analysis, popularity prediction, etc., but less work has been done to find the different trends followed by people living in different cities. This study aims to find such trends across all cities. The trends being considered are the local business trends.

Finding such trends will be beneficial to both the customers as well as the businesses. Let us take some scenarios. Say, tourists want to know what trends the people of a particular city (they are planning to visit) follow. Then, these local business trends will give them an idea as to what local businesses are prevalent in that city. Accordingly, they can plan on visiting those businesses. On the other side of the coin, the businesses can know if their category of business is prevalent in that city or not. On the same terms, they can strategize their business plan to expand their business, and also the startups can know their opportunities in a particular city.

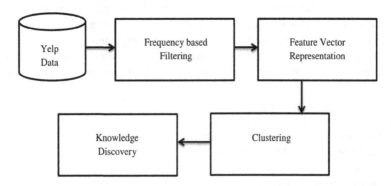

Fig. 1 Proposed workflow

The remainder of this paper is organized as follows. Section 2 presents a review of related research work. Our proposed method is described in Sect. 3. Experimental results are shown in Sect. 4. Finally, the conclusion is summarized in Sect. 5.

2 Related Work

X. Mao et al. have used Yelp to detect popularity of recipes in [3]. Detecting popularity of any resource means predicting how popular that resource is among the Yelpers. Popularity prediction of news, videos, and advertisements has been of interest in recent studies. In this study, researchers have predicted the popularity of recipes by using polynomial regression. Traditional popularity detection schemes only use the statistical measures of the reviews but in this study, the sentiments of reviews were also taken into consideration.

M. Prithivirajan et al. [4] have also found how well the reviews relate to the star ratings the Yelpers have given to a local business. For example, if someone has given a bad rating but has written something good about that business, then the rating becomes irrelevant. The study was motivated by the fact that star ratings are the first to be seen by any person, and bad ratings on social media have the ability to degrade any business by manifolds. So to find the relevance of the ratings, they used LDA (Latent Dirichlet Allocation) [5] for topic modeling; aspect-based sentiment scoring.

Research has also been done on the business reviews provided in the Yelp dataset. A. Salinca [6] proposed several approaches for sentiment classification, which uses various machine learning models and feature extraction models.

Content annotation extraction has also been an interesting area for researchers. Using the reviews on Yelp, one such study by M. Yamamoto et al. [7] focuses on extracting the good and bad points of a service reviewed on Yelp. The content considered are the Yelp reviews, might be food, restaurant, etc. To do this, they used an unsupervised approach of data mining to extract unique aspects and the user opinions on those aspects.

Lot of work has already been done on recommending restaurants, etc., but the study by T. Yamasaki et al. [8] focuses on giving review-based service recommendation. In other words, for a particular restaurant, their system takes a review as a query and provides the user with the services of that restaurant that has been collected from other reviews of a similar kind. This helps the user to not only find the desired restaurant but also the desired service of that restaurant. To achieve this, bag-of-words model and skip-gram-based model have been used.

Another study by I. Varlamis et al. [9] extracts the valuable information from the Yelp reviews regarding the aspects that users prefer in products or services. This aims for personalized advertisement by combining aspect-based opinion analysis and personalized recommendation. This will not only help the users but also aims at benefiting the businesses. For this, the content of the aspect as well as the user opinion on that aspect has been mined.

The study by G. Zhao et al. [10] predicts the user ratings by using the social behavior of Yelper's rating behaviors. A user's daily rating was explored by using aspects like when a user rates an item, what the rating is, on what item did the user give the rating, the user's interest based on his/her previous rating records, and how this rating behavior influences fellow friend Yelpers. Based on these rating behaviors, the study predicts the user ratings by probabilistic matrix factorization model.

As we can see, the work done till now mainly focuses on a single domain, for example, the recommendation of movies, restaurants, and services, determining the popularity of cuisines, news, and businesses, predicting star ratings, analyzing the star ratings of reviewers, sentiment analysis of reviews, etc. Less work has been done to find the trends of local businesses and compare the results across the diverse set of cities provided in the review data. These trends will help to differentiate the cities based on the trends followed by the people living in different cities. So, the study focuses on finding the cultural trends, specifically the trends based on local businesses in different cities using data mining techniques.

3 Proposed Work

The main objective is to find the local business trends followed by people living in different cities. The result intended to obtain is the cities with their prevalent local businesses and finally, show the local business trends followed across cities. To accomplish this, the step-by-step workflow has been shown in Fig. 1 and the steps are detailed below.

3.1 Data Preprocessing

The dataset used might contain reviewers who have reviewed an abnormally high number of businesses. So to remove the impact of these reviewers and free our results from being biased to such users, the study will calculate the standard deviation "s" of the Yelper's reviews. The dataset was then observed for different multiples of "s" (i.e., how many users have reviews greater than several multiples of s). Depending on this observation, the Yelpers having reviews more than a large multiple of "s" will be excluded from further steps. Thus, a frequency-based filtering was carried out in the dataset.

3.2 Vector Representation

After the preprocessing step of removing users having abnormally high reviews, the task is to identify the business categories, like food and restaurants, beauty and spas, nightlife, rafting, party and event planning, etc., that has been provided in the dataset. Each city will then be represented as a normalized vector of the above-determined categories, where each entry in the vector will represent the percentage of presence of the xi-th business category in that city.

3.3 Clustering

There are several data mining techniques available. However, to find the local business trends across cities, clustering has been used in this study. Clustering is a data analysis technique by which objects are grouped into clusters, where each cluster is a collection of objects with similar characteristics. There are many clustering algorithms. But in this study, the clustering technique used is the centroid-based k-means [11] algorithm. The reason for this selection is that it is simpler and has lesser computational complexity such as DBSCAN. K-means requires its distance function to be well defined. This has been made sure by the vector representation of cities as discussed in the above subsection. Based on the selection, k-means clustering was then applied to the above-obtained city vectors. The result would be clusters of cities, each cluster being prevalent for a particular combination of local business categories.

4 Experimental Results

In order to fulfill the objective of identifying the local business trends followed by people living across cities, the proposed work has been discussed in the above section. This section talks about the dataset and the experiments conducted.

4.1 Dataset Description

Yelp is one of the most popular user review sites in the world. Data has been obtained from the Yelp dataset challenge [12] which is a benchmark dataset used by researchers across the world. It has data pertaining to various local businesses, users (Yelpers to be specific), reviews, check-ins, and tips. There is information about 42,153 local businesses, 11,25,458 reviews, and 2,52,898 Yelpers. All this data is in json (JavaScript Object Notation) format. Services included in the dataset are limited to those located in the cities across the United States of America.

4.2 Data Preprocessing

On closely observing the dataset, it was found that there were few Yelpers who had reviewed for an alarmingly huge number of businesses. The mean number of reviews of all the users was four. But there were users whose review count was in hundreds and thousands. There is a good chance that these reviews might be fake or their reviews can impact the final result, thereby biasing it. So, to ignore these Yelpers, the study first found out the standard deviation of all these reviews. This value came out to be 15. As shown in Fig. 2, the line graph plots the number of Yelpers versus the number of reviews that the Yelpers have greater than the number of reviews in multiples of the standard deviation obtained, i.e., 15. In other words, a coordinate on the line graph, say, (40, 18) shows that there are 18 Yelpers whose review count is more than 40 times the standard deviation (15), i.e., 600 reviews. Considering the huge dataset and the number of reviews written by the users, 40 such standard deviations were taken, i.e., 600. As stated, there were 18 Yelpers whose review count was more than 600. Such users were excluded. Ignoring the reviews written by the above-obtained 18 users, the processed data now contains 11,10,452 reviews. This data will now be used for further steps. Table 1 summarizes these preprocessing results.

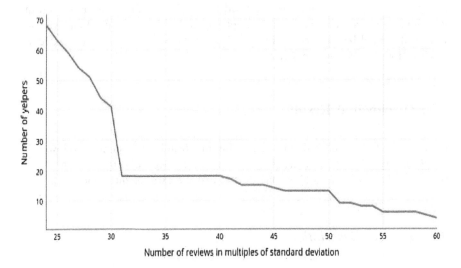

Fig. 2 Line graph of number of Yelpers versus number of reviews in multiples of standard deviation

Table 1 Preprocessing results

	Number of reviews	Number of users
Before preprocessing	11,25,458	2,52,898
After preprocessing	11,10,452	2,52,880

4.3 Vector Representation

Categorizing the 42,153 local businesses, a total of 715 categories of businesses were listed. Then, each city was represented as a vector of 715 categories of businesses. Each entry of the vector determined the percentage of the popularity of that category of business in that city as discussed above.

4.4 Clustering

As discussed, k-means clustering algorithm was applied to these vectors. K value was taken to be 13 by rough estimation. 13 clusters of cities were obtained, prevalent for a particular combination of categories of local businesses. The cities are then mapped to their corresponding prevalent local businesses and a few have been shown in Table 2.

4.5 Discussion

The result thus obtained is the city versus their respective prevalent local business categories. A part of the result is shown in Table 2. For example, Las Vegas is famous for beauty and spas, nightlife, Waunakee is famous for rafting/kayaking, etc. This is just an intercept of the total result. Another observation from this result is that one can see the culture of prevalent local businesses in most of the cities. The pie chart shown in Fig. 3 shows this. In this observation, "Beauty & Spas" is the most prevalent local business trend across most cities. "Jewelry" and "American" food, follow up in line.

Table 2 City versus their prevalent local business categories

City	Local business categories
Florence	American (traditional)
Phoenix	Beauty and spas
Waunakee	Rafting/kayaking
Las Vegas	Beauty and spas, nightlife
Carnegie	Party and event planning

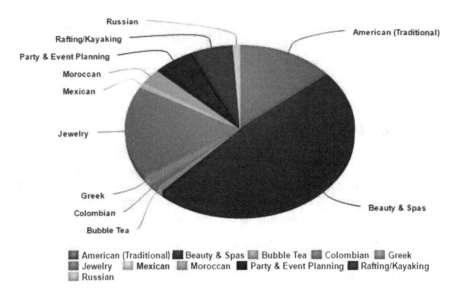

Fig. 3 Contribution of business categories in identifying cultural trends

5 Conclusion and Future Work

In this study, the main aim is to compare cities based on the various local business trends followed by the people living in that city. To accomplish the task, the raw Yelp data was first preprocessed and then, k-means clustering algorithm was applied on each city's normalized vector representation of the different categories of local businesses. The clusters thus formed showed the different local business categories prevalent across cities. The final result was thus obtained as shown in the tabular format (Table 2). Also, visualization was done on the result to see the local business trends across all cities.

The results obtained will help not only the customers for knowing the local business trends of a city and affect their consuming decision but also the local businesses themselves who can then strategize their b-plans accordingly. In future, other trends (say cuisines or recipes, etc.) which have the potential to identify the trends and help distinguish the culture of people living in different cities will also be mined.

References

1. "Ten Things You Should Know About Yelp." Yelp, 2011. [Online]. Available: https://www.yelp.com/about [Accessed 30 12 16]
2. A. Hicks, S. Comp, J. Horovitz, M. Hovarter, M. Miki, J. L. Bevan, "Why people use yelp.com: An exploration of uses and gratifications", Computers in Human Behavior, vol. 28, no. 6, pp. 2274–2279, 2012
3. X. Mao, Y. Rao and Q. Li, "Recipe popularity prediction based on the analysis of social reviews," Awareness Science and Technology and Ubi-Media Computing (iCAST-UMEDIA), 2013 International Joint Conference on, Aizuwakamatsu, 2013, pp. 568–573
4. M. Prithivirajan, V. Lai, K. J. Shim and K. P. Shung, "Analysis of star ratings in consumer reviews: A case study of Yelp," Big Data (Big Data), 2015 IEEE International Conference on, Santa Clara, CA, 2015, pp. 2954–2956
5. Daniel Ramage, David Hall, Ramesh Nallapati, Christopher D. Manning, Labeled LDA: a supervised topic model for credit attribution in multi-labeled corpora, Proceedings of the 2009 Conference on Empirical Methods in Natural Language Processing: Volume 1, August 06–07, 2009, Singapore
6. A. Salinca, "Business Reviews Classification Using Sentiment Analysis," 2015 17th International Symposium on Symbolic and Numeric Algorithms for Scientific Computing (SYNASC), Timisoara, 2015, pp. 247–250
7. M. Yamamoto, T. Yamasaki and K. Aizawa, "Service Annotation and Profiling by Review Analysis," 2016 IEEE Second International Conference on Multimedia Big Data (BigMM), Taipei, 2016, pp. 357–364
8. T. Yamasaki, M. Yamamoto and K. Aizawa, "Review-Based Service Profiling and Recommendation," 2016 Joint 8th International Conference on Soft Computing and Intelligent Systems (SCIS) and 17th International Symposium on Advanced Intelligent Systems (ISIS), Sapporo, Japan, 2016, pp. 377–381
9. I. Varlamis, M. Eirinaki and D. Proios, "TipMe: Personalized advertising and aspect-based opinion mining for users and businesses," 2015 IEEE/ACM International Conference on Advances in Social Networks Analysis and Mining (ASONAM), Paris, 2015, pp. 1489–1494
10. G. Zhao, X. Qian and X. Xie, "User-Service Rating Prediction by Exploring Social Users' Rating Behaviors," in IEEE Transactions on Multimedia, vol. 18, no. 3, pp. 496–506, March 2016
11. Jain, A. K., "Data clustering: 50 years beyond k-means", 2009, Pattern Recognition Letters, 31, pp. 651–666
12. Yelp dataset from the yelp dataset challenge. [Online]. Available: https://www.yelp.com/dataset_challenge

Relative-Feature Learning through Genetic-Based Algorithm

K. Chandra Shekar, Priti Chandra and K. Venugopala Rao

"We take the full responsibility of the ethical issue of the work.
We have taken permission to use the information in our work.
In case of any issue we are only responsible."

Abstract A relative-feature learning system development within the data mining is showing its capability in genetic algorithms to diagnose the normal behavior and misuse features and anomalies in chosen learning environment. Under this contribution research work, we investigated a learning method to diagnose the features in data mining through genetic algorithm. The investigated method is connected with the adaptive propagation planning which controls structuring, computing, and predicting. In this paper, we use relative, a fast and effective behavior, to improve the genetic data feature diagnosis. The authors have proposed Relative-Feature Learning through Genetic-Based Algorithm, which performs genetic data operation, feature fitness evaluation, scalable to large amount of data and effective in the field of soft computing. Our proposed algorithm experimental approach can be used in the developed system for feature learning in more depth.

Keywords Genetic algorithm · Data mining · Soft computing
Feature learning · Fitness function

1 Introduction

Genetic algorithms (GA) represent powerful general purpose search method based on the evolutionary ideas of natural selection and genetics. They simulate natural process based on principles of Lamark and Darwin. The field of genetic and

K. Chandra Shekar (✉)
IIIT – Rajiv Gandhi University of Knowledge Technologies, Basar, India
e-mail: chandhra2k7@gmail.com

P. Chandra
Advanced Systems Laboratory, DRDO, Hyderabad, India
e-mail: priti_murali@yahoo.com

K. Venugopala Rao
G. Narayanamma Institute of Technology & Sciences, Hyderabad, India
e-mail: kvgrao1234@gmail.com

© Springer Nature Singapore Pte Ltd. 2018
V. Bhateja et al. (eds.), *Proceedings of the Second International Conference
on Computational Intelligence and Informatics*, Advances in Intelligent Systems
and Computing 712, https://doi.org/10.1007/978-981-10-8228-3_8

evolutionary computation was first explored by Turing who suggested an early template for the genetic algorithm. Holland performed much of the fundamental work in GEC in 1960 and 1970. His goal of understanding the processes of natural adoption and designing biologically inspired artificial systems led to the formulation of the simple genetic algorithms [1–3].

Genetic algorithms are typically implemented using computer simulations in which an optimization problem is specified. Members of space of candidate solutions, called individuals, are represented as chromosomes. Genetic algorithms work in an iterative manner by generating new populations of strings from old ones. Every string is encoded in binary, real, etc., versions of a candidate solution. An evolution function associates a fitness measure to every string indicating its fitness for the problem [2, 3].

The essence of GAs is an efficient, parallel, and global search method. In the development of different GAs, initial data analysis is made using components of GA, they are (1) selection, (2) crossover, and (3) mutation.

GA [1, 4] is an optimization technique with a scope of selection and genetics, capable of searching effectively large data sets. GAs are made to solve several machine problems [5, 6]. J. H. Holland [7–9] and Bremermann [10, 11] had made usage of the proposed GAs in mining applications. These GAs are capable of solving many hard problems in real time. The applications of Data Mining (DM) focussed on Searching algorithms [12], obtaining fitness values [13], extracting data from large amount of data [14], to detect data intrusion in networks, data preprocessing tasks [15] for data reduction and feature selection over the large amount of data, to optimize the queries, reducing the random search over the large databases, diagnose the best fit data [13], to improve the data coverage and determine the data initial weights [15], optimizes the position of data centers, weights and widths and data mining based on risk factor. Nowadays, GA is extensively used in data clustering [14], data optimization, data cell placement, design plan, hydrocyclone, data cache design, and in data network design.

Many practical GAs are present to perform data feature classification, requiring learning of a classification function that have feature vector values to finite data classes. The usage of data features to represent the data classification is as follows:

1. A learning algorithm or method to perform accuracy in data classification, e.g., a decision tree algorithm. Learning algorithm describes the data patterns explicitly to capture the complete information, which is required for data classification. If the data present is incomplete and inconsistent, then learning algorithm used will be failed in classification accuracy.
2. Data classification function estimates the time of data pattern representation, the attributes of data, defining the classification of relevant and irrelevant attributes, used to describe the search time and data availability. A requirement of data availability is needed to estimate the learning of data accurate classification function.
3. During the learning process, the number of data feature taken is huge for aggregation and is sufficiently taken into account, an accurate learning

Table 1 Algorithm for genetic algorithm

Begin
STEP 1: t=0
STEP 2: initialize population P(t)
STEP 3: compute fitness P(t)
STEP 4: t=t+1
STEP 5: if termination criterion achieved go to step 10
STEP 6: select P(t) from p(t-1)
STEP 7: crossover P(t)
STEP 8: mutate P(t)
STEP 9: go to step 3
STEP 10: Output best and stop
End

algorithm. The data features taken into learning process are subject to interest to improve the classification function.

4. Data mining includes the presence of sophisticated and sensitive features; the data taken from these data sets have valuable features to learn. The algorithm used to learn these kinds of data needs a planning to feature the accurate and relevant data from sample data sets.

The above discussion presents the data feature learning problem, in the design of genetic-based data algorithm. This feature learning problem refers to identifying and selecting a useful subset of data patterns from a large data set [2, 3]. To improve the learning algorithm, we are proposing an adaptive propagation planning algorithm to structure, compute, and predict the data features accurately and to improve the genetic data feature diagnosis explained in Table 1.

2 Relative-Feature Learning Algorithm

There are different learning algorithms presented in literature but none of the method can solve the problem in absence of prior data/information and is not possible to compute the similarities in two different data patterns. Hence, there must be some prior data/information about the problem to solve with similarities. This prior data/information is called relative data feature. Every data/information contains the relative data feature, which classifies the data set correctly. Learning algorithms are classified based on the dependencies: they are supervised learning—need training examples to represent the data—unsupervised learning—no information about the training examples—learning through data process, which results in an uncorrected data classification, and reinforcement learning—need training examples but could not be able to get the correct data classification. Our proposed learning algorithm is based on decision trees under supervised learning method.

2.1   The Adaptive Propagation Planning

When a data instance is initiated, the classification of data, the attributes required to represent the data are acknowledged, the testing of the attribute to specify the data node is performed, each data node corresponding to each attribute results in a data-valued learning function, and data root node performs the corresponding propagation of node paths to next aggregation levels. This process is repeated for every data root node and data subtree node, until total classifications were identified. The decision tree which we have developed has a feature of adaptability by improving the classification of data nodes by variation node analysis and allowing the classified leaf node to modify to a base node. Main features of our proposed adaptive propagation planning method are as follows:

(a) The data instances are regenerative with the data attributes.
(b) Consistent data classification.
(c) Data set considered needs training examples to avoid errors.
(d) The training data including misclassified and unknown data sets can be detectable.

For our training data, we simulated our proposed learning algorithm in C4.5 algorithm using Weka Tool is listed in Table 2.

Table 2 Proposed decision tree algorithm pseudocode	**Input:** instances, attributes, indicators
	rn:root_node(instances, attributes)
	if (instance=available) then
	rn<=rn+1; return rn:
	else if (instance=not available) then
	rn<=rn-1; return rn;
	if (attributes=zero) then
	rn=rn(instance); return rn;
	else
	cn:current_node(instances,current_attribure,indicators)
	d:decision<=true=cn;
	for each can do
	cb:current_branch(rn,cn);
	if d=false then
	ln:leaf_node(indicators)
	endif
	endfor
	endif
	return rn

2.2 Controls Identified in the Adaptive Propagation Planning Method for Feature Learning

The proposed relative-feature learning algorithm has identified the following categories for the adaptive propagation planning method.

(a) *Structuring*: In structuring, a model is created to decide the most probable relational learning classes of data to classify. Best output result is possible using the data recognition during structuring aggregation process. The process of structuring is illustrated below:

1. Divide the data instances into data classes according to the data feature criteria.
2. Label these data instances to recognize as training examples.
3. Use these training examples to detect the situation using anomalous.
4. Map the detected situations with the actual data labels.
5. Now, recognize the data classification.

(b) *Computing*: In computing, after anomaly detection, the data features of input data are identified by training the decision tree for the best values to model the feature aggregative, crossover, and mutation operations. To compute the best hypothesis and fitness function, the process is illustrated below:

1. train the anomaly detected data features using supervised GA;
2. GA, in this case, performs the operation to compute the possibility of outcome as right or wrong.

(c) *Predicting*: In prediction, the computed statement is processed clearly for classification and regression analysis to identify the problem of outcome by relating complex feature learning. During classification and regression analysis, prediction of outcome takes a continuous value to generate latency value. The latency value is used for prediction of actual latency to compare the prediction process with actual latency to improve the future continuous latency values. The future latency values are used as training examples for subsequent actual values. During the optimization problem, the actual value is analyzed for different actions of training examples, and the best will be optimized.

3 A Genetic-Based Algorithm

3.1 Feature Selection

Consider a data set of D_s samples s_i, $i = 0, 1, \ldots, D_s$, and the selected data for analysis is given by criteria as $C_{Ds} = (\min_i < s_i < \max_i)$, where $\min_i$ and $\max_i$ represent the minimum and maximum sets of features to be taken from the given

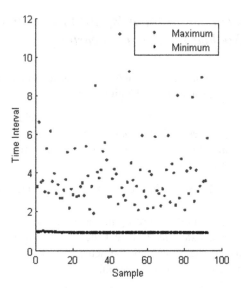

Fig. 1 Set of distance features taken from the data set

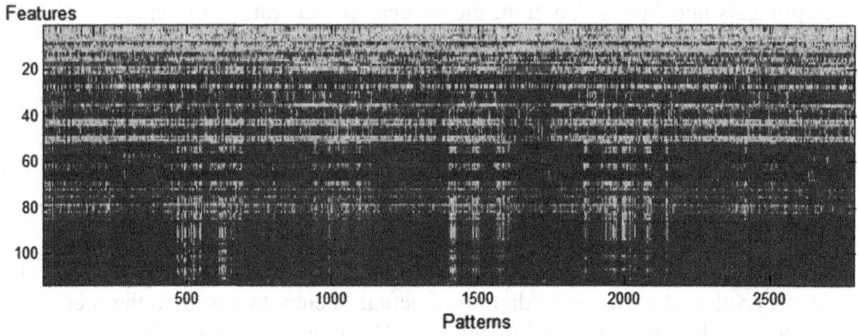

Fig. 2 Pattern features

data samples, respectively. Figures 1 and 2 illustrate the selected features in the data set and their patterns.

3.2 Population

A criteria-based samples were considered for the length of samples,

$L = C_{Ds}(1) \cup C_{Ds}(2) \cup C_{Ds}(3)$, considering three types of features to be extracted.

The features are considered based on the algorithm shown in Table 3.

Table 3 Feature selection algorithm pseudocode

Initialize Ds, CDs, L
Evaluate Ds, CDs
Feature f
f=0
do
parents←feature_set(Ds)
offspring←variable_declare(CDs)
Evaluate CDs
Population p
p(f≠0)←p(f=0) and L(CDs)
f=f+1
return

3.3 Fitness and Crossover

For our GA implementation, we took four feature set samples, which are shown in Table 4. The fitness function features are shown in Figs. 3, 4, 5, and 6, respectively.

During the validation of the fitness function, we used classifiers to estimate the combined performance. Classifiers which we analyzed with our proposed algorithm are 2-classes, 3-classes, and 9-classes. The computation of error rate is based on the accuracies of the above-said classifier and the proposed as shown in Table 5.

The detection of data features is shown in Table 6.

In Table 6, the "Total number of correct answers" and the "Total number of tries" are the pattern features analyzed for the classifications 2-classes, 3-classes,

Table 4 Four feature set fitness and crossover evaluation

	Sample_One	Sample_Two	Sample_Three	Sample_Four
Type	Random	Random	Random	Random
Number of features	10	20	40	60
Population size	50	60	70	80
Generation limit	500	400	300	100
Mutual likelihood	0.1–0.6	0.8–0.9	0.6–0.9	0–1
Crossover likelihood	0.05–0.4	0.1–0.9	0.6–0.9	0.2–0.8
Initial maximum tree depth	2	4	6	8
Maximum number of nodes	32	46	59	62
Crossover type	Swap subtree	Swap subtree	Swap subtree	Swap subtree
Mutation type	Branch mutation	Branch mutation	Branch mutation	Branch mutation
Stop condition	99 generations	99 generations	99 generations	99 generations

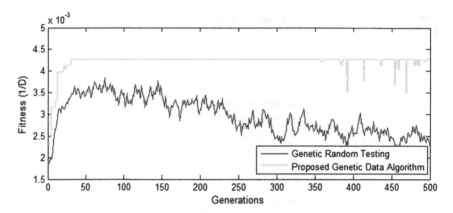

Fig. 3 Fitness: Sample_One feature

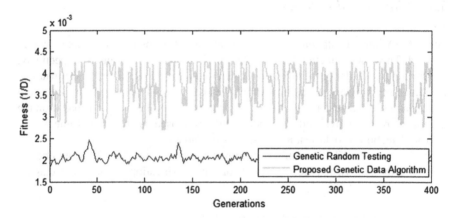

Fig. 4 Fitness: Sample_Two feature

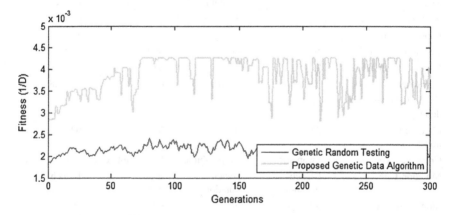

Fig. 5 Fitness: Sample_Three feature

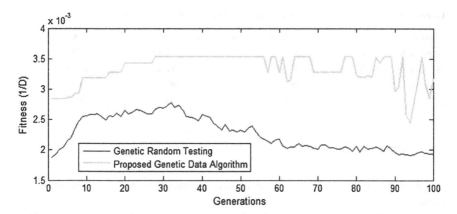

Fig. 6 Fitness: Sample_Four feature

Table 5 Comparing the accuracy of all classifiers

Classifier	Performance		
	2-Classes	3-Classes	9-Classes
C 4.0	80.3	56.8	25.6
kNN	82.3	50.4	28.5
Proposed	84.9	58.9	33.5

Table 6 Feature results in 2-, 3-, and 9-Classes

Feature	Results (%)		
	2-Classes	3-Classes	9-Classes
Total_Correct _Answers	89.25	91.36	95.36
Total_Number_of_Tries	42.56	51.26	53.25

Fig. 7 Fitness condition set with its relative error rate

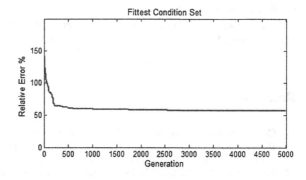

and 9-classes. These comparisons state the feature importance for the data set taken for analysis. The error rate is measured in each classifier by taking the ratio between feature 1 and feature 2 columns (Fig. 7).

Table 7 Comparative analysis of the proposed method with the existing methods

Algorithm	True positive rate (%)	Accuracy (%)	Precision (%)
K-nearest neighbor	59.31	97.93	97.99
N-gram + GHSOM	92.51	99.45	96.21
N-gram + Diffusion Maps	98.72	99.94	100
Proposed method	99.75	99.94	98.25

3.4 Anomaly Detection

In our proposed GA, the anomaly detections metrics were measured as shown in Table 7.

From Table 7, proposed algorithm has shown good results comparatively. The comparative results prove that our proposed genetic-based data algorithm allows a good analysis of the data sets and strive toward an effective prediction with greater accuracy.

4 Conclusion and Future Enhancements

In this paper, a relative-feature learning algorithm is successfully investigated. This paper makes the three contributions to feature learning: (1) it introduces a relational learning class of data structuring aggregation, (2) it presents a hierarchy distribution computing values to model the feature aggregative, and (3) it demonstrates prediction on the genetic data domain that relates complex feature learning. Our proposed algorithm explores strategies that merge (1) automated feature-test-suite diagnose generation technique with long data testing by non-repetitive feature cases for uncovering data from alignment problems, (2) resource leakages, and (3) feature-data ambiguity.

Our further work concentrates on the ensemble methods, leading toward an efficient hybrid model for classification and prediction.

References

1. Chandra Shekar K, Vijay Bhasker G, Laxmi Chaitanya V, "Mining Frequent Itemsets for Non Binary Data Set using Genetic Algorithm", International Journal of Advanced Engineering Sciences and Technologies (IJAEST), Volume No.11, Issue No.1, 143–152, 2011.
2. M.A. Jabbar, B.L. Deekshatulu, Priti Chandra, "An Evolutionary Algorithm for Heart Disease Prediction", Springer-Verlag Berlin Heidelberg, ICIP 2012, CCIS 292, 378–389, 2012.
3. M.A. Jabbar, B.L. Deekshatulu, Priti Chandra, "Intelligent heart disease prediction system using random forest and evolutionary approach", Journal of Network and Innovative Computing, ISSN 2160-2174, Volume 4, 175–184, 2016.

4. Peixian LI, Zhixiang TAN, Lili YAN, Kazhong DENG, "Time series prediction of Mining subsidence based on genetic algorithm neural network", International Symposium on Computer Science and Society, 2011.
5. Haifeng, S., et al., "The Problem of Classification in Imbalanced Data Sets", IEEE, 2010.
6. Mikhail Zolotukhin, Timo Hämäläinen and Antti Juvonen, "Online anomaly detection by using N-gram model and growing hierarchical selforganizing maps", Proceedings of the 8th International Wireless Communications and Mobile Computing Conference (IWCMC), pp. 47–52, 2012.
7. K G Srinivasa, Saumya Chandra, Siddharth Kajaria, Shilpita Mukherjee, "IGIDS: Intelligent Intrusion Detection System Using Genetic Algorithms", IEEE 2011.
8. Chih-Fong Tsai and Jui-Sheng Chou, "Data Pre-Processing by Genetic Algorithms for Bankruptcy Prediction", IEEE, 2011.
9. K G Srinivasa, Jagadish M, K R Venugopal and L M Patnaik, "Data Mining based Query Processing using Rough Sets and Genetic Algorithms", Proceedings of the IEEE Symposium on Computational Intelligence and Data Mining, 2007.
10. D. Goldberg, "Genetic Algorithms in Search, Optimization and Machine learning", Addison Wesley, 1989.
11. Wenxiang Dou, Jinglu Hu, Kotaro Hirasawa and Gengfeng Wu, "Quick Response Data Mining Model Using Genetic Algorithm", SICE Annual Conference, The University Electro-Communications, Japan, 2008.
12. Lijuan Liu and Mingrong Deng, "An Evolutionary Artificial Neural Network Approach for Breast Cancer Diagnosis", Third International Conference on Knowledge Discovery and Data Mining, 2010.
13. Yumin Pan and Weining Xue, Quanzhu Zhang, Liyong Zhao, "A Forecasting Model of RBF Neural Network Based on Genetic Algorithms Optimization", Seventh International Conference on Natural Computation, 2011.
14. Mikhail Zolotukhin and Timo Hämäläinen, "Detection of Anomalous HTTP Requests Based on Advanced N-gram Model and Clustering Techniques", Springer-Internet of Things, Smart Spaces, and Next Generation Networking. Lecture Notes in Computer Science, Vol. 8121, pp. 371–382, 2013.
15. Syed Umar Amin, Kavita Agarwal, Dr. Rizwan Beg, "Genetic Neural Network Based Data Mining in Prediction of Heart Disease Using Risk Factors", Proceedings of IEEE Conference on Information and Communication Technologies(ICT), 2013.

Performance of Negative Association Rule Mining Using Improved Frequent Pattern Tree

E. Balakrishna, B. Rama and A. Nagaraju

Abstract Negative Association Rule (NAR) mining is a task of finding rules which contains data items that are negatively correlated. Many algorithms are implemented to discover the NAR; from the offered approach, the frequent pattern growth (FP-Growth) approach is proficient for finding the item sets, from which we can discover NAR. But in the FP-Growth, it finds numerous Conditional FP Trees (CFP-Tree). Frequent Item Set Mining (FISM) algorithm uses an improved FP-Tree for generating NAR without producing CFP-Tree. In this paper, we presented the overview of three different algorithms: Apriori, FP-Growth, and FISM that can be used in the process of discovering NAR. Finally, we analyze the behavior of these algorithms by considering a simple transactional database.

Keywords Improved FP-tree · Support · Confidence

1 Introduction

NAR is similar to positive association rules (PARs) [1–3] in which either the ancestor or the consequent or both are negated [4]. To discover the NAR, first we have to discover frequent item sets from the database. In this paper, we focus on discovering NAR from frequent item sets. We can find PAR from frequent item sets using a quantify known as "confidence", i.e., $\mathrm{Confd}(X \Rightarrow Y)$, defined as

E. Balakrishna (✉)
JNTUHCEH, Hyderabad, Telangana, India
e-mail: balakrishnakits@gmail.com

B. Rama
Kakatiya University, Warangal, Telangana, India
e-mail: rama.abbidi@gmail.com

A. Nagaraju
Central University of Rajasthan, Bandar Sindri, Rajasthan, India
e-mail: kits.nagaraju@gmail.com

© Springer Nature Singapore Pte Ltd. 2018
V. Bhateja et al. (eds.), *Proceedings of the Second International Conference on Computational Intelligence and Informatics*, Advances in Intelligent Systems and Computing 712, https://doi.org/10.1007/978-981-10-8228-3_9

Table 1 Database

ID	Items
T1	Ink, Pen, Eraser
T2	Pen, Sharpener
T3	Pen, Pencil
T4	Ink, Pen, Sharpener
T5	Ink, Pencil
T6	Pen, Pencil
T7	Ink, Pencil
T8	Ink, Pen, Pencil, Eraser
T9	Ink, Pen, Pencil

$$Confd(X \rightarrow Y) = sup(X \rightarrow Y)/sup(X) \tag{1}$$

Possible NAR can be found from PAR by the following principle: For PAR $A \rightarrow B$, the possible negative rules are

$$\text{(a) } A \rightarrow \sim B \quad \text{(b) } \sim A \rightarrow B \quad \text{and} \quad \text{(c) } \sim A \rightarrow \sim B \tag{2}$$

Valid NAR [5] from set of possible NAR can be generated using the following measures:

$$Confd(X \rightarrow \sim Y) = \{sup(X) - sup(X \cup Y)\}/sup(X) \tag{3}$$

$$Confd(\sim X \rightarrow Y) = \{sup(Y) - sup(X \cup Y)\}/(1 - sup(x)) \tag{4}$$

$$Confd(\sim X \rightarrow \sim Y) = \frac{\{1 - sup(X) - sup(Y) + sup(X \cup Y)\}}{1 - sup(X)} \tag{5}$$

We get confidence of each NAR; the rule is said to be valid NAR, if the confidence of that rule is greater than or equal to user mention confidence value. Consider the below database for generating NAR using Apriori, FP-Growth, and FISM.

Let us assume that minimum support (MS) defined by the user is 2, i.e., 50% (Table 1).

The outline of this paper is as follows: Section 2 includes generating NAR based on Apriori method. Section 3 includes generating NAR based on FP-Growth method. Section 4 includes generating NAR based on FISM and comparative results. Section 5 provides conclusion and future work.

2 Generating NAR Based on Apriori

Procedure for finding NAR is as follows:

(2.1) Find frequent item sets using Apriori method [1].
(2.2) Get PAR from frequent item sets using Eq. (1).
(2.3) Discover possible NAR using Eq. (2).
(2.4) Find valid NAR using Eqs. (3), (4), and (5).

2.1 Frequent Item Sets Generated Using Apriori Method

Minimum support considered here is 20% (Table 2).

2.2 Generating PAR

Based on Eq. (1), below PAR is found using frequent item sets (Tables 2 and 3).

Table 2 Frequent item sets

Frequent 1-item sets	Frequent 2-item sets	Frequent 3-item sets
{Pen: 7}	{Ink, Pen: 4}	{Ink, Pen, Pencil}
{Ink: 6}	{Ink, Pencil: 4}	{Ink, Pen, Eraser}
{Pencil: 6}	{Ink, Eraser: 2}	
{Sharpener: 2}	{Pen, Pencil: 4}	
{Eraser: 2}	{Pen, Sharpener: 2}	
	{Pen, Eraser: 2}	

Table 3 PAR

Minimum confidence = 50%	
Ink → Pencil (confidence = 0.667)	Ink → Pen (conf = 0.667)
Sharpener → Pen (confi = 1.000)	[Pen, Pencil] → [Ink] (conf = 0.500)
Eraser → Pen (conf = 1.000)	[Pen, Ink] → [Pencil] (conf = 0.500)
Eraser → Ink (conf = 1.000)	[Pencil, Ink] → [Pen] (conf = 0.500)
Pen → Pencil (conf = 0.571)	[Pen, Ink] → [Eraser] (conf = 0.500)
Pen → Ink (conf = 0.571)	Eraser → [Pen, Ink] (conf = 1.000)
Pencil → Pen (conf = 0.667)	[Pen, Eraser] → Ink (conf = 1.000)
Pencil → Ink (conf = 0.667)	[Ink, Eraser] → Pen (conf = 1.000)

Table 4 Possible NAR

Ink → ~Pencil	Pen → ~Pencil
~Ink → Pencil	~Pen → Pencil
~Ink → ~Pencil	~Pen → ~Pencil
Sharpner → ~Pen	Pen → ~Ink
~Sharpener → Pen	~Pen → Ink
~Sharpener → ~Pen	~Pen → ~Ink
Eraser → ~Pen	Pencil → ~Pen
~Eraser → Pen	~Pencil → Pen
~Eraser → ~Pen	~Pencil → ~Pen
Eraser → ~Ink	Pencil → ~Ink
~Eraser → Ink	~Pencil → Ink
~Eraser → ~Ink	~Pencil → ~Ink
Ink → ~Pen	[Pen, Ink] → ~[Eraser]
~Ink → Pen	~[Pen, Ink] → [Eraser]
~Ink → ~Pen	~[Pen, Ink] → ~[Eraser]
[Pen, Pencil] → ~[Ink]	Eraser → ~[Pen, Ink]
~[Pen, Pencil] → [Ink]	~Eraser → [Pen, Ink]
~[Pen, Pencil] → ~[Ink]	~Eraser → ~[Pen, Ink]
[Pen, Ink] → ~[Pencil]	[Pen, Eraser] → ~Ink
~[Pen, Ink] → [Pencil]	~[Pen, Eraser] → Ink
~[Pen, Ink] → ~[Pencil]	~[Pen, Eraser] → ~Ink
[Pencil, Ink] → ~[Pen]	[Ink, Eraser] → ~Pen
~[Pencil, Ink] → [Pen]	~[Ink, Eraser] → Pen
~[Pencil, Ink] → ~[Pen]	~[Ink, Eraser] → ~Pen

2.3 Generating Set of Possible NAR from PAR

From PAR generated in the previous step, the following possible NAR is found using the rules (II) (Table 4).

2.4 Generating Valid NAR

The valid NAR is generated by making use of Eqs. (3), (4), and (5) (Table 5).
    In this technique, we got 22 valid NARs.

Table 5 Valid NAR

Minimum confidence = 50%	
~Pen → Pencil	~Eraser → ~[Pen, Ink]
~Ink → Pencil	~Ink → Pen
~Ink → ~Pencil	~[Pen, Eraser] → Ink
~Pen → Ink	[Pen, Pencil] → ~[Ink]
~Sharpener → Pen	~[Pen, Pencil] → [Ink]
~Pencil → Ink	~[Ink, Eraser] → Pen
~Pencil → ~Ink	[Pen, Ink] → ~[Pencil]
~Eraser → Pen	~[Pen, Ink] → [Pencil]
[Pencil, Ink] → ~[Pen]	[Pen, Ink] → ~[Eraser]
~[Pencil, Ink] → [Pen]	~[Pen, Ink] → ~[Eraser]
~Eraser → Ink	~Pencil → Pen

3   Generating NAR Based on FP-Growth

Procedure for finding NAR is as follows:

(3.1)   Find frequent item sets using FP-Growth method [6].
(3.2)   Get PAR from frequent item sets using Eq. (1).
(3.3)   Discover possible NAR using Eq. (2).
(3.4)   Find valid NAR using Eqs. (3), (4), and (5).

3.1   Generating Frequent Item Sets

Minimum support considered here is 20%.

3.2   Generating PAR

Based on Eq. (1), below PAR is generated using frequent item sets (Tables 6 and 7).

Table 6 Frequent item sets

Ink	{Ink: 4}; {Ink, Pen: 4}
Eraser	{Eraser: 2}; {Eraser, Ink: 2}; {Eraser, Pen: 2}; {Eraser, Ink, Pen: 2}
Pencil	{Pencil: 2}; {Pencil, Ink: 2}; {Pencil, Pen: 2}; {Pencil, Ink, Pen: 2}
Sharpener	{Sharpener: 2}; {Sharpener, Pen: 2}

Table 7 PAR

Minimum confidence (MC) = 50%	
[Pen, Pencil] → [Ink] (conf = 0.500)	Ink → Pen (conf = 0.667)
[Pen, Ink] → [Pencil] (conf = 0.500)	Ink → Pencil (conf = 0.667)
[Pencil, Ink] → [Pen] (conf = 0.500)	Sharpner → Pen (conf = 1.000)
[Pen, Ink] → [Eraser] (conf = 0.500)	Eraser → Pen (conf = 1.000)
Pen → Pencil (conf = 0.571)	Eraser → Ink (conf = 1.000)
Pen → Ink (conf = 0.571)	Eraser → [Pen, Ink] (conf = 1.000)
Pencil → Pen (conf = 0.667)	[Pen, Eraser] → Ink (conf = 1.000)
Pencil → Ink (conf = 0.667)	[Ink, Eraser] → Pen (conf = 1.000)

3.3 Generating Set of Possible NAR from PAR

See Table 8.

3.4 Generating Valid Negative Association Rules

The valid NARs are discovered using Eqs. (3), (4) and (5) (Table 9).
   In this technique, we got 22 valid NARs.

Table 8 Possible NAR

[Pen, Pencil] → ~Ink	[Pen, Ink] → ~Eraser	Pencil → ~Pen
~[Pen, Pencil] → Ink	~[Pen, Ink] → Eraser	~Pencil → Pen
~[Pen, Pencil] → ~Ink	~[Pen, Ink] → ~Eraser	~Pencil → ~Pen
[Pen, Ink] → ~Pencil	Pen → ~Pencil	Pencil → ~Ink
~[Pen, Ink] → Pencil	~Pen → Pencil	~Pencil → Ink
~[Pen, Ink] → ~Pencil	~Pen → ~Pencil	~Pencil → ~Ink
[Pencil, Ink] → ~Pen	Pen → ~Ink	Ink → ~Pen
~[Pencil, Ink] → Pen	~Pen → Ink	~Ink → Pen
~[Pencil, Ink] → ~Pen	~Pen → ~Ink	~Ink → ~Pen
Eraser → ~Ink	Eraser → ~[Pen, Ink]	[Pen, Eraser] → ~Ink
~Eraser → Ink	~Eraser → [Pen, Ink]	~[Pen, Eraser] → Ink
~Eraser → ~Ink	~Eraser → ~[Pen, Ink]	~[Pen, Eraser] → ~Ink
Ink → ~Pencil	~Ink → Pencil	~Ink → ~Pencil
Sharpener → ~Pen	~Sharpener → Pen	~Sharpener → ~Pen
Eraser → ~Pen	~Eraser → Pen	~Eraser → ~Pen
[Ink, Eraser] → ~Pen	~[Ink, Eraser] → Pen	~[Ink, Eraser] → ~Pen

Table 9 Valid NAR

Minimum confidence = 50%	
~Pen → Pencil	~Eraser → ~[Pen, Ink]
~Ink → Pencil	~Ink → Pen
~Ink → ~Pencil	~[Pen, Eraser] → Ink
~Pen → Ink	[Pen, Pencil] → ~[Ink]
~Sharpener → Pen	~[Pen, Pencil] → [Ink]
~Pencil → Ink	~[Ink, Eraser] → Pen
~Pencil → ~Ink	[Pen, Ink] → ~[Pencil]
~Eraser → Pen	~[Pen, Ink] → [Pencil]
[Pencil, Ink] → ~[Pen]	[Pen, Ink] → ~[Eraser]
~[Pencil, Ink] → [Pen]	~[Pen, Ink] → ~[Eraser]
~Eraser → Ink	~Pencil → Pen

4 Generating NAR Based on FISM

Procedure for finding NAR is as follows:

(4.1) Find frequent item sets using FISM [7].
(4.2) Get PAR from frequent item sets using Eq. (1).
(4.3) Discover possible NAR using Eq. (2).
(4.4) Find valid NAR using Eqs. (3), (4), and (5).

4.1 Frequent Item Sets Generated Using FISM Method

Frequent item sets are obtained from the transactional database (Table 1) after applying FISM algorithm: Let minimum support considered here is 50%.

4.2 Generating PAR

According to Eq. (1), below PAR is discovered using frequent item sets (Tables 10 and 11).

Table 10 Frequent item sets

Ink	{Ink: 4}; {Ink, Pen: 4}
Eraser	{Eraser: 2}; {Eraser, Ink: 2}; {Eraser, Pen: 2}; {Eraser, Ink, Pen: 2}
Pencil	{Pencil: 2}; {Pencil, Ink: 2}; {Pencil, Pen: 2}; {Pencil, Ink, Pen: 2}
Sharpener	{Sharpener: 2}; {Sharpener, Pen: 2}

Table 11 PAR

Minimum confidence = 50%
[Ink, Eraser] ⇒ Pen
[Eraser, Pen] ⇒ Ink
Eraser ⇒ [Ink, Pen]

Table 12 Possible NAR

[Ink, Eraser] ⇒ ~Pen	~[Eraser, Pen] ⇒ ~Ink
~[Ink, Eraser] ⇒ Pen	Eraser ⇒ ~[Ink and Pen]
~[Ink, Eraser] ⇒ ~Pen	{~Eraser ⇒ [Ink, Pen
[Eraser, Pen] ⇒ ~Ink	~Eraser ⇒ ~[Ink, Pen]
~[Eraser, Pen] ⇒ Ink	

Table 13 Valid NAR

Minimum confidence = 50%
~[Ink, Eraser] ⇒ Pen
~[Eraser, Pen] ⇒ Ink
Eraser ⇒ ~[Ink, Pen]
~Eraser ⇒ ~[Ink, Pen]

4.3 Generating Set of Possible NAR from PAR

The following possible NARs are found using Eq. (2) (Table 12).

4.4 Generating Valid NAR

Valid NARs are generated from set of possible NAR using Eqs. (3), (4), and (5) (Table 13).

In this technique, we got four valid NARs.

5 Comparative Results

See Table 14 and Graph 1

Table 14 Number of rules for given database

	Apriori	FP-growth	IFP-tree
No. of frequent item sets generated	13	12	12
PAR generated	16	16	03
NAR generated	22	22	04

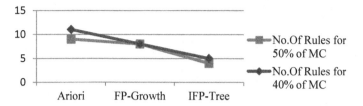

Graph 1 Negative rules are discovered by Apriori, FP-growth and IFP-tree

6   Conclusion and Future Work

In this paper, performance of Apriori, FP-Growth, and FISM is compared by considering sample transactional database. From the results, we can conclude that FISM approach is efficient than Apriori and FP-Growth algorithms. In the future, we will develop the NAR mining tool.

References

1. R. Agarwal, E. R. Srikant, "Fast Algorithms for Mining Association Rules in Large Databases", Proc. of the 20th International conference on very Large Databases, pp. 487– 499, Santiago, Chile, 1994.
2. N. Satyavathi, B. Rama and A. Nagaraju, "Incremental Updating of Mined Association Rules for Reflecting Record Insertions" Proceedings of the First International Conference on Computational Intelligence and Informatics: ICCII 2016 Volume 507 of Advances in Intelligent Systems and Computing. Publisher Springer Singapore, ISBN 9811024707, 9789811024702.
3. P. Yan, G. Chen, C. Cornelis, M. De Cock and E. E. Kerre, "Mining Positive and Negative Fuzzy Association Rules", LNCS 3213, 2004.
4. Wu X., Zhang C., Zhang S.: Mining both positive and negative association rules. In: Proc. of ICML (2002) 658–665.
5. X. Yuan, B. P. Buckles, Z. Yuan and J. Zhang, "Mining Negative Association Rules", Proc. Seventh Intl. Symposium on Computers and Communication, Italy, 2002, pp. 623–629.
6. J. Han, H. Pei and Y. Yin "Mining Frequent Patterns without Candidate Generation". In: Proc. Conf. on the Management of Data (SIGMOD'00, Dallas, TX) ACM Press, New York, NY, USA 2000.
7. E. Balakrishna, B. Rama, A. Nagaraju, "Mining of Negative Association Rules using Improved Frequent Pattern Tree", IEEE, INSPEC Accession Number: 15022153, https://doi.org/10.1109/iccct2.2014.7066748, ICCCT-2014.

Iterative Concept-Based Clustering of Indian Court Judgments

Sumi Mathai, Deepa Gupta and G. Radhakrishnan

Abstract This paper proposes architecture for the legal practitioners to lessen the burden of reading full document. The iterative Latent Semantic Analysis (LSA)-based concept extracting and clustering of legal judgment is proposed here. Headnotes in legal judgment explain about the cases in few sentences which is not enough for the legal practitioners for preparing the arguments. Our approach is to automate text processing for main concepts retrieval of legal judgments. An iterative latent semantic-based concept extraction is used. The Natural Language Processing (NLP) and data mining techniques are used for the comparison of full judgment against: iterative latent semantic analysis based concept retrieval, headnote, and different summarization levels using the existing summarization tool.

Keywords LSA · NLP · Concept retrieval · Hierarchical clustering

1 Introduction

Last few decades have witnessed exponential increase within the use of Information Technology that has resulted in great amount of knowledge being generated, stored, and searched. Information could be structured and kept as records in database management system or may be totally unstructured like weblog posts or plain text documents. With the abundance of data being out there as text documents, the

S. Mathai (✉) · G. Radhakrishnan
Department of Computer Science & Engineering, Amrita School of Engineering,
Amrita Vishwa Vidyapeetham, Amrita University, Bengaluru, India
e-mail: suminelson@gmail.com

G. Radhakrishnan
e-mail: g_radhakrishnan@blr.amrita.edu

D. Gupta
Department of Mathematics, Amrita School of Engineering,
Amrita Vishwa Vidyapeetham, Amrita University, Bengaluru, Karnataka, India
e-mail: g_deepa@blr.amrita.edu

© Springer Nature Singapore Pte Ltd. 2018
V. Bhateja et al. (eds.), *Proceedings of the Second International Conference on Computational Intelligence and Informatics*, Advances in Intelligent Systems and Computing 712, https://doi.org/10.1007/978-981-10-8228-3_10

problem of retrieval of data from such unstructured dataset is sitting new challenges to the analysis community [1]. Court judgment is such an unstructured large dataset.

Judgment is a decision by a court or other tribunals that resolves a controversy and determines the rights and obligations of the parties. Judgment is that the final part of a court case. An applicable judgment includes all the contest issue and concludes the lawsuit, since it is regarded as the courts official pronouncement of the law on the action that was pending before it. It states what remedies the winner is awarded and who wins the case [2]. Legal Judgments which is large in size has an ever-increasing need of automatic text processing. The main challenge legal practitioners are facing is to find judgment that is similar to his current case from the huge dataset. It is not an easy task to read each and every document and decide whether it is a relevant case to be referred for his current case. Lawyers cannot take the risk of missing an appropriate case that will be available to the opposing lawyer.

The headnotes provided in the judgment are giving what the case is about; it is not providing the details for legal practitioners to prepare the argument or get a better review about the case. Legal practitioners have to read all related judgments to prepare his argument. It is an onerous task for the legal practitioners to read the whole judgments and find the facts from each document. The complexness of legal domain datasets requires better and more refined methods to process legal judgments to satisfy information need of legal practitioners.

This paper describes an approach for the legal practitioners to retrieve the main concepts, for the argument preparation and to decide whether the case is similar to lawyer's current case with less reading using the iterative latent semantic analysis based concept retrieval method. The automated concepts retrieval can be beneficial for the headnote preparation too, which are prepared manually. Existing summarization tool is used to summarize the document in different levels, and the accuracy is verified using clustering techniques.

The outline is the paper is as follows. Section 2 is having a brief survey of the works related to this paper. Section 3 gives a description about the court judgment dataset. Section 4 gives the details about clustering approach. Section 5 gives details of the experimental methods. Section 6 gives details about the baseline model. Section 7 discusses about the experiment results and the result analysis. Section 8 discusses about the conclusion.

2   Literature Survey

Using the Indian court judgments, researchers have implemented different information retrieval systems and summarization methods. Topic retrieval, tabular summarization, and some of the works which have done in this area are discussed below.

In [1], document cluster analysis is presented as a tool to improve search in legal domain. Cosine similarity method is used to group the case notes (abstracts) of legal document dataset for better grouping. Latent Dirichlet Allocation (LDA) is a generative probabilistic model used to group the documents by topic. Using LDA, the topics are created by breaking down the documents. Here, the number of cluster is equal to the number of topics. Cosine similarity between topic and document is used to decide to which cluster or topic the document belongs to [3]. Automatically, preparing the catchphrases with respect to precedents is meant to identifying the important legal points of a document. Several statistical extraction approaches to extract the catch phrases have been proposed and tested it using Rouge [4]. Finding argumentative roles and thematic structures of the legal document and preparing a tabular structure with the important features like decision data, introduction, context, juridical analysis, and conclusion can be done to give a quick review of the legal document [5]. A hybrid approach in which a number of different summarization techniques combined is a rule-based system for automatic summarization of legal case reports [6].

Identification of rhetorical roles present in the sentences of a legal document for automatic summarization using Conditional Random Field (CRF) has been done in [7]. Accuracy of 80% for automatic summarization when compared it with expert's summarization was obtained. Cosine similarity has been used to find similar judgments using different features like all-term, co-citation, in citation, paragraph link, etc. Paragraph link is found to be the better way to find the similar documents [2]. Anna Huang used five different distance measures like cosine similarity, Euclidean distance, Jaccard coefficient, Pearson correlation coefficient to perform clustering using k-means clustering [8]. Majority of the work done here are using latent Dirichlet allocation, cosine similarity, conditional random field, etc. to find tabular structure or to find the topics about which legal document represents or clustering between the topics. Minority work is done in summarization of the legal documents.

Recent year's works have been reported on text mining, document similarity, clustering, summarization, etc. Different studies about information retrieval on the different useful pattern, classification, and clustering methods can be found [9, 10]. The concept of correlated term vector to represent a document has been used in three dynamic document algorithms: Term frequency based Maximum Resemblance Document Clustering (TMARDC), Correlated Concept based Maximum Resemblance Document Clustering (CCMARDC), and Correlated Concept based Fast Incremental Clustering Algorithm (CCFICA) [11]. Document similarity judgment for interactive document clustering using snippet, topic term, and original text has been proposed in [12]. Using Euclidean and Manhattan as similarity measures, Rakesh Chandra Balabantaray et al. have used k-means and k-medoid algorithms for clustering [13]. Different approaches of LSA method to summarize Turkish and English document using sentence selection approaches have been reported by Gong and Liu [14], Steinberger and Jezek [15], and Murray et al. [16]. Cross method and topic method have been implemented by Makbule Gulcin Ozsoy et al. [17]. They found that cross method was performed better than other methods. LSA [18] method is used to summarize the document and SVD method for capturing and modeling interrelationships between terms to semantically cluster terms and sentences. The evaluation is done using precision and recall F-measure giving different weighting to the sentences. Preprocessing, feature extraction, and Support Vector Machine (SVM) techniques to extract the opinions from review are explained in [19].

The above papers discuss the different methods for the topic retrieval and the tabular form of summarization of the court judgment. Concept retrieval from Legal judgments few research works is there. The proposed work investigates the possibilities of concept retrieval from the legal judgment for the legal practitioner's argument preparation. When considering the legal document, concept retrieval is an important aspect. This paper addresses the main concept retrieval of the legal document for the legal practitioners, for their argument preparation, which can also be used for the headnote preparation also.

3 Court Judgment Dataset

The Indian legal court judgments are collected from International Environmental Law Research Centre (Water and Sanitation www.ielrc.org). The International Environmental Law Research Centre is an independent research organization focusing on international and comparative environmental law issues, with a particular emphasis on India and East Africa. Supreme Court and State High Court judgments from India are collected. Statistical information about the dataset is given in Table 1. Legal judgments used for the study are related to different water and sanitation cases.

Legal judgments are very multifarious, in the sense that there is no particular structure to express the opinion of the judges pertaining to different information about a case. A sample of court judgment is given in Fig. 1 The content of the legal judgment document will have the following details.

1. Headnote about the case.
2. Citation number that gives the details of which all cases the current case has referred.
3. Date.
4. Unique name for the judgment.
5. Detail about the judges.
6. The judgment part that gives the opinion given by the judges for the case in the form of free text.

Table 1 Court judgment dataset statistics

Documents description	Statistics
Maximum size of a document	132 KB
Minimum size of a document	2 KB
Total number of sentences	33542
Total number of words	871842

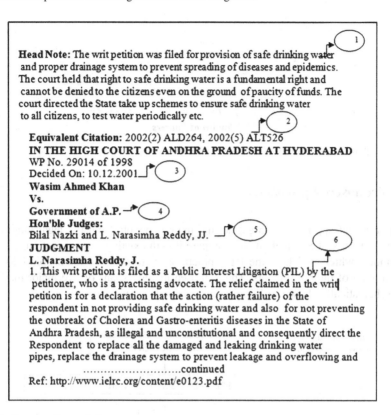

Head Note: The writ petition was filed for provision of safe drinking water and proper drainage system to prevent spreading of diseases and epidemics. The court held that right to safe drinking water is a fundamental right and cannot be denied to the citizens even on the ground of paucity of funds. The court directed the State take up schemes to ensure safe drinking water to all citizens, to test water periodically etc.

Equivalent Citation: 2002(2) ALD264, 2002(5) ALT526
IN THE HIGH COURT OF ANDHRA PRADESH AT HYDERABAD
WP No. 29014 of 1998
Decided On: 10.12.2001
Wasim Ahmed Khan
Vs.
Government of A.P.
Hon'ble Judges:
Bilal Nazki and L. Narasimha Reddy, JJ.
JUDGMENT
L. Narasimha Reddy, J.
1. This writ petition is filed as a Public Interest Litigation (PIL) by the petitioner, who is a practising advocate. The relief claimed in the writ petition is for a declaration that the action (rather failure) of the respondent in not providing safe drinking water and also for not preventing the outbreak of Cholera and Gastro-enteritis diseases in the State of Andhra Pradesh, as illegal and unconstitutional and consequently direct the Respondent to replace all the damaged and leaking drinking water pipes, replace the drainage system to prevent leakage and overflowing and
...........................continued
Ref: http://www.ielrc.org/content/e0123.pdf

Fig. 1 Sample of court judgment

4   Decision of Number of Clusters

The number of clusters, judgments form, is difficult to decide as we do not know exactly how many types of legal documents there are and how they are classified initially. Elbow method is one of the suitable methods to decide the number of clusters. Here, we have n court judgments to decide how many clusters they form. Experiment is initially conducted to decide the hierarchical clustering cut; steps like data preprocessing, vector creation, and distance matrix formation are explained later. The hierarchical clustering of documents is drawn using the distance matrix taking the distance on the y-axis and documents on the x-axis. Using the inter-cluster distance, the optimal number of cluster is determined by the elbow method [20] as shown in Fig. 2 with the distance in the y-axis and number of clusters in the x-axis where the elbow shape is taking around the three clusters. For analysis, four clusters are also taken for the evaluation of accuracy.

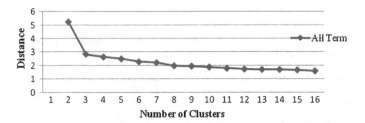

Fig. 2 Elbow shape curve to identify the number of clusters

5 Proposed Approach

Our proposed approach is to extract the main concepts from the legal judgments through iterative LSA method and compare it with existing summarization tool and headnotes which will help the legal practitioners to find the main points for the preparation of their arguments. The proposed method spans of five major modules are schematically shown in Fig. 3 and explained below.

Fig. 3 Schematics diagram of the proposed approach

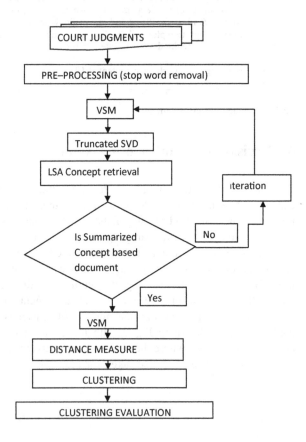

5.1 Preprocessing

In raw data, preprocessing is done to remove the noisy and the inconsistent data which are discussed here. The stop words that are present in the document have to be pruned. Natural Language Processing (NLP) toolkit is used to prune the stop word. The number of words can be further reduced by lemmatizing (the different infected forms of a word are grouped to a single term and for the analysis considered as a single word), so that there will not be any repetition of the same word in different forms in one document.

5.2 Documents Vector Space Model

Court judgment documents are represented in a vector form using document vector space models. After preprocessing, the occurrence of each word in the legal document is counted to weigh the frequent terms. Then, the term frequency of legal document is calculated by dividing the number of times term t appeared in that document by the total number of terms in that document. Then, the term frequency is calculated using Eq. (1).

$$TF(t) = \frac{\text{Number of times term } t \text{ appear in a document}}{\text{Total number of terms in a document}} \tag{1}$$

The court judgments will be having varying length. Mostly, a term will appear more frequently in longer documents than a shorter document. The term frequency for each term in the document is normalized. For normalization, the term frequency is constantly divided by the document length [21].

$$IDF(t) = \log\_e\left(\frac{\text{Total number of documents}}{\text{Number of documents with term } t \text{ in it}}\right) \tag{2}$$

Term frequency and inverse document frequency [21] are calculated, to produce a composite weight for each term in each document. The *tf-idf* weighting scheme for term t in document d is calculated.

$$tf - idf = tf_{td} * idf_{td} \tag{3}$$

5.3 Latent Semantic Analysis (LSA)

The LSA method is chosen to find the important concepts from the set of legal documents. A matrix X is created with row as the document and column as the terms from the documents. Singular Value Decomposition (SVD) is an algebraic model. Here, the given input matrix X is decomposed into three new matrices U, S, and V [15].

Table 2 Concept selection

Full document	5 concepts	10 concepts	15 concepts
3 cluster	69.02	47.00	59.00
4 cluster	59.80	38.40	44.00

$$X = USV^T \tag{4}$$

Here, X has a rank n. The SVD is trying to reduce the matrix n to k where k $<$ n by truncated SVD. This means that it takes a list of n base unique vectors and approximates them to linear combination k unique vectors. The k is decided by giving different values to k till we get an equal representation as n. The documents are simplified into k linearly independent base unique vectors which represent the main concepts of the document. The concepts are arranged in sorted form so that to get the concept of maximum weight on the top. The number of concepts breakpoint identification is verified using different number of concepts given in Table 2. Five concepts are selected when taking more concepts; the accuracy is varying. So the five concepts are used for the analysis of full document against headnotes, summarization tool, and iterative methods. The concept with maximum weight which is on the top is similar, when going down the similarity decreases. Then, the sentences are extracted from the original document which is having the above concepts in order.

5.4 Cosine Similarity

After deciding the number of concepts, the cosine similarity of each document is calculated using *tf-idf*.

$$cos(D_1, D_2) = \frac{\sum (D_1 \cdot D_2)}{\sqrt{\sum (D_1)^2} * \sqrt{\sum (D_2)^2}} \tag{5}$$

The dot product of document represented in vector form D1 and D2 is calculated. Court judgments which are not sharing any single word get the similarity value as zero because of orthogonality of the vector; else it will get greater values. The cosine similarity for information retrieval of two documents will range from 0 to 1, since the term frequencies (tf-idf weights) cannot be negative. The angle between two term frequency vectors cannot be greater than 90°. The distance measure is calculated as

$$Distance\ Measure = 1 - Cosine\ Similarity \tag{6}$$

5.5 *Hierarchical Clustering*

Hierarchical cluster similarities of each document with every other document are found. There are "n" court judgments to be clustered, an "n × n" judgment distance (or similarity) matrix are created. Hierarchical clustering starts by assigning each item to its own cluster, such that if there are n court judgments, and n clusters are formed. The steps of hierarchical clustering are as follows:

1. Join together the most similar pair of judgment by finding the least distance measure and join them as a single cluster. Now one cluster is less.
2. Compare the distances between the new cluster and each of the old clusters.
3. Continue steps 2 and 3 until all judgments are clustered into a single cluster of size n.

6 Experimental Setup of Baseline Model

In the experimental setup, headnote given in the document and the existing summarization tool are used as the baseline models against our proposed approach.

- Full document—Full legal judgments are considered as baseline model using which the clusters are formed and compared with proposed approach, the iterative steps are not included in baseline model.
- Headnote—Headnote given in the legal court judgment is considered as another baseline model where the clusters are formed as explained in proposed approach; the iterative step is not included in baseline model.
- Summary 25 and 50%—The existing summarization tool which is having the option to select the summary according to user's need is taken to analyze the cluster, where the steps for clustering are explained in proposed method, where the iterative step is not included in baseline model.

The proposed method results are then compared with the baseline models to verify the concepts which are shown in the graphical form in Fig. 4. The accuracy rate in percentage is taken in y-axis for the comparison of full document against baseline features in x-axis. The accuracy rate in percentage is calculated using the clustering techniques keeping the full document as the comparison document against other features. The number of sentences in each method is shown in Table 3 for the comparison of it with the proposed model.

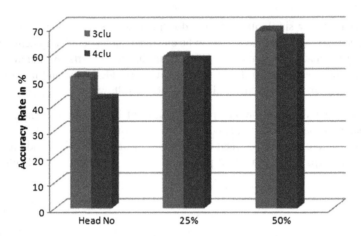

Fig. 4 Baseline model

Table 3 Number of sentences

Full doc	Headnote	25% Summarization	50% Summarization
33542	396	6556	14381

7 Experiment Results and Analysis

In the experiment shown in Fig. 5 where the y-axis shows the accuracy rate in percentage and the x-axis shows the comparison features against full document. Table 4 shows the comparison of the full document against the other entire feature. This shows that our iterative method is giving better accuracy than judgments existing headnotes and existing summarization tools. When comparing the proposed iterative method clustering with summarization tool, iterative method is showing accuracy similar to 50% summarizations; the advantage of iterative method is that they have less number of sentences than the 50% summarization given in Table 5, where PA-1 denotes our proposed approach. Proposed approach is showing a slight decline after third iteration, so that we can stop the iteration after third. The headnote which is given in the document is giving just a glance about what the case is about, which is less informative for the legal practitioners checked manually. Using our method, they will able to retrieve important concepts with less reading for the argument preparation, which can also be used for the headnote preparation which is manually done.

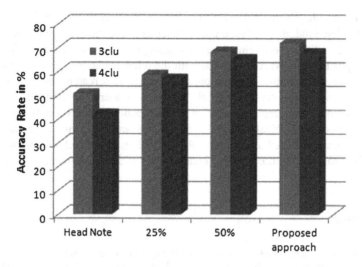

Fig. 5 Analysis of all-term against other features

Table 4 Comparison with full document

Full doc	Headnote	25%	50%	PA-1	PA-2	PA-3	PA-4	PA-5
3clu	50.6	58.5	68.3	69.2	70.1	71.9	69.5	68.9
4clu	42.1	56.9	65.2	59.8	64.6	67.7	61.6	61

Table 5 Number of sentence count in each method

Full doc	Headnote	25%	50%	PA-1	PA-2	PA-3	PA-4	PA-5
33542	396	6556	14381	604	356	322	297	279

8   Conclusions and Future Work

Iterative Latent Semantic Analysis (LSA)-based concept retrieval method has been developed for concept retrieval of the court judgment using Natural Language Processing (NLP) and data mining techniques. Headnotes in legal judgment explain the cases in few sentences which is not enough for the legal practitioners for preparing the arguments. The onerous challenge legal practitioner are facing is to find judgment that are similar to his current case from the huge dataset and the important points for the argument preparation. Reading each and every document and deciding whether it is a relevant case to be referred for current case is a tedious work. In the same way, the headnotes which are available in the document are also created manually which is also a monotonous task. The automated text processing using latent semantic analysis method is showing better results, when the results are compared using clustering. Full document is compared against headnotes, different summarization levels used by existing summarization tool, iterative concepts retrieval of legal judgments. The number of sentences in the iterative method is less and at the same time, it is having the major concepts which are more informative for the legal practitioners when compared with the other methods. These concepts can be useful for the headnote preparation also. Rather than reading the whole document, with use of the iterative model, legal practitioners can find that the main concept is what the current research on iterative concept retrieval anticipates.

Here, basic hierarchical clustering techniques are used; in future, advanced technologies will be used to improvise the technology. LSA method can be further improvised to find the more refined concepts of the court judgment.

References

1. Rupali Sunil Wagh.: Exploratory Analysis of Legal Documents using Unsupervised Text Mining Techniques. In: International Journal of Engineering Research & Technology (IJERT), Vol. 3 Issue 2, February 2014.
2. Sushanta Kumar.: Similarity Analysis of Legal Judgments and applying 'Paragraph-link' to Find Similar Legal Judgments. IIIT/TH/2014/9, 2014.
3. Ravi Kumar V, K. Raghuveer.: Legal Documents Clustering using Latent Dirichlet Allocation. In: International Journal of Applied Information Systems (IJAIS) 2(6): 27–33, May 2012.
4. Filippo Galgani, Paul Compton, and Achim Hoffmann.: Towards Automatic Generation of catchphrases for Legal Case Reports. In: Springer-Verlag Berlin Heidelberg 2012 LNCS 7182, pp. 415–426 (2012).
5. Atefeh Farzindar and Guy Lapalme.: Legal Texts Summarization by Exploration of the Thematic Structures and Argumentative Roles. RALI, Département d'Informatique et recherche opérationnelle. Université de Montréal, Québec, Canada, H3C 3J7(2004).
6. Filippo Galgani, Paul Compton, Achim Hoffmann.: Combining Different Summarization Techniques for Legal Text. Proceedings of the Workshop on Innovative Hybrid Approaches to the Processing of Textual Data (Hybrid2012), EACL 2012, pp. 115–123, (2012).
7. M, Saravanan and Ravindran, Balaraman and Raman.: Improving Legal Document Summarization using GraphicalModels. In: Nineteenth Annual Conference on Legal Knowledge and Information Systems, pp. 51–60. JURIX (2006).
8. Anna Huang.: Similarity Measures for Text Document Clustering. In. *Research gate* NZCSRSC, April 2008.
9. Bharati M. Ramageri.: Data Mining Techniques and Applications. In: Indian Journal of Computer Science and Engineering Vol. 1 No. 4 301–305, Dec 4, 2010.
10. Mohammed. J. Zaki.: Data Mining Techniques. In: ResearchGate, August 9, 2003.
11. Jayaraj Jayabharathy, Selvadurai Kanmani.: Correlated concept based dynamic document clustering algorithms for newsgroups and scientific literature. Springer (2014).
12. Yasufumi Takama, Minghuang Chen, Seiji Yamada.: Document Similarity Judgment for Interactive Document Clustering. SCIS & ISIS 2010, Dec. 8–12, 2010.
13. Rakesh Chandra Balabantaray, Chandrali Sarma, MonicaJha.: Document Clustering using K-Means and K-Medoids. In: *International Journal of Knowledge Based Computer System, Vol. 1, Issue 1, June 2013.*
14. Gong Y, Liu X.: Generic text summarization using relevance measure and latent semantic analysis. ACM 1-58113-331-6/01/0009, SIGIR'01, 2001.
15. Steinberger J, Jezek K.: Using Latent Semantic Analysis in text summarization and summary evaluation: pp. 93–100, ISIM'04, 2004.

16. Murray G, Renals S, Carletta J.: Extractive summarization of meeting recordings. 9th European conference on speech communication and technology. pp. 593–596, 2005.
17. Makbule Gulcin Ozsoy, Ferda Nur Alpaslan, Ilyas Cicekli.: Text summarization using Latent Semantic Analysis. In: JIS (2011).
18. Y. Gong, X. Liu.: Generic Text Summarization Using Relevance Measure and Latent Semantic Analysis. In: Proceedings of the 24th annual international ACM SIGIR conference on Research and development in information retrieval, New Orleans, Louisiana, pp. 19–25, United States (2001).
19. C. Priyanka, Deepa Gupta.: Identifying the best feature combination for sentiment analysis of customer reviews. In. Second International Conference on Advances in Computing, Communications and Informatics. (ICACCI-2013), IEEE (2013).
20. Ravi Kumar V, K. Raghuveer.: Legal Documents Clustering using Latent Dirichlet Allocation. In: International Journal of Applied Information Systems (IJAIS) 2(6): 27–33, May 2012.
21. Kevin P. Murphy.: A Probabilistic Perspective. Text book.

Improving the Spatial Resolution of AWiFS Sensor Data Using LISS III and AWiFS DataPair with Contourlet Transform Learning

K. S. R. Radhika, C. V. Rao and V. Kamakshi Prasad

Abstract Acquiring satellite images having high spectral, spatial, temporal, and radiometric resolutions simultaneously is very difficult. In this work, data from two sensors, viz., LISS III and AWiFS (DataPair) of Resourcesat-1 satellite (ISRO), are fusioned for improving the spatial resolution of AWiFS data. Best temporal resolution available from Resourcesat-1 is at 5-day repetivity by AWiFS. Using high spatial temporal resolution images, more information can be extracted from the images. Non-subsampled contourlet transform (NSCT) is used for the fusion of LISS III and AWiFS data. Through this fusion, even fine details can be extracted from the surface of the earth. Fusion of one or more images can be used, to determine the information content more accurately in our unambiguous manner. Image fusion process deals with two images with different characteristics and combines the merits of both. Here, an image with high spatial resolution (HSR) is used for a corresponding low spatial resolution (LSR) image. The quality of the output data are checked with the quality of the original image. The results of the desired high-resolution image through this current method are found to be satisfactory.

Keywords AWiFS data · LISS III data · NSCT · LSR · HSR quality assessment

K. S. R. Radhika (✉)
DMS SVH College of Engineering, Machilipatnam, India
e-mail: kammilisrr@gmail.com

C. V. Rao
National Remote Sensing Centre, Hyderabad, India
e-mail: cvrao909@gmail.com

V. Kamakshi Prasad
JNTUHCEH, Hyderabad, India
e-mail: kamakshiprasad@gmail.com

© Springer Nature Singapore Pte Ltd. 2018
V. Bhateja et al. (eds.), *Proceedings of the Second International Conference on Computational Intelligence and Informatics*, Advances in Intelligent Systems and Computing 712, https://doi.org/10.1007/978-981-10-8228-3_11

1 Introduction

Agriculture is the main source of revenue and is the prime occupation in India. Periodical variation of the crops, percentage of yields, damage due to various reasons, and several other factors pose a challenge.

LISS III, LISS IV (Linear Imaging Self Scanners), and AWiFS (Advanced Wide Field Sensor) on board are the sensors available with Resourcesat-1 satellite mission. Temporal resolution (TR), radiometric, spectral, and spatial resolutions (SR) are the main properties that describe remotely sensed data. Temporal resolution is the satellite's revisit time. Radiometric resolution represents sensitivity to the magnitude of the electromagnetic energy when an image is acquired. Spectral resolution describes the number of bands in the satellite. Spatial resolution is the smallest discernable detail in the image.

LISS III and AWiFS are two sensors of the same satellite Resourcesat-1 of ISRO, having identical spectral bands for both of them. Swath width (SW) and temporal resolution are high for AWiFS data, while spatial resolution is high for LISS III. Thus, images AWiFS and LISS III can be fusioned. The resulted image formed will have high TR, high SR along with a wider swath. Thus, more information can be derived through this technique [1].

To derive fine details from the RS data, HSR is needed. High spatial resolution is a need for accurate boundary delineation of smaller features. To obtain area details over wide range, high swath is recommended. High temporal resolution is required to study periodical changes in the crops (change detection analysis). In Resourcesat-1, sensor parameters of LISS III and AWiFS are given in Table 1.

1.1 Non-subsampled Contourlet Transform

The objective is to use an effective transform aiming the applications in which redundancy is not of significant concern such as denoising. Being completely shift-invariant and multiscale expansion, NSCT has fast implementation. In addition

Table 1 Resourcesat-1 sensor parameters

	Swath width (km)	Temporal resolution	Spatial resolution (m)	Spectral bands	Wavelength ranges (μm)
LISS3	740	5 days	56	Band 2	0.52–0.59
				Band 3	0.62–0.68
				Band 4	0.77–0.86
				Band 5	1.55–1.70
AWiFS	140	24 days	23.5	Band 2	0.52–0.59
				Band 3	0.62–0.68
				Band 4	0.77–0.86
				Band 5	1.55–1.70

to this, NSCT is highly successful in applications such as image denoising and enhancement techniques [2].

NSCT transform is composed of two parts. One is a multiscale property involved a non-subsampled pyramid (NSP) structure and the another is a non-subsampled directional filter bank (NSDFB) structure that results in directionality.

NSP: NSCT has multiscale property acquired from shift-invariant filtering which gives a subband decomposition. This is possible by two-channel non-subsampled 2D filter bands.

NSDFB: DFB of Bamberger and smith [3] combines sampled two channels for filter bands with resampling operations. The resultant tree-structured filter bank decomposes the 2D-frequency plain to directional wedges.

Image is decomposed into multi-direction subbands at multiple levels using NSCT. With the available subbands in the image, seven contourlet coefficients are possible which are further processed. Coefficients of both the sensors in the over-lapped region are used for creating the training database. This in turn is used for predicting HR data for the corresponding LR data. In simultaneous capturing according to their swath widths of the two sensors, LISS III sensor data are overlapped at the center of AWiFS image.

In simultaneous coverage of an area with both the sensors, LISS III image overlapped at the center of AWiFS data. Thus, an overlapping area occurs at the center (Figs. 1 and 2). In Fig. 1, the SW of both sensors LISS III and AWiFS is

Fig. 1 Swath of LISS III and AWiFS

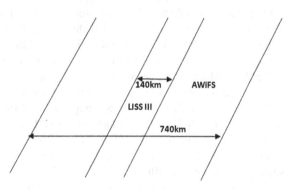

Fig. 2 LISS III image overlapped at the center of AWiFS image

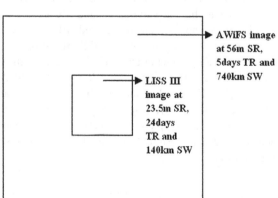

shown. Figure 2 shows the coverage area of both sensors. In the overlapped region corresponding to the low-resolution AWiFS data, high-resolution LISS III data will be predicted.

2 Background

AWiFS has TR, SR, and SW of 5 days, 56 m, and 740 km, respectively. LISS III on other hand has TR, SR, and SW of 24 days, 23.5 m, and 140 km. Inverse relationship exists between SW and revisit time, and direct proportion lies between ground sampling distance (GSD) and SW [4].

The main important describing factors of remotely sensed data are resolutions of the image. They are TR, SR, radiometric resolution, and spectral resolution. The discriminable smallest linear distance separation that exists between any two objects on the surface of earth represents SR of RS system [5]. Each bandwidth and the responsive bands number of the EM spectrum describe spectral resolution of the RS system [6]. The revisit time of the imagery of a specific geographic area by the system is called temporal resolution [7]. Remote sensed data at periodical data are required to observe study certain agricultural variables.

Radiometric resolution is ability/sensitiveness of an RS detector to detect variations in signal strength while it captures the radiant which is either emitted, backscattered, or reflected, from the area/location [8]. Countable number of distuinshable signal levels are represented through it. Resourcesat-1 satellite has three onboard sensors, viz., LISS IV (SR is 5.8 m), LISS III (SR is 23.5 m) of AWiFS (SR is 56 m). Parameters of various sensors in the satellite Resourcesat-1 are given in Table 1. More than one number of sensors data are fusioned to significantly enhance the information for decision-making application and analysis.

For improving SR of AWiFS image using LISS III dataset, single image super-resolution (SISR) [9] technique with NSCT is proposed. Various applications in RS such as precision agriculture, flood forecasting, and land-use–land-cover changes require images with the greater swath along with increased temporal and spatial resolution. Identification of vegetation-type habitat derivation improves land-cover classification with high spatial data [10].

Sensor data having varying TR and SR are combined resulting in images with simultaneous improvement in TR and SR. This technique of sensor data combination is observed by fusioning LISS III data (23.5 m of SR) with AWiFS data (5 days of TR) producing an image having 23.5 m of SR and 5 days of TR. This method hence derives the name LISS III SR and AWiFS TR data combination (LSAT) [11]. For Landsat and MODIS data, the SR and TR are 30 m, 16 days and 250 m–1000 m, 1 day, respectively. Location regularized sparse representation technique is used to predict the fine image corresponding to coarse image using a pair of Landsat and MODIS images. Later, optimal solution is found using FISTA algorithm [12].

SISR method tries to find priori correspondence between low-resolution and high-resolution image blocks and in the subsequent stages, these results are used to get high-resolution data from low resolution data. This method uses support vector regression (SVR) that accepts vector input and outputs scalar quantity. Corresponding pairs in AWiFS and LISS III identical in geometry are made to train the SVR model. High-resolution image pixel along with its corresponding block (patch) in LR image constitutes pair. LR image patch is used by the trained model of getting HR image pixel [13].

3   Methodology

Input images: As input images, two datasets dataset I and dataset II are considered. Each dataset constitutes two images: one is LISS III and another is AWiFS data. This pair of images is used in the entire work for processing. The datasets I and II image pairs (AWiFS image and LISS III image) are acquired at the same area on the same date, by Resourcesat-1 and this data has been collected from BHUVAN database provided by NRSC. These datasets are widely used by research community in India and abroad [14]. The input image datasets are selected such that they are covering uniform features.

Input DataPair 1: *One pair images acquired on November 27, 2009. These are the original images used as input DataPair* (Figs. 3 and 4).

Input DataPair 2: *One pair images acquired on October 31, 2009. These are the original images used as input DataPair* (Figs. 5 and 6).

Proposed method obtains HR image through the given steps, to prone the concept.

3.1   Image Registration

Image-to-image registration is the translation, scale, and rotation alignment process by which two images of similar geometry and of same geographic area are positioned coincident with respect to one another that corresponding elements of the ground area appear in the same place on the registered images. This is used to obtain exact pixel-to-pixel matching of two images LISS III and AWiFS, which is obtained.

3.2   Normalization

LR images through LISS III are possible by weighted averaging of neighboring so to compare the corresponding AWiFS radiance. SR equal to AWiFS (56 m) is the same as LISS III data being down-sampled 2.4 times [15]. Direct digital number

(DN) comparision is not suggestable as LISS III image is 8-bit data and AWiFS is 10-bit data. So radiometric normalization in radiance domain is performed. The DN of down-sampled LISS III (L_3) is calculated using radiance values through "(1)" and "(2)" [16].

$$\text{Radiance } L_3 = (\text{DN of } L_3/\text{Max. DN of } L_3) \times \text{LISS III Saturation Radiance} \quad (1)$$

Using these radiance values, pixel values are estimated just as AWiFS image using the following equation.

$$\text{DN} = \text{Radiance } L_3 \times (\text{Maximum DN of AWiFS}/\text{Saturation} - \text{Radiance of AWiFS}) \quad (2)$$

3.3 NSCT Training

Using NSCT, for one band say B2 in AWiFS image, seven contourlet coefficients of which one constitutes low-frequency information, two belonging to middle-level frequency information, and remaining four of high-frequency information are extracted. In the similar way, using band B2 of LISS III image, seven coefficients are obtained. Subsequently, corresponding coefficients of the two images are combined to form the training database. The process thus yields seven different databases corresponding to seven different frequency information.

3.4 NSCT Prediction

Using the training database and AWiFS information for each contourlet coefficient, the respective coefficients of LISS III are predicted. So seven coefficients will be predicted in the same manner of band B2. These seven NSCT coefficients of the predicted image are combined to get the total NSCT information. Inverse transformations result in the predicted band B2 information of LISS III image in spatial domain. The procedure is adopted for all the remaining three bands. Thus, training database of step (c) is used to predict the HR equivalent of the original LR AWiFS data in the overlapped area.

3.5 Quality Assessment

The quality of the output image is tested with original data. Comparison of assessment parameter such as RMSE, CC, SAM, R^2, and SSIM of the predicted HR image of original HR image is done in the overlapped area [17].

4 Results

In this work, a pair of remote sensing images from two sensors AWiFS and LISS III of Resourcesat-1 satellite are considered. Experiments are conducted on two datasets (Figs. 3 and 4) and (Figs. 5 and 6) comprising of 300 × 300 pixels image size. These two input images are acquired on same date and at the same

Fig. 3 AWiFS LR image

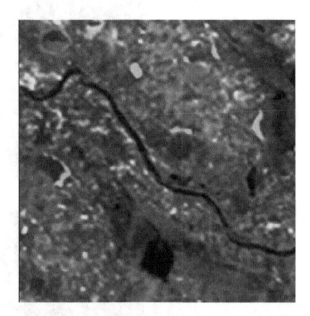

Fig. 4 LISS III HR image

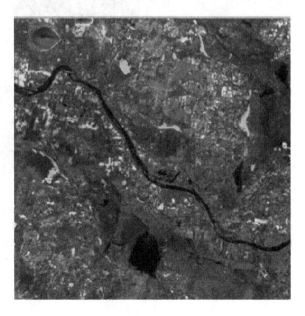

Fig. 5 AWiFS LR image

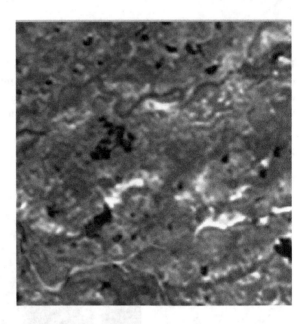

Fig. 6 LISS III HR image

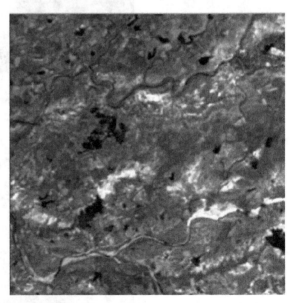

geographical location. The AWiFS and LISS III sensor images are having four spectral bands. The first band is green band denoted by B2, second is red band denoted by B3, third is the near infrared band denoted as B4, and the fourth is shortwave infrared band denoted by B5. The methodology is implemented on all

Fig. 7 Band 2

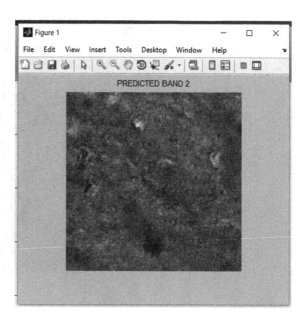

the above-mentioned bands separately. Later, four predicted HR bands are obtained for the corresponding LR bands.

DataPair 1:

Figures 7, 8, 9, and 10 are the bands of LISS III image that are predicted using the dataset pair on November 27, 2009 of the respective AWiFS bands 2, 3, 4, and 5, respectively. Multispectral HR image of the predicted bands is obtained corresponding to the low-resolution AWiFS data (Fig. 11).

DataPair 2:

Figure 12 shows the predicted LISS III band 2 image of the respective AWiFS band 2 that is obtained. Similarly, Figs. 13, 14, and 15 predicted LISS III bands 3, 4, and 5, respectively, for the corresponding LR AWiFS bands. Image pair of LISS III and AWiFS on October 31, 2009 is used for the prediction of HR image corresponding to the LR AWiFS image. Above four bands are layer stacked to obtain the multispectral predicted LISS III image. Predicted multispectral image with band combination of band 2, band 3, and band 4 is given in Fig. 16.

The proposed method using NSCT has been implemented to predict high-resolution LISS III data in overlapping area of DataPair. The output image quality is checked with the original LISS III image in the overlapping area. The quality assessment parameters are computed for the predicted image which is found to be satisfactory. There may be similar earth features in the nonoverlapping area as that of overlapping area. The NSCT database of the overlapping region can be used to predict the features in the nonoverlapping region. If the nonoverlapping region has similar features, then the proposed method can be extended even for the

Fig. 8 Band 3

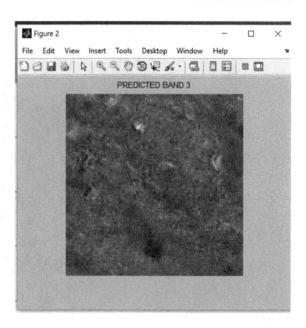

Fig. 9 Band 4

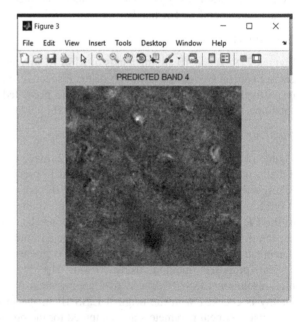

application to predict HR LISS III image. Thus, the spatial resolution of AWiFS can be improved equal to that of L3. Later, the LISS III swath width and temporal resolution can be improved with as that of AWiFS sensor data.

Fig. 10 Band 5

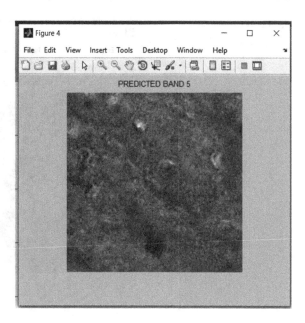

Fig. 11 Predicted HR image

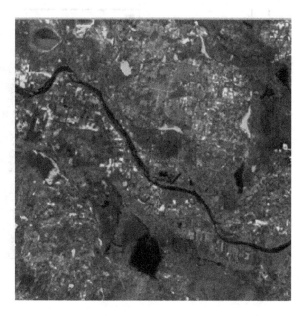

Fig. 12 B2

Fig. 13 B3

Fig. 14 B4

Fig. 15 B5

Fig. 16 Predicted HR image

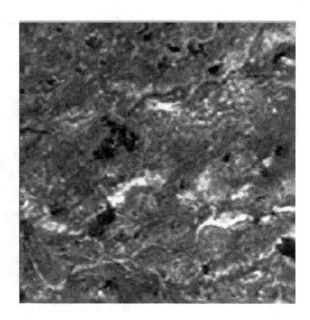

5 Conclusion

The predicted HR image for the LR image is obtained in the overlapped area of DataPair imagery. The predicted HR image is satisfactory in the overlapped area. Quality assessment is done on the predicted HR image and found to be high accurate, in terms of all the assessment parameters described above.

The proposed work can even be extended for predicting the HR image nonoverlapping regions since in simultaneous capture of the DataPair with LISS III patch at the center of AWiFS patch the method predicts the HR data corresponding to the LR data in the overlapping region. This can be extended even to the nonoverlapping region where LISS III (HR data are not available). This uses the fact that classes (say water bodies, vegetation, etc.) have similar characteristics either in overlapping or nonoverlapping regions. So it is possible to improve resolutions for the entire AWiFS swath, which will enable us to create the images with spatiotemporal image fusion in the nonoverlapping regions of the image, with sufficient large training database.

Acknowledgements The images used in this work are downloaded from Bhuvan website of NRSC, Hyderabad. Sincere thanks and gratitude by authors to NRSC for providing these datasets.

References

1. Radhika, K.S.R., Rao, C.V., Kamakshi Prasad, V.: Enhancement of AWiFS Spatial Resolution with SVM Learning. In 6th International Advanced Computing Conference, 978-1-4673-8286-1/16 IEEE https://doi.org/10.1109/iacc.2016.42, pp. 178–183 (2016).
2. Arthur L. da Cunha, Jianping Zhou, Minh N. Do.: The Nonsubsampled Contourlet Transform: Theory, Design, and applications. In: IEEE Transactions on Image processing, 15: 3089–3101. https://doi.org/10.1109/tip.2006.877507 (2006).
3. Bamberger, R.H., Smith, M.J.T.: A Filter Bank for the Directional Decomposition of Images: Theory and design. In: IEEE Trans. Signal Process., vol. 40, no. 4, pp. 882–893, Ar. (1992).
4. Coops, N.C., Johnson, M., Wulder, M.A., White, J.C.: Assessment of Quickbird High Spatial Resolution Imagery to Detect Red Attack Damage Due to Mountain Pine Beetle Infestation. In: Remote Sensing of Environment, 103(1): 67–80, https://doi.org/10.1016/j.rse.2006.03.012 (2006).
5. Jensen, J.R.: Remote Sensing of the Environment: An Earth Resource Perspective. 2/e. Pearson Education India (2009).
6. Landgrebe, D.A.: Signal theory methods in multispectral remote sensing. John Wiley & sons (2005).
7. Abrams, M.: The Advanced Spaceborne Thermal Emission and Refection Radiometer (ASTER): data products for the high spatial resolution images on NASA's Terra platform. In: International Journal of Remote Sensing, 21(5): 847–859 (2000).
8. Lambin, E.F. Strahlers, A.H.: Change - vector analysis in multi temporal space: a tool to detect and categorize land-cover change processes using high temporal-resolution satellite data. In: Remote sensing of environment, 48(2): 231–244, (1994).
9. Baker, S., Kanade, T.: Limits on Super-Resolution and How to Break Them. In: IEEE Transactions on Pattern Analysis and Machine Intelligence, 24: 1167–1183. https://doi.org/10.1109/tpami.2002.1033210 (2002).
10. Gillespie, T.W., Foody, G.M., Rocchini, D., Giorgi, A.P., Saatchi, S.: Measuring and Modelling Biodiversity from Space. In: Progress in Physical Geography, 32; 203–221, https://doi.org/10.1177/0309133308093306 (2008).
11. Rao, C.V., MalleswaraRao, J., Senthil Kumar, A., Dadhawal, V.K.: Fast spatiotemporal data fusion: merging LISS III with AWiFS sensor data. In: International Journal of Remote Sensing, 35(24): 8323–8343, https://doi.org/10.1080/01431161. 2014. 985396 (2014).
12. Xun Liu, Chenwei Deng, Baojun Zhao.: Spatiotemporal Reflectance Fusion Based on Location Regularized Sparse Representation. In: IGARSS, 978-1-5090-3332-4/16/$31.00 ©2016 IEEE, pp. 2562–2565 (2016).
13. Rao, C.V., MalleswaraRao, J., Senthil Kumar, A., Lakshmi, B., Dadhwal, V.K.: Expansion of LISS III swath using AWiFS wider swath data and contourlet coefficients learning. In: GI Science and Remote Sensing, 52(1): 78–93, https://doi.org/10.1080/15481603.2014, 983370 (2015).
14. National Remote Sensing Centre, https://bhuvan.nrsc.gov.in.
15. Zhang, H., Huang, B.: Support Vector Regression-Based Downscaling for Inter Calibration of Multiresolution Satellite Images. In: IEEE Transactions on Geoscience and Remote sensing, 51: 1114–1123. https://doi.org/10.1109/tgrs.2013.2243736 (2013).
16. Rao, C.V., MalleswaraRao, J., Senthil Kumar, A., Manjunath, S.: Restoration of High Frequency Details while Constructing the High Resolution Image. In: Annual IEEE India conference on Engineering sustainable solutions (INDICON), (2011).
17. Wang, Z., Bovik, A.C., Sheikh, H.R., Simoncelli, E.P.: Image Quality Assessment: From Error Visibility to Structural Similarity. In: IEEE Transactions on Image Processing, 13 (4): 600–612, https://doi.org/10.1109/tip.2003.819861 (2004).

Regular Expression Tagger for Kannada Parts of Speech Tagging

K. M. Shiva Kumar and Deepa Gupta

Abstract Part of speech tagging for Indian languages in general and Kannada in particular is not a very widely explored territory. There have been many attempts at developing a good POS tagger for Kannada, but the morphological complexity of the language makes it a hard nut to crack. Some of the best taggers available for Indian languages employ hybrids of machine learning or stochastic methods and linguistic knowledge. Though the results achieved using such methods are good, their practicability for other inflective Indian languages is reduced due to their heavy dependence on linguistic knowledge. Even though taggers can achieve very good results if provided good morphological information, the cost of creating these resources renders such methods impractical. In this paper, we present regular expression parts of speech tagger for Kannada. We apply 100 patterns incorporating the TDIL tags for Kannada and tested for accuracy with manual tagged corpus.

Keywords POS tags · NLTK · Morphology · Regular expressions
Pattern · NLP · Semantics · Syntax

1 Introduction

Grouping the words into their respective classes is widely known as tagging. The part of speech tagging is one of the stages in natural language processing. An NLP system needs more information about each word in a corpus. The syntax-level information is the words parts of speech, and the semantic level information is its meaning and also a context. In this paper, we use the word tagging in the context of

K. M. Shiva Kumar (✉)
Department of Computer Science, Amrita University, Mysore, India
e-mail: kh_shivakumar@asas.mysore.amrita.edu

D. Gupta
Department of Mathematics, Amrita School of Engineering,
Amrita University, Coimbatore, India
e-mail: g_deepa@blr.amrita.edu

© Springer Nature Singapore Pte Ltd. 2018   121
V. Bhateja et al. (eds.), *Proceedings of the Second International Conference
on Computational Intelligence and Informatics*, Advances in Intelligent Systems
and Computing 712, https://doi.org/10.1007/978-981-10-8228-3_12

parts of speech tagging for Kannada words. Indian languages are well organized in their syntax semantics and lexical information. We find more scripts about the grammar rules of creating meaningful and syntactically correct sentences. The process of framing the sentence with proper order in a language is acquired by human beings by their parents and the surrounding environment very easily but to train a computer that can process natural language like Kannada needs a well-framed model and a good annotated corpus. The English language is resource rich in providing language-related corpus and language processing tools on Internet. Kannada corpus and linguistic data are resource poor compared to other languages. So we need to create these resources manually or create software to automatically create these resources. The free order languages like Kannada require tagging information to reduce the ambiguity and get the context information from the given sentence. Two broad approaches are followed by researchers for tagging the corpus. (1) Rule-based approach which purely based on linguistic knowledge and (2) statistical approach and a hybrid approach is used to create a tagged corpus or a tagged sentence. Statistical tagging techniques use either supervised or unsupervised or hybrid methods. The tagging process is termed as the sequence labelling problem using the noisy channel model in which every word in the input sequence is to be assigned with a tag from the given tags. Technology development for Indian languages (TDIL) has released a standard tags for each official language in India. We use the 32 official TDIL tags for Kannada in this work. Statistical approaches require training data. The resource-poor languages like Kannada do not have online annotated corpus. We need to create the corpus manually. A huge quantity of tagged corpus creation is a difficult task for languages like Kannada because of their agglutinative feature and morphological richness. The data sparseness and smoothing techniques used in statistical methods lead other problems of omitting important words in the different contexts. Statistical taggers use machine learning algorithms to train their systems. In Kannada, inflections attachment to the main word decides its meaning at lexical level as well as at sentence level. Because of this feature, the number of unique words in the corpus is very high compared to other Indian languages and English. Each lemma word in Kannada can be attached with 13 inflections, and these inflections are orthographically attached to the root word. The morphological richness makes the statistical tagger to treat the lemma words as different if it is attached with variable inflections. Kannada consonants are attached with vowels in a word to make them agglutinative. This leads to increase in number of unique words in the training corpus, and their frequency is also less compared to English or Hindi corpus. This is the major challenge which makes the statistical taggers fail to give good results for Kannada.

In this paper, we present a regular expression-based parts of speech tagging for Kannada textual data. We explore the morphological feature associated with Kannada word glyphs to identify the POS tags and tag with NLTK tagger. We got a promising result without using statistical or machine learning algorithms to tag the Kannada words.

2 Related Work

The need of Indian language scripts recognition and processing task is in demand as the number of online users is increasing day by day. The phonetic keyboard input is enabled for all Indian languages with the effort of TDIL. In Kannada, we can represent any pronunciation with their equivalent script and the vernacular representation of the given word. The languages with fixed word order follow certain order of words in a sentence. Any change in this order gives a different meaning. The SVO order is strictly followed in representing English sentences. The free word order languages do not follow any fixed order in arrangement of words in a sentence because in these languages each word acts as a functional or nominal word. In English language, a phrase or a group of words represent the partial meaning of a sentence [1]. The combination of prepositions and nouns or verbs gives a partial meaning of the sentence. Such chunking may be achieved using different methods. For fixed word order languages, phrase structure grammars and parse tree adjoining grammars are used for modelling the sentence structures [2] (Joshi and Schabes 1997). These techniques cannot be applied for free word order languages. In free order languages, the position of phrases can be altered just by changing the attached inflectional affixes. The position of inflectional suffix attached with each word in free order languages decides the meaning of the sentence. This leads to device a constraint-based model to process these languages adhering to the grammar rules. The free order language grammar and their computational representation are proposed by Bharati et al. [3] (Bharati and Sangal [4]; Tapanainen and Jarvinen [5]). These modelling methods are also been used to parse fixed order languages (Bharati et al. 1997). They proposed the rules and methods to parse Hindi sentences. 'Unknown Word Guessing and Part-of-Speech Tagging Using Support Vector Machines' [6]. They were able to predict POS tag for mysterious words using support vector machines. They achieve high accuracy in POS tag calculation using substrings and surrounding context. This method does not depend on exacting characteristics of English. They can be applied to other languages. Drawback of this method is computational cost. It is difficult to train for a large amount of training data, and testing time increases in more complex models. Karthik Kumar et al. [7] proposed the 'Comparative study of various Machine Learning methods For Telugu Part of Speech tagging' and they introduced the Smoothing the flexibility of an information and Probability of transition between depends current observations, past and future observations. The drawback of this paper is low accuracy of Telugu POS tagging, when it is compared to other Indian languages. Kumar and Josan [8] had proposed 'Part of Speech Taggers for Morphologically Rich Indian Languages'. This paper tells about the part of speech (POS) taggers proposed for various Indian languages like Hindi, Punjabi, Malayalam, Bengali and Telugu. This paper introduces the high generalization performance independent of the dimension of feature vectors in support vector machines. The limitation of this paper is that low accuracy leads to less size of lexicon. Rajeev et al. [9] proposed the 'Part of Speech Tagger for Malayalam'. They mentioned about the statistical approach with hidden Markov model and TnT tagger. The taggers are used for parsing, for recognition in message extraction system, for speech recognition, and for text-based information retrieval. The system works fine with Unicode-8 bit format to train and test data and provides basic building block for constructing statistical models for automatic processing of natural languages with hidden Markov model. Dhanlaxmi et al. [10] developed an SVM-based POS tagger and chunker for Tamil language. They got POS tag accuracy of 95.64% and chunker accuracy of 95.82%. They used a customized Amrita tags to tag the corpus of 165000 training set and 60000 test set Tamil words. Shambhavi and Kumar [11] proposed the 'Kannada Part-Of-Speech Tagging with Probabilistic Classifiers'. This paper has a process of assigning POS tags for each word in sentences using a supervised machine learning classification algorithms, second-order hidden Markov model (HMM) and conditional random fields (CRF) in Kannada language. They have taken training data of 51,269 words, and test data consists of 2932 tokens. This experiment illustrates the accuracy of the tools based on HMM and CRF, which are 79.9% and 84.58%, respectively. Badugu [12] proposed 'Morphology Based POS Tagging on Telugu'; in this work, they presented the morphological feature of functional words in Telugu and used the morphological analyzer to tag the words in Telugu. Pallavi and Pillai [13] developed a Kannada POS tagger using conditional random fields with 80000 words corpus. They used the TDIL tags to train and test the system. The rules or arrangement of words in Kannada is a free form in nature; due to this, the system requires more number of rules for specifying a single part of speech. They got 92.4% accuracy in POS Tagging.

3 Challenges in POS Tagging for Kannada

Attachment of Kannada inflections to the root words makes the number of unique words much more larger than English corpus. Figure 1 presents the comparison between the number of words and unique words in English and Kannada Bible text corpus of 3000, 6000 and 9000 sentences. The analysis of parallel corpus of English Kannada Bible text which is a parallel corpus reveals that the number of words used to represent the information is twofolds more compared to Kannada words. The number of unique words between English and Kannada is fivefolds more. We observe more number of repeated words in English than Kannada. The main reason

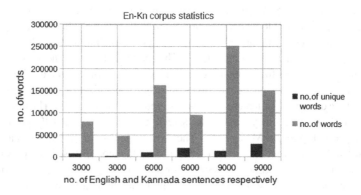

Fig. 1 English–Kannada corpus details

Table 1 Kannada case markers

English case name	Kannada case name	Characteristic affixes
Nominative	Prathama vibhakti	rama*Nu*, (masculine) seetha*Lu* (*feminine*)
Accusative	Dvitiya vibhakti	Annu, (colloquial)
Instrumental	Trutiya vibhakti	Inda
Dative	Chaturthi vibhakti	Ge, ige, akke, kke
Genitive	Shashti vibhakti	Aa
Locative	Saptami vibhakti	Alli
Vocative	Sambodhana vibhakti	Ae

for this feature for Kannada textual representation is Kannada words are attached with inflectional suffixes which includes the gender number and case markers.

In Kannada language the declinable words have seven cases as shown in Table 1.

A phrase or chunk in English is a group of two or more words but in Kannada the phrase can be an inflected word with number gender or case markers. This morphological richness makes the task of tagging more complex with statistical methods as in the case of English.

4 Kannada Morphology-Based Parts of Speech Tagging

The proposed approach differs from all other work on tagging for Indian languages in the way we process words. In English, a functional word is a combination of two or three words. The meaning of the phrase is spread over all the parts of the phrase but in Kannada the morphological affixes attached to the main words decide the meaning, gender, number and case. A word or token in Kannada is a complex functional word inflected with morphological suffix, and the position of these suffixes will decide the meaning of the given sentence. This leads to less number of words in a given sentence of Kannada. Because of this feature, Kannada tokens are to be seen beyond the space separated group of characters. Kannada words can be classified based on meaning and morphological and syntactic properties. When a Kannada word attached with morphological suffix its phonetic pronunciation slightly varies reflecting in a different meaning. Each word in Kannada sentence comprises a definite meaning and syntactic role in the sentence. The arrangement of words in the sentence follows the well-defined syntactic rules especially with respect to gender number and case. The case markers in Kannada are attached to noun and verb categories. We need to define a set of syntactic relations that are applicable to all domain-specific data which is available in Kannada.

Statistical approaches for tagging assume that the information necessary for tag assignment comes from the associated tokens in the sentence. This is true for English but the same approach cannot be applicable for Kannada. In many cases, only the combination of two or three words together that come before the current word is taken for consideration. In contrast, the information required for assigning the correct tag for Kannada word comes from within the word that is in morphology. This is true in the case of morphologically rich languages. The majority of the words in Kannada can be tagged correctly by looking at the main word and the inflections attached to the word. The case, gender and number attachment in Kannada words make the language rich with more number of unique words of the language. This feature is a threat for statistical processing of the language because we cannot derive the pattern of arrival of words in the sentences. In Kannada, the morphology assigns 8–10 affixes to the main word. Because of this feature, the syntax modelling for Kannada becomes more complex compared to English or Hindi languages. The Kannada sentence syntax follows a free form arrangement of words. In the present context we use interrelationship between main tags, such as nouns and verbs that will not model the syntax correctly. The complex functional dependencies between the words of the sentence are decided by the inflectional suffixes in Kannada. Each noun and verb in Kannada is attached with any of the above-mentioned inflectional suffixes; it is required to refine the POS tags to represent different forms of nouns and verbs. This makes the statistical techniques to capture and utilize complex functional dependencies between words in a sentence. The syntactic parser with improved of tags will be able to eliminate most of the tag ambiguities. Tagging is intended only to reduce tag ambiguities, not necessarily to eliminate all ambiguities.

5 Tags and Kannada Tagging

Kannada words are inflected with more than ten affixes which decide the meaning and context of word in the sentence. The question here is whether we need to tag the words with inflectional suffixes or without consideration of the suffixes. If we do not consider the suffix attachment to the words in Kannada sentences, we will be successful in good tagging but we may lose inherent meaning attached with the inflected words. In this work, we present the impact of using the modified POS tags by considering the associated inflections. We are using the variants of noun and verb main category parts of speech to inclusively tag the Kannada words. We retain the other main POS tags provided by TDIL. Since the inflections attached to the main parts of speech category decide the unique meaning of the word in the sentence, we believe in creating more variants of main category parts of speech such as noun and verb will improve the tagging accuracy and to extract the linguistic pattern of Kannada sentence. The attachment of case markers in Kannada sentence follows a particular pattern as shown in Table 2.

The free order arrangement of words in Kannada sentences follows a predefined pattern of arrangement of words which are inflected by the number and case markers. In the above example, all three variations of the sentence give the same meaning but the word order is different in each sentence. We can see the difference in case markers position in each of these sentences. Hence, in Kannada the inflectional suffixes decide the meaning associated with their position in a given sentence. Hence, to extract the syntactic information for Kannada sentences, we need to analyse the pattern of arrangement of these number, gender and case markers positions in the sentences.

Table 2 Free form word arrangement in Kannada sentences

	Kannada sentence forms	English phrase forms
Kn-sentence	ಮೋದಿಯವರು ದೆಹಲಿಯಿಂದ ಬೆಂಗಳೂರಿಗೆ ಬಂದರು.	Modiji from Delhi to Bangalore came
Transliterated sentence	Modiyavaru Dehaliyinda bengaLurige bandaru	
	1 2 3 4	
PoS	NN_1 NN_3 NN_4 VB_1	
Kn-sentence	ದೆಹಲಿಯಿಂದ ಬೆಂಗಳೂರಿಗೆ ಮೋದಿಯವರು ಬಂದರು.	From Delhi to Bangalore Modiji came
Transliterated sentence	Dehaliyinda bengaLurige modiyavaru bandaru	
PoS	NN_3 NN_4 NN_1 VB_1	
Kn-sentence	ಬೆಂಗಳೂರಿಗೆ ದೆಹಲಿಯಿಂದ ಮೋದಿಯವರು ಬಂದರು.	To Bangalore from Delhi Modiji came
Transliterated sentence	bengaLurige Dehaliyinda modiyavaru bandaru	
PoS	NN_4 NN_3 NN_1 VB_1	
English sentence	Modiji came to Bangalore from Delhi	

Table 3 Proposed tags for Kannada

POS tag	Description	Example	
NN_Pr	Proper noun	ಬೆಂಗಳೂರು, BengaLuru	Bangalore
NN_S	Noun singular	ರಾಮ, Rama	Rama
NN_P	Noun plural	ಪೆನ್ನುಗಳು, pennugaLu	Pens
NN_M	Noun masculine	ಸೇವಕ,sevaka	Servant
NN_F	Noun feminine	ಸೇವಕಳು, ಸೇವಕಿ sevakaLu, sevaki	Maid
NN_N	Noun neutral	ಕುರ್ಚಿಯ, Kurchiya	Chair
NN_1	Noun nominative	ಸೇವಕನು, Sevakanu	Servant
NN_2	Noun accusative	ಸೇವಕನನ್ನು, Sevakanannu	Servant's
NN_3	Noun instrumental	ಸೇವಕನಿಂದ, Sevakaninda	From servant
NN_4	Noun dative	ಸೇವಕನಿಗೆ, Sevakanige	To servant
NN_5	Noun genitive	ಸೇವಕನ, Sevakana	Servant's
NN_6	Noun locative	ಸೇವಕನಲ್ಲಿ, Sevakanalli	In servant
NN_7	Noun vocative	ಸೇವಕನೆ, sevakanE	The servant
VB_M	Verb masculine	ಬಂದನು, Bandanu	He came
VB_F	Verb feminine	ಬಂದಳು, bandaLu	She came
VB_N	Verb neutral	ಬಂದಿತು, Banditu	That come
VB_S	Verb singular	ಬಂತು, Bantu	Came
VB_P	Verb plural	ಬಂದವು, Bandavu	Many have come
VB_1	Verb nominative	ಕೊಡು, koDu	Give
VB_2	Verb accusative	ಕೊಡಲು, koDalu	To give
VB_3	Verb instrumental	ಕೊಟ್ಟಿದ್ದರಿಂದ, Kottiddarinda	Given
VB_4	Verb dative	ಕೊಡಲಿಕ್ಕೆ, koDalikke	To give
VB_5	Verb genitive	ಕೊಟ್ಟ, Kotta	He gave
VB_6	Verb locative	ಕೊಟ್ಟಲ್ಲಿ, Kottalli	Where he gave?
VB_7	Verb vocative	ಕೊಟ್ಟನೆ Kottane, ಕೊಟ್ಟಳೆ kottaLe	Did he gave? Did she gave?

Nouns and verbs are treated as universal categories. Since the inflections attached to these main categories do not alter their respective categories in Kannada sentence but carries a high impact on deciding the meaning of the sentence, we need to tag the words with all possible variants of the main POS tag types. In this regard, we present the following POS tags for Kannada nouns and verbs. We retain the other POS categories specified in TDIL tags for Kannada is shown in Table 3.

6 Experiments and Results

In availability of a standard corpus for Kannada, textual data is the major challenge for Kannada language processing. We downloaded the publicly available Bible text corpus which is in xml format. We converted the xml form corpus into textual form and obtained 20000 sentences. It is observed that number of unique words in Kannada is more compared to English.

6.1 Algorithm Kannada RegEx Tagger

Input: Raw Kannada sentences or a text document.
Step 1: Tokenize sentences using NLTK tokenize module.
Step 2: Design the RegEx patterns.

Patterns=   [(r"(.*ಲಿಂದ)$",   "NN_F"),(r"(.*ಗಳ)$",   "NN_P"),(r"(.*ವನ್ನು)$",   "NN_2"),(r"(.*ಯಲ್ಲಿ)$",
"NN_6"),(r"(.*ನ್ನು)$",   "NN_2"),(r"(.*ಯನ್ನು)$",   "NN_S"),(r"(.*ಗ)$",   "NN_S"),(r"(.*ಗಿ)$",   "NN_S"),(r"(.*ಳು)$",
"NN_S"),   (r"(ದೇವರು)$",   "NN"),(r"(.*ದನು)$",   "VB_M"),(r"(.*ದಳು)$",   "VB_F"),   (r"(.*ವಾಗಿ)$",
"ADJ"),(r"(ತುಂಬಾ)$",   "ADJ"),(r"(ಒಳ್ಳೆಯ)$",   "ADJ"),(r"(ವಿಶಾಲ*)",   "ADJ"),(r"(ದೊಡ್ಡ*)$",
"ADJ"),(r"(.*ವಾಗಿಯೂ)$",   "ADJ"),   (r"(ಮೇಲೆ)$",   "ADV"),(r"(.*ದ)$",   "ADJ"),(r"(.*ದೊತಹ)$",
"ADJ"),(r"(.*ದೆ0ತ)$",   "ADJ"),(r"(ಇತ್ತು)$",   "VB"),(r"(ಅದು)$",   "PRN"),(r"(ಇದು)$",   "PRN"),(r"(ಅವರು)$",
"PRN"),(r"(ಅವರು)$",   "PRN"),(r"(ಇವರು)$",   "PRN"),(r"(ಇವರ)$",   "PRN").......]

Step 2.1: Pass above patterns to nltk RegEx tagger.
Step 2.2: For each token in sentence or file.
Step 2.3: Assign POS tags using the RegEx tagger.

Step 3: Copy tagged words into json dump as a file.
Step 4: Read or print Tagged sentences.
Step 5: Close file.

6.2 RegEx Tagger Results

The Kannada RegEx pos tagger performance for randomly selected sentences is as follows in Table 4.

Since we used the morphology-based regular expression patterns to identify all possible variants of noun and verb forms, number of unknown words has been reduced to maximum extent and the tag accuracy is more. The precision and recall with respect to identifying the noun and verb forms in the Kannada text are more compared to other statistical approaches of tagging.

Table 4 RegEx tagger results for Kannada

# sentences	# tokens	Unique words	Identified by RegEx tagger	Unknown (%)
3000	47324	1274	1210	5
6000	94942	20248	19788	2.27
9000	150899	28859	27959	3.11

7 Conclusion and Future Work

In this paper, we presented morphology-based regular expression tagger for Kannada parts of speech tagger according to the complex morphological feature of Kannada words. We have shown that the performance and accuracy are high in the proposed system. Manual verification of the results shows that the system is flexible to tag any domain-specific data. We are planning to extend the same for developing a speaker identification system for Kannada textual data using the gender and case markers associated with verbs. The proposed work can also be used to model a more flexible syntax and semantic analyser for Kannada sentences.

References

1. Christopher Manning., Herbert Schutze: *Foundations of Statistical Natural Language Processing*, pp. 407–409; MIT Press, 1999
2. Charniak E: Statistical parsing with a context-free grammar and word statistics, Proceedings of the 14th AAAI, Menlo Park, 1997
3. Bharati A., Chaitanya V., Rajeev Sangal: *Natural Language Processing: A Paninian Perspective*; Prentice Hall India, 1995
4. Bharati, A., and Sangal, R., (1993). Parsing Free Word Order Languages in the Paninian Framework. In Proceedings of Annual Meeting of Association for Computational Linguistics, pp. 105–111
5. Pasi Tapanainen and Timo Järvinen A non-projective dependency parser, ANLC 1997 Proceedings of the fifth conference on Applied natural language processing Pages 64–71
6. Tetsuji, Nakagawa, T, Kudo, T., Matsumoto, Y; Unknown Word Guessing and Part-of-Speech Tagging Using Support Vector Machines. In *NLPRS* (pp. 325–331). 2001
7. Karthik Kumar G, Sudheer, K., Avinesh, "Comparative study of various machine learning methods for telugu part of speech tagging". In *Proceedings of the NLPAI contest workshop during NWAI '06, SIGAI Mumbai*. 2006
8. Dinesh Kumar., Josan, G. S: Part of speech taggers for morphologically rich Indian languages: a survey". *International Journal of Computer Applications*, 6(5), 32–41. 2010
9. Rajeev R R, Jisha P jayan, Elizabeth Sherly, Parts of speech Tagger for Malayalam, IJCSIT International Journal of Computer Science and Information Technology, vol 2, No.2, December 2009, pp 209-213
10. Dhanlaxmi V., Anand Kumar M., Rajendran S., Soman K P: PoS Tagger and Chunker for Tamil Language, Proceedings of the 8th Tamil Internet Conference, Cologne, Germany. 2010
11. Shambhavi B R., Kumar P R: Kannada Part-Of-Speech Tagging with Probabilistic Classifiers. *International Journal of Computer Applications*, 48(17), 26–30, 2012
12. Badugu, Srinivasu: Morphology Based POS Tagging on Telugu, International Journal of Computer Science Issues (IJCSI), 2014
13. Pallavi KP., Pillai A.S KannPOS—Kannada Parts of speech Tagger using conditional Random Fields, IN: Shetty N., Prasad N., Nalini N (eds), Emerging Research in computing, Information, Communication and Applications, Springer, New Delhi, pp 479–491(2016)

Design of Conservative Gate and their Novel Application in Median Filtering in Emerging QCA Nanocircuit

Bandan Kumar Bhoi, Neeraj Kumar Misra and Manoranjan Pradhan

Abstract In this era of emerging technology, fault-tolerant logic is applied for circuit design. As the deep submicron and scaling, a number of pitfalls are the faces of the CMOS technology. So a lot of constraints related to CMOS have shorted with the quantum-dot cellular automata (QCA) technology. In this work, a new parity conservative gate referred as parity-QCA (P-QCA) is proposed. The gate is simulated with QCADesigner and compared with existing parity preserving logic gates. By the comparative outcomes, it is found that the proposed design achieved higher efficiency as compared to the counterparts. We achieved 100% stuck-at fault coverage.

Keywords Parity preserving logic · Testability · Quantum-dot cellular automata · Nanotechnology

1 Introduction

Advances in CMOS technology and deep submicron scaling have made reliability a prime concern in advance semiconductor technology. In fact, it is worth mentioning that QCA technology is much used in research because of the smaller area, high computation speed, and low power concern [1, 2]. Electric potential-based charge flow occurs in QCA cell. This approach is advantage for both no leakage current and high computation; however, as with any CMOS technology, it suffers from leakage current as for the reliability issue. In VLSI circuits, parity preserving method is preferable for fault detection. The primary abstract behind the preserving

B. K. Bhoi (✉) · M. Pradhan
Department of Electronics & Telecommunication, Veer Surendra Sai University
of Technology, Burla 768018, India
e-mail: bkbhoi_etc@vssut.ac.in

N. K. Misra
Department of Electronics Engineering, Institute of Engineering and Technology,
Lucknow, India

© Springer Nature Singapore Pte Ltd. 2018
V. Bhateja et al. (eds.), *Proceedings of the Second International Conference on Computational Intelligence and Informatics*, Advances in Intelligent Systems and Computing 712, https://doi.org/10.1007/978-981-10-8228-3_13

logic of the observance of testability is to match the parity of inputs with parity of outputs [3]. In fact, it is the hamming weight used in both inputs and outputs to ensure the observance of testability. More formally, momentary incorrect logic output in the circuit can cause fault to occur, which effect circuit functionality. Hence, the parity preserving feature is of importance to circuits' reliability. Since the parity preserving approach is not suited of error correction, the testing engineerings usually accept the testing by means of stuck-at fault to be able to correct the fault. So using parity preserving circuits faults can be easily detected which increases the reliability and performance. All these above factors motivate us to design a new logic gate which is proving more efficient than previously published parity-preserving gates.

This work is structured as follows. Section 2 gives a glance of QCA logic. Section 3 provides related work on parity preserving circuits. Section 4 introduces the proposed design, and Sect. 5 evaluates the performance of P-QCA. A novel efficient median filter architecture is reported in Sect. 6. Section 7 deals with the test evaluation of the P-QCA for all possible defects. The conclusion is presented in Sect. 8.

2 Preliminaries

The state-of-the-art technology was driven towards smaller area and low power. QCA technology was all geared towards succeeding these two aims. In a QCA, quantum cell is computed by the logic bits. Each cell is composed of four quantum-dot and two free electrons [2, 4]. Two extra electrons are present in the cell and because of tunneling effect they move around the quantum dots. Due to Coulomb repulsion, the electrons are repealed to the corner locations to maximize their space. The QCA cell has values, binary 1 and binary 0, according to the different positions of two free electrons as shown in Fig. 1a.

QCA devices can be designed by various physical arrangements of QCA cells. The two primary logic gates in QCA are the MAJ and INV [5, 6]. MAJ gate can be realized by five QCA cells as shown in Fig. 1b. The inverter is shown in Fig. 1c, where 45 displacements in the two lines of the merging cells produce the complementary action of the input signals. QCA operates by the Columbic interaction that connects the state of one cell to the state of the neighbors, unlike the conventional logic circuits in which information is transferred by electrical current.

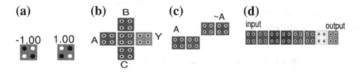

Fig. 1 Basic QCA designs **a** cell, **b** majority gate (MAJ), **c** inverter (INV), **d** wire

Figure 1d shows binary wire. In the binary wire, a signal propagates from the input to the output due to the Columbic interactions between the cells. An external clock signal is used in CMOS-based system for designing sequential circuits and controlling the timing operations. However, timing in QCA is necessary for both combinational and sequential circuits, and is achieved by clock zones [7, 8]. Clocking zone satisfied the power gains in QCA as well as the control of the information flow between the cells.

3 Related Works

Reliability of the devices is always a major concern. To ensure the reliability of devices, parity checking is the preferred method used in fault-tolerant circuits due to its low overhead for storage as well as interconnect. Researchers have designed testable logic circuits using parity preserving logic gates such as Fredkin gate, CQCA gate, MX-gate, t-QCA, PPRG gate, and TPC-QCA gate. Fredkin gate [9] is a universal gate that is shown in Fig. 2a. But its disadvantage is that it requires six majority gates and four clocking zones in QCA implementation, so that it is difficult to build a large circuit using Fredkin gates. MX-QCA gate [6] shown in Fig. 2b has five majority gates and four clocking zones. The structures of CQCA gate [10] and t-QCA gate [11] are shown in Fig. 2c and d, respectively. These two gates have majority logic functionality at the primary outputs which are useful for circuits having that functionality. But their QCA implementation requires more number of cells. TPC-QCA gate [12] structure is shown in Fig. 2e. QCA implementation of this gate is better than other parity-preserving gates in terms of cell count and area. But the disadvantage is that it requires five majority gates and majority logic functionality implementation requires three numbers of such gates. PPRG gate [13]

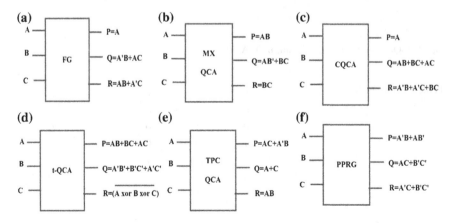

Fig. 2 a Fredkin gate, b Mx-QCA, c CQCA, d t-QCA, e TPC-QCA, f PPRG

is shown in Fig. 2f. The disadvantage of PPRG gate is that it requires a large number of QCA cells for QCA implementation. All these factors motivate us to design a new parity-preserving gate having less cell count, area, and delay.

4 Proposed Gate

In this section, we proposed a self-testable parity-preserving gate called P-QCA. There are three inputs and three outputs having input, output mapping as P = AB + BC + AC, Q = B′, R = A′B + BC′ + A′C′. The schematic of the P-QCA gate is shown in Fig. 3a. The logic block of the proposed P-QCA gate is illustrated in Fig. 3b. Truth table is shown in Table 1. The truth table of the proposed P-QCA gate confirms that A XOR B XOR C = P XOR Q XOR R. It requires two majority gates and three inverters. Two possible ways of QCA realization of P-QCA gate are shown in Fig. 4a and b. The first implementation has coplanar structure. In coplanar design, 90° cells are used for wire crossing. The second implementation has multilayer structure. This structure uses crossover cells for wire crossing. The multilayer layout has the advantages of reduced number of cells and area.

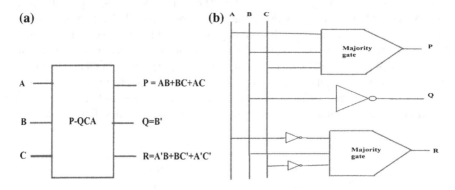

Fig. 3 P-QCA gate **a** block diagram, **b** QCA schematic

Table 1 Truth table of P-QCA gate	A	B	C	P	Q	R
	0	0	0	0	1	1
	0	0	1	0	1	0
	0	1	0	0	0	1
	0	1	1	1	0	1
	1	0	0	0	1	0
	1	0	1	1	1	0
	1	1	0	1	0	1
	1	1	1	1	0	0

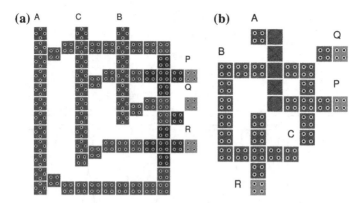

Fig. 4 **a** Coplanar QCA layout of P-QCA, **b** multilayer QCA layout of P-QCA

5   Simulation Results and Comparison

The P-QCA gate is simulated using QCADesigner tool [14]. QCA results of P-QCA gate are shown in Fig. 5. These results confirm that our proposed QCA layout has achieved desired functionality. The P-QCA gate is compared with Fredkin, CQCA, MX-QCA, t-QCA, and TPC-QCA gates. Table 2 reports the implementation results of the proposed design and previous parity preserving logic gate designs. The P-QCA design has 0.06 µm^2 and 0.03 µm^2 area for coplanar and multilayer design techniques, respectively, as shown in Table 2. A comparative performance of P-QCA and some of existing QCA primitive gates implementation of thirteen standard function is presented in Table 2. It indicates that the multilayer implementation of a P-QCA gate is most efficient in terms of QCA design metrics. It requires 33 QCA cells compared to 71 cells required in coplanar structure. Clocks zone is reduced to 1 in multilayer QCA layout which is highly efficient to design larger circuits. It also covers a minimal area of 0.03 µm^2. Table also reported that the multilayer layout of proposed P-QCA gate is more efficient than existing parity-preserving gates in all QCA design metrics. P-QCA gate has reduced cell count, clock zones, and area compared to previous most efficient TPC-QCA gate. It is because five majority gates having two-level implementation are needed in TPC-QCA gate compared to two majority gates and one-level implementation in the proposed design. In order to have the comparison with existing parity-preserving gates for logic realization, we have realized 13 standard three input-variable Boolean operations in [15] for QCA based. These thirteen operations have 256 Boolean functions for three input variables. Table 3 presented the comparative analysis results with existing gates by synthesizing these 13 standard functions.

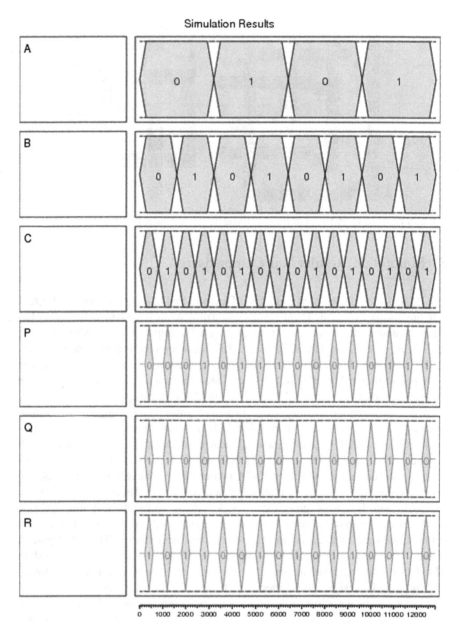

Fig. 5 Simulation result of P-QCA gate

Table 2 Comparison of P-QCA and existing parity-preserving gates

Parity-preserving gates	Cell count	Clock zone	Area (μm^2)
Fredkin [9]	246	5	0.36
MX-QCA [6]	218	4	0.35
PPRG [13]	171	5	0.19
CQCA [10]	117	2	0.11
t-QCA [11]	113	4	0.12
TPC-QCA [12]	34	3	0.04
Coplanar P-QCA	71	2	0.06
Multilayer P-QCA	33	1	0.03

Table 3 Comparison of P-QCA and existing parity-preserving gates

Standard function	Logic gate							
	CQCA		t-QCA		TPC-QCA		P-QCA (Multilayer)	
	# Gate	# Clk	# Gate	# Clk	# Gate	# Clk	# Gate	# Clk
1. F = ABC	2	4	2	8	2	6	2	2
2. F = AB	1	2	1	4	1	3	1	1
3. F = ABC + AB′C′	3	4	2	8	3	9	3	3
4. F = ABC + A′B′C′	6	8	8	16	3	9	4	4
5. F = AB + BC	2	4	2	8	2	6	2	2
6. F = AB + A′B′C	5	8	6	16	2	6	3	3
7. F = AB + A′BC′ + AB′C′	6	6	6	16	4	9	5	5
8. F = A	1	2	1	4	1	3	1	1
9. F = AB + BC + AC	1	2	1	4	3	9	1	1
10. F = AB + B′C	3	4	4	12	1	3	2	2
11. F = AB + BC + A′B′C′	6	8	2	8	3	9	4	4
12. F = AB + A′B′	4	6	1	4	2	6	2	2
13. F = ABC + A′B′C + AB′C′ + A′BC′	3	4	2	8	5	9	5	5
Total	43	62	38	116	32	87	35	35
Improvement (%)	18.6	43.5	7.8	69.8	–	59.7		

6   Logic Synthesis with P-QCA

The primary purpose of this P-QCA gate is to design fault-tolerant logic devices with low design cost. In this section, a novel median filter is designed using the proposed P-QCA gates.

Median filtering is basically improving the interpretability of information in image processing for noise reduction. It is well known that the noise reduction is a

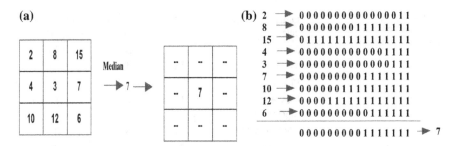

Fig. 6 **a** Median finding of 3 × 3 matrix, **b** median finding using one-hot encoding

challenging task and the goal of the image processing. Many methods such as nonlinear to reduce the noise; another method is linear filtering. Nonlinear technique is the very first step performed for the noise reduction as the algorithm is applied. To apply this algorithm, first find the middle element of the group numbers. In this group number, we are applying the shorted technique either ascending or descending. However, applying the standard median filter may yield replace the pixel value as a result of algorithm apply. Figure 6 shows the median finding and encoding. Here, the median finding algorithm is designed using one-hot encoding method which is shown in Fig. 6b. The advantage of this encoding method is that no decoders are needed to decode the input data.

Figure 6a shows an example of 3 × 3 size filter. In Fig. 6b, all the numbers are decoded using one-hot encoding method. Here, every column has nine digits. From this nine digits, either 1 or 0 will be taken which is maximum among the nine digits to calculate the median. In this example, 0000000001111111 is the value for each column, where the value is 7 in one-hot decoding. So the median is 7. As we are calculating the majority of nine numbers in a column, our proposed P-QCA gate is more suitable for it. The proposed design is shown in Fig. 7. Here, four P-QCA gates are used to calculate the majority number among the input nine numbers of a single column. Here, I0 to I9 are nine inputs to find median value, Y is final median output, and g1 to g8 are unused or garbage outputs. Above example shows a median value calculation of four-bit size numbers, so that here 16 columns are required to design. Similarly, for 8-bit numbers, 256 columns are required. For this size, 256 × 4 = 1024 P-QCA gates are required to design the median filter. From Table 3, it is shown that majority number finding functionality $F = AB + BC + AC$ is achieved by a single gate in CQCA and t-QCA. So a comparative study of the proposed gate with these gates to design an 8-bit median filter is shown in Table 4.

Fig. 7 Median finding using P-QCA gate

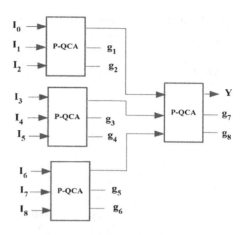

Table 4 Comparison for implementing median finding architecture of 3 × 3 matrix

Gate	No. of gates	Area (μm²)	Clock zones
CQCA [10]	1024	112.64	2048
t-QCA [11]	1024	122.88	4096
P-QCA (multilayer)	1024	30.72	1024

7 Testing and Fault Coverage

This section explores the fault testing methodology of the proposed P-QCA gate. In parity preserving logic gates, the parities of input and output vectors are same. So that faults can be easily detected by comparing the parity of input test vectors with output vectors. Table 5 illustrates the testability of proposed P-QCA gate for stuck-at faults. Table 5 illustrates the different single stuck-at fault patterns of the P-QCA gate. The outputs corresponding to test vector 000 and 111 are 011 and 100, respectively. If any input/output is stuck-at 1, the output corresponding to test vector 000 yields an output different from 011. Similarly, if any, input/output is stuck-at 0, and the output corresponding to test vector 111 yields an output different from 100.

$$\text{Fault coverage of} <000, 111> \ = 100\%$$

$$\text{E-output} = \text{Expected output, F-output} = \text{Faulty output}$$

Table 5 illustrates the different multiple stuck-at fault patterns of the P-QCA gate. In both single stuck-at fault and multiple stuck-at fault, the test vector set {000, 111} can detect all faults with 100% fault coverage. It also indicates that there

Table 5 Stuck-at fault characterization

I/O	Fault type	Test vector ABC	E-output PQR	F-output PQR
Single stuck-at faults A	s-a-0	111	100	101
A	s-a-1	000	011	010
B	s-a-0	111	100	110
B	s-a-1	000	011	001
C	s-a-0	111	100	101
C	s-a-1	000	011	010
P	s-a-0	111	100	000
P	s-a-1	000	011	111
Q	s-a-0	000	011	001
Q	s-a-1	111	100	110
R	s-a-0	000	011	010
R	s-a-1	111	100	101
Multiple stuck-at faults AB	s-a-0	111	100	010
AB	s-a-1	000	011	101
BC	s-a-0	111	100	010
BC	s-a-1	000	011	101
AC	s-a-0	111	100	001
AC	s-a-1	000	011	110
ABC	s-a-0	111	100	011
ABC	s-a-1	000	011	100
PQ	s-a-0	111	100	000
PQ	s-a-1	111	100	110
QR	s-a-0	000	011	000
QR	s-a-1	111	100	111
PR	s-a-0	111	100	000
PR	s-a-1	000	011	111
PQR	s-a-0	111	100	000
PQR	s-a-1	000	011	111

is a parity mismatch between inputs and faulty outputs. This led us to conclude that P-QCA gate can concurrently detect a permanent fault by matching the parity.

8 Conclusion

In this paper, a new parity-preserving gate is designed using QCA. The design is implemented and verified in the QCADesigner tool. We compared the proposed gate with existing parity-preserving gates. We have also implemented 13 standard functions. The comparison results illustrate that our proposed gate has significant improvements in area, delay, and power consumption. We also designed a new median finding architecture using majority logic. Finally, stuck-at fault analysis of the proposed gate indicates that with minimum test vectors all stuck-at faults can be detected.

References

1. ISCAS, IEEE International Symposium on Circuit and Systems, QCA: A Promising Research Area for CAS Society, (2004)
2. Lent, C.S., Tougaw, P.D., Porod, W., Bernstein, G.H.: Quantum cellular automata. Nanotechnology. 4, 49–57 (1993)
3. Parhami, B.: Fault-tolerant reversible circuits. In: Fortieth Asilomar Conference on Signals, Systems and Computers. ACSSC' 06. pp. 1726–1729 (2006)
4. Orlov, A.O., Amlani, I., Bernstein, G.H., Lent, C.S., Snider, G.L.: Realization of a Functional Cell for Quantum-Dot Cellular Automata. Science, 277, 928–930 (1997)
5. Tougaw, P.D., Lent, C.S.: Logical devices implemented using quantum cellular automata. J. Appl. Phys. 75(3), 1818–1825 (1994)
6. Thapliyal, H., Ranganathan, N., Kotiyal, S.: Design of testable reversible sequential circuits. IEEE Transactions on Very Large Scale Integration (VLSI) Systems. 21, 1201–1209 (2013)
7. Tóth, G., Lent, C. S.: Quasi adiabatic switching for metal-island quantum-dot cellular automata. J. Appl. Phys. 85, 2977–2984 (1999)
8. Lent, C.S., Liu, M. and Lu, Y.: Bennett clocking of quantum-dot cellular automata and the limits to binary logic scaling. Nanotechnology. 17, 4240–4251 (2006)
9. Fredkin, F., Toffoli, T.: Conservative logic. Springer. (2002)
10. Thapliyal, H., Ranganathan, N.: Conservative QCA gate (CQCA) for designing concurrently testable molecular QCA circuits. In: 22nd International Conference on VLSI Design VLSID, pp. 511–516 (2009)
11. Sen, B., Dutta, M., Sikdar, B.K.: Efficient design of parity preserving logic in quantum-dot cellular automata targeting enhanced scalability in testing. Microelectronics Journal. 45 (2), 239–248 (2014)
12. Heikalabad, S.R., Karkaj, E.T.: A testable parity conservative gate in quantum dotcellular automata. Superlattices and Microstructures. (2016) https://doi.org/10.1016/j.spmi.2016.08.054
13. Roohi, Arman, et al.: A Parity-Preserving Reversible QCA Gate with Self-Checking Cascadable Resiliency. IEEE Transactions on Emerging Topics in Computing. (2016)
14. Walus, K., Dysart, T.J. Jullien, G., Budiman, A.R.: QCADesigner: A rapid design and simulation tool for quantum-dot cellular automata, Nanotechnology, IEEE Transactions on. 3(1), 26–31 (2004)
15. Momenzadeh, M., Huang, J., Tahoori, M.B., Lombardi, F.: Characterization, test, and logic synthesis of and-or-inverter (AOI) gate design for QCA implementation. IEEE Transaction on Computer-Aided Design of Integrated Circuits and Systems, 24(12), 1881–1893, December (2005)

A Brief Survey: Features and Techniques Used for Sentiment Analysis

P. N. V. S. Pavan Kumar, N. Kasiviswanath and A. Suresh Babu

Abstract Today, people are more passionate in using websites, social media, and e-shopping. They are also more eager to express and share their opinions or feedbacks on web regarding day-to-day activities and global issues. But most of the reviewers are not genuine in giving reviews and opinions. Some people may create false reviews to promote or disparage products and services. This practice of making fake, untruthful, or deceptive reviews is known as opinion spam. Over the past few decades, research communities, academia, public, and industries are meticulously working on sentiment analysis (opinion mining) to extract and predict the authentic interests and usage requirements of customers. Using natural language processing, meaningful features can be extracted from the text and are possible to conduct review spam detection using various machine learning techniques. In this paper, we emphasized and analyzed the various feature selection methods in sentiment analysis by studying the various papers of different authors.

Keywords Feature selection · Opinion mining · Sentiment analysis
Machine learning · Supervised learning · Natural language processing (NLP)

P. N. V. S. Pavan Kumar (✉) · N. Kasiviswanath
Department of CSE, G. Pulla Reddy Engineering College (Autonomous),
Kurnool, Andhra Pradesh, India
e-mail: pegatraj@gmail.com

N. Kasiviswanath
e-mail: gprechodcse@gmail.com

A. Suresh Babu
Jawaharlal Nehru Technological University, Anantapur, Andhra Pradesh, India
e-mail: asureshjntu@gmail.com

© Springer Nature Singapore Pte Ltd. 2018
V. Bhateja et al. (eds.), *Proceedings of the Second International Conference on Computational Intelligence and Informatics*, Advances in Intelligent Systems and Computing 712, https://doi.org/10.1007/978-981-10-8228-3_14

1 Introduction

In this internet world, share marketing, opinion sharing and checking of opinions, e-ticket bookings, e-shopping, etc., are all increasing in day-to-day life of people. The size and importance of these reviews is also increasing. Hence, in taking decisions for purchasing products and in share value predictions in share markets, etc., people are depending on these reviews (opinions, feedbacks, or ratings). These reviews, made by the reviewers, will be in different types that include comment reviews (positive or negative), ratings, color or graph indications, good and bad indications, likes, etc.

Some companies or organizations write reviews by themselves for promoting their own products or disparaging other companies or organization products. Sometimes reviewers, usually common people, give blind opinions on the products that are purchased. Definitely, online reviews are helpful to the user but predicting the genuineness of the reviews is very difficult for the user and trusting of these reviews blindly is perilous for seller and buyer. Hence, review spam [1] is grievous enough that have enticed to heed by both government and media. Hence, it is a challenge for researchers to develop efficient methods in order to help sellers and buyers for distinguishing genuine reviews from fake reviews.

Sentiment analysis or opinion mining is a natural language processing (NLP) application. It depends on feature selection methods and approaches to extract the feature sets from the review corpus. For performing this, certain statistical measures are used. Here, features are chosen based on empirical evidence. Hence, features are very important in the task of sentiment analysis. Text analytics is the most commonly used method to find and extract polarity (positive or negative) of a subjective information in the review documents. This review paper presents the survey report on the techniques and approaches for feature selection and extraction in sentiment analysis. In this paper, we have done a literature review process by studying various papers of different authors to identify areas that are thoroughly observed by researchers.

The paper is organized into four sections. Section 1 is an introduction; Sect. 2 discusses about the feature selection algorithms and methods used for the review process; Sect. 3 gives a survey report on the features and techniques used in the process of feature selection and extraction for sentiment analysis, and Sect. 4 concludes and reports opportunities for further research.

2 Feature Selection

Feature selection (FS) and extraction methods are needed in sentiment analysis especially for improving learning performance, minimizing computational complexity, building generalizable models in a better way, and reducing required storage. Actual feature space is mapped to a new feature space, with reduced dimensions in the feature extraction method (FE). But this transformation and further analysis of transformed features are difficult because transformed features obtained have no physical meaning. But without any transformation, FS chooses a subset of features from the actual feature set and the physical meanings of actual features are preserved. As FS is better readable and interpretable than FE, it is considered to be superior. Certain FE methods can also be transformed into FS methods [2].

In terms of label availability, FS methods can be broadly classified into supervised, unsupervised, and semi-supervised methods [3]. In terms of different selection strategies, FS can be categorized as filter, wrapper, embedded, and hybrid models.

2.1 Feature Selection Algorithms

The characteristics of the FS algorithms [4, 5] in terms of their output types and feature redundancy handling capability characteristics of some FS algorithms are listed in Table 1.

2.2 Feature Selection Methods

FS methods are classified into offline and online methods [6]. Offline FS methods are classified into four types: filter, wrapper, embedded, hybrid, and ensemble

Table 1 Feature selection algorithms

S no	FS algorithm	Characteristics of FS algorithm	Feature redundancy capability
1	BLogReg	Embedded, supervised, multivariate, feature set	Cannot handle
2	CFS	Filter, supervised, multivariate, feature set	Cannot handle
3	FCBF	Filter, supervised, multivariate, feature set	Cannot handle
4	mRMR	Filter, supervised, multivariate, feature set	Cannot handle
5	SBMLR	Embedded, supervised, multivariate, feature set	Cannot handle
6	Gini index	Filter, supervised, univariate, feature weighting	Can handle
7	Information gain	Filter, supervised, univariate, feature weighting	Can handle

(continued)

Table 1 (continued)

S no	FS algorithm	Characteristics of FS algorithm	Feature redundancy capability
8	Kruskal–Wallis	Filter, supervised, univariate, feature weighting	Can handle
9	Fisher score	Filter, supervised, univariate, feature weighting	Can handle
10	ReliefF	Filter, supervised, univariate, feature weighting	Can handle
11	Chi-square	Filter, supervised, univariate, feature weighting	Can handle
12	t-test	Filter, supervised, univariate, feature weighting	Can handle
13	Spectrum	Filter, unsupervised, univariate, feature weighting	Can handle

feature selection (EFS) methods. These methods are called as flat FS methods [6]. The working of each flat FS method is depicted in Fig. 1.

The structured FS methods [7] are categorized into a tree, graph, and a group. Traditional FS is the offline method. Supervised FS methods use labeled variables and find the correlation among features. Unsupervised methods do not use labeled data but usually preserve the data similarity or manifold structure. Thus, this method has gained an increasing attention since many years. Semi-supervised methods utilize a manifold structure corresponding to both the label and unlabeled data.

Features from various web documents arrive in by streams which are different from classical online learning that allows samples to flow dynamically. Out of all the samples available, only one feature descriptor fd_i is assumed by the online FS methods. Thus, the ultimate task of online FS is to evaluate whether the feature with fd_i should be accepted by their arrival or not using the proposed methods like Grafting [8], Alpha-investing [9], and online streaming feature selection (OSFS) [10, 11].

The features of various reviews exhibit certain intrinsic structures [12] in many real-world applications, e.g., trees, graphs, spatial or temporal smoothness, and disjoint or overlapping groups [4, 13, 14]. Hence, online and offline feature structures should be analyzed and understood well in reviews in order to significantly improve classification performance and help identify the important features.

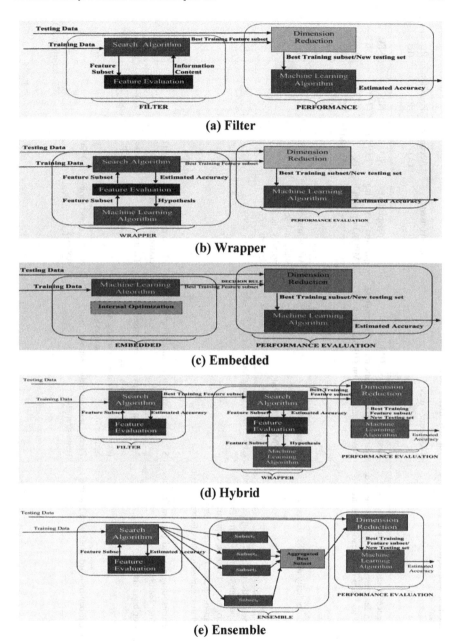

(a) Filter

(b) Wrapper

(c) Embedded

(d) Hybrid

(e) Ensemble

Fig. 1 Flat FS Models: **a** Filter, **b** wrapper, **c** embedded, **d** hybrid, **e** ensemble

Table 2 Survey on features, techniques, and datasets used for sentiment classification

Reference number and year	Features used	Techniques used	Datasets used	Accuracy (A)	Precision (P)	Recall (R)
[15] (2005)	Word, sentence, document structure, modification features. Positive, negative, and both binary and non-modification features	Boostexter AdaBoost.HM for neutral-polar classification (28 features), polarity classification (10 features)	Multi-perspective question answering (MPQA) opinion corpus	NPC (75.9%) PC (65.7%)	NPC (71.6–77.7%) PC (28.4–72.9%)	NPC (56.8–87%) PC (11.2–82.2)
[16] (2005)	Item-specific polarity and PSP features	Naive Bayes SVM OVA SVM OVA + PSP SVM Regression	Movie review of scale and sentence polarity, author datasets	66.3%		
[17] (2005)	Syntactic, semantic, and orthographic lexical features, dependency parse features, and opinion recognition features, capitalization features, part-of-speech features, opinion lexicon features, semantic class features, basic features	CRF Model using the MALLET Code	MPQA corpus		79.3–81.2	59.5–60.6
[18] (2006)	Binary features, unigram and trigram features, discriminating features	SVM and sequential minimal optimization	Movie review and customer feedback	72–91%		
[19] (2011)	Unigrams or bigrams	Maximum entropy, Feature-based sentiment analysis, augmented lexicon-based method, learning and lexicon-based sentiment analysis, Lms	Twitter	67.9–85.4%	31.8–68.7%	35–82.7%
[20] (2012)	Product and bag-of-word (unigram) features	Psenti, lexicon only, learning only, senti-strength	Software and movie reviews	89.64%		

(continued)

Table 2 (continued)

Reference number and year	Features used	Techniques used	Datasets used	Accuracy (A)	Precision (P)	Recall (R)
[21] (2013)	Positive and negative	Df, Df-Tf, Mi, Mi-Df, Ig, Ig-Tf, Svm, Nb	Chinese hotel online reviews	88.05% (SVM using MI-DF)	88%	79–87%
[22] (2014)	True positive (TP), true negative (TN), false positive (FP), false negative (FN), false positive (FP)	SVM-based, information gain, relief, principal component analysis (PCA), chi-square, Gini index, uncertainity-based term, SVM, KNN, NB	Opinion corpus for Malay (OCM)	87% (SVM)	63.40–84.19%	
[23] (2015)	Binary features	Optimized swarm search-based feature selection (OS-FS), clustering-by-coefficient-of-variation (CCV), Feature evaluation method, FAST Algorithm	100 sample news from CNN.com	98%	98%	98%
[24] (2015)	N-gram features	SVM with linear and radial basis function	Google Financial News	96–97%		
[25] (2015)	Optimum features	Genetic algorithm and rough set theory, meta-heuristic-based algorithms	Customer reviews			
[26] (2016)	Positive and negative features	Bootstrapping and sampling, multinominal naïve Bayes classifier (MNB), support vector machine (SVM)	Twitter Amazon Beauty products, IMDB Movie review, Yelp restaurant datasets			
[27] (2016)	Movie review features	IDF (Features separated) PCA (Features reduction) CART, naïve Bayes and LVQ with decision forest-based (Features arranged) decision backwoods and decision forest-based feature extraction.	Movie data from Twitter dataset	75–81%	75–81% (average precision)	75–81%

3 Survey on Features, Techniques, and Datasets Used

In Table 2, we specified the year and reference number, features used, techniques, and datasets used in various papers from 2005 to 2016. Accuracy (A), precision (P), and recall (R) values are also specified in the table above.

4 Conclusion

FS is still an active field and is perpetually rejuvenating itself to answer new challenges. In this paper, we provide a brief survey report from 13 reference papers (2005–2016) based on the features used, techniques applied for extracting features for the sentiment classification, and performance evaluation of the techniques applied on the datasets or corpus used in terms of accuracy, precision, and recall. The datasets or corpus used in Table 1 are Movie Reviews, MPQA, News, Twitter, Hotel and restaurant, customer opinion corpus, etc.

From the survey report, the goals in 2005 papers are polarity classification, rating inference, information extraction, text categorization, and scalable pattern evaluation. Various methods used are augmented lexicon-based methods (2011), proposed hybrid approach (2012), pair-wise combination of techniques (2013), an extensive study on the effect of seven FS methods on the three machine learning classifiers (2014), swarm-based FS methods and meta-heuristic algorithms (2015), bootstrapping, sampling, machine learning techniques, and decision forest-based methods (2016).

From the 13 papers surveyed, the accuracy ranges from 67 to 98% when N-gram and binary features are used, whereas for other features like syntactic and semantic features, polarity features, and lexicon features, the accuracy ranges from 65% to 88% only. Hence, N-gram based approach of FS outperforms than the other approaches. From the survey report, it is observed that SVM-based approaches and pair-wise combination of techniques like DF-TF, MI-DF, etc. gives better performance than other techniques.

Very few papers are on social media web sources like Facebook, Twitter, YouTube, etc. Most of the reviews are from movies and news and less research is there on regional movie reviews and the regional news. There is a vast scope for research in a well-designed sentence structure, regional, and language-centric sentiment analysis.

References

1. Lau RY., Liao SY., Kwok RCW., Xu K., Xia Y., Li Y.: Text mining and probabilistic language modeling for online review spam detecting. ACM Trans Manage Inf Syst Vol 2(4), pp: 1–30 (2011).

2. Masaeli M., Dy J.G., and Fung G.: From transformation-based dimensionality reduction to feature selection. In: 27th International Conference on Machine Learning, pp: 751–758 (2010).
3. Ang, Jun Chin, et al.: Supervised, Unsupervised and Semi-supervised Feature selection: A Review on Gene Selection. In: IEEE/ACM Transactions on Computational Biology and Bioinformatics IEEE Transactions On Computational Biology And Bioinformatics, Manuscript Id 1 (2015).
4. Zhao Z., Morstatter F., Sharma S., Alelyani S., Anand A., and Liu H.: Advancing feature selection research asu feature selection repository. Technical report (2013).
5. Sahu, Tirath Prasad, and Sanjeev Ahuja:Sentiment analysis of movie reviews: A study on feature selection & classification algorithms. In: International Conference on Microelectronics, Computing and Communications (MicroCom), IEEE (2016).
6. Tang J., Alelyani S., Liu H.: Feature Selection for Classification: A Review. In: Data Classification: Algorithms and Applications, CRC Press (2013).
7. Jundong Li, Kewei Cheng, Suhang Wang, Fred Morstatter, Robert P Trevino, Jiliang Tang, and Huan Liu. Feature selection: A data perspective. arXiv preprint arXiv:1601.07996 (2016).
8. Perkins S. and Theiler J., Online feature selection using grafting. In: ICML, pp. 592–599 (2003).
9. Zhou J., Foster D.P., Stine R., and Ungar L.H.: Stream wise feature selection using alpha-investing. In: KDD, pp. 384–393 (2005).
10. Wu X., Yu K., Ding W., Wang H., and Zhu X.: Online feature selection with streaming features. IEEE Transactions on Pattern Analysis and Machine Intelligence, vol. 35, no. 5, pp. 1178–1192 (2013).
11. Wu X., Yu K., Wang H., and Ding W.: Online streaming feature selection. In: 27th International conference on machine learning, pages 1159–1166 (2010).
12. Zou H.: The adaptive lasso and its oracle properties. Journal of the American statistical association, vol. 101, no. 476, pp. 1418–1429 (2006).
13. Yuan L., Liu J., and Ye J.: Efficient methods for overlapping group lasso. In: IEEE Transactions on Pattern Analysis and Machine Intelligence, vol. 35, no. 9, pp. 2104–2116 (2013).
14. Jing Wang, Meng Wang, Peipei Li, Luo qi Liu, Zhongqiu Zhao, Xu gang Hu, and Xindong Wu.: Online feature selection with group structure analysis. In: IEEE Transactions on knowledge and Data Engineering, 27(11):3029–3041 (2015).
15. Wilson T., Wiebe T. and Hoffman P.: Recognizing contextual polarity in phrase-level sentiment analysis. In: Human Language Technology Conference and Conference on Empirical Methods in Natural Language Processing; British Columbia, Canada (2005).
16. Pang B. and Lee L.: Seeing stars: exploiting class relationships for sentiment categorization with respect to rating scales. In: 43rd annual meeting of the Association for Computational Linguistics (ACL), University of Michigan, USA, (2005).
17. Choi Y., Cardie C., Riloff E., and Patwardhan S.: Identifying sources of opinions with conditional random fields and extraction patterns. In: Conference on empirical methods in natural language processing (EMNLP 2005); Vancouver, B.C., Canada.
18. Koenig A.C. and Brill E.: Reducing the human overhead in text categorization. In: 12th ACM SIGKDD conference on knowledge discovery and data mining; Philadelphia, Pennsylvania, USA (2006).
19. Zhang L., Ghosh R., Dekhil M., Hsu M. and Liu B.: Combining Lexicon-based and Learning-based Methods for Twitter Sentiment Analysis. In: Technical report, HP Laboratories, (2011).
20. Mudinas A., Zhang D., Levene M.: Combining lexicon and learning based approaches for concept level sentiment analysis. In: First International Workshop on Issues of Sentiment Discovery and Opinion Mining; New York, NY, USA (2012).
21. Chen, Xian, Jing Ma, and Yueming Lu.: Feature selection for Chinese online reviews sentiment classification. In: International Conference on Computational Problem-solving (ICCP), IEEE (2013).

22. Alsaffar A., Omar N.: Study on feature selection and machine learning algorithms for Malay sentiment classification. In: 2014 International Conference on Information Technology and Multimedia (ICIMU). IEEE (2014).

23. Fong, Simon, Elisa Gao, and Raymond Wong: Optimized Swarm Search-Based Feature Selection for Text Mining in Sentiment Analysis. In: International Conference on Data Mining Workshop (ICDMW), IEEE (2015).

24. Foroozan S., et al.: Improving Sentiment Classification Accuracy of Financial News Using N-Gram Approach and Feature Weighting Methods. In: 2nd International Conference on Information Science and Security (ICISS), IEEE (2015).

25. Ahmad S.R., Bakar A.A. and Yaakub M.R.: Metaheuristic algorithms for feature selection in sentiment analysis. Science and Information Conference (SAI), pp. 222–226, London (2015).

26. Gary Goh S.W., Andy Ang J. L., Allan Zhang N. S.: Optimizing Performance of Sentiment Analysis through Design of Experiments. In: IEEE International Conference on Big Data (Big Data) (2016).

27. Jeevanandam Jotheeswaran and Koteeswaran S.: Feature Selection using Random Forest method for Sentiment Analysis. In: Indian Journal of Science and Technology, Vol 9(3), https://doi.org/10.17485/ijst/2016/v9i3/75971, January, 2016.

Content-Centric Global Id Framework for Naming and Addressing for Smart Objects in IoT

Prasad Challa and B. Eswara Reddy

Abstract In the Internet of Things (IoT), things include variety of objects like computers, sensor nodes, RFID tags, vehicles, medicines, books, etc., and these things can be uniquely identified for the addressing and communicating each other. As the key part of the Internet architecture, naming scheme is crucial change brought by the vision of the IoT. Classical naming schemes like IP and URI are facing great challenges due to adopted features and have been altered and broaden to endure with all the aspects. Other identification schemes are also offered for distinct causes. The building of new naming schemes within future Internet architecture projects has brought fresh thoughts and fresh answers. In this paper, we propose a naming scheme called content-centric global Id for addressing the problems of mobility, scalability and energy conservation challenges effectively. The advantage of the proposed method is delivering simple routing approach that exclusively depends on content and efficient content distribution.

Keywords IoT · 6LoWPAN · URI · AAID · GS1 · Sensor UID
NDN · MF

1 Introduction

The Internet of Things (IoT) connects billions of devices. In such a vast network of different interconnected objects, the issue of naming and identification of objects plays an important role and it affects the other aspects such as architecture, privacy characteristics, governance, etc.

P. Challa (✉)
Department of CSE, VFSTR University, Guntur, India
e-mail: chprasad18@gmail.com

B. Eswara Reddy
Department of CSE, JNTU Kalikiri, Anantapur, India
e-mail: eswarcsejntua@gmail.com

© Springer Nature Singapore Pte Ltd. 2018
V. Bhateja et al. (eds.), *Proceedings of the Second International Conference on Computational Intelligence and Informatics*, Advances in Intelligent Systems and Computing 712, https://doi.org/10.1007/978-981-10-8228-3_15

To identify many numbers of low-power smart devices, it is required effective naming and addressing policies. Soon, the TCP/IP network protocol identifies each device using 4-byte IPv4 address in which IPv4 we can address 2^{32} devices only. Therefore, other alternative solutions used for the addressing policy. In this context, IETF group proposed IPv6 addressing that has been advised for low-power wireless communication nodes within the 6LoWPAN context. IPv6 addresses are expressed by means of 128 bits and therefore it should be enough to identify any object which is worth to be accosted. Therefore, we may look at assigning an IPv6 addressing policy for identifying the objects in the cyberspace world.

Radio Frequency Identification (RFID) tags are embedded into IoT objects and are interconnects the objects using standard communication protocols. In the literature, various solutions are proposed [1–3] for mapping the RFID into IPv6 addressing. Due to the limitations of RFID technology, the solutions can address the mobility of objects and security issues in the networks.

IoT devices are constrained in terms of memory, processing capability and power supply, and to support mobility to these constrained devices, several IP-enabled mobility protocols are proposed [4, 5]. Mobility management protocols evolved by host-centric network or network-centric each of them having their own importance in handover delay, signal strength and packet loss to improve the QoS in real-time applications. In IPv6 networks, auto-configuration supports the mobility management.

In IP-based networks, all the host systems are identified by querying the Domain Name Server (DNS) which cannot sufficient to address the scalability and adaptability for object to object communication in IoT. However, instead of querying DNS servers in hierarchical, Object Name Service (ONS) [6, 7] is flat naming scheme to resolve names of objects.

In order to achieve a true identity management in the IoT, an essential feature is to provide support for finding the IoT devices in order to addressable, named and discovered. The taxonomy of IoT identifiers is categorized into object Ids, communication Ids and application Ids. The sensor data is effectively distributed over the core network with routing on GUIDs [8]. It can also inherit the nice mobility and security features of GUID.

In [9], IoT applications are classified into four types: Energy management application, supply chain management and logistics, urban mobility applications and defence and Intelligence applications. Each application has its own trade when it comes to naming issues.

The rest of the article is organized as follows. Section 2 reviews the related work on naming and addressing of IoT objects. Section 3 gives the details of problem definition and system model. Section 4 presents the simulation results and discussion.

2   Related Work

Today's Internet is mainly based on IPv4 and uses 32-bit addresses which limit the address space to 4,294,927,296 different addresses. In future, according to the estimate, there are 50 billion connected devices by the year 2020. The limitation of IPv4 stimulated the growth of IPv6 that covers the address space of 128 bits, that is, $3.4 * 10^{38}$ unique addresses. Based on IPv6 addressing, 6LoWPAN (IEEE 802.15.4) is designed to typically constrain devices in terms of storage and processing capability which uses the features like header compression and machine configuration, neighbouring discovery, security and interoperability [10]. 6LoWPAN is lacking RFID mobility support and likewise a large number of information generated and having different traffic characteristics due to heterogeneity.

The header size of 6LoWPAN (IEEE 802.15.4) is 21–41 bytes, to reduce the header size and for supporting non-IP technologies like Bluetooth Low Energy. In [11], global IP protocol, which introduces Access Address Identifier (AAID) that translates the in-node frame to out-node frame by mapping the heterogeneous devices where header size is reduced to 5 bytes, improves the performance but when nodes are moving it cannot address the problems of home agent relocation issues and security is weak.

Electronic Product Code (EPC) global and global standard 1 are the solutions for identifying the objects in supply chain management; it can identify the objects and provide the strong security [12]. In [13], it is not suitable for smart objects in IoT due to lack of interoperability among the protocol standards. In [14, 15], Open Sensor Web Architecture (OSWA) focuses on providing a platform to discover, access heterogeneous sensor networks and then process the information collected from those resources. In addition to this, a service-oriented prototypes of sensor collection service, sensor planning service, web notification service and sensor repository service have been designed and implemented targeting the sensor applications running on top of TinyOS.

In [9], for naming and addressing smart objects in IoT, NDN is a future Internet architecture project which addresses the problem of interoperability and mobility but it uses name-based routing that depends on dynamic binding of network address which creates a bottleneck at gateway nodes.

ICN-NDN (Information Centric Network-Named Data Network) and ICN-MF (Mobility First) are projects under future Internet architecture in which routing is based on content instead of depending on network address; it uses the context to transmitting packets. In [16–21], name-based routing and mobility first routing address the network congestion in pervasive computing. These are lacking in interoperability and flexibility.

The summary of classification on related work which addresses the naming and addressing of smart object in IoT is given in Table 1.

Table 1 Summary of methods of naming and addressing adopted by various schemes

	6LoWPAN	Global IP	GS 1	SWE	IoT@Work	ICN-NDN	ICN-MF
Network independent	✗	✗	✗	✗	✗	✓	✓
Scalability	✓	✗	✓	✗	✓	✓	✓
Efficient	✗	✓	✓	✓	✗	✓	✓
Preserving of privacy	✗	✗	✓	✓	✗	✓	✓
Interoperability	✗	✗	✗	✓	✗	✗	✓
Reliability	✗	✗	✓	✗	✓	✓	✗
Flexibility	✗	✗	✗	✗	✗	✗	✓
Mobility	✗	✗	✗	✗	✗	✗	✓

3 Content-Centric Global Id Framework

The architecture of content-centric global Id framework is shown in Fig. 1. The main idea is based on content and identity-based naming scheme for device-to-device communications in IoT. Figure 1 describes how things (sensors, RFID tags, actuators, etc.) of heterogeneous system are accessed by various applications through the IoT middleware.

IoT middleware serve platform is associated with content store database which employs an XML encoding scheme to encode data packets and decode the names stored in content store. The key functionalities of an IoT middleware are discovery, security, interoperability, data management and content redistribution, and those are summarized into three groups as

Fig. 1 Architecture of CCGId

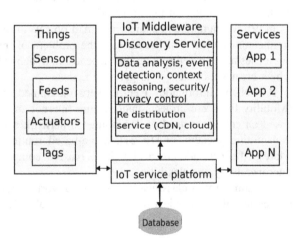

(1) Discovery,
(2) Data processing and
(3) Data redistribution.

- **Discovery service**:
  Discovery service provides the applications for finding things through the Object Name Service (ONS) which is similar to Domain Name Service (DNS) in the traditional TCP/IP Internet architecture for resolving the names. The IoT object connects to CCGId network and makes itself available to other applications.
- **Data Processing**:
  The applications access sensor data through the CCGId network and generate new data from an existing knowledge by applying semantic rules.
- **Data Redistribution:**

The CCGId host distributes sensor data through the network and delivers data from their source to different locations for quick access.

3.1 Protocol Stacks of CCGId

The protocol stack of CCGId is shown in Fig. 2 which elucidates the CCGId core network functionalities to achieve flexible communication among the IoT objects. The main components of CCGId are Object Name Service (ONS) which registers all object's global Ids using the two components: one is *public key* and another one is *sequence number* which is shown in Fig. 3. ONS is a distributed and dynamic service which updates in real time.

Name Certification Service (NCS) is a component for supporting access of data and/or services of IoT objects; NSC verifies the objects global Id for providing the access of data/service that ensures the privacy communication endpoints.

Fig. 2 Protocol stack of CCGId

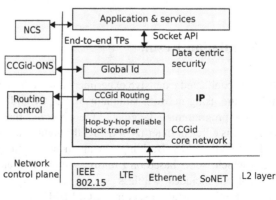

Public key	Sequence Number

Fig. 3 Structure of global Id

In contrast to TCP/IP, which leaves a responsibility for security to endpoints, CCGId secures the data itself by requiring data producers to cryptographically sign on every data packet.

The gateway device can identify the service type such as light control, temperature, sensing, etc. Routing makes use of both global Id and Network Address (NA) from ONS for hop-by-hop reliable block transfer of data packets. In CCGId core network, each router has enough storage for caching to provide Delay Tolerant Network (DTN).

Applications are in the following sequence of steps.

- **Name Registration:** The user of device publishes global Id and service type through the NAS function, which can available on both device and router of CCGId.
- **ONS Refresh**: When the application service connects to CCGId network, the access router calls ONS update to provide a global Id to connect to network address mapping.
- **Service Discovery**: Whomever wish provide service to the corresponding application can find global Id through query for requesting service with proper keywords.
- **Service Request**: Once discovering, the service-seeking device next makes a service request with an RDF query to join the service membership.
- **Routing and ONS lookup**: The service request towards god is routed to the corresponding network address which is obtained from an ONS lookup.
- **Service Request Redirection**: User application device selects one of the service providers by making use of GPS location.

In the process of routing, router storage is useful for processing location-based services. All the routers in the core network are connected in a distributed fashion.

4   Simulation Result

We have conducted extensive experiments of the ICN-building management system architecture using network simulator-3. The CCGId framework is implemented in the C++ programming language. We observed that the CCGId framework performance is efficient in terms of delay, throughput and energy consumption metrics.

The CCGId-BMS (Building Management System) system topology is shown in Fig. 4 which works based on Building Automation and Control network (BACnet) protocol. It includes wireless sensor nodes, sink nodes, actuator nodes and CCGId-BMS application server. CCGId-BMS server receives the data from all the

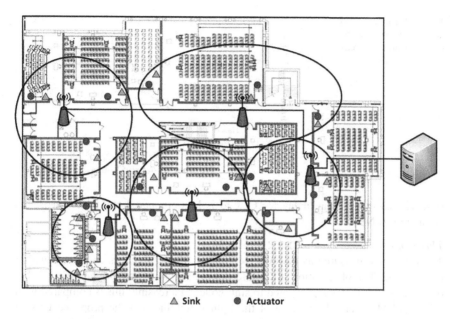

△ Sink   ● Actuator

Fig. 4 BMS topology based on flood plan

smart objects via sink nodes corresponding to application user requests. The assumption is that every actuator is associated with the sink node.

Delay:

We measure the delay metric as the average time for a processing a request from the application server. We show the average delay for the application server requests in Fig. 5. In the experiment, there is a sink for every 10 sensor nodes to report the queries. Every sensor periodically reports data for every 500 ms. We observed that the delay is more when the number of hops increases from the application server to the sink nodes.

Compared with existing frameworks like NDN and MF, the proposed CCGId average delay is improved because of reducing per-packet overhead. Therefore, we

Fig. 5 Average reporting delay between the sink and the BMS application server

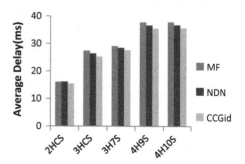

Fig. 6 Throughput at BMS
application server as number
of sensor nodes increasing

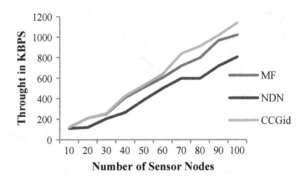

observe that the average delay for CCGId is less compared to MF and it is less compared to NDN.

Throughput:

Throughput is measured as the average number of user requests completed with a unit of time. The observation of CCGId throughput is improved when the number of sensor nodes that are associated to sink node is increasing; this is compared with NDN and MF methods because of the light weightiness of the proposed CCGId framework. In case of MF, the GUID->NA mapping entry requires higher number of routing updates. In case of NDN interest packets carries sensor data from nodes increases the Pending Interest Table (PIT) on the sink node should be enough to accommodate the maximum rate of Interest load. It is reduced by CCGId decentralized NA mechanism. The results are presented in Fig. 6.

Energy Consumption:

The energy consumption is the amount of energy consumed by each sensor node. Initially, all the batteries of sensor nodes are fully charged and only sink node is provided by active power supply. There is a significant improvement in the proposed CCGId framework in terms of energy consumption because of the periodic-per hop scheduling and multilevel data aggregations which reduce the average transmission rate of data packets. In addition to this, the data packets are encoding and decoding information in present in Content Store (CS) database. Due

Fig. 7 Residual energy as
increasing sensor nodes

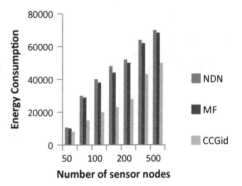

to number of bytes, per-packet is reduced which inturn improves the residual energy of passive devices is explained in Fig. 7.

5   Conclusion

In this report, we presented naming and addressing of smart objects in future Internet architectures. Classical solutions in the Internet of Things (IoT) like 6LoWPAN and global IP are inefficient to support networking and privacy characteristics. Other naming and identification methods are also offered for specific purpose. The proposed content-centric global Id (CCGId) framework provides better solution for all the issues to support IoT devices. We presented the naming scheme called content-centric global Id for addressing the problems of mobility, scalability and energy conservation challenges effectively. The advantage of the proposed method is delivering that simple routing exclusively depends on content and efficient data distribution. In the future, it can be extended to support more security and bulk data transferring.

References

1. Lee, Sang-Do, Myung-Ki Shin, and Hyoung-Jun Kim. "EPC vs. IPv6 mapping mechanism." *The 9th international conference on advanced communication technology.* Vol. 2. IEEE, 2007.
2. Yoon, Dong Geun, et al. "RFID networking mechanism using address management agent." *Networked Computing and Advanced Information Management, 2008. NCM'08. Fourth International Conference on.* Vol. 1. IEEE, 2008.
3. Ghaleb, Safwan M., et al. "Mobility management for IoT: a survey." *EURASIP Journal on Wireless Communications and Networking* 2016.1 (2016): 1–25.
4. Akyildiz, Ian F., Jiang Xie, and Shantidev Mohanty. "A survey of mobility management in next-generation all-IP-based wireless systems." *IEEE Wireless Communications* 11.4 (2004): 16–28.
5. C. Perkins, IP Mobility Support for IPv4, IETF RFC 3344, August 2002.
6. Atzori, Luigi, Antonio Iera, and Giacomo Morabito. "The internet of things: A survey." *Computer networks* 54.15 (2010): 2787–2805.
7. Jara, Antonio J., Miguel A. Zamora, and Antonio Skarmeta. "Glowbal IP: An adaptive and transparent IPv6 integration in the Internet of Things." *Mobile Information Systems* 8.3 (2012): 177–197.
8. Li, Jun, et al. "Supporting efficient machine-to-machine communications in the future mobile internet." *Wireless Communications and Networking Conference Workshops (WCNCW), 2012 IEEE.* IEEE, 2012.
9. Syrus, Publilius. "Internet of Things."
10. Jinghua Ding, Navrati Saxena: "Applications And Open Issues Of Internet Of Things: A Brief Overview", International Journal of Advance Computational Engineering and Networking (IJACEN), Volume-2, Issue-5, pp 36–41, 2014.
11. "GS1", http://www.gs1.org/.

12. Mealling, Michael. "Auto-ID object name service (ONS) 1.0." *Auto-ID Center Working Draft* 12 (2003).
13. EPCglobal.Inc., "GS1 Object Name Service (ONS) Version 2.0.1", 2013.
14. "SWE", http://www.opengeospatial.org/ogc/markets-technologies/swe.
15. Hu, Chuli, Nengcheng Chen, and Chao Wang. "Remote sensing satellite sensor information retrieval and visualization based on SensorML." *Geoscience and Remote Sensing Symposium (IGARSS), 2011 IEEE International.* IEEE, 2011.
16. Zhu, Zhenkai, Alexander Afanasyev, and Lixia Zhang. "A new perspective on mobility support." *Named-Data Networking Project, Tech. Rep* (2013).
17. Baid, Akash, Tam Vu, and Dipankar Raychaudhuri. "Comparing alternative approaches for networking of named objects in the future Internet." *Computer Communications Workshops (INFOCOM WKSHPS), 2012 IEEE Conference on.* IEEE, 2012.
18. Zhang, Lixia. "Evolving Internet into the Future via Named Data Networking." *Presentation, APRICOT-APAN* (2011).
19. Jacobson, Van, et al. "Named Data Networking (NDN) Project 2012–2013 Annual Report." (2014).
20. Yuan, Haowei, and Patrick Crowley. "Experimental evaluation of content distribution with NDN and HTTP." *INFOCOM, 2013 Proceedings IEEE.* IEEE, 2013.
21. Jeff Burke,"Named Data Networking: Cyberphysical Applications Research", PPT, UCI, 2012.

Road Traffic Management System with Load Balancing on Cloud Using VM Migration Technique

Md. Rafeeq, C. Sunil Kumar and N. Subhash Chandra

Abstract The collection of traffic data using multiple sensors and other capture devices are been addressed in multiple researches deploying the mechanism using geodetically static sensor agents. Nevertheless to avoid the congestion, the parallel research works have proposed frameworks based on cloud-based data centers. Those approaches do not propose any technique to reduce the cost and improve the service-level agreements to match with the current industry and research demands. Thus, this work proposes a cloud-based automatized framework for virtual machine migration to increase the SLA without compromising the cost for storage and energy. The major achievement of this work is to minimize the SLA violation compared to existing virtual machine migration techniques for load balancing. The extensive practical demonstrations of virtualization and migration benefits are also carried out in this work. With the extensive experimental setup, the work furnishes the comparative analysis of simulations for popular existing techniques and the proposed framework.

Keywords Load balancing · Virtual machine · Migration · Performance

Md. Rafeeq (✉)
CMR Technical Campus, Kandlakoya, Medchal, Hyderabad 501401, Telangana, India
e-mail: rafeeqmail@gmail.com

C. Sunil Kumar
SNIST, Yamnampet, Ghatkesar, Hyderabad 501301, Telangana, India
e-mail: ccharupalli@gmail.com

N. Subhash Chandra
HITS, Keesara, Bogaram, Rangareddy 501301, Telangana, India
e-mail: subhashchandra.n.cse@gmail.com

© Springer Nature Singapore Pte Ltd. 2018
V. Bhateja et al. (eds.), *Proceedings of the Second International Conference on Computational Intelligence and Informatics*, Advances in Intelligent Systems and Computing 712, https://doi.org/10.1007/978-981-10-8228-3_16

1 Introduction

Load balancing techniques on cloud computing is the generic framework-based process where the generated workloads are distributed over multiple data center resources. The load balancing techniques bring the advantage of lower response time [1]. However, the cost of replication of resources is also to be taken care of an additional cost. The cloud data center-based load balancing is distinguished from the domain name service-based load balancing. The domain name service load balancers deploy the hardware and software components to balance load for the hardware resources, whereas the cloud-based load balancing techniques deploy the software algorithms or protocols to distribute the load over multiple data center nodes. Also, it is to be understood that the cloud-based load balancing techniques allow the customers to use the global or geodetically distributed services based on geodetically distributed servers. Multiple parallel researches are been carried out to demonstrate the benefits of load balancing on cloud-based data centers as handling the high unexpected traffic generally referred to Cyber Spikes [2]. Making the application scalable based on demand without degrading the performance increases the reliability at the cost of VM migration and load balancing on cloud data centers using proposed three-phase optimal virtual machine migration technique.

2 Proposed Framework for Road Traffic Data Management Framework

Henceforth with the detailed understanding of the generic traffic data monitoring and management framework, we propose the novel framework for road traffic data management control with replication control on cloud storage. In the proposed framework, we have considered the layer-based approach for better controlling and management of the agent-based components. The agents in the wireless network are single function oriented but the collective network is multipurpose. In this study, we propose the framework consisting of deployed network layer, monitoring layer, application management layer, storage layer, and finally the server-based server layer virtualization benefits for cloud data centers. This work also highlights the benefits of virtual machine migrations and also evaluates the parameters influencing the performance and productivity [2, 3].

2.1 Open Access Control

The virtual machines come with a reduced abstraction in the system level and allow the provider, customer, and researchers to access more properties of the system. The access to computing environment data, system-level codes, hardware utilization

Table 1 Parameters for open access control [4]

Parameter type	Parameter name	Access permissions	
		Traditional	Virtual machine migration
Processing	CPU type	Not allowed	Allowed
	Allocation	Allowed	Allowed
	Priority	Allowed	Allowed
Memory	Size	Allowed	Allowed
	Buffer	Not allowed	Allowed
Storage	Access IDE bus	Not allowed, physical	Allowed, logical
	Capture mode	Not allowed	Allowed
	Library group	Allowed, physical	Allowed, logical
Network	IP address	Allowed	Allowed
	MAC address	Not allowed	Allowed
	Internal network	Partially allowed	Allowed

statistics, traces of the active application, failing and down timing component configurations, and the guest operating system configuration parameters and the ability to control them independently helps to understand the performance perimeters [4] (Table 1).

2.2 Optimal Hardware Control

Virtual machines come with a flexibility to change or alter the operating system and hardware components seamlessly. After the initial cost for setting up a virtual environment, the users are free to modify the computing system including the operating system, libraries, tools, and other supporting patches without investing the full time needed for computing system change or upgrade [5] (Table 2).

2.3 Optimal Replication Control

The replication of the virtual machines using the snapshot feature allows the users to take timely and on-demand backups of the virtual machine images. Thus, the backups help to quickly reproduce the same computing environment without investing the complete setup time (Table 3).

Table 2 Reduced hardware upgrade constraints [5]

Parameter type	Parameter name	Accessibility	
		Traditional	Virtual machine migration
Operating system	Version	Available	Available
	Interoperability	No continuous availability	Available
	Patch	Available	Available
Development environment	Patch	Available	Available
	Device driver	No continuous availability	Available
	Version control	Available	Available
Configuration	Configuration delay	Very high	Low

Table 3 Reduced replication duration [5]

Parameter type	Replication time	
	Traditional (min)	Virtual machine migration
Windows server	50–90	Just in time
MAC servers	40–60	Just in time
Linux servers	30–40	Just in time

2.4 Service Provider Support for Virtual Machine Migration

The virtual machines are hosted by all service providers with similar configurations but with added advantages. Hence, adopting to virtual machine computing is the best choice to avoid the lack of support and facility availability (Table 4).

Table 4 Service provider support for migration [6]

Server type	Amazon Cloud	Microsoft Azure Cloud	Google App Engine Cloud	IBM Bluemix Cloud	Private Hosted Cloud
Windows server	YES	YES	YES	YES	NO
MAC servers	YES	YES	YES	YES	NO
Linux servers	YES	YES	YES	YES	NO

2.5 *Optimal Manageability of Updates*

Application on virtual machines hosted on cloud is always liable for automatic and regular updates from the service provider without any extra cost. However, on the other side, hosting the traditional system demands the cost and time implications for updates.

3 Proposed Optimal Migration Framework

This work deploys a cost evaluation function to determine the most suitable virtual machine to be migrated considering the least SLA violation [6]. The framework for optimal migration is presented here (Fig. 1).

The proposed framework is classified into three major algorithm components as VM identification, VM migration, and cost function. Algorithms for all three phases are been discussed here.

3.1 *Virtual Machine Identification*

The first phase of the algorithm analyzes the highest loaded node and migrates the virtual machine to the available less loaded node. After identifying the source and destination, the algorithm identifies the virtual machine to be migrated. The outcome of this algorithm is to obtain optimal load balanced condition for the data center after virtual machine migration. The detail of the algorithm is explained here:

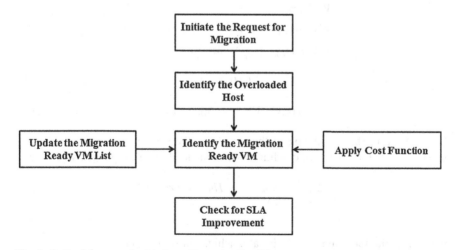

Fig. 1 Optimal framework virtual machine migration [6]

Step-1.1. Calculate the load on each node in the data center

$$Phy_{CPUCapacity} = \sum_{i=1}^{n} VM(i)_{CPUCapacity} \tag{1}$$

$$Phy_{MemoryCapacity} = \sum_{i=1}^{n} VM(i)_{MemoryCapacity} \tag{2}$$

$$Phy_{IOCapacity} = \sum_{i=1}^{n} VM(i)_{IOCapacity} \tag{3}$$

$$Phy_{NetworkCapacity} = \sum_{i=1}^{n} VM(i)_{NetworkCapacity} \tag{4}$$

$$\Pi = (Phy_{CPUCapacity} + Phy_{MemoryCapacity} + Phy_{IOCapacity} + Phy_{NetworkCapacity}) \tag{5}$$

Step-1.2. In the second step, the algorithm identifies the highest and lowest loaded node in the data center

$$\Pi_{MAX} = \begin{cases} If \ \Pi_i > \Pi_j, then \ \Pi_{MAX} = \Pi_i \\ Else \ \Pi_j > \Pi_i, then \ \Pi_{MAX} = \Pi_j \end{cases} \tag{6}$$

$$\Pi_{MIN} = \begin{cases} If \ \Pi_i < \Pi_j, then \ \Pi_{MIN} = \Pi_i \\ Else \ \Pi_j < \Pi_i, then \ \Pi_{MIN} = \Pi_j \end{cases} \tag{7}$$

Step-1.3. Once the source and destination are identified as MAX and MIN, respectively, the identification of virtual machine to be migrated is carried out. During the identification, the optimal load balanced condition is identified.

$$VM(i) = $$
$$VM(i)_{CPUCapacity} + VM(i)_{MemoryCapacity} \tag{8}$$
$$+ VM(i)_{IOCapacity} + VM(i)_{NetworkCapacity}$$

$$\Pi_{MAX} - VM(i) = \Delta_{Source} \tag{9}$$

$$\Pi_{MIN} + VM(i) = \Delta_{Destination} \tag{10}$$

Step-1.4. After the calculation of the new load, the source and destination nodes must obtain the optimal load condition, where the loads are nearly equally balanced.

$$\begin{cases} If \ \Delta_{Source} \approx \Delta_{Destination}, Then \ Migrate \ VM(i) \\ Else \ i = \in (n) \end{cases} \tag{11}$$

where n is the total number of virtual machines in source node.

3.2 Virtual Machine Allocation

During the second phase of the algorithm, this work analyzes the time required for VM allocation for the selected virtual machine with other parameters like energy consumption, number of host shutdowns, execution time—VM selection time, execution time—host selection time and execution time—VM reallocation time. These parameters will help in generating the cost function

Step-2.1. Calculate the energy consumption at the source before migration:

$$E_{Source} = \sum_{i=1}^{t} (E_{CPU} + E_{NETWORK} + E_{IO} + E_{MEMORY})_i \qquad (12)$$

Step-2.2. Calculate the energy consumption at the destination after migration:

$$E_{Destination} = \sum_{i=1}^{t} (E_{CPU} + E_{NETWORK} + E_{IO} + E_{MEMORY})_i \qquad (13)$$

Step-2.3. Calculate the difference in energy consumption during migration:

$$E_{Diff} = |E_{Source} - E_{Destination}| \qquad (14)$$

Step-2.4. Calculate the number of host shutdowns, execution time—VM selection time, execution time—host selection time and execution time—VM reallocation time during migration:

$$\begin{pmatrix} Host_{Down} & VM_{SelectionTime} \\ Host_{SelectionTime} & VM_{ReallocationTime} \end{pmatrix} \qquad (15)$$

Henceforth, the comparative analysis is been demonstrated in the results and discussion section.

3.3 Cost Analysis of Migration

The optimality of the algorithm focuses on the SLA. During the final phase of the algorithm, the migrations are been validated with the help of the cost function to measure the optimality of the cost. The final cost function is described here:

$$Cost(VM) = E_{Diff} + \begin{pmatrix} Host_{Down} & VM_{SelectionTime} \\ Host_{SelectionTime} & VM_{ReallocationTime} \end{pmatrix} + SLA_{Violation} \qquad (16)$$

4 Results and Discussion

This work has performed extensive testing to demonstrate the improvement over the existing migration techniques [7]. The various considered migration techniques are listed with the used acronyms here (Table 5).

The simulation of the algorithm is based on CloudSim, which is a framework for modeling and simulation of cloud computing infrastructures and services. The experimental setup used for this work is been explained here (Table 6; Fig. 2).

Finally, the proposed technique is been tested for the load balancing with the below-furnished simulation setup (Table 7).

Table 5 List of techniques used for performance comparison [8]

Used name in this work	Selection policy	Allocation policy
IQR MC	Maximum correlation	Inter quartile range
IQR MMT	Minimum migration time	Inter quartile range
LR MC	Random selection	Local regression
LR MMT	Minimum migration time	Local regression
LR MU	Minimum utilization	Local regression
LR RS	Rom selection	Local regression
LRR MC	Maximum correlation	Robust local regression
LRR MMT	Minimum migration time	Robust local regression
LRR MU	Minimum utilization	Robust local regression
LRR RS	Rom selection	Robust local regression
MAD MC	Maximum correlation	Median absolute deviation
MAD MMT	Minimum migration time	Median absolute deviation
MAD MU	Minimum utilization	Median absolute deviation
MAD RS	Rom selection	Median absolute deviation
THR MC	Maximum correlation	Static threshold
THR MMT	Minimum migration time	Static threshold
THR MU	Minimum utilization	Static threshold
THR RS	Rom selection	Static threshold
OPT ALGO	Proposed algorithm Part-1	Proposed algorithm Part-2

Table 6 Experimental setup [7]

Setup parameters	Number of physical hosts	Number of virtual machines	Total simulation time (s)
Values	800	1052	86400.00

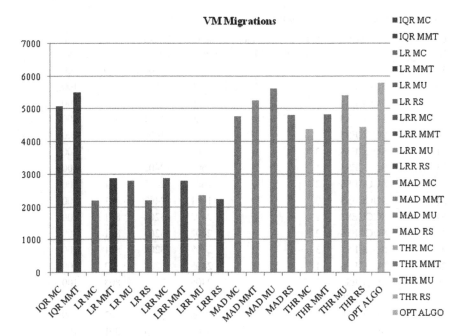

Fig. 2 Number of VM migration comparison [8]

Table 7 Load balancing simulation setup [8]

Simulation duration (s)	Requests per user	Data size (bytes)	Avg. users	Virtual machines	Memory	CPU
216000	120	2000	2000	5	512	2.4 GHz

The CPU utilization achieved during the simulation is furnished below (Table 8) and 100% of the CPU utilization is achieved during load balancing.

5 Conclusion

Load balancing can be achieved through virtual machine migration. However, the existing migration techniques constraints to improve the SLA and often compromise to a higher scale on the other performance evaluation factors. However, the proposed technique is independent of the virtual machine image format and demonstrates the same improvement. The comparative analysis is been done with the proposed technique with the existing techniques like IQR MC, IQR MMT, LR MC, LR MMT, LR MU, LRR MC, LRR MMT, LRR MU, LRR RS, LR RS, MAD MC, MAD MMT, MAD MU, MAD RS, THR MC, THR MMT, THR MU,

Table 8 Load balancing simulation [9]

Cloudlet ID	STATUS	Data center ID	VM ID	Start time	Finish time	Time	Utilization (%)
1	SUCCESS	1	0	0	800	800	100
2	SUCCESS	2	0	0	800	800	100
3	SUCCESS	3	0	0	800	800	100
9	SUCCESS	1	0	800	1601	801	100
10	SUCCESS	2	0	800	1601	801	100
11	SUCCESS	3	0	800	1601	801	100
25	SUCCESS	1	0	1601	2402	801	100
28	SUCCESS	2	0	1601	2402	801	100
31	SUCCESS	3	0	1601	2402	801	100
37	SUCCESS	1	0	2402	3203	801	100
40	SUCCESS	2	0	2402	3203	801	100
43	SUCCESS	3	0	2402	3203	801	100
26	SUCCESS	1	3	2405	3208	803	100
29	SUCCESS	2	3	2405	3208	803	100
32	SUCCESS	3	3	2405	3208	803	100
35	SUCCESS	1	3	2405	3208	803	100
49	SUCCESS	2	0	3203	4004	801	100
52	SUCCESS	3	0	3203	4004	801	100
55	SUCCESS	1	0	3203	4004	801	100
293	SUCCESS	2	3	20071	20874	803	100
296	SUCCESS	3	3	20071	20874	803	100

and THR RS. The work also furnishes the practical evaluation results from the simulation to retain the improvement of the other parameters at least to the mean of other techniques during SLA improvement. Also, this proposed technique for virtual machine migration demonstrates no loss in existing CPU utilization during load balancing [9, 10].

References

1. Beloglazov and R. Buyya, "Managing overloaded hosts for dynamic consolidation of virtual machines in cloud data centers under quality of service constraints", IEEE Trans. Parallel Distrib. Syst., vol. 24, no. 7, pp. 1366–1379, 2013
2. H. Xu and B. Li, "Anchor: A versatile and efficient framework for resource management in the cloud", IEEE Trans. Parallel Distrib. Syst., vol. 24, no. 6, pp. 1066–1076, 2013
3. S. Di and C.-L. Wang, "Dynamic optimization of multi-attribute resource allocation in self-organizing clouds", IEEE Trans. Parallel Distrib. Syst., vol. 24, no. 3, pp. 464–478, 2013
4. J. Zhan, L. Wang, X. Li, W. Shi, C. Weng, W. Zhang and X. Zang, "Cost-aware cooperative resource provisioning for heterogeneous workloads in data centers", IEEE Trans. Comput., vol. 62, no. 11, pp. 2155–2168, 2013

5. D. Carrera, M. Steinder, I. Whalley, J. Torres and E. Ayguad, "Autonomic placement of mixed batch and transactional workloads", IEEE Trans. Parallel Distrib. Syst., vol. 23, no. 2, pp. 219–231, 2012

6. T. Ferreto, M. Netto, R. Calheiros and C. De Rose, "Server consolidation with migration control for virtualized data centers", Future Generation Comput. Syst., vol. 27, no. 8, pp. 1027–1034, 2011

7. K. Mills, J. Filliben and C. Dabrowski, "Comparing vm-placement algorithms for on-demand clouds", Proc. IEEE 3rd Int. Conf. Cloud Comput. Tech. Sci., pp. 91–98, 2011

8. W. Zhao and X. Tang, "Scheduling Data Collection with Dynamic Traffic Patterns in Wireless Sensor Networks," Proc. IEEE INFOCOM '11, pp. 286–290, Apr. 2011

9. J. C. Herrera, D. B. Work, R. Herring, X. J. Ban, Q. Jacobson and A. M. Bayen "Evaluation of traffic data obtained via GPS-enabled mobile phones: The Mobile Century field experiment", Transp. Res. C, Emerging Technol., vol. 18, no. 4, pp. 568–583 2010

10. A. Adya, W.J. Bolosky, M. Castro, G. Cermak, R. Chaiken, J.R. Douceur, J. Howell, J.R. Lorch, M. Theimer, and R. Wattenhofer, "Farsite: Federated, Available, and Reliable Storage for an Incompletely Trusted Environment," Proc. Fifth Symp. Operating System Design and Implementation (OSDI), pp. 1–14, 2002

Energy Constraint Service Discovery and Composition in Mobile Ad Hoc Networks

P. Veeresh, R. Praveen Sam and C. Shoba Bindhu

Abstract In recent years, processing capability of mobile devices had greatly increased. Utilization of service-oriented architecture (SOA) in mobile ad hoc networks will increase efficiency of the running application. Single service cannot solve the complex requirements of the user; hence, in service composition, multiple services are combined to handle complex requirements. The nature of MANETs like heterogeneity in resources, dynamic network topology, and distributed nature makes the service composition as a complex task. Most of the existed service discovery and composition techniques consider one service in one node and QoS constraints are not considered. In this paper, a new node model is proposed where a node maintains multiple services by considering minimum energy constraint associated with the services during service discovery and composition which reduces the number of nodes involved in composition path and thereby reduces the failure rate in composition. The performance of proposed method is measured in ns3 simulation tool, where it outperforms the existing protocols by reducing energy consumption and failure rate of composition.

Keywords MANET · Energy constraint · Service discovery
Service composition

P. Veeresh (✉)
Jawaharlal Nehru Technological University Anantapur, Anantapur, Andhra Pradesh, India
e-mail: veeresh_kaly@yahoo.co.in

R. Praveen Sam
G. Pulla Reddy Engineering College, Kurnool, India
e-mail: praveen_sam75@yahoo.com

C. Shoba Bindhu
JNTU College of Engineering, Anantapur, India
e-mail: shobabindhu@gmail.com

© Springer Nature Singapore Pte Ltd. 2018
V. Bhateja et al. (eds.), *Proceedings of the Second International Conference
on Computational Intelligence and Informatics*, Advances in Intelligent Systems
and Computing 712, https://doi.org/10.1007/978-981-10-8228-3_17

1 Introduction

In recent years, a rapid development occurred in processing capabilities of mobile devices such as smartphones and personal digital assistants (PDA), growing needs of the users, and it is essential to communicate with each other. MANET is an infrastructure-less, temporary, spontaneous, and multi-hop communication network with collection of independent nodes [15]. The challenges in MANETs are dynamic topology, limited bandwidth, packet loss, battery constraints, and security.

Service-Oriented Architecture (SOA) has attractive benefits, where it can be implemented by services. Services are self-descriptive, self-encapsulated, dynamic discovery, loosely coupled, heterogeneous, machine interaction, and dynamic loading components.

Service-oriented methodology retains the benefits of component-based development. A lot of service-oriented research is presented in [6, 10, 18]. It has an interface portrayed in a machine-processable format (especially Web Service Description Language—WSDL). In SOA, other systems can interact seamlessly with each other using SOAP (Simple Object Access Protocol) messages. Implementing services into MANET increases taste of technology and solves complex tasks of the users [5].

Service discovery means identifying required services in the vicinity of network. Service discovery in wired networks uses centralized registries, i.e., Universal Description Discovery and Integration-UDDI. The approaches defined for wired networks are not suitable for wireless networks.

Service composition means combining individual services into a large and complex distributed service to satisfy the user's needs. Service composition includes combining and planning an arrangement of services. This process can be performed either automatically or manually [17] but requires cooperation from other mobile nodes.

Wired networks are stable networks but MANETs are not stable networks. In MANET, energy constraints are very important because nodes in the network purely depend on limited energy sources. Service discovery and composition methodologies defined for wired networks not suitable for MANET due to time-varying and dynamic in nature [2]. This may enable great challenges in MANETs for service composition.

To counter the problems of existing approaches, the proposed method has the following features.

1. Service description is maintained in decentralized registries.
2. Energy-aware service composition for MANET finds least energy constraint services in composition path.
3. Multiple services can be accommodated in a single node.

Service composition success rate will depend on number of nodes involved in composition [16]. The proposed method increases the system performance with minimum number of nodes in the service composition.

The paper is organized as follows. Section 2 specifies the related work, i.e., work done so far. Section 3 represents system model and Sect. 4 represents energy constraint service discovery and composition in MANETs. Section 5 represents simulation and performance evaluation. Section 6 presents conclusion and future work.

2 Related Work

In [3], JINI and SLP architectures are proposed for service discovery. In this approach, resource and service discovery are based on either broadcasting of requests or service advertisement propagation. In [9], the authors have proposed a group-based service discovery (GSD) process. This protocol works based on two methods. One is intelligent selective group-based request forwarding and another is peer-to-peer caching of advertisements.

In [8], Chakraborty et al. proposed a protocol, where the service discovery is the part of routing protocol. It reuses the path, which is discovered in both requests and advertisements at the routing layer. But this method lacks of modularity and scalability features.

In [4], service composition is proposed by constructing hierarchical task graph using smaller components. In graph, a node represents logical services and edge represents data flow between corresponding nodes. In [7], a distributed broker-based service composition protocol is proposed for MANET. The composition is accomplished with localized broadcasting and intelligent selective forwarding technique to minimize composition path length. In [1], service composition will not depend on the central repositories for service registration. Here, each node maintains the table with network information. To achieve the service composition process, table update messages are sent to neighbor nodes until it reaches to maximum propagation distance in the network.

A method named service composition by simple broad casting has been proposed in [19]. After initiating service composition by the starting node, it will broadcast the requirement to the neighbor nodes. The neighbor nodes will use the output of the previous service as the input to find the appropriate service.

In [11], minimum disruption index is evaluated using dynamic programming for optimal solution for service composition. They used minimum disruption service composition and recovery (MDSCR) algorithm for uncertain node mobility. This algorithm predicts the service link lifetime by approximating the node location and velocity with one-hop look-ahead prediction. But they do not consider the arbitrariness of node mobility. The prediction result may be inaccurate.

In [14], they proposed service field, and when the service is executed in a node, the neighbor nodes calculate the field strength of the next service. Based on service field strength, the node which is executing service selects the next service for composition.

In [12], they present QASSA, an efficient service selection algorithm which provides ground for QoS-aware service composition in ubiquitous environments. They formulate global QoS requirements for service selection as a set-based optimization problem.

To counter the problems in the existed system, in this paper, a new node model proposed where more than one service in a node is considered and also energy constraint associated with each service. In the results of energy constrained service discovery and composition, the proposed method reduces the overall failure rate and increases throughput.

3   System Model

Here, the node architecture in terms of services and service response packet structure is discussed, which are used in ESDC.

3.1   Node Architecture of a Service in MANET

Node description with more than one service in MANET can be defined as

$$N = (N_{id}, \{S_1, S_2, \ldots S_n\}, E_{Node}),$$

where

N_{id} = Node ID, node can be identified uniquely, i.e., IP address of a node.
$\{S_1, S_2, \ldots S_n\}$ Services present in a node.

Each service $Sj = (S_{id}, S_{ip}, S_{op}, E_{Ser}, C)$, where S_{id}, service ID, service can be identified uniquely, S_{ip}, service input, S_{op}, service output, E_{Ser} energy index value for service, be the amount of energy needed to process the service. C, other constraints. E_{node}, node energy.

A node defined as triplet, node id N_{id}, services in the node $\{S_1, S_2, \ldots, S_n\}$ and node energy E_{node}. E_{node} specifies the amount of energy needed to maintain its properties like bandwidth, mobility, and reliability. Calculation of E_{node} value is discussed in next section. Each service can be uniquely identified by its service id S_{id}. A service can be machine interoperable with its input S_{IP} and output S_{OP}. E_{Ser} represents energy required to process a service. A service can be uniquely identified by Ni. Sj, Ni represents node, and Sj represents J-th service on that node.

Figure 1 represents example scenario of a node structure in MANET with respect to services. Each node contains one or more services. For example, a node N_2 contains s_{10}, s_{18}, and s_4 services.

Fig. 1 Nodes in MANET
with multiple services

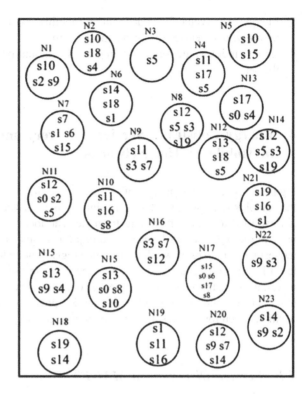

3.2 Response Packet

After receiving request from the service composition initiator, if the node contains related services, it initiates response packet for each service separately. Figure 2 shows the response packet format.

Response packet format is explained as follows:

1. Service composition initiator address represents source node ip address.
2. Service provider address represents target node ip address.
3. Sequence number counters stale use of old packets.
4. Hop count represents number of hops between sources to destination.
5. Timeout represents the packet alive time.
6. ServiceID represents unique identification of service.

Service Composition Initiator Address	Service Provider Address	Sequence Number	Hop Count	Timeout	ServiceID	Energy Index Value for Node	Energy Index Value for Service

Fig. 2 Response packet format

7. Energy index value for node includes mobility constraints like bandwidth, node mobility, and service discovery mechanism.
8. Energy index value for service represents amount of energy needed to process the service.

4 Energy Constraint Service Discovery and Composition in Mobile Ad Hoc Networks

Figure 3 visualizes the process of service composition. Composition initiator issues the complex request and in turn splits into atomic services. Atomic services request is forwarded to local service registry node. The local registry node gives response back to the initiator and forwards the request to other registry nodes. The service request packet keeps forwarding from one registry node to another registry node till the TTL value reaches to zero. After receiving responses from multiple registry nodes, the initiator node runs the energy constraint service composition algorithm to trace out minimum energy constraint service providers. After completion of composition execution, the results are to be transmitted to service composition initiator.

4.1 Service Discovery in MANET

Service discovery in MANET uses decentralized mechanism due to dynamic behavior of nodes. The service discovery in the MANET is done by categorizing the nodes participating as service registry, service providers, and service request nodes. The topology contains the nodes belonging to one of the categories stated. The discovery process is executed repeatedly whenever a new node gets into the network or else an existing one leaves or expires.

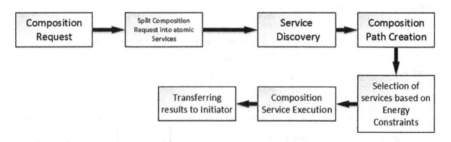

Fig. 3 Service composition process

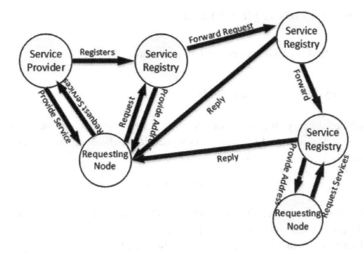

Fig. 4 Process of service discovery

Service registry node takes the responsibility of registering and maintaining the services available in the network. Service provider node provides services. Usually, they are registered with the service registry. Service request node is the consumers of the services.

Service Discovery Process:

Here, we explain the service discovery process in brief as shown in Fig. 4.

(1) Service providers' nodes register at the service registry which is at minimum hop distance.

(2) On receiving the request from service provider, service registry node registers the list of service details which are provided by provider.

(3) The service registry node maintains the information of all the services provided including service description and provider information. Additionally, a service registry maintains the information regarding the energy constraint of each service and node.

(4) A node requests a service it needs to query the service registry node.

(5) Service registry node sends response to the requester and forwards requested packet to the next registry node until TTL value reaches to zero.

4.2 Energy Consumption Evaluation

We proposed two indexed values for energy evaluation.

4.2.1 Energy Index Value for Node

This includes energy required for mobility constraint and service discovery. Equation (1) measures cost of energy consumed for each packet in the network for each node.

$$E_{Pkt} = m * size + b, \tag{1}$$

where b is the channel acquisition overhead, m is incremental component proportional to packet size and size of the packet.

Total energy consumed by each node is calculated as

$$E_{Node} = E_{ack} + \sum_{i=1}^{n} Cost_{Ei} \tag{2}$$

Energy cost for transmission and reception of control packets is

$$E_{ack} = n \times E_{Pkt} \tag{3}$$

n is the number of control packets, where Ei = {Node Movement, Band Width, Resources, Service Discovery, etc.,}. $Cost_{Ei}$ Cost incurred for various mobility constraints.

4.2.2 Energy Index Value for Service

The calculation of energy consumption of a single service is as follows:

$$E_{Ser} = E_{pre} + E_{com} + E_{loc} + E_{dis} + E_{use}, \tag{3}$$

where E_{ser} is the energy consumption of a single service. E_{pre} E_{loc} E_{dis} and E_{use} is the energy consumption at preparation stage, composition stage, the logistic transportation stage and service configuration, service waste disposal, and execution stage of a service, respectively. Total energy consumption of a service for each node is (E_{CN})

$$E_{CN} = E_{node} + E_{Ser}, \tag{4}$$

where E_{Node} is the energy index value for a node and E_{Ser} is the energy index value of a service.

4.3   Energy-Aware Service Composition

In energy-aware service composition, minimum energy constraint services are selected. Composition initiator will get all nodes energy index value E_{Node} and energy index value for service E_{Ser}. A matrix is established between nodes and its services with E_{CN} as shown below:

$$\begin{pmatrix} E_{11} & E_{12} & E_{13} & \ldots & E_{1n} \\ E_{21} & E_{22} & E_{23} & \ldots & E_{2n} \\ E_{31} & E_{32} & E_{33} & \ldots & E_{3n} \\ \ldots & \ldots & \ldots & \ldots & \ldots \\ E_{m1} & E_{m2} & E_{m3} & E_{m4} & E_{mn} \end{pmatrix}$$

We maintain a matrix, where the columns represent the nodes and rows represent their services. An entry in the matrix E_{mn} represents the energy required for the node n to provide the service m. We select the nodes that provide the required set of services with minimum energy power consumption.

Algorithm 1: ServiceCompositionAlgorithm(N, m, n)

Procedure ServiceCompositionAlgorithm(N, m, n)
//NList of nodes which provides services for composition, Ener[1 : m; 1 : n]is the
//energy of node and its services, MinEnrgy[i; j] minimum energy consumed services.
//n is number of nodes in composition. m is the number of services
for *i ← 1 to m*
{
MinEnrgy[i]→a;
  for *j ← 1 to n*
    if *N[j].S[i].E_{ser} < MinEnrgy[i] then*
       MinEnrgy[i] ← j;
}
for *i←1 to m*
       TotEner←TotEner + N[MinEnrgy[i]].S[i].Eser;
CompositionpathisMinEnrgy[1 : m];
AddpathtoCompositionExcuteP acket;
SendCompositionExcutePacket;

For example, as shown in Fig. 5b, matrix consists of total energy index values of each service in Jowls. Node N_1 contains three services S_1, S_3, and S_4. Node N_2 contains two services S_2 and S_4. Node N_3 contains two services S_1 and S_2. Node N_4 contains two services S_3 and S_4. Node N_5 contains only one service S_4. Among these nodes, for Service S_1 node N_3 is selected because of minimum energy index value, for Service S_2 node N_2 is selected, for Service S_3 node N_4 is selected, and for Service S_4 node N_4 is selected. The minimum energy constraint service composition path is $N_3 \rightarrow N_2 \rightarrow N_4 \rightarrow N_4$.

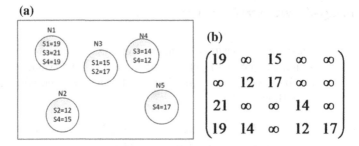

Fig. 5 **a** Nodes with multiple services, **b** energy constraints matrix

5 Simulation and Performance Evaluation

In this section, the performance of proposed method is compared with traditional AODV in service composition. The performance is measured with metrics path failure rate, throughput, and energy consumption. For implementation of services, we utilized the tool specified in [13], which is the extension framework of network simulator NS-3.

Table 1 summarizes simulation parameters.

First, we organized an infrastructure-less MANET with 100 mobile devices with wireless capabilities, and each device can communicate within their proximity with other devices.

Figures 6, 7, and 8 represent (a) path failure rate, (b) throughput, and (c) energy consumption. Using visual trace analyzer tool, we obtained the number of times path failed in duration of time as shown in Fig. 6. The simulation is run for 150 s, where the proposed method has shown an enhanced performance when compared with the traditional AODV approach. In the service composition, the proposed method chooses nodes with good energy levels; hence, it could reduce the path failures.

It has shown a rapid increase in the throughput when compared with the traditional AODV approach as shown in Fig. 7.

Table 1 Simulation setup

Parameter	Value
Number of nodes	100
Simulation time	150 s
Wifi standard	802.11b
Wifi rate	Dsss rate 1 Mbps
Transmission range (R)	45 m
Routing protocol	AODV
Number of concrete services	180
Size of composition plan	5 (abstract services)

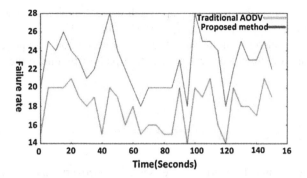

Fig. 6 Service composition failure rate

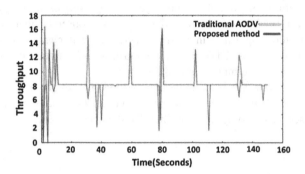

Fig. 7 Throughput

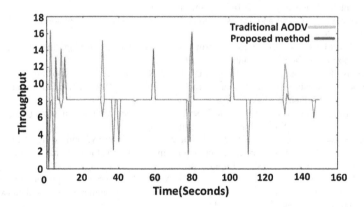

Fig. 8 Energy consumption

Figure 8 represents that traditional approach will consume more energy than our proposed methodology. Energy consumption reduces which means that network lifetime increases thereby failure rate of service consumption may be reduced. The proposed method defines the service execution path to minimize the total energy consumption.

6 Conclusion and Future Work

The proposed approach energy-aware service discovery and composition in MANET is designed for highly dynamic environment applications and requires multiple participants; a node provides multiple services and distributes components in an ad hoc manner. Node mobility, bandwidth, energy constraints, etc. are the important constraints in MANET. Due to course, we proposed energy constraint service discovery and composition in this paper. We observed that our proposed service composition algorithm in MANET performs better results through simulation than traditional AODV-based compositions in terms of failure rate, throughput, energy consumption, and delay.

In our future work, we would like to add QoS metrics for a node as well as metrics for services are into the consideration to get better results. We also planned to test proposed methodology in real-time disaster management.

References

1. Unai Aguilera and Diego L´opez-de Ipina. Service composition for mobile ad hoc networks using distributed matching. In International Conference on Ubiquitous Computing and Ambient Intelligence, pages 290–297. Springer, 2012.
2. Mia Backlund Norberg and Ter´ese Taaveniku. A web service architecture in mobile ad hoc networks, 2005.
3. Michel Barbeau. Service discovery protocols for ad hoc networking. In CASCON 2000 Workshop on ad hoc communications, 2000.
4. Prithwish Basu, Wang Ke, and Thomas DC Little. Scalable service composition in mobile ad hoc networks using hierarchical task graphs. In Proc. 1st Annual Mediterranean Ad Hoc Networking Workshop, 2002.
5. David Booth, Hugo Haas, Francis McCabe, Eric Newcomer, Michael Champion, Chris Ferris, and David Orchard. Web services architecture. 2004.
6. Fabio Casati, Ski Ilnicki, LiJie Jin, Vasudev Krishnamoorthy, and Ming- Chien Shan. Adaptive and dynamic service composition in eflow. In International Conference on Advanced Information Systems Engineering, pages 13–31. Springer, 2000.
7. D CHAIGKABORTY, Y Yesha, and A Joshi. A distributed service composition protocol for pervasive environments. In Proceedings of IEEE Wireless Communications and Networking Conference: Mar, pages 21–25, 2004.
8. Dipanjan Chakraborty, Anupam Joshi, and Yelena Yesha. Integrating service discovery with routing and session management for ad-hoc networks. Ad Hoc Networks, 4(2):204–224, 2006.

9. Dipanjan Chakraborty, Anupam Joshi, Yelena Yesha, and Tim Finin. Gsd: A novel group-based service discovery protocol for manets. In Mobile and Wireless Communications Network, 2002. 4th International Workshop on, pages 140–144. IEEE, 2002.

10. Xiaohui Gu and Klara Nahrstedt. Distributed multimedia service composition with statistical qos assurances. IEEE transactions on multimedia, 8(1):141–151, 2006.

11. S. Jiang, Y. Xue, and D. C. Schmidt. Minimum disruption service composition and recovery over mobile ad hoc networks. In Mobile and Ubiquitous Systems: Networking Services, 2007. MobiQuitous 2007. Fourth Annual International Conference on, pages 1–8, Aug 2007.

12. N. B. Mabrouk, N. Georgantas, and V. Issarny. Set-based bi-level optimization for qos-aware service composition in ubiquitous environments. In Web Services (ICWS), 2015 IEEE International Conference on, pages 25–32, June 2015.

13. Petr Novotny and Alexander L Wolf. Simulating services-based systems hosted in networks with dynamic topology. Technical report, Technical Report DTR-2016-2, Department of Computing, Imperial College London, 2016.

14. Sun Qibo, Wang Wenbin, Zou Hua, and Yang Fangchun. A service selection approach based on availability-aware in mobile ad hoc networks, 8(1):87–94, 2011.

15. R. Ramanathan and J. Redi. A brief overview of ad hoc networks: Challenges and directions. Comm. Mag., 40(5):20–22, May 2002.

16. R. Ramanathan and J. Redi. A brief overview of ad hoc networks: challenges and directions. IEEE Communications Magazine, 40(5):20– 22, May 2002.

17. Biplav Srivastava and Jana Koehler. Web service composition-current solutions and open problems. In ICAPS 2003 workshop on Planning for Web Services, volume 35, pages 28–35, 2003.

18. Mea Wang, Baochun Li, and Zongpeng Li. sflow: Towards resource efficient and agile service federation in service overlay networks. In Distributed Computing Systems, 2004. Proceedings. 24th International Conference on, pages 628–635. IEEE, 2004.

19. Qing Zhang, Huiqiong Chen, Yijun Yu, Zhipeng Xie, and Baile Shi. Dynamically Self-Organized Service Composition in Wireless Ad Hoc Networks, pages 95–106. Springer Berlin Heidelberg, Berlin, Heidelberg, 2005.

Classifying Aggressive Actions of 3D Human Models Using Correlation Based Affinity Propagation Algorithm

Binayak Gouda and Satyasai Jagannath Nanda

Abstract Unlike traditional clustering techniques like K-means, K-nearest neighborhood where the number of initial clusters is known apriori, in affinity propagation (AP) based approach the numbers of clusters present in a given dataset are chosen automatically. The initial cluster centers in AP are termed as exemplars which are initialized randomly, and their numbers are automatically set and refined with the progress in the iteration. The AP has improved accuracy than K-means and has much lower computational time than the evolutionary based automatic clustering approaches and density based clustering like DBSCAN. The original AP algorithm was based on Euclidean distance. Here a Correlation based affinity propagation (CAP) algorithm is introduced considering Pearson Correlation as affinity measure. Simulation results reveal that the CAP provides effective clusters than that achieved by AP for datasets having a large number of attributes. The proposed algorithm is applied to categorize aggressive and regular actions of 3D human models. Extensive simulation studies on fifteen cases reveal the superior performance of CAP on finding the exact number of clusters as well as numbers of points in each cluster close to that of the true partition.

Keywords Density based clustering · Affinity propagation algorithm
Correlation coefficient · 3D human models · Aggressive activities

B. Gouda · S. J. Nanda (✉)
Department of Electronics and Communication Engineering, Malaviya National
Institute Technology Jaipur, Jaipur 302017, Rajasthan, India
e-mail: nanda.satyasai@gmail.com

B. Gouda
e-mail: binayak4431@gmail.com

© Springer Nature Singapore Pte Ltd. 2018
V. Bhateja et al. (eds.), *Proceedings of the Second International Conference
on Computational Intelligence and Informatics*, Advances in Intelligent Systems
and Computing 712, https://doi.org/10.1007/978-981-10-8228-3_18

1   Introduction

Frey and Dueck [1] introduced the Affinity Propagation (AP) clustering algorithm in 2007. The algorithm becomes popular in the research community due to automatic determination of number of clusters, insensitive to initialization of cluster centre, accurate allocation of data points to clusters and lower computational time requirements. It has been applied for analyzing functional magnetic resonance images [2], shadow detection in remote sensing images [3], gesture recognition in video frames [4], routing in wireless sensor networks [5], creating energy-efficient base station network for green cellular networks [6], cluster formation in Vehicular Ad Hoc Networks while managing node mobility [7], oral conversation texts [8]. Based on applications the original algorithm has been modified in several aspects.

Wang et al. [9] proposed a multi-exemplar affinity propagation (MEAP) algorithm which uses a number of exemplars in every cluster to associate them to a super exemplar. They have reported superior performance of MEAP for unleveled image categorization and clustering of handwritten digits. Gan and Ng [10] developed a subspace clustering algorithm by combining AP with attribute weighting which is effective for analysis on Human liver cancer, Breast and colon tumor datasets. Hang et al. [11] reported a Transfer affinity propagation-based clustering algorithm (TAP) which is helpful when there are insufficient data to develop model.

It is observed that the accuracy of original AP algorithm [1, 12] is not promising for applications when real life datasets have large number of attributes. Therefore in the present article instead of distance minimization from the points to the exemplars, the maximization of correlation between them is exploited as the affinity measure. The primary intuition is that when a dataset has a large number of attributes there may be a similarity between them and correlation may be a suitable statistic to explore it. Interestingly simulation results reveal that the assumption worked on five benchmark datasets, and clustering accuracy of CAP is better than AP.

Theodoridis et al. [13] developed an experimental set up in University of Essex to record the physical activities of human being in the 3D environment. The aim is to develop an automatic recognition system which can categories the aggressive and regular behaviour of human models. Originally Theodoridis et al. [13] viewed it as a classification problem and proposed an artificial neural network (ANN) model trained with genetic programming (GP) to solve it. Later on Nanda and Panda [14] have formulated it as an automatic clustering problem and solve it using multi-objective immunized PSO algorithm. Though the authors have achieved good accuracy in obtaining the desired number of clusters and the samples in them but computational time involve in the process is enormous. The proposed CAP is able to achieve similar accurate clusters at lower computational time than immunized PSO [14].

The rest part of the manuscript is organized as follows. Section 2 highlights the significance of using Pearson Correlation Coefficient as affinity measure. The proposed algorithm step wise implementation is also narrated in this section.

Section 3 describes the simulation environment and performance of the proposed CAP and original AP algorithm on five benchmark datasets. The real life application of the proposed algorithm to classify the aggressive and normal actions of human models is demonstrated in Sect. 4. The concluding remarks are given in Sect. 5.

2   Proposed Correlation Based Affinity Propagation Algorithm

The proposed algorithm is motivated by Nanda et al. recently published computationally efficient density based algorithm [15] which uses Pearson correlation coefficient (PCC) as fitness measure to group the samples. In [15] the density based algorithm DBSCAN is suitably modified with PCC as affinity measure. Here the maximization of PCC is introduced as the affinity measure to improve the performance of affinity propagation algorithm. The PCC between data samples x, y ∈ P_ ((M × N)) is given by

$$\rho(x, y) = \frac{COV(x, y)}{s_x \times s_y} = \frac{\sum_{j=1}^{M} (x_{1,j} - \mu_x)(y_{1,j} - \mu_y)}{\sqrt{\sum_{j=1}^{M} (x_{1,j} - \mu_x)^2 \times \sum_{j=1}^{M} (y_{1,j} - \mu_y)^2}} \tag{1}$$

where COV(x, y), μ, s stands for covariance, mean, standard deviation respectively. The range of ρ(x, y) is [−1, 1]. The essential properties of PCC which make it an effective function for clustering (like symmetric, invariant with change in scale and location of variables etc.) are discussed in details in [15].

The step wise implementation of the proposed CAP algorithm is outlined here

Step 1: Formation of Correlation Matrix: For any given input dataset D of dimension M × N compute the correlation matrix c(x, y) of dimension M × M. As, c(x, y) is a symmetric matrix so its upper diagonal elements are computed by taking the PCC of the samples in D as given in (1).

Step 2: Selection of Shared Preference: After correlation matrix c(x, y) is obtained, replace the diagonal elements of this matrix by the minimum value of c(x, y). This process is termed as 'Selection of shared preference'. This works fine for the benchmark datasets like Iris, Wine, Small Soyabean, Lung Cancer and Image Segmentation. But, for 3D human modelling dataset the sample values lie pretty close to each other. Therefore the correlation values lie between 0.7 and 1. So, in this case for proper clustering the diagonal elements are replaced by a much lower value i.e. −2.2.

Step 3: Formation of Responsibility Matrix: A responsibility message r(x, y) is sent from data point x to exemplar y, which indicates how suitable the data sample y is to serve as a role model for point x.

$$r(x, y) = c(x, y) - \sum_{y' \neq y} \left\{ a(x, y') + c(x, y') \right\} \tag{2}$$

Step 4: Formation of Availability Matrix: Availability a(x, y) is a message sent from data point y to data point x which indicates how pertinent it would be for data point x to choose point y as an exemplar.

$$a(x, y) = \min \left\{ 0, r(y, y) + \sum_{x' \neq (x, y)} \max\{0, r(x', y)\} \right\} \tag{3}$$

The message between all the exemplar points to the data points, create matrix a (x, y).

Step 5: Self Availability Matrix: Self-availability a(y, y) reflects the collective evidence that y is an exemplar, based on the positive responses (responsibilities) sent to it from other points.

$$a(y, y) = \sum_{x' = y} \max\{0, r(x', y)\} \tag{4}$$

Step 6: Selection of Exemplar: The exemplar is identified from criterion matrix which is the combination of responsibility and availability matrices is given by

$$Cr(x, y) = R(x, y) + A(x, y) \tag{5}$$

For any data point $x \in Cr$ the value of $y \in Cr$ which maximizes $Cr(x, y)$ is identified as an exemplar for the point x.

Step 7: Damping of Matrices: After messages are updated, a small random noise is added to correlation matrix for avoiding numerical oscillations as suggested in [5].

$$C(x, y) = c(x, y) - 10^{-12} \cdot randn(M, M)(\max(c(:)) - \min(c(:))) \tag{6}$$

Each message is assign to λ times of its past value and (1−λ) times of its updated value. The value λ = 0.9 is taken for simulation.

The steps 2–7 are repeated for a number of iterations till the exemplar points or corresponding clusters do not change for certain number of iterations (taken as 10).

3   Simulation Results on Benchmark Datasets

In order to validate the performance of proposed CAP algorithm simulation is carried out on five benchmark datasets taken from UCI repository [16]. The datasets and their size (samples × dimensionality) used for testing are: Iris (150 × 4),

Small Soyabean (47 × 35), Wine (178 × 14), Lung cancer (32 × 57) and Image segmentation (210 × 19). The simulations of AP and proposed CAP are carried out in the MATLAB R2014.

In order to evaluate the performance of original AP [1, 12] and proposed CAP algorithm the percentage of classification accuracy is taken as a validation measure given by

$$PA = \frac{Number\ of\ misclassified\ events}{Total\ number\ of\ events} \times 100\% \tag{7}$$

The PA and run time achieved by both the algorithms for the five datasets are presented in Table 1. The best outcomes are highlighted in bold letters. It is observed that except the IRIS dataset in rest four dataset the proposed CAP provides superior classification accuracy. The IRIS dataset has only 4 dimensions and thus correlation is not very effective. But in other datasets as the dimensionality is higher correlation is a suitable candidate.

4 Classification of Aggressive Actions of 3D Human Models

This real life application set up is originally developed by Theodories et al. [13], which is now standardized in UCI repository [6] as 'Vicon physical action database'. The database has records for 10 human beings (three females and seven males). The physical actions of the human beings are recorded with 9 cameras and each one records the moment in 3D (i.e. X, Y and Z coordinate). Therefore, each action is recorded in 27 dimensions. A model for recording the actions are shown in Fig. 1. The actions associated with 'Subject 1' are investigated over here. The numbers of samples present in each type of actions are presented in Table 2.

Five case studies are carried out to classify the various actions of 'subject 1' are highlighted below. In each case study, there are three subcases. In first subcase (a) the dataset comprises of 100 samples from the aggressive and regular action. In subcase (b) the dataset consists of 10% of the data from aggressive and normal gesture. In subcase (c) the dataset composes of 20% of the data for aggressive and normal actions for case 1–3 and 15% of the data for case 4–5.

Case 1: Separation of Body movements for normal and aggressive activities: The objective is to separate body launching actions for regular (class 1) and aggressive (class 2) gestures. Class 1 has a dimension of 2014 × 81 which is obtained by taking the first 2014 × 27 samples for three actions bowing, standing, seating and placing them side by side. Similarly, class 2 has dimension 1511 × 81, taking 1511 × 27 samples from headering, pushing and pulling.

Case 2: Discrimination of Arm movements for normal and aggressive activities: The goal is to classify arm launching for regular actions (class 3) and aggressive actions (class 4). The class 3 having dimension 1716 × 81, consists of

Table 1 Comparative results of AP and CAP on benchmark datasets

Name of the database	Ideal cluster division	AP clustering	PA (%)	CT (s)	CAP clustering	PA (%)	CT (s)
Iris	3[50 + 50 + 50]	3[50 + 54 + 46]	94.67	7.11	3[50 + 55 + 45]	93.33	7.43
Wine	3[59 + 71 + 48]	3[23 + 50 + 105]	61.80	10.7	4[56 + 60 + 14 + 48]	84.27	11.8
Soya small	4[10 + 10 + 10 + 17]	3[10 + 10 + 27]	57.45	0.35	4[10 + 10 + 13 + 14]	87.23	0.39
Lung cancer	3[9 + 13 + 10]	2[23 + 9]	37.50	0.181	3[14 + 9+9]	93.75	0.21
Image segment	7[30 × 7]	2[100 + 110]	42.86	11.54	7[34 + 2+36 + 38 + 48 + 22 + 30]	65.71	12.51

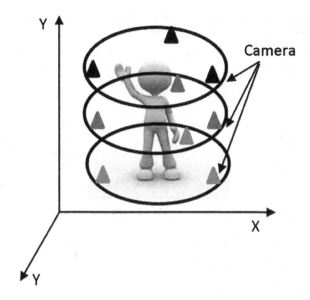

Fig. 1 A model representing recording the physical actions of human beings with nine cameras and each one records in 3D coordinate. Original setup developed by Theodories et al. [13] at University of Essex, UK

Table 2 Classification of aggressive and normal gestures based on different body parts movement for 'subject 1'

Movement of body parts	Class	Normal action	Num of sample	Class	Aggressive action	Num of samples
Body launching	1	Bowing	2014	2	Headering	1511
		Standing	2102		Pushing	1873
		Seating	2274		Pulling	2096
Arm launching	3	Clapping	1716	4	Hammering	1902
		Hugging	1973		Slapping	2137
		Hand shak.	2559		Punching	2486
Leg launching	5	Walking	1721	6	Side kicking	1445
		Jumping	1902		Front kicking	2092
		Running	2242		Kneeing	2461

1716 × 27 samples taken from each of aggressive actions hammering, slapping and punching. Class 4 has dimension 1902 × 81, with 1902 samples and 27 attributes collected from each of normal action clapping, hugging and handshaking.

Case 3: Classification of Leg movements for normal & aggressive actions: The aim is to differentiate leg launching for normal (class 5) and aggressive activities (class 6). Class 5 has dimension 1721 × 81 consists of 1721 samples from

Table 3 Comparative results of cluster analysis on 3D human actions modelling using subject 1 of 'Vicon physical action database' using AP & CAP algorithm

Case	Subcase	Ideal cluster division	AP clustering	PA (%)	CT (s)	CAP clustering	PA (%)	CT (s)
1	(a)	2[100, 100]	2[100, 100]	100.00	9.78	2[100, 100]	100.0	10.45
	(b)	2[201, 151]	2[201, 151]	100.00	55	2[201, 151]	100.0	59.24
	(c)	2[400, 300]	3[400, 217, 83]	76.29	460	**2[400, 300]**	100.0	497
2	(a)	2[100, 100]	2[100, 100]	100.00	9.27	2[100, 100]	100.0	9.77
	(b)	2[190, 171]	3[86, 104, 171]	52.35	54.23	**2[190, 171]**	100.0	56.77
	(c)	2[380, 342]	6[85, 119, 26, 150, 211, 131]	–	473	**3[204, 176, 342]**	51.25	484.7
3	(a)	2[100, 100]	2[100, 100]	100.00	9.32	2[100, 100]	100.0	9.87
	(b)	2[172, 144]	3[108, 64, 144]	59.49	38.84	**2[172, 144]**	100.0	40.32
	(c)	2[344, 288]	4[160, 120, 64, 288]	41.77	345	**3[184, 116, 252]**	50.63	351.8
4	(a)	3[100, 100, 100]	4[100, 64, 36, 100]	76.00	33.9	**3[100, 100, 100]**	100.0	36.45
	(b)	3[200, 171, 172]	5[200, 102, 69, 108, 64]	51.01	211	**3[200, 171, 172]**	100.0	227.73
	(c)	3[300, 257, 257]	8[300, 120, 69, 68, 64, 98, 25, 70]	27.27	739.9	**4[300, 93, 164, 257]**	77.15	752.31
5	(a)	3[100, 100, 100]	4[89, 11, 100, 100]	92.67	33.3	**3[100, 100, 100]**	100.0	34.65
	(b)	3[151, 190, 144]	5[140, 11, 87, 103, 144]	59.59	143.78	**3[151, 190, 144]**	100.0	147.64
	(c)	3[225, 285, 216]	9[116, 109, 87, 100, 18, 53, 27, 101, 115]	–	510	**3[225, 285, 216]**	100.0	533

jumping, running, walking. The aggressive actions class 6 has 1445 × 81 includes 27 attributes from each gesture side kicking, front kicking and kneeing.

Case 4: Segregation of normal gestures: The goal is to classify the collective actions caused by three different type of body movements. The data set consists of regular actions due to body launching (2014 × 81), arm launching (1716 × 81) and leg launching (1721 × 81).

Case 5: Segregation of aggressive actions: The need is to separate the offensive gestures due to different types of body movements. The data set comprises of aggressive actions due to body launching (1511 × 81), arm launching (1902 × 81), leg launching(1445 × 81).

The comparative results of cluster analysis obtained in the five case studies are presented in Table 3. The accurate clusters obtained with CAP over AP are highlighted in bold letters. It is observed that the obtained number of clusters and samples present in them with CAP closely resembles with the ideal division. This is also validated with the obtained percentage of accuracy. It is also observed that the computational time of CAP is bit higher due to the more computation involve in calculating the correlation coefficient than Euclidean Distance. However this computational time is much lower than that by immunized PSO in [14].

5   Conclusions

In this manuscript a new CAP clustering algorithm is proposed considering Pearson correlation coefficient as affinity measure. The performance of the proposed algorithm is validated on five benchmark datasets obtained from UCI repository. A real life application of the proposed algorithm is demonstrated to classify the regular and aggressive actions of 3D human models. Simulations on fifteen case studies established the accurate classification of human actions by the proposed algorithm over the original affinity propagation algorithm. The computational time of the proposed algorithm is slightly higher than the original due the additional computation involves in the Pearson correlation coefficient. Therefore it is concluded that the proposed approach is an effective tool to classify datasets which have similarity in patterns rather than closeness among them.

References

1. Frey, B.J. and Dueck, D.: Clustering by passing messages between data points, Science Express, 315, 5814, 972–976 (2007).
2. Ren, T., Zeng, W., Wang, N., Chen, L. and Wang, C.: A novel approach for fMRI data analysis based on the combination of sparse approximation and affinity propagation clustering. Magnetic resonance imaging, 32, 6, 736–746 (2014).
3. Xia, H., Chen, X. and Guo, P.: A shadow detection method for remote sensing images using affinity propagation algorithm. In: IEEE International Conference on Systems, Man and Cybernetics, pp. 3116–3121 (2009).

4. Kokawa, Y., Wu, H. and Chen, Q.: Improved affinity propagation for gesture recognition. *Procedia Computer Science,* In: 17th International Conference in Knowledge Based and Intelligent Information and Engineering Systems, 22, pp. 983–990 (2013).
5. Ying, Z. and Changgang, J.: A kind of routing algorithm for heterogeneous wireless sensor networks based on affinity propagation. In: 26th Chinese Control and Decision Conference, pp. 2481–2485 (2014).
6. Lee, S.H. and Sohn, I.: Affinity propagation for energy-efficient BS operations in green cellular networks. IEEE Trans. on Wireless Communications, 14, 8, 4534–4545 (2015).
7. Hassanabadi, B., Shea, C., Zhang, L. and Valaee, S.: Clustering in vehicular ad hoc networks using affinity propagation. Ad Hoc Networks, 13, 535–548 (2014).
8. Liu, D. and Jiang, M.: Affinity propagation clustering on oral conversation texts. In: IEEE 11th International Conference on Signal Processing, 3, pp. 2279–2282 (2012).
9. Wang, C.D., Lai, J.H., Suen, C.Y. and Zhu, J.Y.: Multiexemplar affinity propagation. IEEE Trans. on Pattern Analysis and Machine Intelligence, 35, 9, 2223–2237 (2013).
10. Gan, G. and Ng, M.K.P.: Subspace clustering using affinity propagation. Pattern Recognition, 48, 4, 1455–1464 (2015).
11. Hang, W., Chung, F.L. and Wang, S.: Transfer affinity propagation-based clustering. Information Sciences, 348, 337–356 (2016).
12. Dueck D.: Affinity propagation: clustering data by passing messages, Doctoral Thesis. University of Toronto, (2009).
13. Theodoridis T. et al.: Ubiquitous robotics in physical human action recognition: a comparision between dynamic ANNS and GP, Proc. of IEEE international conference on robotics and automation, pp. 3064–3069 (2008).
14. Nanda, S. J. and Panda G.: Automatic clustering algorithm based on multi objective immunised PSO to classify actions of 3D human models, Engineering Application of Artificial Intelligence, 26, 1429–1441 (2013).
15. Nanda, S. J. and Panda G.: Design of computationally efficient density-based algorithms, Data & Knowledge Engineering, 95, 25–38 (2015).
16. UCI Repository of Machine Learning: https://archive.ics.uci.edu/ml/datasets.html.

How Safe Is Your Mobile App? Mobile App Attacks and Defense

Kireet Muppavaram, Meda Sreenivasa Rao, Kaavya Rekanar
and R. Sarath Babu

Abstract Usage of mobile phones has increased over the years, all over the world. With the extensive usage of smartphones, it has been confirmed by that most of the smartphone users work in an Android platform as it is an open platform which can be used by anyone to develop applications or introduce malicious activities; Reports suggest that up to 85% of the mobile phones used are in Android platform (Portal, Statistics. Number of monthly active Facebook users worldwide as of 1st quarter 2015 (in millions), 2015) [1]. At the least, a smartphone user uses a minimum of 12 applications, which arises the question, "how safe is your mobile?". This paper focuses on the said problem and provides an insight about some common types of attacks that use the vulnerability in applications and some defense methods and techniques that could be followed to prevent them.

Keywords Android smartphones · Vulnerabilites · Attacks · Defences

1 Introduction

Technology is day-by-day getting really advanced and it is becoming an essential part of life. Compared to the mobile phones, which mainly provide telephony functions, smartphones are handheld communication devices that support multimedia communications and applications for work and entertainment. Due to this

K. Muppavaram (✉) · M. Sreenivasa Rao · K. Rekanar · R. Sarath Babu
JNTUHCEH, JNTUSIT, BTH, Hyderabad, India
e-mail: kireet04@gmail.com

M. Sreenivasa Rao
e-mail: srmeda@gmail.com

K. Rekanar
e-mail: kare15@student.bth.se

R. Sarath Babu
e-mail: sharath.rakki@gmail.com

© Springer Nature Singapore Pte Ltd. 2018
V. Bhateja et al. (eds.), *Proceedings of the Second International Conference on Computational Intelligence and Informatics*, Advances in Intelligent Systems and Computing 712, https://doi.org/10.1007/978-981-10-8228-3_19

fast-growing functionality, the rate of usage at which updating traditional mobile phones to smartphones is tremendous. As per reports of statistics portal, it almost crossed more than 100 billion mobile applications had been downloaded from the Apple App Store and the number of free mobile apps downloads crossed more than 92 billion [1].

The expeditious growth of the global smartphone market in the upcoming years will also be escalated by the increasing business use of smartphones. Besides, the basic traditional corporate-liable model, the new BYOD (bring your own device) model is gaining much acceptance in enterprises throughout the world at present. According to these studies, International Data Corporation believes that in 2013, 132.3 million were used as employee-liable and 61.4 million smartphones were used as corporate-liable devices [2]. Most popular mobile properties are mainly being accessed using mobile apps instead of mobile browsers. Some of the popular examples are mobile social-networking properties, led by Facebook with 727 million monthly active mobile-only users [2]. Growing market is mobile retail via shopping apps as well as mobile gaming and other possibilities of app monetization.

Another growing market is mobile retail via shopping apps, as well as mobile gaming and other possibilities of app monetization.

One of the major distinct features of smartphones is that they give provision for the users to install and run third-party application programs which are usually in generalized term called as apps. According to different standard app reports all over the world, the usage of Android apps is more when compared with the other mobile operating systems and also Android gives easy provision for the users to develop and run their own apps as it is open platform [1]. This openness gave the malware writers easy directions to introduce the malware using mobile applications and extract the user sensitive information (Fig. 1).

Fig. 1 How safe is your mobile app?

2 Vulnerability of Smart Phones

Smartphones are vulnerable to security attacks due to many factors. Some of those are as follows:

- Users tend to store a lot of personal data in their smartphones, particularly as many people these days prefer to use online banking services for many financial purposes and choose the feature of saving their details on the phone itself; such data is sensitive [3]. Hackers gather such data from the smartphone to use it to their own advantage of gaining financially in a substantial manner, thus making a smartphone remunerative [4].
- Most of the smartphones used are developed on the Android platform, and with the platform that encourages open-source kernel, malware writers have the chance to gain a better understanding of the platform. Google's marketing strategy, which encourages the development of third-party applications and publishing them, has always been easy to gain a pro table market share [5]. Thus, hackers are presented with an opportunity to create and publish malware of their choice. And, as the never-ending usage of smartphones by owners' increases, so does the installation of malware which cannot be controlled very easily.
- Most smartphone users are not aware of the fact that their phone is a handheld computer that is vulnerable to any kind of cyber attacks [4]. They assume that their smartphones are just mobile phones that have many applications installed for communication and entertainment purpose [4]. Hence, there is not much attention that is paid to security measures.

Additionally, the origin of hardware and operating system of the smartphones has had the malware writers less constrained about the implementation of their actions. Besides, it is always easy to migrate a computers malware to any of the smartphone platforms.

3 Malware Behavior and Threats

Malicious attack behavior, remote control behavior, and propagation behavior have always been characterizing mobile malwares [6]. Attack behavior concerns about how the mobile malware will exploit the infected mobile device further by infecting all the other victims devices using different communication channels, e.g., Bluetooth [6]. Remote control behavior is about how the mobile malware utilizes a remote server to exploit a mobile device further after infecting it [6]. Propagation behavior is about the how malware can be transmitted to the victims [6].

A malware would try to acquire access to the data stored in the devices, and meddle with the functionality of the device, and possibly open up more security vulnerabilities like enabling unauthorized remote access to the hacker. There are many threats which can be launched using malware, which have been listed in the next subsection.

3.1 Attacks

The most typical attacks due to malware include the following:

i. **Mobile Application Permission Leakage Attacks**: The mobile application permission leakage attacks are of three types: Confused deputy attacks, intent spoofing, and permission collusion [7]. Confused deputy attacks are completely depending on misconfigured mobile applications. Intent spoofing is similar form of confused deputy attack which effects the applications that are not meant to communicate with other applications [7]. Collusion attacks are the attacks which use overt and covert channels and aggregates the permissions from different mobile applications and releases user mobile sensitive data [7]. Collusion attacks are quite di cult to detect and which causes a great deal of research in the mobile applications.

ii. **Spyware Attacks**: Spyware is malware that conveniently collects information from an infected device. A user stores a lot of information on a smartphone which attracts the hackers towards it. Besides, there are many channels that can be used in smartphones to collect information. For instance, an application that shows the weather conditions would have a permission to send the location data to servers, which can be used by the hacker to acquire the location of that user in a malicious manner. Sometimes, when an application might seem legitimate in most senses, the permission settings in a users' phone might not be secure enough to prevent such abuse of a permission given.

iii. **Phishing Attacks**: Malware needs to have a database of fake URLs that are capable of taking a users' personal information like credit card details, and any other personal information that can be misused as the websites have been cleverly masquerade as trusted websites. Studies prove that 25 percent of attacks use such techniques [8]. As this type of attack does not have a requirement of attacking the user directly in any manner due to its platform-independent nature, they are mostly applicable to smartphones.

There are numerous reasons for a hacker to prefer this attack. Some of which are as follows:

- The smaller screen of a mobile device (in this case, smartphone when compared to a personal computer) tends to enhance the chances of disguising the trust sign which any user would rely on to make a decision of submitting credentials, for instance, if a site is enabled by secure sockets layer.
- A smartphone provides a large variety of channels to use phishing attack on, due to wide scale of applications that are used. These channels include short message service or any other messenger applications.
- It is facile to mask the infected application as a legitimate application and cleverly distribute them in the market for usage.
- Many users do not have awareness on the fact that a smartphone gives the same risk as any other personal computer and tend to trust it more than a computer which obviously makes the computers work easier.

iv. **Diallerware Attacks**: A hacker can cause a financial loss to a smartphone user by diallerware attacks, which send premium rate SMS without a user being aware of it [9]. The premium rate SMS service was created to provide value-added services like news and stock quotations periodically with the price of it being charged to the users' phone bill. This can be used by the hacker to financially deteriorate the user without them being aware of it.

v. **Worm-Based Attacks**: A worm has the capability to compromise on the security of smartphones [9]. Besides, its nature of duplicating itself and propagating from one device to another without the user's notice helps the case [9]. And as network function virtualisation has been introduced lately in the next generation of mobile phones, worm-based attacks have a chance of increasing exponentially.

vi. **Botnets**: A set of zombie devices that are remotely controlled by a hacker when infected by malware is called a botnet [9]. When there are many mobile devices in the network, it is called a botnet [9]. They impose serious security threats to the Internet and most of them would be likely used in any organized crime, to launch attacks and gain profit financially.

vii. **Financial Malware Attacks**: This attack aims at stealing credentials from a smartphone using buffer-overflow techniques or man-in-the-middle attacks on financial applications that are run by the user [4]. As we have established earlier, a smartphone is equally prone to risks as much as a computer is. Financial malware can come in many forms, for instance, it could just be a key-logger application which collects credit card numbers, or in a more sophisticated format, it could be an application that impersonates a real banking application that can launch man-in-the-middle attacks when a user is performing a bank transaction.

3.2 Malicious Activities Through Mobile Apps

Some of the ways where malicious activities can be introduced through apps are as follows:

  i. Most of the app developers tend to request permissions which are irrelevant to the app; this makes the easy provision for malware writers to introduce their malicious activities.
 ii. If two apps were developed by the same developer, then there may be a chance of collusion attacks where in the case of user ID are shared using API calls which is more dangerous.

4 Defense Mechanisms

There are different strategies to defend against malware, some of which have been listed below:

a. Proactive detection can be done using tools or softwares.
b. Creating honeypots so that can catch hold of malware writers easily.
c. Updating the applications within short span every time.
d. System cleanup everyday.
e. Using most trusted applications by verifying the signature of the used applications.

4.1 Preventive Measures

Listed below are a few preventive measures that can be taken in order to avoid malware attacks.

- **Application developers**: Most of the malware attacks are introduced using mobile applications. The application developers should follow the secured policies strictly by not giving any sort of the loopholes to the malware attackers. The Android consists of about 135 permissions out of which by different researches at about 30 permissions are considered as malicious and Android itself has categorized 23 permissions as the most dangerous permissions, so developers should restrict using these malicious, dangerous permissions in their application so that scope of introducing the malware decreases.
- **Smartphone users**: Smartphone users should strictly restrict app installation from untrusted mobile markets or websites. Most of the malwares have been introduced by Chinese and Russian developers, so when any smartphone user who is installing an app should check the developers by whom app is ordered

by; reviews should be checked; users should not blindly accept all the permissions; if the user feels that there are some dangerous permissions which are not required for the app, they should be disabled; even the app runs by disabling some permission which can be done after installation.

- **Mobile market administrators**: Mobile market administrators should follow strict policies to ensure that mobile market is secure and highly trusted mobile market. If an app is found suspicious that should be removed and developer should be blocked, apart from different scanners like bouncer which scans to find malicious activities, it is necessary to provide few more scanners that enhances the usage of apps through scanners.

4.2 Detection Techniques

Most of the malware can be detected in either of two ways, i.e., using signature-based detection or anamaly-based detection. Signature-based detection solely concentrates on detection through signatures; it has some limitations like if the particular signature of app has not been found in the database detection fails. Anamoly-based detection is done based on the behavior of the system, and it mainly checks whether there are any inconsistencies in the behavior. These anamoly-based and signature-based detections are basically techniques which are usually come under network-based detection. Detection will also done on client side which is called as client-side detection. Client-side detection is again classified into static client detection and dynamic client detection. Static and dynamic client detection performs detection in terms of signature-based detection and anamoly-based detection.

5 Common Attacks on a Mobile Phone/Smartphone

There are the three most common ways where the smartphone/mobile phone most likely to be breached:

- i. Poor security decision by users,
- ii. Flaws in operating systems, and
- iii. Usage of malicious apps.

Poor security decision by users: Most of the present smartphones at present have good built-in technology which warns the users with different alerts like, for example, when unsecure Wi-Fi is used, the present smartphones give the alert "a warning will come up saying that the server identity cannot be verified and asking if user still want to connect". User will be prompted to click & quot; continue & quot;

before you can join the Wi-Fi. Such kinds of warnings are usually ignored by the users by giving their poor security decision to continue the harmful activity.

Flaws in operating systems: Developers initially unknowingly may give some privileges which are vulnerable to extract the most sensitive information. Frequent updates are being released by fixing the bugs by later stages but most of the users will not install the updates as a result of the bugs or errors which have been developed by the developer in the initial stages that may not be fixed and that leads to extract of the data.

Malicious apps downloaded from website: Most of the apps which were downloaded from the website or the links increase the risk of data breach; apps which are downloaded from the Playstore somehow have been protected but still need careful evaluation in giving the required permissions for the app by giving good security decision.

6 Algorithm: CURD—Confidential Unrestricted and Restricted Data

The algorithm proposed is used to check the safety level of the application being used, i.e., to know, "How safe is the app?"

Step 1. Dividing available the mobile user data into three sensitive levels: confidential, unrestricted, and restricted.

- Confidential—High sensitive data which cannot be given permission to any app or other sources.
- Unrestricted—Data can be accessed by app or any source.
- Restricted—Sensitive data can be accessed only by special permissions through users' acceptance.

Step 2. Considering the app apk and extracting permissions through apk tool and calculating the percentage of confidential, unrestricted, and restricted data required for a particular app for installation.

Step 3. Based on the calculation, the app seeks more than 25% of confidential data in the form of permissions acceptance by the user when the app is treated as not safe.

This algorithm gives only the basic checking of the app that is safe or not. There are different tools which are available for further enhanced detection of malicious applications.

Available tools are TaintDroid, FireDroid, MockDroid, and RecDroid. Further, there are many methods which use permission-based detection and explores the permissions for detection and gives rating to the permissions in terms of risky permissions, and not-risky permissions to calculate whether the app is malicious or not.

7 Conclusion

In this paper, we presented a summarized view on different types of attacks by which a smartphones' vulnerability is exploited. Most smartphone users are dependent on using different apps for their daily usage. For all e-transactions, the respective merchants are providing their own applications, as Android is an open platform and it is not secure enough to use untrusted apps. In near future, almost every transaction can be done through mobile applications, for instance, Paytm. By summarizing the attacks that would give a smartphone user, an idea about the types of attacks which can be introduced in different ways into an application make it malicious. By this, we conclude that every user should have awareness if the apps used by the user are safe enough or not.

References

1. Portal, Statistics. "Number of monthly active Facebook users worldwide as of 1st quarter 2015 (in millions)." Luettavissa: http://www.statista.com/statistics/264810/number-ofmonthly-active-facebook-users-worldwide/. Luettu 26 (2015).
2. Ballagas, Rafael, et al. "Byod: Bring your own device." Proceedings of the Workshop on Ubiquitous Display Environments," Ubicomp. Vol. 2004. 2004.
3. Kang, Joon-Myung, Sinseok Seo, and James Won-Ki Hong. "Usage pattern analysis of smartphones." Network Operations and Management Symposium (APNOMS), 2011 13th Asia-Pacific. IEEE, 2011.
4. Kulkarni, Shakuntala P., and Sachin Bojewar. "Vulnerabilities of Smart Phones." (2015).
5. Rekanar, Kaavya. "Text Classification of Legitimate and Rogue online Privacy Policies: Manual Analysis and a Machine Learning Experimental Approach." (2016).
6. D. Guo, A. Sui, and T. Guo, "A Behavior Analysis Based Mobile Malware Defense System," Proc. ICSPCS, pp. 16, 2012.
7. M. Kireet, Dr. Meda Sreenivasa Rao. "Investigation of Collusion Attack Detection in Android Smartphones." International Journal of Computer Science and Information Security, (IJCSIS) Vol. 14, No. 6, June 2016.
8. Garera, Sujata, et al. "A framework for detection and measurement of phishing attacks." Proceedings of the 2007 ACM workshop on Recurring malcode. ACM, 2007.
9. He, Daojing, Sammy Chan, and Mohsen Guizani. "Mobile application security: malware threats and defenses." IEEE Wireless Communications 22.1 (2015): 138–144.

An Agile Effort Estimation Based on Story Points Using Machine Learning Techniques

Ch. Prasada Rao, P. Siva Kumar, S. Rama Sree and J. Devi

Abstract Nowadays, many software companies face the problem of predicting the accurate software effort. Most of the software projects are failed due to over budget and over schedule as well as under-budget and under-schedule. The main reason for the failure of software projects is inaccurate effort estimation. To improve the accuracy of effort estimation, various effort estimation techniques are introduced. Functional points, object points, use case points, story points, etc., are used for effort estimation. Earlier, traditional process models like waterfall model, incremental model, spiral model, etc., are used for developing the software, but none of them have given the successful projects to the customers. Now, 70% of the application software's have been developed by the agile approaches. The success rate of the projects developed by using agile methodologies has been increased. The major objective of this research is to estimate the effort in agile software development using story points. The obtained results have been optimized using various Machine Learning Techniques to achieve an accurate prediction of effort and compared performance measures like MMRE, MMER, and PRED.

Keywords Adaptive Neuro-Fuzzy Interface System (ANFIS)
Effort estimation · Generalized Regression Neural Networks · Mean Magnitude of Error Relative (MMER) · Mean Magnitude of Relative Error (MMRE)
Prediction Accuracy (PRED) · Radial Basis Function Networks (RBFNs)

Ch. Prasada Rao (✉) · P. Siva Kumar
K L University, Vaddeswaram, Guntur, India
e-mail: prasadarao.chatla@aec.edu.in

P. Siva Kumar
e-mail: spathuri@kluniversity.in

S. Rama Sree · J. Devi
Aditya Engineering College, Surampalem, India
e-mail: ramasree_p@rediffmail.com

J. Devi
e-mail: charandevi85@gmail.com

© Springer Nature Singapore Pte Ltd. 2018
V. Bhateja et al. (eds.), *Proceedings of the Second International Conference on Computational Intelligence and Informatics*, Advances in Intelligent Systems and Computing 712, https://doi.org/10.1007/978-981-10-8228-3_20

209

1 Introduction

Project management in the software development is the prominent responsible activity. The major objectives of software project management are to plan, establish the scope document, determine the business case, estimating the effort accurately, providing the development and testing environments to feature teams, identify the probable risks, etc. The contract between the product owner and organization purely depends on the project scope. The contract includes estimation of the budget, deadlines for the project to be delivered and it is defined as a business case. The estimation of the effort is measured in terms of the size of the project, the time required to complete the project and how much budget is needed to complete the project on time. The accurate prediction of the effort in the software development leads to the success of software project otherwise it leads to failure of the project. The main reason to get a failure of the software projects is poor management, inaccurate estimation, continuous change of requirements, incomplete require- ments, lack of communications among developers, unable to identify the risks in the early stage of development, adopting the relevant process model, etc. There is a statement in software development "Good Management can't assure the success of the project but Poor Management leads to failure of the projects." The effective and efficient process model plays a major role in developing the successful software projects. The Standish Group reported in 2015, the success rate of the projects developed by waterfall model is only 11% and when coming to the projects developed by using the agile methodologies: 39% of the projects delivered on time, on a budget, with high quality and satisfy all the customer needs and expectations. Now, 80% of the software industries are transforming from conventional process models (Waterfall, Spiral, Increment, RUP, Prototyping, RAD, etc.) to agile process models (Scrum, XP, FDD, etc.) [1]. The scrum and XP are very popular and widely recognized.

In 2001, some of the experts from top software industries form a team and establish an "Agile Manifesto" [2]. This is called as a bible for software organi- zation to develop the successful software projects. The advantages of agile process models over conventional process models are:

- Deliver some usable software to the Product Owner (Customer) in a reasonable time
- Good chemistry among the team members to improve the productivity
- Involvement of Product Owner in the entire development to give better feedback or he can also change some features
- Agile teams are self-organized as well cross-functional for effective progress of other software development
- Regular Assessment by feature teams and so on

Either in Conventional or Modern Approaches, the accurate effort estimation is essential for developing the successful projects. To predict the effort in software projects, many metrics are introduced like KLOC, Functional Points, Class Points, Use Case Points, Object Points, etc. None of them have given the accurate results. In the agile process model, Story Points are derived from the user stories [3]. User stories can be written in a very general language where everyone can able to understand. User stories are of the form

As a <Designation>
I want to <some task to be done>
So that <goal to be achieved>
Ex: As a parent
I want to know the percentage of my son
So that I can take necessary actions to improve their studies.

Let us take an example to estimate the time and cost using User stories:

Total User Stories	20
10 of them can be measured in 3 Story Points each	10 * 3 = 30
Remaining can be measured in 5 Story Points each	10 * 5 = 50
Total Story Points	30 + 50 = 80
Velocity	10
Total number of sprints (total Story Points/Velocity)	80/10 = 8
Each sprint may take	4 weeks
Estimated time for development	32 weeks
Each sprint cost	50,000
Estimated cost for development	8 * 50,000 = 400,000

There are many techniques available for effort estimation like Delphi Estimation, Planning Pocket, Use Case Points, etc., [4]. But they are mainly designed for conventional process models. We can apply these conventional estimation techniques to agile methodologies, but they will not give accurate results. To improve the accuracy of effort estimation, we have to use story points in agile methodologies. The implementation of effort estimation is carried out using 21 project data set and evaluated using different machine learning techniques. Most of the researchers are considered the performance metrics like MMRE, MMER, and PRED(X). Estimation of the effort has been implemented using some soft computing techniques like ANFIS, RBFN and evaluated and compared using the three performance metrics that are mentioned above.

2 Related Work

A. Schmietendorf et al. [5] provided the definition and characteristics of agile methodologies and described different possibilities of effort estimation for different types of agile methods, especially for XP projects and provided an analysis on the importance of prediction of software effort and agile software development. S.M. Satapathy et al. [6] described various Adaptive Regression Techniques such as multi-layer perceptron, projection pursuit regression, multivariate adaptive splines, K nearest neighbor regression, constrained topological mapping and radial basis function networks for predicting the software effort and provided the cost estimation Class Point approach.

S. Keaveney et al. [7] proposed different cost estimation techniques for both conventional estimation techniques as well as for agile software development approaches and examined the causes of inaccurate estimates and steps to improve the process. Shashank Mouli Satapathy et al. [8] described various SVR kernel methods for improving the estimation accuracy that helps in getting optimal estimated values. This estimation is carried out using Story Point approach and performance of different SVR methods were compared in terms of MMRE and PRED.

Hearty et al. [9] provided an investigation about Bayesian Networks, which can combine sparse data, prior assumptions, and expert judgment into a single model. Shows how XP can learn from project data in order to make quantitative effort predictions and risk determinants and also described project velocity. Donald F. Specht et al. [10] described Generalized Regression Neural Network which is a one-pass learning algorithm and also explains neural network implementation with one-dimensional example and adaptive control systems.

3 Evaluation Criteria

The performance of various methods discussed in the proposed approach is determined by the following criteria

- The Mean Magnitude of Error Relative can be calculated as

$$\text{MMRE} = \frac{100}{N} \sum_1^N \frac{|P_i - A_i|}{A_i}. \tag{1}$$

- The Mean Magnitude of Error Relative can be calculated as

$$\text{MMER} = \frac{100}{N} \sum_1^N \frac{|P_i - A_i|}{P_i}. \tag{2}$$

- Prediction Accuracy can be calculated as

$$\text{PRED}(X) = \frac{100}{N} \sum_1^N \begin{cases} 1 & \text{if } \frac{|P_i - A_i|}{A_i} < \frac{N}{100} \\ 0 & \text{otherwise.} \end{cases} \tag{3}$$

where

N   The total observations or projects in test set
P_i Predicted effort of ith project in the test set
A_i Actual effort of ith project in the test set
X   MMRE value of the test set for which we are calculating the prediction.

The model with high PRED value and low MMRE and MMER values will give accurate estimation results.

4   Proposed Approach

The proposed approach is based on the data collected from 21 projects' dataset [11]. The dataset is used to evaluate software effort in terms of time and cost. In this paper, Story Points and Project Velocity are taken as inputs.

4.1   Story Points

Story Points are a unit of measure that is used to implement a User Story. When estimating based on Story Points, we will assign a size to each User Story [12]. Generally, Story Points follow Fibonacci like sequence

$$\text{i.e., } 0, \frac{1}{2}, 1, 2, 3, 5, 8, 13, 20, 40, 100, \propto, ?$$

While predicting the size of each user story, most of the industries will prefer the either 3 or 5 as size for the user story. The entire team who are involved in developing the sprints in scrum will be estimating the total story points of all the user stories. While predicting the story points for each user story from all team persons there will be some different opinions and weights for user stories. In this case, the people with different opinions will be explained their opinions and re-estimate it. Still, it has ambiguity, then scrum master will be considered the average weights of the entire team. Later these have been partitioned into sprints. Generally, each sprint may take 2–4 weeks of time with the help of 5–8 people of the team with one scrum master.

4.2 Project Velocity

The Project Velocity is the deliveries made per sprint. Sprint is time-boxed iteration. Velocity may differ from initial sprint to next sprints, i.e., velocity may be less
for initial sprints and may increase in subsequent sprints [13].

4.3 Soft Computing

In recent years soft computing gains popularity over hard computing because of its
easiness, availability in Computer Science, Machine Learning, Artificial Intelligence, etc. By applying soft computing we can easily find better solutions to
complex problems at a reasonable cost. The techniques provided by soft computing
are Fuzzy Systems (FS), Neural Networks (NN), Probabilistic Reasoning (PR), etc.
We can get more accurate results by using soft computing than hard computing.
Here, in this paper, we have implemented the Machine Learning Techniques such
as Adaptive Neuro-Fuzzy Modeling, Generalized Regression Neural Network and
Radial Basis Function Networks (RBFNs) using MATLAB software.

4.3.1 Adaptive Neuro-Fuzzy Interface System

In this, we need to build Adaptive Neuro-Fuzzy Interface System (ANFIS) to train
Sugeno systems. An ANFIS is a fuzzy system and the training is done using
neuro-adaptive learning methods as in the case of neural network training.
MATLAB provides a Fuzzy Logic Toolbox which provides a number of
command-line functions to train Sugeno-type Fuzzy Interface System (FIS) using
the data provided for training. First, we need to generate FIS structure using genfis
algorithm, because anfis algorithm takes genfis (genfis1/genfis2/genfis3) function as
one of the argument. For this we have 3 methods available in MATLAB (Table 1).
    In the proposed approach we have used genfis2.
    The syntax used in the proposed approach is

Table 1 Different functions of genfis available in MATLAB

Function name	Description
genfis1	It will generate FIS structure from data (training set) using grid partition
genfis2	It will generate FIS structure from data (training set) using subtractive clustering
genfis3	It will generate FIS structure from data (training set) using FCM clustering

$$\text{Fis} = \text{genfis2(Input, Output, Rad, Bounds, Options)},$$

where

Input	Input from training set
Output	Output from training set
Rad	In this research method Rad value is [0.5 0.4 0.3]
Bounds	Specifies how to map the training set data i.e., The bounds argument can be empty matrix or need not be provided In the proposed approach, it is an empty matrix
Options	It is also an optional vector which specifies algorithm parameters to override the default values. The default value for this argument is [1.25 0.5 0.15 0] and this value is only used in the proposed approach.

The anfis uses genfis (genfis1/genfis2/genfis3) to create initial FIS. Now we have to give this genfis2 (in the proposed approach we have used genfis2 algorithm) output to anfis function which acts as a training routine for Sugeno-type FIS. The syntax of anfis that is used in this research method is

$$\text{fis1} = \text{anfis(Traindata, Fis, Dispopt)},$$

where

Traindata	Training set
Fis	Output of genfis2 is used in our approach
Dispopt	If 1 is specified, training information will be displayed and if 0 is specified information won't be displayed. In the proposed approach 1 is specified.

In the Adaptive Neuro-Fuzzy Modeling, the last step is performing fuzzy interface calculations. For this we need evalfis algorithm. The syntax used in the proposed approach is:

$$\text{estimated} = \text{evalfis(Input, fis1)}$$

where

Input	Input data from test set
fis1	FIS structure generated by anfis function.

4.3.2 Generalized Regression Neural Network

To implement this algorithm MATLAB provides a function known as newgrnn. GRNN which is a type of Radial basis network. GRNN is used to approximate functions [10]. They are quick to design. The general form of newgrnn function is

$$net = newgrnn(Input, Output, Spread),$$

where

Input   Input vector (training set inputs)
Output   Target vector (training set outputs)
Spread   Constant (default 1.0). In the Proposed approach default value is taken.

4.3.3   Radial Basis Function Networks (RBFNs)

RBFNs are similar to neural networks. Input to the RBFNs can be one or more numeric values and output from RBFNs is also one or more numeric values. RBFN is also known as a Radial net which can be used for classifying the data and to make predictions. It contains three layers:

1. Input layer: Takes input vector
2. Hidden layer: Contains neurons
3. Output layer: Contains linear combinations of neurons and It will form network outputs by taking a weighted sum of outputs from the second (hidden) layer.

If there are N neurons then the number of categories will be N + 1. There are some Radial Basis Functions such as Gaussian, Multiquadrics, and Inverse Multiquadrics [14]. There are two types of functions available in MATLAB to implement BBFNs. They are as follows:

1. newrb: It is used to design a radial basis network. This radial basis function is also used for function approximation. The newrb will set a Mean Square Error (MSE) goal and it adds neurons to hidden layer until it reaches the goal. This function stops either when the goal is reached or when maximum number of neurons is reached. The syntax used in the proposed approach:

$$net = newrb(Input, Output, Goal, Spread).$$

   Arguments are discussed later.
2. newrbe: It is used to design exact radial basis network. These are also used for function approximation. This function is used to design radial basis network very quickly. If there are n input vectors then there will be n neurons.
   Syntax:

$$net = newrbe(Input, Output, Spread)$$

Let us discuss the arguments of all 3 functions.

Input   Input vector (training set inputs)
Output   Target vector (training set outputs)

Spread   Constant (default is 1.0). In the Proposed approach default value is taken
Goal   Mean Square Error goal (default is 0.0). In this research method, default
          value is taken.

Note: For all three functions are simulated for new input to get the predicted
values.
Syntax:

$$sim(Net, Newinput)$$

where

Newinput   Input from test set
Net   Net is the output of newgrnn/newrb/newrbe function.

Most of us try to give the outputs directly using matrix, but when we are using
different number of columns in input and output arguments it will show error. So
that we have to complement all the variables which we are passing as arguments in
newgrnn, newrb, newrbe, and also for sim function we need to complement test
input set but not the radial basis function (net variable) which is going to be
simulated. As we took complement to the inputs, we need to complement the final
output, i.e. the output from sim function.

5   Experimental Results

Table 2 shows the results of our approach. We have taken three machine learning
techniques: Adaptive Neuro-Fuzzy Modeling, Generalized Regression Neural
Network, and Radial Basis Function Networks. The obtained results are based on
the training set and test set which we have used in the proposed approach. We can
observe that all functions differ in terms of MMRE and MMER, but almost all
functions gave similar PRED.

Table 2 Comparision among anfis, newgrnn, newrb, newrbe functions

Model	Function	MMRE		MMER		PRED	
Adaptive Neuro-Fuzzy Modeling	Anfis	*Cost*	*Time*	*Cost*	*Time*	*Cost*	*Time*
		3.9079	8.4277	3.9587	19.5922	57.1429	76.1905
Generalized Regression Neural Networks	Newgrnn	4.8335	2.7864	5.8700	3.2042	76.1905	76.1905
Radial Basis Function Networks	Newrb	9.9604	8.0909	16.2868	6.6430	76.1905	76.1905
	Newrbe	10.6099	8.0909	11.3974	6.6430	76.1905	76.1905

6 Threats to Validity

- We have taken only 21 projects' data and examined with few projects. Maybe it gives better results from small set of data set, so we should consider the larger data set and examine
- Automation of User Stories to Story Points is not possible.
- Experienced and skill set of scrum team may also be considered for the success of software projects.

7 Conclusion

In this paper Story Point approach is used for predicting the software effort. Story Point approach is one of the most popular effort estimation techniques that can be applied to agile software projects. In this paper, three Machine Learning Techniques are chosen for predicting the software effort. The three techniques are Adaptive Neuro-Fuzzy Modeling, Generalized Regression Neural Networks and Radial Basis Function Network. Adaptive Neuro-Fuzzy Modeling is implemented using anfis function and GRNN is implemented using newgrnn function and RBFNs are further implemented using two algorithms namely newrb, newrbe which are provided by MATLAB software. The results are analyzed using MMRE, MMER, and PRED parameters. This study can also be expanded using Fireworks Algorithm (FA), Random Forest, etc.

References

1. D. Cohen, M. Lindvall and P. Costa, "An introduction to agile methods," Advances in Computers, vol. 62, pp. 1–66, 20/Jan/2004.
2. M. Flower and J. Highsmith, "The Agile Manifesto", Software Development, vol. 9, no. 8, pp. 28–35, August 2001.
3. "Writing User Stories, Examples and Templates in Agile Methodologies", Jan 8-2016.
4. Tutorialpoint.com, "Estimation Technique".
5. A. Schmietendorf, M. Kunz, and R. Dumke, "Effort Estimation for Agile Software Development Projects", in 5th Software Measurement European Forum, 2008.
6. S.M. Satapathy, M. Kumar, and S.K. Rath, "Fuzzy-class Point approach for software effort estimation using various adaptive regression methods", CSI Transactions on ICT, vol. 1, no. 4, pp. 367–380, 31 December 2013.
7. S. Keaveney and K. Conboy, "Cost estimation in Agile Development Projects", in ECI, 2006, pp. 183–197.
8. Shashank Mouli Satapathy, Aditi Panda, Santanu Kumar Rath, "Story Point Approach based Agile Software Effort Estimation using Various SVR Kernel Methods", International Conference on Software Engineering, 2014 July 1–3.

9. P. Hearty, N. Fenton, D. Marquez, and M. Neil, "Predicting Project Velocity in XP Using a Learning Dynamic Bayesian Network Model," Software Engineering, IEEE Transactions on Software Engineering, vol. 35, no. 1, pp. 124–137, 2009.

10. Donald F. Specht, "A General Regression Neural Network", IEEE Transactions on, vol. 35, no. 1, pp. 124–137, November 2009.

11. Z.K. Zia, S.K. Tipu and S.K. Zia, "An Effort Estimation Model for Agile Software Development", Advances in Computer Science and its Applications, vol. 2, no. 1, pp. 314–324, July 2012.

12. E. Coelho and A. Basu, "Effort Estimation in Agile Software Development using Story Points", Development, vol. 3, no. 7, August 2012.

13. Monty Python and Holy Grail, "Velocity: Measuring and Planning an Agile Project", Dec 10–2007.

14. Shafinaz Elinda Saadon, Zarita Zainuddin, "Radial Basis Function and Multilayer Perceptron Neural Networks in Function Approximation".

OFS-Z: Optimal Features Selection by Z-Score for Malaria-Infected Erythrocyte Detection Using Supervised Learning

Md. Jaffar Sadiq and V. V. S. S. S. Balaram

Abstract Analyzing microscopic image of blood smears has a pivotal role in analysis characterization of erythrocytes in the screening of malarial parasites. Characteristics feature of erythrocyte undergo changes if it infected by malaria parasite infection. Morphology, texture and intensity are key features of erythrocytes. Numerous solutions have proposed in improving the microscopic image analysis. An image processing algorithm to automate and accurate the diagnosis of malaria parasite scope in microscopic images of the blood smears is proposed in this paper. The experimental study indicating that identification of malaria parasite scope in erythrocytes by the proposed OFS-Z (Optimal Feature Selection by Z-Score) is significant and robust. The renowned classifiers such as naive Bayes, SVM, and AdaBoost classifiers are used to assess the performance of the proposed model. The metrics cross-validation and miss-classification rate rarely used to estimate the optimality of the classifiers under the proposed feature selection approach.

Keywords Malaria detection · Microscopic image processing
Erythrocyte · Feature selection · Z-score · P-value · Morphological
features · Texture features · Classifiers

1 Introduction

Malaria is one of the major epidemic disease risk envisaged across the countries, and globally thousands of deaths are reported annually resulting from malarial infection. The protozoan parasites like the falciparum, knowlesi, vivax, malaria, and

Md. Jaffar Sadiq (✉) · V. V. S. S. S. Balaram
Department of IT, Sreenidhi Institute of Science & Technology, Ghatkeser, Hyderabad, Telangana, India
e-mail: jaffer_610@yahoo.com

V. V. S. S. S. Balaram
e-mail: vadrevu_kinnera@yahoo.com

© Springer Nature Singapore Pte Ltd. 2018
V. Bhateja et al. (eds.), *Proceedings of the Second International Conference on Computational Intelligence and Informatics*, Advances in Intelligent Systems and Computing 712, https://doi.org/10.1007/978-981-10-8228-3_21

ovale of genus Plasmodium lead to the critical disease of malaria, which is transmitted from one to other by mosquito bites. As the global statistics reflect, it estimated that globally, for every 30 s one death is being resulted due to malaria [1].

Numerous studies are carried out to develop techniques and solutions for early diagnosis of malarial symptoms in humans, which can support in early treatments and reduce the ratio of deaths resulting from malaria. The clinical strategies for diagnosis of many diseases that affect the health based on computer-aided visual and analysis systems. Substantial research that has taken place in the field of biomedical and engineering reflects significant developments in the domain. Computer-aided visual systems adaptation for medical image selection from a clinical database, disease diagnosis can be very resourceful provided if the accuracy rates of results from image processing are high.

The clinical strategy for diagnosing malaria infection is to identify parasite scope in microscopic images of blood smears, carried out by the clinical experts. Among the key factors of malaria diagnosis and treatment, the challenge lies in the need for early and accurate detection. Manual processes of microscopic image detection are time consuming, prone to errors and require significant expertise from clinical experts. Effective solutions based on computer-aided and artificial intelligence models for performing clinical diagnosis under divergent and sensitive standards can lead to more improvements in the diagnosis.

To mitigate the challenges of the manual process, vivid range of analysis techniques proposed to enhance timely diagnosis of malaria parasite infection in humans. Varied features are adapted in different kinds of image analysis solutions for differentiating the infected and noninfected erythrocytes. For instance, solutions like the morphology, texture, and intensity are certain features used in the microscopic analysis of erythrocytes. Structural changes to the erythrocytes take place because of malaria parasite infection. In certain malarial species, morphological features do not affect the textural changes and such factors could lead to inaccurate results.

2   Related Work

Malaria parasite scope detection using image processing methods has been discussed in [2–4] many of the contemporary literature. In [3], automatic thresholding method is contributed which relies on morphological approach for segmenting the cells in the blood images. The proposed model in [3] considers disk-shaped flat structuring element for distinguishing overlapping cells.

In [4] the closed boundaries formation technique is adapted, in which the closed boundaries formed around blood smears. Upon application of such method, initially, the image contrast enhances resulting from Histogram Equalization [5] and is further broken to contours upon connectivity. In [6] the novel method is proposed, in which the relations observed amidst radiometric representation and other such related clinical observations for automating malaria parasite scope in erythrocytes.

K-means clustering discussed in [7], which used for distinguishing the erythrocytes and parasite of chosen blood smear image. Such an approach used with varied color models.

Profoundly, in the majority of the contemporary models, the detection accuracy of parasite scope directly impacted by the region detection accuracy and it delivers the least significance in terms of detecting parasite scope at the premature levels.

To ensure better levels of accuracy and diagnosis, some of the current researches are also focusing on anomaly-based models that trained on identifying infected erythrocytes. In addition, the knowledge that gained in the training phase shall use for identifying parasite scope over the thin blood smear microscopic images. Feature extraction is an effective method of quantitative measurement of images that typically used in identifying objects or regions, and analyzing pathology of a tissue or structure in the pathology slides. Upon computing the features, selection appropriate subgroup of significant and sound features is essential for improving the classification accuracy and minimizing the overall complexities.

For distinguishing the infected and noninfected erythrocytes, varied features used from the array of image and new variables computed for classifying the information into separate classes. Features set consist features that lead to a wider distance of class and within-class variance in the feature vector space.

Morphological features decide the resulting overall shape and size of erythrocytes even without considering the density into account. It is imperative from the earlier study that the size of an erythrocyte enlarged in the instance of infection in *P. vivax* and provable infection. In the case of falciparum, the size of erythrocyte does not change.

Feature of morphology comprises area, compactness ratio, perimeter, the minor axis of best-fit ellipse, eccentricity, Hu's moment, bending energy, area granulometry, and roundness ration [8].

Texture features implicitly reflect the spatial distribution of intensity over a specific region. GLCM, Gray level run-length matrix, entropy, histogram of color channel, local binary pattern, saturation histogram [2, 9, 10] are part of intensity and texture features. Researchers have chosen numerous feature sets for classifying infected and noninfected erythrocytes.

In [2] automated image processing method proposed, which uses features like relative size and eccentricity over the thin blood smears. In the solution proposed in [11] a semiautomatic method for detection, features like color histogram, grayscale histogram, Sobel histogram, Tamura texture histogram, saturation-level histogram are used.

In the other automated diagnosis solution proposed in [12], flat texture, run-length matrix, gradient features, intensity histogram, Hu set of the invariant moment, co-occurrence matrix, and Laplacian features chosen for classification of erythrocytes.

In [13], digital analysis of changes by Plasmodium vivax was investigated using features like perimeter, form factor, and area. In [14], content-based image retrieval approaches proposed in which features like histogram and Hu moment are chosen.

In [15], the textural approach quantitative characterization of Plasmodium vivax for infected erythrocytes was proposed. Automatic screening of malaria parasite depending on machine learning comprising 96 features proposed in [16]. In [17], the method of Malaria Parasite Detection in Giemsa–Stained Blood Cell Images that adapts features like, area granulometry, flat texture, gradient, color histogram had been studied [17].

In [18], researchers have discussed a sound diagnostic tool for automatic classification of diverse blood diseases. The solution relies on digital image processing methods. The solution on computer-based screening and visualization of Plasmodium falciparum candidate areas in blood smears digitalized image. The method obtains a diagnostic system that adapts feature set of local binary pattern-rotation, scale-invariant feature transform, invariant local contrast.

The solution proposed in this manuscript is an optimal morphological and texture features selection by statistical covariance assessment scale called z-score [19].

3   Optimal Features Selection by Z-Score

The devised Optimal Feature Selection by z-score includes multiple phases and those are preprocessing, feature abstraction, and feature set reduction by selecting optimum features of the given microscopic thin blood smear images. The preprocessing phase includes the conversion of given RGB images to grayscale, contrast adaption, noise abstraction, and determining the edges.

The feature abstraction phase includes the process of image segmenting and extracting texture and morphological features.

The feature set reduction by optimum feature selection done by the z-score of the values found in normal and infected erythrocytes in respect to each feature.

3.1   Methods and Materials

This section explores the methods used in image preprocessing, feature selection, and classification, which also includes the exploration of the features adapted.

3.1.1   RGB to Grayscale by PCA [20]

Firstly, the degree of angles amidst axes denoting Red, Green, and Blue obtained and cosine values of the respective degrees changed to grayscale values.

$$xr = \sum_{i=0}^{|P|} x(r_i) \qquad yg = \sum_{i=0}^{|P|} y(g_i) \qquad zb = \sum_{i=0}^{|P|} z(b_i)$$

$$mw_{\{x,y,z\}} = \frac{xr + yg + zb}{\sqrt{x^2 + y^2 + z^2}} \tag{1}$$

In the above equation (Eq. 1), the mean of the absolute weights $mw_{\{x,y,z\}}$ embraced to Red, Green, and Blue of each pixel of the given image. The notations xr, yg, zb are the aggregate weights I in the respective order of x, y, z given to corresponding r, g, b of the all pixels.

3.1.2 Gamma Equalization

Poor illumination is a phenomenal constraint for optimal segmenting in microscopic images. Thus, gamma equalization is adapted for improving the contrast of the chosen image [21, 22], which is carried out as follows:

The gamma equalization to the image $g(x, y)$ is performed using the following set of equation (Eq. 2):

$$f(x, y) = g_{\max}(x, y) * \frac{(g(x, y) - g_{\min}(x, y))^\gamma}{(g_{\max}(x, y) - g_{\min}(x, y))^\gamma}. \tag{2}$$

3.1.3 Median Filter

The key aim of the median filter [23] is to run towards replacing each entry with a median of neighboring entries, using the "Window" (sequence of neighbors). For a 1D signal, the most optimal window is the preceding and following entries, but for 2D signals or more dimensional ones, there is the scope for more complex window sequencing.

In the instance of a window comprising odd number of entries, defining the median shall be a simple process, as just the middle value of the entries in the window are sorted in a numeric manner. However, in the case of an even number of entries, the scope of more than one possible median.

3.1.4 Gaussian Filter

The Gaussian-smoothening operator [24] is a 2D convolution operator adapted for "blurring" images and eliminating any kind of noise and detail. In simple terms, it is defining like mean filter, but a different kernel used for denoting the shape of the Gaussian hump. To ensure point-spread function, the idea of Gaussian smoothening is used, and achieved by convolution. As the image is stored as an integration of discrete pixels.

Technically, the Gaussian distribution is a nonzero anywhere, and it requires convolution kernel that is infinitely large, however, in practical terms it is effectively zero more than any of the three standard deviations from mean and hence kernel can be truncated.

3.1.5 Canny Filter-Based Edge Detection

Canny edge detection [25] is a method for extracting resourceful information from vivid visionary objects and significantly reduces the quantum of data that requires processing. The phenomena are adapted vividly in numerous computer vision systems. It can effectively implement in edge detection solution for addressing the requirements in a wide range of scenarios.

3.1.6 K-Means Clustering by Optimal Centroids

The k-means algorithm is widely used clustering algorithm. The considerable constraint of the k-means is an optimal selection of initial centroids, which discussed and limited by a novel binary search technique in [26]. The k-means clustering by optimal centroid selection that proposed in [26] used here in this manuscript to cluster the given input image. The optimal centroid selection is done as follows:

$$ICR(A_i) = \frac{\max(A_i) - \min(A_i)}{K}. \tag{3}$$

Here in the (Eq. 3) $ICR(A_i)$ indicates the range of initial cluster center of attribute A_i.

Such that initial cluster center for each attribute shall be assessed and represented as

$$ICR = \{ICR(A_1), ICR(A_2), \ldots, ICR(A_n)\}$$

Further, the initial centroids for the k clusters can assess as follows:

step 1. *for* $k = 1, 2, \ldots, K$ //for each value from 1 to K
step 2. $C_k = \phi$ //an empty set representing centroid k
step 3. *for* $i = 1, 2, \ldots, n$ //for each attribute
step 4. $C_k \rightarrow \min(A_i) + (k-1) * ICR(A_i)$

      End //of step 3
      End //of step 1

3.1.7 Z-Score

The z-score that denotes the given two vectors are distinct or not assessed as follows:

- The notation a is the occurrence of feature f in positive set
- The notation b is the occurrence of feature f in negative set
- The notation c is the occurrence of all other features (other than feature f) in the positive set
- The notation d is the occurrence of all other features (other than feature f) in the negative set

Then the z-score Z of feature f can be assessed as follows:

$$Z(f) = \frac{a - (a+c) * \frac{(a+b)}{a+b+c+d}}{\sqrt{(a+c) * (a+b) * (1 - (a+b))}}. \tag{4}$$

In the above (Eq. 4), we can find the degree of probability (p-value) of the z-score z (f) from the z-table. If the degree of probability is less than the given degree of probability threshold then the feature f is optimal.

3.1.8 Classifiers

The contemporary classifiers widely used in research are adapted for classifying the given unlabeled microscopic images such that malaria infected and not infected, and they are SVM [27], naïve Bayes [28], and AdaBoost [29].

The classifier called Support Vector Machines (SVM) is proven to be near optimal to the classification process. The other contemporary classifier is naïve Bayes, which is easy to build and robust if given input is having features of independent classes. The AdaBoost is another benchmark classifier that boosts the performance of the decision trees, hence the miss-classification rate is very low that compared to SVM and naïve Bayes classifiers. The process completion time observed for SVM, naïve Bayes is approximately stable, and low that compared to Adaboost.

3.1.9 Features

In GLCM [30], the texture information explored in a significant manner. GLCM is an n × n matrix wherein n denotes varied gray shades that found in the chosen image. Further, this matrix is adapted for detailing the metrics defined below.

LBP (Local Binary Pattern) [31, 32] is the features of texture reflecting similarities of neighbor regions of the grayscale image. LBP computation is primarily reliant on two key factors such as circular neighborhood and bilinear interpolation.

The features proposed in [22] in addition to the features of invariant moments [33, 34] are termed as morphometric information features that are extensive for illustrating anomalous erythrocytes recognition.

3.2 Preprocessing

The conversion of RGB to grayscale can do any of the conversion techniques [35] found in contemporary literature. In the context of the model proposed here, the principal component analysis strategy that devised in [20] is used. Further, the contrast adaption has done using gamma equalizer [21, 22]. The next phase of the preprocessing is noise reduction, which done by hybridizing the Median [23] and Gaussian filters [24]. The objective of combining these two filters is to remove the salt and pepper noise and noise reflected by overlapping. The median filter is found to be robust to remove salt and pepper noise and Gaussian filter is more optimal to remove noise of overlapping. Further, Canny filters [25] are used to retain the continuity of the edges of the erythrocytes found in processed grayscale images.

Figure 1 depicts the structure of the given microscopic blood smear at different phases of the preprocessing.

3.3 Segmenting Erythrocytes by K-Means with Optimal Centroids

The preprocessed erythrocyte image is clustered into 2 segments such that normal, infected erythrocytes represented by respective clusters. In regard to this the k-mean with optimal centroids [26]. The expected clusters are 2, which is since the pixels to be clustered in given input image are fall in either normal or infected erythrocyte scope. Hence, the superfluous and traversed regions of the given input image eradicated and holes observed if any filled by morphological destruction technique [22]. The input and resultant images depicted in Fig. 2.

Fig. 1 Structure of the image at different phases of preprocessing

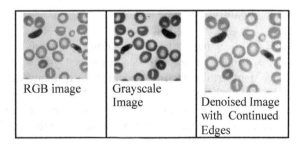

RGB image | Grayscale Image | Denoised Image with Continued Edges

Fig. 2 The input and resultant image of the k-means clustering with binary search

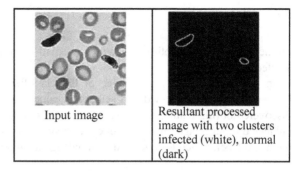

| Input image | Resultant processed image with two clusters infected (white), normal (dark) |

3.4 Features Extraction

The texture and morphological features of the normal and infected erythrocytes considered and extracted. The 19 features explored by gray level concurrence matrix and the all-possible local binary patterns [32, 36] considered under texture features category. The features explored in [22, 33, 34] are considered under morphological features category.

3.5 Feature Optimization

A feature said to be optimal if that feature coverage for both infected and normal erythrocytes are highly covariant. If feature challenges to this property such that coverage for both infected and normal erythrocytes is not diverse then that feature discarded. The feature optimization is done as follows:

The values observed for all considered features in corresponding normal and infected erythrocytes are represented in a matrix format of the respective order.

The corresponding matrix of the normal erythrocytes is represented as follows:

Each row is the values observed for all of the considered texture and morphological features corresponding to a normal erythrocyte.

Similarly, the matrix will be built for infected erythrocytes, such that each row represents the values observed for all the features considered in respective of infected erythrocyte.

Further, the significance of each feature is explored by estimating the z-score between the values observed for the corresponding feature in respective to normal and infected erythrocytes. If z-score is found to be significant at a given degree of probability threshold then the respective feature will be considered as optimal.

3.6 Classification

Further, the selected optimum features used to train the classifier on the given labeled training set.

In order to test the performance of the classifier, the given unlabeled input images will be preprocessed and select the values observed for optimum features discovered during feature selection process. Then these values obtained from each unlabeled input image given submitted to the classifier and then classifier allocates predicted label, which is done by the knowledge acquired during the training phase.

3.7 Validation

Further, these predicted labels will categorize as true positives (truly labeled as infected erythrocytes), false positives (falsely predicted as infected erythrocytes), True negatives (truly predicted as normal erythrocytes) and false positives (falsely predicted as normal erythrocytes).

4 Experimental Study and Results Analysis

In order to perform the experimental study, the labeled (malaria-prone, normal) microscopic images were collected under statistical guidelines [37].The microscopic images of the infected and normal erythrocytes considered from different clinical diagnostic centers under statistical guidelines for contributors to medical journals [37]. Peripheral blood smear slides were collected from a total 1600 (134 normal, 1322 *P. vivax* and 144 *P. falciparum*) patients. Among the collected 800 samples, only 1127 (925 *P. vivax*, 101 *P. falciparum* and 101 normal) samples were considered based on microscopically visible factors. All samples were attested and diagnosed by clinical experts (pathologists). The dataset statistics are depicted in Table 1.

The experiments conducted on i5 generation Intel processor with 4 GB ram and windows operating system. The implementation of the proposed model was carried out using FIJI [38], which is a medical image processing tool. The overall process includes preprocessing (as depicted in Sect. 3.2), segmenting that depicted in

Table 1 Dataset statistics

	Normal	Infected	Total
Training	213	576	789
Testing	91	247	338
Total	304	823	1127

Sect. 3.3, feature extraction (see Sect. 3.4), and feature optimization (see Sect. 3.5) implemented using relevant Java APIs provided in FIJI.

All samples were attested and diagnosed by clinical experts (pathologists).

4.1 Performance Analysis

The significance of the feature selection by OFS-Z as explored using benchmark classifiers like naïve Bayes, SVM, and AdaBoost. The 70% of the samples (213 normal and 576 infected) among the total 1127 microscopic images were used to train the classifiers. The rest 30% (91 normal and 247 infected) images were used to test the performance of the classifiers that concludes the significance of the feature selection strategy devised. The results obtained for the metrics called as accuracy, sensitivity, and specificity was evincing that the optimum features obtained from OFS-Z played a significant role to achieve the significant disease detection accuracy.

The values obtained for the above-stated metrics from the classification results of the selected classifiers explored in Table 2.

The results depicted in Table 2 evincing that the overall performance of classifiers, which is in terms of prediction accuracy is more than 84%, which is substantial that compared to the present average accuracy, which is approx. 81% that depicted in contemporary benchmark feature extraction models explored in Sect. 2. The other critical advantage of the proposed feature selection strategy is a minimal prediction as false positives, which is since the average specificity observed from experiments is 79% that depicts 4% more than the specificity observed in contemporary models reviewed. Concerning the results obtained from an experimental study, among the all three selected classifiers, AdaBoost is best toward classification performance. However, the process time observed for AdaBoost is more than the other two classifiers. The results obtained for the other metric specificity for

Table 2 Results obtained for validation metrics

	SVM	NB	AB
TP	225	210	244
FP	20	17	19
TN	71	74	72
FN	22	37	3
Precision	0.880832081	0.884222779	0.895013401
Accuracy	0.875739645	0.840236686	0.934911243
Sensitivity	0.871700305	0.808295055	0.947040757
Specificity	0.78021978	0.813186813	0.791208791
F-measure	0.914634146	0.886075949	0.956862745
Process time in milliseconds	8195 ± 27	8539 ± 32	9924 ± 40

SVM is lesser than that compared to other two classifiers, which indicates that false negative rate is high. On medical grounds, false negative rate (diseased records are claimed as normal) is certainly intolerable. The specificity is significantly high in the case of AdaBoost. Hence, it is obvious to claim that the AdaBoost performed best among the classifiers considered. The process time evinced for AdaBoost is considerably more among all three classifiers, which is negligible in the context of classification accuracy and specificity.

5 Conclusion

The optimal feature selection for machine learning classifiers is considerable research objective and it is a most critical factor for computer-aided disease pre-diction models. In this context, an optimal feature selection by z-score proposed here in this manuscript. The devised model aimed to detect the optimal texture and morphological features from microscopic images of the thin blood smears to predict the scope of malaria. The experiments were conducted on three classifiers using 1127-labeled microscopic images. The results obtained from the experiments indicated that the feature selection strategy is optimal since the classification accuracy of the all three classifiers considered is more than 84%. Among these classifiers, AdaBoost outperformed other two classifiers called SVM and naïve Bayes. The disease prediction accuracy and specificity are two critical metrics that were used to explore the performance of these classifiers. The process time observed for AdaBoost is more than the other two, which can be negligible as its classification performance is higher than the other two classifiers. The contribution motivating us to define hybrid feature selection models with heuristic and meta-heuristic approaches found in contemporary literature.

References

1. World Health Organization. What is Malaria? and What is Leukemia? http://apps.who.int/gb/ebwha/pdf_files/EB118/B118_5-en.pdf.
2. Ross, N. E., Pritchard, C. J., Rubin, D. M., and Dusé, A. G., Automated image processing method for the diagnosis and classification of malaria on thin blood smears. Med. Biol. Eng. Comput. 44(5): 427–436, 2006.
3. Ruberto, C. D., Dempster, A., Khan, S., and Jarra, B., Analysis of infected blood cell images using morphological operators. Image Vis. Comput. 20(2): 133–146, 2002.
4. Sio, S. W. S., Sun, W., Kumar, S., Bin, W. Z., Tan, S. S., Ong, S. H., Kikuchi, H., Oshima, Y., and Tan, K. S. W., Malaria Count: An image analysis-based program for the accurate determination of parasitemia. J. Microbiol. Methods 68(1): 11–18, 2007.
5. Zuiderveld, K., Contrast limited adaptive histogram equalization. In: Heckbert, P. (Ed.) Graphics gems IV, Academic Press, pp. 474– 485, 1994.

6. Le, M. T., Bretschneider, T. R., Kuss, C., and Preiser P. R., A novel semi-automatic image processing approach to determine plasmodium falciparum parasitemia in giemsa-stained thin blood smears. BMC Cell Biol. 9(15), 2008.

7. Nasir, A. S. A., Mashor, M. Y., and Mohamed, Z., Colour image segmentation approach for detection of malaria parasites using various colour models and k-means clustering. WSEAS Trans. Biol. Biomed. 1(10): 41–55, 2013.

8. C. Di Ruberto, A. Dempster, S. Khan, and B. Jarra, "Morphological Image Processing for Evaluating Malaria Disease", Springer, 2001.

9. F.B. Tek, A.G. Dempster, I. Kale, "Malaria Parasite Detection in Peripheral Blood Images", In: Proceedings of the British Machine Vision Conference, UK, pp. 347–356, 2006.

10. F. Boray Tek, "Parasite detection and identification for automated thin blood film malaria diagnosis", Computer vision and Image Understanding, pp. 21–31, 2007.

11. Diaz, G. & Gonzalez, A. & Romero, E., "A semi-automatic method for quantification and classification of erythrocytes infected with malaria parasites in microscopic images," Journal of Biomedical Informatics, pp. 296–307, July 2009.

12. V. Springl, "Automatic Malaria Diagnosis through Microscopic Imaging", Thesis, Faculty of Electrical Engineering. Prague, 2009.

13. M. Edison, J.B. Jeeva, M. Singh, " Digital analysis of changes by plasmodium vivax malaria in erythrocytes", Indian Journal of Experimental Biology, Vol. 49, pp. 11–15, 2011.

14. Khan, M. & Acharya, B., "Content Based Image Retrieval Approaches for Detection of Malarial Parasite in Blood Images," International Journal of Biometrics and Bioinformatics (IJBB)", Vol. 5, pp. 97–110, 2011.

15. M. Ghosh, D. Das, C. Chakraborty, and A. K. Ray, "Quantitative characterization of Plasmodium vivax in infected erythrocytes: a textural approach," Int. J. Artif. Intell. Soft Co., vol. 3, no. 3, pp. 203–221, 2013.

16. D. K. Das, M. Ghosh, M. Pal, A. K. Maiti, C. Chakraborty, "Machine learning approach for automated screening of malaria parasite using light microscopic images", Micron, vol. 45, pp. 97–106, 2013.

17. L. Malihi, K. Ansari-Asl, A. Behbahani, "Malaria parasite detection in giemsa-stained blood cell images", In Proceedings of 8th Iranian Conference on Machine Vision and Image Processing (MVIP), Zanjan, Iran, pp. 360–365, 2013.

18. N. Linder, R. Turkki, M. Walliander, A. Martensson, V. Diwan, E. Rahtu, M. Pietikainen, M. Lundin, J. Lundin, "A malaria diagnostic tool based on computer vision screening and visualization of plasmodium falciparum candidate areas in digitized blood smears", PLoS One 9, e104855.

19. http://www.sjsu.edu/faculty/gerstman/StatPrimer/z-table.PDF.

20. John, C., and J. Christian Russ. "Introduction to image processing and analysis." CRC Pr I Llc, ISBN (2008): 978-0.

21. Lai, Ching-Hao, et al. "A protozoan parasite extraction scheme for digital microscopic images." Computerized Medical Imaging and Graphics 34.2 (2010): 122–130.

22. Gonzalez, R. C. Digital image processing. Prentice Hall (2002).

23. Wei, Zhouping, et al. "A median-Gaussian filtering framework for Moiré pattern noise removal from X-ray microscopy image." Micron 43.2 (2012): 170–176.

24. Chokkalingam, Sp, K. Komathy, and M. Sowmya. "Performance Analysis of Various Lymphocytes Images De-Noising Filters over a Microscopic Blood Smear Image." International Journal of Pharma and Bio Sciences (2013): 1250–1258.

25. Gonzalez, Rafael C. Eddins, et al. Digital image processing using MATLAB. No. 04; TA1637, G6. 2004.

26. Hatamlou, Abdolreza. "In search of optimal centroids on data clustering using a binary search algorithm." Pattern Recognition Letters 33.13 (2012): 1756–1760.

27. Suykens, Johan AK, and Joos Vandewalle. "Least squares support vector machine classifiers." Neural processing letters 9.3 (1999): 293–300.

28. Murphy, Kevin P. "Naive Bayes classifiers." University of British Columbia (2006).

29. An, Tae-Ki, and Moon-Hyun Kim. "A new diverse AdaBoost classifier." Artificial Intelligence and Computational Intelligence (AICI), 2010 International Conference on. Vol. 1. IEEE, 2010.
30. Galloway, Mary M. "Texture analysis using gray level run lengths." Computer graphics and image processing 4.2 (1975): 172–179.
31. Ojala, Timo, Matti Pietikainen, and Topi Maenpaa. "Multi-resolution gray-scale and rotation invariant texture classification with local binary patterns." IEEE Transactions on pattern analysis and machine intelligence 24.7 (2002): 971–987.
32. Krishnan, M. Muthu Rama, et al. "Textural characterization of histopathological images for oral sub-mucous fibrosis detection." Tissue and Cell 43.5 (2011): 318–330.
33. Hu, Ming-Kuei. "Visual pattern recognition by moment invariants." IRE transactions on information theory 8.2 (1962): 179–187.
34. Das, Devkumar, et al. "Invariant moment based feature analysis for abnormal erythrocyte recognition." Systems in medicine and biology (ICSMB), 2010 international conference on. IEEE, 2010.
35. Orozco-Morales, Rubén. "Image Color Dimension Reduction. A comparative study of state-of-the-art methods." (2016).
36. Choras, Ryszard S. "Image feature extraction techniques and their applications for CBIR and biometrics systems." International journal of biology and biomedical engineering 1.1 (2007): 6–16.
37. Altman, Douglas G., et al. "Statistical guidelines for contributors to medical journals." British medical journal (Clinical research ed.) 286.6376 (1983): 1489.
38. Schindelin, J.; Arganda-Carreras, I. & Frise, E. et al. (2012), "Fiji: an open-source platform for biological-image analysis", Nature methods 9(7): 676–682, PMID 22743772 (on Google Scholar).

An Effective Hybrid Fuzzy Classifier Using Rough Set Theory for Outlier Detection in Uncertain Environment

R. Kavitha and E. Kannan

Abstract Data mining is the process that is used to extract the meaningful information from the large size dataset. The effective dataset utilization depends on the proper classification of outliers. The main objective of the outlier detection is to extract the abnormal data with inconsistency. The information collection from the different mechanisms is uncertain in nature. The data uncertainty causes the knowledge imperfections namely, vagueness and indiscernibility. The proposed research work implements an efficient fuzzy–rough set classifier for an outlier detection with less computational complexity. The fuzzy logic utilization and the fix of abnormal data easily determine the outliers in the large size database. The proposed classifier is compared with the existing classification methods with performance parameters like average running time, average execution time, execution time, false negative, false positive, true negative, true positive, precision, recall, and accuracy. The comparative analysis depicts the effectiveness of classifier in outlier detection.

Keywords Fuzzy logic · Rough set · Outlier detection · Classification

1 Introduction

Fuzzy logic [1] is expressed as the degree to which an element belongs to a set. The characteristics function of a fuzzy set is allowed to have values between 0 and 1, which denotes the degree of membership. Fuzzy logic [2] combined with rough set theory [3] is used to detect the outliers. As the medical database contains a large amount of data, the data contains noises that are categorized as attribute noise or class noise. Outliers [4] are considered as an errors or noise.

R. Kavitha (✉) · E. Kannan
Vel Tech Dr. RR & Dr. SR University, Chennai, India
e-mail: rkavitha1984@gmail.com

E. Kannan
e-mail: ek081966@gmail.com

© Springer Nature Singapore Pte Ltd. 2018
V. Bhateja et al. (eds.), *Proceedings of the Second International Conference on Computational Intelligence and Informatics*, Advances in Intelligent Systems and Computing 712, https://doi.org/10.1007/978-981-10-8228-3_22

Rough set theory [5] is the soft computing technique used to represent vagueness or incomplete knowledge. It has been applied in health science industry in the recent years. It is used for the heart disease diagnosis and classification. Using the rough set theory, discussions on information system, lower approximation, upper approximation, indiscernibility, rule generation, accuracy, certainty, rule coverage, and the strength are computed with the improved results. It is also the useful way of finding knowledge from the medical disease dataset to estimate the imprecise concept. The rough set theory is different from fuzzy since it provides the vagueness due to the lack of information on attributes present in the dataset. It also acts as a mathematical tool.

This paper explores the use of fuzzy logic [6] in the classification of data and suggests a method that can determine outliers in medical dataset and predict them by fixing the abnormal data. The outlier detection from the large size database is extensively used in various applications namely, reliability analysis, fault detection, and the medical diagnosis. Rough set theory [7] based on lower and upper approximation classifies the elements accurately belonging to the set and the elements possibly belonging to the set respectively. The hybrid fuzzy and rough set theory based data classification reduces the computational complexities and improves the outlier detection capability.

2   Related Work

In year 2016, Siva et al. proposed work on "A fuzzy logic based method for outlier detection". The work is carried out to find the anomalous and erroneous data. The mathematical statistical techniques are used for detecting and removing the outliers. A real steel industry data is considered for outlier detection using fuzzy inference systems. In year 2016, Jiao Shi et al. performed work on "Enhanced rough-fuzzy c-means algorithm with strict rough sets properties". The author proposed work by integrating fuzzy sets and rough sets; a hybrid algorithm is designed to find the patterns from large datasets. Also, fuzzy weighted factor is designed to find the patterns. An enhanced version of rough-fuzzy C-means clustering algorithm is used in computing a new centroid. Each cluster is calculated with its own lower and upper approximation. Objects are partitioned into corresponding regions for updating their centroids. The centroids point is calculated based on the weighing average calculated for lower and upper approximations of its own boundary region.

In year 2015, Jun Meng et al. proposed a work on "Gene Selection Integrated with Biological knowledge for plant stress response using neighborhood system and Rough set Theory". In this work, the preprocessing algorithm is used to calculate the threshold. The gene selection is carried on by selecting the top-ranked genes, building a neighborhood system, using reduction algorithm based on rough set theory, applying the attribute reduction and finally using classifier to achieve the accuracy. The attribute reduction algorithm is used to select significant genes, based on the neighborhood rough set model.

In year 2014, Ahmad Taher Azar et al. performed a work on "Inductive learning based on rough set theory for medical decision making". This work is performed with two approaches. They are based on past historical records including previous symptoms and treatments and rule-based expert system. The work is carried out in finding the upper, lower, and boundary approximation from the dataset. The rough set exploration system, an open source rough set tool, is used for the experimental results. In year 2013, Aquil Burney et al. performed a work on "Application of rough set in health sciences and disease diagnosis". The proposed work discusses on rough set information systems, approximation sets, indiscernibility, rule generation, reducts, accuracy, and discernibility matrix. It mainly focuses on reduction of the attributes after finding the discernibility matrix. The results are calculated using the Rosetta rough set tool.

3 Proposed Work

The hybrid fuzzy and rough set theory based data classification [8] reduces the computational complexities and improves the outlier detection capability. The lower approximation consists of the set that is certainly contained in the set from the universe set. Universe is called the dataset. The upper approximation contains the data that is possibly contained in the observed features. The boundary region consists of those objects that cannot be classified from the dataset. The relevant attribute is identified by evaluating the attribute set called as reducts. It is the technique used to remove the redundant attributes which provides the effective way for disease diagnosis. This technique can be used for classifying any type of medical data as low risk, moderate risk, and high risk.

3.1 Training Pseudocode

The training dataset is given as the input. The fuzzy classification is used as the preprocessing step to classify the data into lower, medium, and higher categories. After this step, the probability is computed for the individual attributes, which is carried out to find the probability for all the attributes. Then, apply the rough set classification to classify the data into lower approximation, upper approximation, and boundary approximation. After finding the records in the boundary region, the sensitive or critical attributes are extracted. With the results of the sensitive attributes, the rules are extracted which are applied to the testing data for the prediction of the medical disease dataset. For the experimental purpose, a sample dataset is considered from UCI Machine Learning repository. It consists of data which are used by the machine learning researchers for their empirical analysis. The heart disease dataset with 14 test attributes is considered.

Input: D_{tr}—Training dataset
Output: R—Rules

Step 1: Extract training dataset and load into database
Step 2: Separate the numerical values as records (x) and attributes (a) in the dataset
Step 3: Select the threshold value (th_{hr}) for different ages as per following table:

Age a	Heart rate threshold limit th_{hr}
$a \le 20$	200
$20 < a \le 25$	195
$25 < a \le 30$	190
$30 < a \le 35$	185
$35 < a \le 40$	180
$40 < a \le 45$	175
$45 < a \le 50$	170
$50 < a \le 55$	165
$55 < a \le 60$	160
$60 < a \le 65$	155

Step 4: Split the stages based on the membership functions for heart rate, blood pressure, and cholesterol

$$st_{hr} = \begin{cases} 1 & if(hr < th_{hr}) \\ 2 & if(hr > th_{hr}) \end{cases}$$

$$st_{bp} = \begin{cases} 1 & if(bp < 130) \\ 2 & if(100 < bp \le 155) \\ 3 & if(130 \le bp < 220) \\ 4 & if(145 \le bp < 200) \end{cases}$$

$$st_{ch} = \begin{cases} 1 & if\ (ch < 160) \\ 2 & if(160 < ch \le 280) \\ 3 & if(ch \le 280) \end{cases}$$

Step 5: Compute the individual probability $(P(x_i))$

$$H(a_i) = maxValue[D_{tr}(a_i)]$$

$$V(a_i) = currentValue[D_{tr}(a_i)]$$

$$P_{ind}(a_i) = H(a_i)/V(a_i)$$

$$P(x_i) = \{P_{ind}(a_1), P_{ind}(a_2) \ldots \ldots P_{ind}(a_n)\}$$

Step 6: Compute the overall probability $(P_{ov}(x_i))$

$$P_{ov}(x_i) = 1/n \sum_{i=1}^{n} P(x_i)$$

Step 7: Getting lower limit and upper limit of overall probability $P_{ov}(x_i)$ for each status

$L_1 \rightarrow$ *Lower Limit for status* 1
$U_1 \rightarrow$ *Upper Limit for status* 1
$L_2 \rightarrow$ *Lower Limit for status* 2
$U_2 \rightarrow$ *Upper Limit for status* 2

Step 8: Rough set classification:
Lower approximate region $R_l(x) \rightarrow L_1 \leq R(x) < L_2$
Upper approximate region $R_u(x) \rightarrow U_1 < R(x) \leq U_2$
Boundary region $R_b(x) \rightarrow R_l(x) \cap R_u(x) \rightarrow L_2 \leq R(x) \leq U_1$

Step 9: Getting sensitive attributes from R_l and R_u records

$aL_l \rightarrow$ *frequent Count* $[R_l(a)]$
$aL_u \rightarrow$ *frequent Count* $[R_u(a)]$
$S_{attri} \rightarrow aL_l \cap aL_u$

Step 10: Frame rules based on sensitive attributes S_{attri}

$S_{attri}(a) = S_{t1}, S_{t2} \ldots, S_{tN}$ and value of each attribute is $\{1, 2, 3\}$
Assign set of rules are $r = \{r_1, r_2 \ldots, r_Z\}$

Step 11: Calculate index for the result of fuzzy formulation

Set the value -1 for the attribute status $a_{status} = 2$ *for each* r_i
$a_s \rightarrow match(sum\ of\ a_{status} from\ R_b(x), r_i)$
$if(a_s > 0)a_s \rightarrow 1$
$if(a_s = 0)a_s \rightarrow 2$
$else\ a_s \rightarrow 3$
$Ru_i \rightarrow r_i, a_s$

Step 12: Output the fuzzy rule set

$$R = \{Ru_1, Ru_2, Ru_3 \ldots \ldots Ru_i\}$$

3.2 Testing Pseudocode

In the testing algorithm, test dataset is collected and loaded in the database. The fuzzy classification is applied to categorize the data into low, medium, and higher.

Then, the individual probabilities of each attribute are computed to find the overall probability of each attribute. Based on the overall probability, the records are classified. It is then checked with the threshold. If the value is less than the threshold, the values are declared as the low-risk records. If not they are recorded as high-risk records. These records are in turn verified with the boundary regions. It is been checked with the rules generated. And finally, the rules are categorized as low, moderate, and high-risk records.

Input: D_{ts}—Testing dataset
Output: $D_{ts}(X_i)$

Step 1: Extract testing dataset and load into database

Step 2: Separate the dataset with the records and attributes as in training algorithm

Step 3: Select the threshold value (th_{hr}) for different ages as per table

Step 4: Split the stages based on the value of heart rate, blood pressure, and cholesterol

$$st_{hr} = \begin{cases} 1 & if \ (hr < th_{hr}) \\ 2 & if \ (hr > th_{hr}) \end{cases}$$

$$st_{bp} = \begin{cases} 1 & if \ (bp < 130) \\ 2 & if \ (100 < bp \leq 155) \\ 3 & if \ (130 \leq bp < 220) \\ 4 & if \ (145 \leq bp < 200) \end{cases}$$

$$st_{ch} = \begin{cases} 1 & if \ (ch < 160) \\ 2 & if \ (160 < ch \leq 280) \\ 3 & if \ (ch \leq 280) \end{cases}$$

Step 5: Compute the individual probability $(P(x_i))$

$$H(a_i) = maxValue[D_{tr}(a_i)]$$
$$V(a_i) = currentValue[D_{tr}(a_i)]$$
$$P_{ind}(a_i) = H(a_i)/V(a_i)$$
$$P(x_i) = \{P_{ind}(a_1), P_{ind}(a_2) \ldots \ldots P_{ind}(a_n)\}$$

Step 6: Compute the overall probability $(P_{ov}(x_i))$

$$P_{ov}(x_i) = 1/n \sum_{i=1}^{n} P(x_i)$$

Step 7: Compute the status value using the lower, upper limit, and overall probability

$$if \ L_1 \leq P_{ov}(x_i) < L_2 - set \ sT = 1$$
$$if \ U_1 \leq P_{ov}(x_i) < U_2 - set \ sT = 3$$
$$if \ L_2 \leq P_{ov}(x_i) \leq U_1 - follow \ rule \ matcher$$

Step 8: Update the records in the dataset

$$D_{ts}(X_i) = \{D_{ts}(x_i), sT$$

3.3 Rule Matcher

The rule matcher algorithm is used to match the rule with the original dataset.

Step 1: Check whether the rules matched with the records of original dataset

$$M(x_i) \rightarrow matched\,R\,with\,D_{ts}(x_i)$$

Step 2: Assign the status values

$$s_i = valueOf[M(x_i), a_{status}]$$
$$if\,s_i > 0\,sT = 1$$
$$if\,s_i = 0\,sT = 2$$
$$if\,s_i < 0\,sT = 3$$

4 Experimental Results

4.1 Accuracy

The accuracy of the classifier is based on the number of instance that is correctly classified. The accuracy of the method is shown in Fig. 1.

$$Accuracy = no.of\,predicted\,outliers/Total\,Outliers$$

4.2 Area Under Curve

The AUC metric measures the performance of a binary classifier. It is the probability that randomly selected positive patterns will be ranked higher than a randomly selected negative pattern. Figure 2 shows the comparison with the other classification methods.

Fig. 1 Accuracy

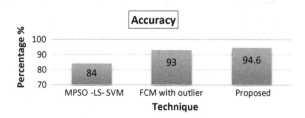

Fig. 2 Average area under curve

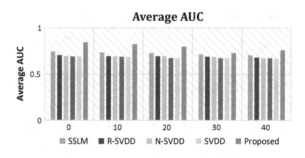

Table 1 Average running time

Average running time in sec							
Method	ε NR-SVDD	SSLM	R-SVDD	N-SVDD	DW-SVDD	SVDD	FRSC
Time in Sec	4.8	3.9	3.9	2.6	2.4	2.4	2.2

4.3 Running Time

Table 1 shows the average running time of the classifier compared to existing classification methods.

$$\text{Average Running Time} = 1/\text{no. of process} \times \sum(\text{time taken to finish} - \text{starting time of the process})$$

5 Conclusion

The medical database contains large amount of data related to the patient's medical conditions. These data contain the outlier called as data noise or attribute noise. These outliers are identified and are removed using the fuzzy and rough set theory. They are the mathematical tool to deal with decision system in providing the uncertain and vague decisions. The paper presents the efficient fuzzy–rough set classifier for an outlier classification with less computational complexity [1, 9]. The proposed classifier is compared with the existing classification methods regarding the performance parameters like average running time, average execution time, execution time, false negative, false positive, true negative, true positive, precision, recall, and accuracy.

References

1. Ali. Adeli, Mehdi. Neshat: A Fuzzy Expert System for Heart Disease Diagnosis. Proceedings of the International MultiConference of Engineers and Computer Scientists, vol I, pp. 134–139, IMECS 2010, March 17–19, Hong Kong (2010)
2. Sanjeev Kumar, Gursimranjeet Kaur: Detection of Heart Diseases using Fuzzy Logic. International Journal of Engineering Trends and Technology (IJETT), vol. 4, Issue 6, pp. 2694–2699 (2013)
3. Sudhakar, K., Manimekalai, M.: Study of Heart Disease Prediction using Data Mining. International Journal of Advanced Research in Computer Science and Software Engineering, vol. 4, Issue 1, pp. 1157–1160 (2014)
4. Vembandasamy, K., Karthikeyan. T.: Novel Outlier Detection In Diabetics Classification Using Data Mining Techniques. International Journal of Applied Engineering Research ISSN 0973–4562, vol. 11, Number 2, pp. 1400–1403 (2016)
5. Ahmad Taher Azar, Nidhal Bouaynaya, Robi Polikar.: Inductive Learning based on Rough Set Theory for Medical Decision Making. Fuzzy Systems, FUZZY-IEEE (2015)
6. Guijun Chena, Xueying Zhanga, Zizhong John Wanga, FenglianLia: Robust support vector data description for outlier detection with noise or uncertain data. Knowledge Based Systems 90, pp. 129–137 (2015)
7. Jun Meng, Jing Zhang., Yushi Luan.: Gene Selection Integrated With Biological Knowledge For Plant Stress Response Using Neighborhood System And Rough Set Theory. IEEE Transactions On Computational Biology And Bioinformatics, vol. 12, No. 2 (2015)
8. Deepali Chandna: Diagnosis of Heart Disease Using Data Mining Algorithm. International Journal of Computer Science and Information Technologies. vol. 5 (2), pp. 1678–1680 (2014)
9. Nidhi Bhatla Kiran Jyoti: A Novel Approach for Heart Disease Diagnosis using Data Mining and Fuzzy Logic. International Journal of Computer Applications (0975 – 8887), vol. 54, No.17, pp. 16–27 (2012)

CSES: Cuckoo Search Based Exploratory Scale to Defend Input-Type Validation Vulnerabilities of HTTP Requests

S. Venkatramulu and C. V. Guru Rao

Abstract Web application servers are prone to attacks that are more vulnerable and thousands of security breaches that are taking place everyday. Predominantly, the hackers to breach the web application systems security use the method of SQL injections and XSS models. IDS systems play a pivotal role in identifying the intrusions and alerting about the attacks. Despite that, there are numerous models of IDS systems in place; one of the commonly approached systems is the syntax analyzers. However, the limitations in terms of programming language dependency and the related issues drop the performance levels of syntax analyzer based strategies. To ensure the right kind of http request vulnerabilities, detection methods are in place; the Cuckoo Search based Exploratory Scale (CSES) to defend input-type validation vulnerabilities of HTTP requests is proposed here in this paper. The key objective of CSES is to magnify the speed and accuracy of input-type validation of web applications. The programming language dependency and server level process overhead issues do not impact the performance of CSES. In addition, the other key benefit of CSES model is optimal speed in search related to vulnerability scope detection. The experimental studies that are carried out on a dataset that contains the records prone to cross-site scripting, SQL injection alongside the normal records, depict better performance of the model, when compared to the other benchmarking model of DEKANT. CSES model has delivered improved accuracy levels in identifying the attacks.

Keywords XSS attack · Web application · SQL injection attack
CVE · CSES

S. Venkatramulu (✉)
Department of CSE, K.I.T.S, Warangal, Telangana, India
e-mail: venkatramulu10@gmail.com

C. V. Guru Rao
SR Engineering College, Department of CSE, Warangal, Telangana, India
e-mail: guru_cv_rao@hotmail.com

© Springer Nature Singapore Pte Ltd. 2018
V. Bhateja et al. (eds.), *Proceedings of the Second International Conference on Computational Intelligence and Informatics*, Advances in Intelligent Systems and Computing 712, https://doi.org/10.1007/978-981-10-8228-3_23

1   Introduction

Internet-based solutions have become an integral part of our daily activities. For instance, right from professional networking to personal communication, banking, transaction management, and ecommerce, web-based application solutions have become an integral part of our business process. As significant quantum of personal and critical data of users is stored in such applications, data security has become integral element in handling the application systems. In addition, the security challenges that are creeping up are fast emerging, and everyday new kinds of security challenges are emerging in terms of impact the security of the web-based application systems.

It is imperative from the report of OWASP that SQLI (SQL injections), XSS (cross-site scripting), FI (File Inclusion), and RCET (Remote Code Execution) are among the key techniques that are used by hackers to exploit the security vulnerabilities of the web application system. Such vulnerability exploitation leads to data breach, comprising security of clients and applications. Many contemporary solutions are emerging as a result of extensive research carried out toward improving the web vulnerability detection, like the static taint analysis, dynamic taint analysis, concolic testing, and model checking solutions.

Though static taint analysis approaches are scalable, one of the significant challenges envisaged in the process is about the false positive rates [1] that are generated. Dynamic taint analysis [1], symbolic [2], model checking [3], and concolic testing methods are being more accurate in detecting real attack values but envisage scalability issues for large systems resulting from path explosion problem [4].

In [5], Shin et al. have proposed scalable vulnerability prediction but the granularity of current predictions is profoundly based on coarse-grained solution to identify the vulnerabilities at varied level of software components. There is integral need for more effective, scalable, accurate, and robust solutions that can benefit web developers.

It is evident from the review of literature that input validation and input sanitization shall be two key coding techniques that can be resourceful in securing the programs from varied kind of vulnerabilities. Input validation certainly inspects the inputs for defined properties like the length of data, range sign, and type. Input sanitization ensures the process of cleansing, by accepting only the predefined characters and terminating any other kind of requests comprising special meaning. Innately an application could be vulnerable if the developers do not adapt contemporary techniques effectively.

Henceforth, a new vulnerable detection strategy "Cuckoo Search based Exploratory Scale (CSES) is proposed in this paper, for input-type validation vulnerabilities detection and also prevention of HTTP requests". CSES model comprises two phases as (i) training and (ii) testing.

The approach is that the required dataset generated using open source tools called XSS proxy [6] and SQL power injector [7] for evaluating the proposed model. Significance of the model estimates the statistical metrics [8] called precision, specificity, accuracy, and sensitivity.

Statistical metrics are used for assessing the prediction accuracy of CSES and DEKANT. As per the metrics, the accuracies of CSES and DEKANT are about 95% and 80%, respectively. The performance of the model is high as the results of specificity and sensitivity observed for CSES are higher than that of DEKANT.

Review of related work and the contemporary solutions from the literature is discussed in Sect. 2; in Sect. 3, the proposed solution is discussed, followed by Sect. 4 in which the experimental process is carried out; and the study results are demonstrated and discussed in Sect. 5. Section 6 depicts the results obtained from experiments and the analysis, and the research work concluded in Sect. 6.

2 Related Work

Usually, the static tools for analysis look for vulnerabilities in the applications typically at source code level [9–12]. Majority of such tools conduct taint analysis like tracking the user inputs for verifying if they are targeting any sensitive link. For identifying such links in PHP coding, Pixy [9] has been one of the first of its kind tools proposed. PhpSafe [10] is the another contemporary tool adapted in taint analysis for searching vulnerabilities in CMS plugins (for instance in the case of WordPress Plugins) without analyzing at CMS source code levels. Hence, it can be configured with certain functions of CMS, which is considered as entry points and sensitive sinks.

In the case of static analysis tools, because of complexity related to coding knowledge about vulnerabilities, the challenge of more number of false positives and false negatives getting triggered takes place. As discussed in [13, 14] even in WAP, the taint analysis is carried out but the data mining process supports in predicting false positives apart from correcting any kind of vulnerabilities that are detected. In [14], authors have presented an approach for more precise static analysis that depends on contemporary data structure for representing source code. An alternative approach has also been proposed wherein, unlike the words do not comprise coding knowledge about vulnerabilities and emphasis is on extracting knowledge from annotated code samples.

Machine learning is used in some recent works for measuring the quality of software, by collecting attributes that disclose any kind of defects in the software [15]. However, many works use machine learning models for predicting vulnerabilities in the source code, which is different from identifying their existence as depicted in [16–18]. Certain attributes like the vulnerabilities and function calls [16], code complexity, and developer activities [17] are used for code-metric analysis with metadata that is generated from code repositories [18].

In [19, 20], machine learning methods are used for predicting the vulnerabilities in PHP programs. The tools extract set of attributes, which end in sensitive sink but do not start at an entry point level. Set of annotated slices are used for training the tool before using them for identifying vulnerabilities that are present in the source code. WEKA tools are used for analysis, as the detection tools do not perform directly any kind of data mining process [21]. The authors have enhanced the detection ability, which is based on traces of detection and program executions [22].

WAP relies more on machine learning and data mining for predicting whether the vulnerability predicted by taint analysis is a false positive or a real vulnerability [13, 14]. Unlike the solution proposed in [19], the WAP solutions identify the location of vulnerabilities in the source code as required for eliminating them. Php Miner tools and WAP use certain standard classifiers like Naïve Bayes, multilayer perceptron, and sequence classifiers.

Some of the static analysis tools even adapt the machine learning techniques for contexts apart from web applications too. In [23], it is used for discovering vulnerabilities by identifying missing checks in source code of C programming. The tool performs taint analysis for identifying checks between entry points and sensitive sinks, and applies the text mining for discovering neighbors of such checks. Accordingly, a model is built for identifying the missing checks. In [24], the method of text mining is used for predicting software components that are vulnerable in Android applications [24]. Text mining techniques are used for getting the terms present in files (software components) and its related frequencies. In addition, a static code analyzer is used for checking the software components to identify them as vulnerable or non-vulnerable. Accordingly, the term frequencies are coordinated for term frequencies of vulnerable software components and building a model for predicting the vulnerability if any in a software component.

Another contemporary model is called DEKANT [25], which is a static analysis tool that detects HTTP request vulnerabilities. The DEKANT is conceptually similar to the model proposed in this manuscript. The DEKANT characterizes the vulnerabilities based on the knowledge gained from the sequence of annotated source code slices. The efficacy of this tool is proportionate to the degree of modularity of the source code, which is often a considerable constraint.

In this paper, the model comprising "Cuckoo Search based Explorative Scale (CSES)" is proposed to defend web vulnerabilities in HTTP requests. The proposed model is at the network level and the process load at web application servers superseded. In the proposed model, syntax analyzers are not used and hence the issues of imprecise code analysis and programming language dependency do not occur in CSES. The further section discusses the proposed model of CSES in detail.

3   Cuckoo Search Based Exploratory Scale (CSES) to Defend Input-Type Validation Vulnerabilities

Proposed CSES model is an evolutionary computation based on network-level heuristic scale for vulnerability detection in http requests. CSES discovers the impact of varied parameter types and the values at the network levels. The proposed solution is program independent, as it does not use any kind of code analyzers. CSES can connect dynamically and could use for monitoring vulnerabilities in http requests that are targeted for any web application. Analysis of http requests to possible vulnerabilities is carried out using cuckoo search based learning of varied request parameter values and the types of such parameters.

The key objective of CSES is to deploy an evolutionary computing-based exploratory scale that scales the vulnerability scope of association of input types and values given for request parameters. At preprocessing level, the given labeled http requests will be partitioned into two sets *vrs*, *frs* representing vulnerable prone and normal records, respectively. At learning phase, nests will build from the record sets *vrs*, *frs* as two different hierarchies *vnh*, *fnh*, respectively. The formation of these nests hierarchies *vnh*, *fnh* is intended to perform cuckoo search, such that the divergent patterns of input types as nests and the pattern of values observed for corresponding input types in record sets *vrs*, *frs* will be placed as eggs in respective nests of hierarchies *vnh*, *fnh*. Further, these nests searched to track the vulnerability scope toward the attacks such as SQL injection and XSS attack. The overall process of CSES is dual-fold strategy of training the model through nest hierarchy formation for vulnerable prone and normal record sets *vrs*, *frs* and assessing the vulnerability scope of the unlabeled http request given as input for testing. The process flow of CSES is discussed in detail from Sect. 3.1 onward.

3.1 Cuckoo Search

Cuckoo Search is a search technique stimulated by the holoparasite act of some cuckoo birds. The species of type cuckoo is unable to complete its reproduction cycle without proper host (nest of the other types of birds that contains eggs which resemble to cuckoo bird egg). The cuckoo bird places its egg(s) in the nest of host. The strategy of search followed by a cuckoo bird is adapted in numerous fields [26, 27]. Cuckoo search executes under three traditional rules [28], and they are cuckoo selects a host nest randomly to place the egg, the nest contains most compatible eggs that are compared to cuckoo egg which enables the reproduction of the cuckoo, and the finite number nests (usually the 15) [29] are adapted for cuckoo search.

The probability factor to notify the cuckoo egg as artifact by a host bird is $\{P(a) \exists a \in (0, 1)\}$. In order to optimize the initial nest of the search, usual search follows the techniques such as Levy flights and random walks [29]. In regard to the proposed model, the features are deterministic and ordered by size in descending order; hence, the levy flights technique is adapted for perch search in the proposal.

3.2 Feature Selection

Let *itV*, *itF* be the sets representing all input types observed in vulnerable and normal record sets *vrs*, *frs*, respectively. The criteria that opted to select optimal features (input type) are as follows.

"If the values observed for an input type in records represented by set *vrs* are significantly distinct from the values observed for corresponding input type in records represented by set *frs*, then that input type will be considered as optimal feature." This is since the input type with similar values for vulnerable and normal records is insignificant to diversify the vulnerability scope from normal. In order to find the optimal features (input types) as set f_{opt}, we adapt hamming distance to estimate the distance between values observed for an input type in respective vulnerable and normal sets *vrs*, *frs*.

3.2.1 Assessing Hamming Distance is as Follows

Let $V = \{v_1, v_2, \ldots, v_m\}$, $W = \{w_1, w_2, \ldots, w_n\}$ be the two vectors representing values observed for an input type t in sets *vrs*, *frs*, respectively. Then, the Hamming distance between these two vectors can be measured as depicted in (Eq. 1):

$$hd_{V \leftrightarrow W} \leftarrow \frac{\sum_{i=1}^{\max(m,n)} \begin{cases} 0 & if(v_i - w_i \equiv 0) \\ 1 \end{cases}}{\max(m,n)} \tag{1}$$

// $hd_{V \leftrightarrow W}$ is the hamming distance between V and W.

3.3 Learning Process (the Nest Formation)

In order to prepare the nest hierarchy, possible unique subsets $N = \{\{n_1 \subseteq f_{opt}\}, \{n_2 \subseteq f_{opt}\}, \ldots, \{n_{|N|} \subseteq f_{opt}\}\}$, according to set theory [30], which is of $2^{|f_{opt}|-1}$ number of subsets from optimal features (input types) f_{opt} of size $|f_{opt}|$ will be defined. The set with all optimal features also been considered as subset; hence, the subsets exist with $|f_{opt}|$ count as maximum length and minimum length as 1.

Further, nest hierarchies *vh* and *fh* will be formed that are representing respective vulnerable *vrs* and fair *frs* record sets. Each hierarchy will be formed such that a subset maximum length is in root level of the hierarchy, subsets with length $|f_{opt}| - 1$ will be in first level (next to root level) of the hierarchy, subsets with length $|f_{opt}| - i$ will be in ith level of the hierarchy, and finally the subsets with length 1 will be in last level of the hierarchy.

The values observed for each subset $\{n_i \exists n_i \subseteq f_{opt}\}$ of input types in vulnerable records set *vrs* will be considered as eggs and placed in a nest, which is represented by respective subset $\{n_i \exists n_i \subseteq f_{opt}\}$ in hierarchy *vh*. Similarly, the values observed for each subset $\{n_i \exists n_i \subseteq f_{opt}\}$ of input types in normal record set *frs* will be considered as eggs and placed in a nest, which is represented by respective subset $\{n_i \exists n_i \subseteq f_{opt}\}$ in hierarchy *fh*.

3.4 Testing Phase (Process of Cuckoo Search)

In order to assess, the given http request is vulnerable prone or, initially not all the input types and their respective values extracted as sets sf, sv in respective order from the http request. Then, obtain the optimal input types as set sf_{opt}, such that the input types exist in both sets sf and f_{opt}. Further, define a set tN that contains $2^{|sf_{opt}|-1}$ unique subsets from sf_{opt}, such that $tN \subset N$. The values observed in given http request for each subset $\{tn \exists tn \in tN\}$ will be considered as a map $E = \{tn_1 \Rightarrow e(tn_1), tn_2 \Rightarrow e(tn_2), \ldots, tn_{|tN|} \Rightarrow e(tn_{|tN|}), \}$ that maps each subset $\{tn_i \exists tn_i \in tN\}$ and the respective value set $e(tn_i)$. The corresponding value sets of the each subset tn that exists in set tN are considered further as cuckoo eggs to be placed in appropriate nests in both hierarchies vh, fh.

3.4.1 Cuckoo Search on Both Hierarchies

Arrange all subsets found in set tN in descending order of their size.

Let $s_{vh} = 0$ be the similarity score observed between the eggs placed in the nests of hierarchy vh and eggs to be placed.

Let $s_{fh} = 0$ be the similarity score observed between the eggs placed in the nests of hierarchy fh and eggs to be placed.

Step 1: $\overset{|tN|}{\underset{p=1}{\forall}} \{tn_p \exists tn_p \in tN\}$ Begin

Step 2: $e(tn) = E\{tn_p\}$ //extracting the values represented by tn_p in corresponding map E

Step 3: $\overset{|N|}{\underset{q=1}{\forall}} \{n_q \exists n_q \in N \wedge n_q \equiv tn_p\}$ Begin //for each subset n_q as a nest that find in both hierarchies, which is identical to the subset tn_p

Step 4: $ls_{\max} = 0$ //max local similarity that set initially to 0.

Step 5: $\overset{|n_q|}{\underset{i=1}{\forall}} \{e_i \exists e_i \in n_q \wedge n_q \in N \wedge n_q \in fh\}$ Begin //for each egg e_i found in nest n_q of hierarchy vh

Step 6: $if \left(\frac{e_i \cap e(tn)}{e(tn)} > ls_{\max} \right)$ Begin

Step 7: $ls_{\max} = \frac{e_i \cap e(tn)}{e(tn)}$

Step 8: End //of condition

Step 9: End //of loop

Step 10: $s_{vh} + = ls_{\max}$

Step 11: //Similar process continued further in regard to hierarchy fh.

Step 12: $ls_{\max} = 0$ //max local similarity that set initially to 0.

Step 13: $\overset{|n_q|}{\underset{i=1}{\forall}} \{e_i \exists e_i \in n_q \wedge n_q \in N \wedge n_q \in fh\}$ Begin //for each egg e_i found in

nest n_q of hierarchy fh

Step 14: $if \left(\frac{e_i \cap e(tn)}{e(tn)} > ls_{\max} \right)$ Begin

Step 15: $ls_{\max} = \frac{e_i \cap e(tn)}{e(tn)}$

Step 16: End //of condition

Step 17: End //of loop

Step 18: $s_{fh} += ls_{\max}$

Step 19: End

Step 20: End //of loop

3.5 Assessing the State of the Http Request

The minimum similarity score for both hierarchies will be 0, and the maximum similarity score for both hierarchies will be $|tN|$, which is since the max similarity for each nest is 1 and the eggs to be placed are at most compatible to all nests.

Henceforth, the difference between the similarity scores observed for both hierarchies taken as metric to estimate the given request is vulnerable prone or not, which is as follows:

$s_{fh} - s_{vh} \geq \frac{|tN|}{2}$	The given http request is surely not prone to vulnerable
$s_{fh} - s_{vh} \geq \frac{|tN|}{3}$	The given http request is possibly not prone to vulnerable
$s_{vh} - s_{fh} > 0$	The given http request is surely prone to vulnerable
$s_{fh} - s_{vh}$ is $\left(> 0 \&\& < \frac{|tN|}{3} \right)$	The given http request is possibly prone to vulnerable

4 Experimental Study

In this section, the emphasis is on experimentation that performed on the proposed model. Experimental study conducted on synthetic dataset that includes http requests intended to cross-site scripting and SQL injection attacks alongside normal requests.

4.1 The Dataset Preparation

The open source tools XSSPROXY [6] and SQL POWER INJECTOR [7] are used to synthesize cross-scripting and SQL injection attacks, respectively. The number of records accumulated for experiments is 7539. Among these, 5265 records are labeled as attack prone and rest 2274 records are labeled as normal.

4.2 Implementation Process

In the initial phase, the preprocessing of the dataset that loaded into the application environment was performed. Further, the segmentation of the data as attack prone and normal records as per the label is carried out. The cross-validation and mis-classification rate of the proposed model were carried out. In order to this, the labeled records partitioned into 70 and 30%. The 70% of the records were used to train the model (nest formation) and rest 30% were unlabeled and used to test the proposed model. The records considered for training phase were preprocessed initially to obtain optimal features (input types) as depicted in Sect. 3.2. Further, these optimal features are used to form the nest hierarchy as depicted in Sect. 3.3. The leftover 30% records were unlabeled and used to perform the testing process as depicted in Sect. 3.4. The following section explores the results obtained from testing phase.

5 Results and Analysis

In this section, the emphasis is on results obtained from the experiments. Input parameters of the experimental study are detailed in Table 1. Using the statistical metrics [8], the performance of the model is discussed in Table 2.

The performance of the model is compared to the benchmarking model of DEKANT [25] that intended to prevent the vulnerabilities in web applications. From the experimental studies, it is imperative that the accuracies of CSES are 93% and 89%, respectively; and the CSES model is robust in terms of attack detection as the miss-classification rate, detection missing rate, and detection fallout rate that observed in CSES is comparatively much lower than DEKANT (See Table 2). The another critical advantage is minimal detection missing rate (false negative rate) that denoted by specificity, which is 11% more for CSES compared to DEKANT.

Table 1 Input records used
for experimental study

Total number of records	7539
Number of attack-prone records	5265
Number of normal records	2274
Attack-prone records used for training	3685
Attack-prone records used for testing	1580
Normal records used for training	1592
Normal records used for testing	682

Table 2 Input records and
prediction statistics of the
CSES and DEKANT [25]

	CSES	DEKANT
Total number of records	2262	2262
The number of attack-prone records	1580	1580
The number of normal records	682	682
Records predicted as attack prone	1583	1580
Records predicted as normal	679	682
True positives	1577	1498
False positives	120	149
True negatives	559	533
False negatives	6	82
Precision	0.883	0.904
Sensitivity	0.966	0.948
Specificity	0.870	0.782
Miss-classification rate	0.063	0.102
Detection missing rate	0.034	0.052
Detection fallout rate	0.13	0.218
Accuracy	0.937	0.898

6 Conclusion

In this paper, a new model of anomaly-based input-type validations for web-based
attacks like cross-site scripting and SQL injection detection is denoted. The model
of CSES for attack-prone http request detection and prevention is evaluated.
The CSES system is used for avoiding the constraints like the process overhead at
the host servers and programming language dependency emerging because of
syntax analyzers. Such constraints are used in analyzing the performance of existing
benchmarking models. Based on experimental study results carried out, the results
emphasize performance improvement and detection accuracy, when compared to
benchmarking strategy of DEKANT [25], which is one of the effective models of
web vulnerability detection. Accuracy levels of vulnerability request have resulted
high with fewer false alarms. The outcome of CSES motivates the future research
for defining the novel heuristic scales through renowned machine learning
strategies.

References

1. D. Balzarotti et al., "Saner: composing static and dynamic analysis to validate sanitization in web applications," Proc. IEEE Symposium on Security and Privacy, pp. 387–401, 2008
2. X. Fu and C.-C. Li, "A string constraint solver for detecting web application vulnerability," Proc. International Conference on Software Engineering and Knowledge Engineering, pp. 535–542, 2010
3. M. Martin and M.S. Lam, "Automatic generation of XSS and SQL injection attacks with goal-directed model checking," Proc. USENIX Security Symposium, pp. 31–43, 2008
4. K.-K. Ma, K. Y. Phang, J.S. Foster, and M. Hicks. "Directed Symbolic Execution," Proc. International Conference on Static Analysis, pp. 95–111, 2011
5. Y. Shin, A. Meneely, L. Williams, and J.A. Osborne, "Evaluating complexity, code churn, and developer activity metrics as indicators of software vulnerabilities," IEEE Transactions on Software Engineering, vol. 37, no. 6, pp. 772–787, 2011
6. Rager, Anton. "XSSProxy'." (2005); http://xss-proxy.sourceforge.net/
7. Larouche, Francois. "SQL Power Injector." (2011); http://www.sqlpowerinjector.com
8. Powers, David Martin. "Evaluation: from precision, recall and F-measure to ROC, informedness, markedness and correlation." (2011).
9. Jovanovic, N., Kruegel, C., Kirda, E.: Precise alias analysis for static detection of web application vulnerabilities. In: Proceedings of the 2006 Workshop on Programming Languages and Analysis for Security. pp. 27–36 (Jun 2006)
10. Nunes, P., Fonseca, J., Vieira, M.: phpSAFE: A security analysis tool for OOP web application plugins. In: Proceedings of the 45th Annual IEEE/IFIP International Conference on Dependable Systems and Networks (Jun 2015)
11. Son, S., Shmatikov, V.: SAFERPHP: Finding semantic vulnerabilities in PHP applications. In: Proceedings of the ACM SIGPLAN 6th Workshop on Programming Languages and Analysis for Security (2011)
12. Yamaguchi, F., Golde, N., Arp, D., Rieck, K.: Modeling and discovering vulnerabilities with code property graphs. In: Proceedings of the 2014 IEEE Symposium on Security and Privacy. pp. 590–604 (May 2014)
13. Medeiros, I., Neves, N.F., Correia, M.: Detecting and removing web application vulnerabilities with static analysis and data mining. IEEE Transactions on Reliability 65(1), 54–69 (March 2016)
14. Medeiros, I., Neves, N.F., Correia, M.: Equipping WAP with weapons to detect vulnerabilities. In: Proceedings of the 46th Annual IEEE/IFIP International Conference on Dependable Systems and Networks (2016)
15. Arisholm, E., Briand, L.C., Johannessen, E.B.: A systematic and comprehensive investigation of methods to build and evaluate fault prediction models. Journal of Systems and Software 83 (1), 2–17 (2010)
16. Neuhaus, S., Zimmermann, T., Holler, C., Zeller, A.: Predicting vulnerable software components. In: Proceedings of the 14th ACM Conference on Computer and Communications Security. pp. 529–540 (2007)
17. Shin, Y., Meneely, A., Williams, L., Osborne, J.A.: Evaluating complexity, code churn, and developer activity metrics as indicators of software vulnerabilities. IEEE Transactions on Software Engineering 37(6), 772–787 (2011)
18. Perl, H., Dechand, S., Smith, M., Arp, D., Yamaguchi, F., Rieck, K., Fahl, S., Acar, Y.: VCC Finder: Finding potential vulnerabilities in open-source projects to assist code audits. In: Proceedings of the 22nd ACM SIGSAC Conference on Computer and Communications Security. pp. 426–437. CCS '15 (Oct 2015)
19. Shar, L.K., Tan, H.B.K.: Mining input sanitization patterns for predicting SQL injection and cross site scripting vulnerabilities. In: Proceedings of the 34th International Conference on Software Engineering. pp. 1293–1296 (2012)

20. Shar, L.K., Tan, H.B.K.: Predicting common web application vulnerabilities from input validation and sanitization code patterns. In: Proceedings of the 27th IEEE/ACM International Conference on Automated Software Engineering. pp. 310–313 (2012)
21. Witten, I.H., Frank, E., Hall, M.A.: Data Mining: Practical Machine Learning Tools and Techniques. Morgan Kaufmann, 3rd edn. (2011)
22. Shar, L.K., Tan, H.B.K., Briand, L.C.: Mining SQL injection and cross site scripting vulnerabilities using hybrid program analysis. In: Proceedings of the 35th International Conference on Software Engineering. pp. 642–651 (2013)
23. Yamaguchi, F., Wressnegger, C., Gascon, H., Rieck, K.: Chucky: Exposing missing checks in source code for vulnerability discovery. In: Proceedings of the 20th ACM SIGSAC Conference on Computer Communications Security. pp. 499–510 (Nov 2013)
24. Scandariato, R., Walden, J., Hovsepyan, A., Joosen, W.: Predicting vulnerable software components via text mining. IEEE Transactions on Software Engineering 40(10), 993–1006 (2014)
25. Medeiros, Ibéria, Nuno Neves, and Miguel Correia. "DEKANT: a static analysis tool that learns to detect web application vulnerabilities." Proceedings of the 25th International Symposium on Software Testing and Analysis. ACM, 2016
26. Baik, Nam-Kyun, et al. "Analysis and design of an intrusion tolerance node for application in traffic shaping." Control, Automation and Systems, 2008. ICCAS 2008. International Conference on. IEEE, 2008
27. Garg, Aman, and AL Narasimha Reddy. "Mitigation of DoS attacks through QoS regulation." Microprocessors and Microsystems 28.10 (2004): 521–530
28. Ranjan, Supranamaya, et al. "DDoS-shield: DDoS-resilient scheduling to counter application layer attacks." IEEE/ACM Transactions on Networking (TON) 17.1 (2009): 26–39
29. Das, Debasish, Utpal Sharma, and D. K. Bhattacharyya. "Detection of HTTP flooding attacks in multiple scenarios." Proceedings of the 2011 International Conference on Communication, Computing & Security. ACM, 2011
30. Jech, Thomas. Set theory. Springer Science & Business Media, 2013

Stream Preparation from Historical Data for High-Velocity Financial Applications

K. S. Vijaya Lakshmi, K. V. Sambasiva Rao and E. V. Prasad

Abstract Event processing is playing a progressively more important part in building inventiveness in the application that can instantaneously answer to the business acute event. Several technologies have been anticipated in recent years, for example, event processing and data streams. Live Data Analysis [1] had bagged lots of investigation attention nowadays, and an ample range of novel techniques and applications have been evolved based on business needs. In financial data analysis [2] however, analysts still generally are dependent on statistical performance parameters blended with conventional line charts to facilitate the consideration of assets and to make decisions. For the evolution of better trading techniques, a live data or event stream is mandatory requirement asset which is highly costly. The proposed technique that creates a live stream of data from historical data sets eliminating the inadequacies, offering a complete ease of working with live streams. This technique can be further used for the creation of a new trend of data analysis [3] and visualization mechanisms [4, 5].

Keywords Live stream · Historical data · Event processing

1 Introduction

As the live data is highly costly which is in terms of lakhs and it is highly impractical to test on live data, we were in a thought to prepare a virtual live stream from the historical data which is to be used for further analysis. For that, we had

K. S. Vijaya Lakshmi (✉)
V R Siddhartha Engineering College, Vijayawada, India
e-mail: vijaya@vrsiddhartha.ac.in

K. V. Sambasiva Rao
N R I Institute of Technology Agiripalli, Vijayawada, India
e-mail: kvsambasivarao@rediffmail.com

E. V. Prasad
G V P College of Engineering for Women, Visakhapatnam, India
e-mail: profevprasad@yahoo.com

© Springer Nature Singapore Pte Ltd. 2018
V. Bhateja et al. (eds.), *Proceedings of the Second International Conference on Computational Intelligence and Informatics*, Advances in Intelligent Systems and Computing 712, https://doi.org/10.1007/978-981-10-8228-3_24

subscribed 1-year Features and Options (F&O) historical data from National Stock Exchange (NSE), which costs around 17 thousand for 1-year data. The data contains five folders.

- *Masters*: Contains all the contracts as on the month end including the contracts that expired on the last Thursday of the month.
- *Circulars*: Contains comprehensive set of circulars released in that month to build a formal communication between NSE and Brokerage firms.
- *Bhavcopy*: Contains one file for each working day, each line in it is one observation.
- *Trades*: Contains details about every trade that took place on that day (nearly 4 lakh entries a day).
- *Snapshot*: Contains order book snapshots at 11 am, 12 a.m., 1 p.m., 2 p.m., and 3 p.m. for each working day with 110000, 120000, 130000, 140000, and 150000 folders. For example, the 110000 folder contains the orders on various stocks from 11.00 a.m.–11.59 a.m.

Any system has to be checked for various types of data. So that, taken data from TrueFx. TrueFx [6] is the member of Integral Corporation. This company provides historical foreign exchange data like the currency conversions for free of cost. The general format of data is conversion symbol (ex: USD/JPY), timestamp (until milliseconds), bid, ask. This data is used for creating virtual live streams based on time and the conversion symbol between respective currencies.

1.1 Scope and Objective

Creation of Virtual Live Stream from historical data. The virtual live stream created from this data is used for training the online machine learning models in order to train an efficient and robust ML trading strategy model. This further continues to end goal visualizing the interactive graphs [7] based on this trained models in real time. As in a big picture this data plays an important role in the further idea of an end to end product beginning from preprocessed historical data to completely analyzed and visualized graphs.

2 Literature Survey

The Technology has altered the investment and trading over the past 30 years. In previous days, it was carried out by computer networks and stockbrokers, which was now replaced by continuous access and algorithm trading. Reporters are disinter interceded when speculators approach on essential sources in the meantime they do. An ever bigger perspective of exploitable monetary and business activity can found on the web. Professional interests concentrate on how the latest information technologies are most excellent useful in trading and investing [8].

The statistical measures for the financial data analysis have several problems and inadequacies. The yearly performance of an asset, for example, is simply determined by the growth or decline of an asset within a specific time interval [9].

In recent times data reduction techniques were introduced to store historical stream data [10]. The common approach of this work is to build random samples, sketches, or other defeat summaries of the historical data, sinking its volume by a considerable quantity. This volume decrease has been shown to be competent for maintenance data in the core, regulating query response time, and dropping update costs. Though, techniques based on data reduction generate fairly accurate answers. The inaccuracy limits on these answers can be moderately loose in practice, principally in issues where the system desires to carry out grouped aggregation or to retrieve a tiny piece of historical data.

Stream processing is fetching a novel and important computing paradigm. Inventive streaming applications are being developed in areas ranging from scientific applications (e.g., atmosphere monitoring), to business intelligence (e.g., scam revealing and trend analysis), to financial markets (e.g., algorithmic trading strategies). Developing, understanding, debugging, and optimizing streaming applications is nontrivial for the reason that of the adaptive and active nature of these applications. The absolute density and the distributed nature of a huge figure of cooperating components hosted on a distributed setting further make difficult matters [11].

2.1 Disadvantages

These observations show that the three most widely used performance measures (1-, 3- and 5-year performance) in the financial analysis represent just three single points from a large set of performance measure, and may provide unstable and unreliable information about the characteristics of an asset which makes them inappropriate for making important investment decisions.

3 Requirements

3.1 Functional

- This work needs to deal with data equivalent to and that can be represented as a real-time data.

- This involves data to be sent to several applications clients which need to be free from noisy and redundant data.
- As it deals with the preparation of live streams from historical data, the data should be preprocessed for seamless streaming.
- It involves web application which receives live stream data and passes its data to remote clients.
- This work is coded in non-i/o blocking programming procedures to tackle the on-field latencies that affect the live streams.

3.2 Nonfunctional

- Also known as quality requirements (performance, security and reliability), which supports functional requirements.
- Dependency on other parties is one of the major nonfunctional requirements of our system as it needs to deal with multiple servers and web applications servers.
- Deployment is another major nonfunctional requirement where two different servers will be deployed simultaneously.
- As our system deals with real-time stream preparation efficiency is also a major nonfunctional requirement.
- Sufficient network bandwidth required for the project involves publishing data through sockets and many of the libraries were predefined that are to be imported in real time for execution of code and deployment of servers.
- Extensibility, where the components are to be scaled to increase traffic and increase in a number of subscription requests.

4   Block Diagram for Creating Virtual Live Stream

See Fig. 1.

4.1 Description

- Financial data regarding the currency transactions of several countries is our initial raw historical data present in files with .csv extension.

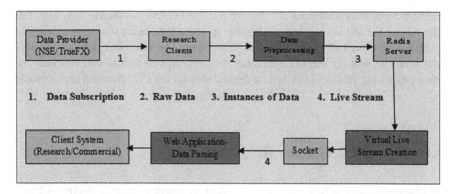

Fig. 1 The research clients subscribe data from data providers, preprocess it, and transfer it to Redis server. Redis server converts it into the virtual live stream. Web application server parses the data to the client through non-i/o blocking socket programming and finally the client receives the stream

- This, raw data will be available for the general public for a particular amount and subscription, if it is NSE or it can also be acquired for free from some websites like TrueFx.
- This, raw data will be preprocessed using shell scripting and our required data will be extracted from the data.
- This data is further set to be published to the socket using Redis server.
- The Redis server will generate virtual live stream from the historical data.
- The web application server receives the stream from Redis server.
- The stream generated by the researcher can be used by another researcher for analysis or an ordinary client that utilizes this stream to predict.

5 Proposed System of Live Stream Preparation

Event: An **event** is an action or occurrence caused by the change in state of a physical or conceptual entity.

Event Stream Processing (ESP): Event Stream is the continuous collection of events and event stream processing involves methods and techniques to process the event streams for processing, analysis, storing, etc.

Sockets: Socket is an endpoint of an interprocess communication flow across the computer network. Now a day the communication between computers is based on the Internet Protocol; therefore most network sockets are internet sockets.

Non I/O Blocking Programming: Permits other processing to carry on prior to the transmission has been completed. It is possible to start the communication and then perform processing, that does not require that the I/O has completed. Any task that depends on the I/O having completed (this includes both using the input values and critical operations that claim to assure that a write operation has been completed) still needs to wait for the I/O operation to complete, and thus is still blocked, but other processing that does not have a dependency on the I/O operation can continue [12].

5.1 Description of Data Set

As discussed previously 1-year historical data was subscribed from National Stock Exchange which was proprietary. In addition to that, the currency exchange data was downloaded from TrueFX. The data downloaded from TrueFX [6], was formulated into a table to present clearly (Table 1). The S. No. was synthesized to show the number of possible instances per a day. It shows that more than 5 lakh instances per a day. The attributes in this data set were Conversion Symbol, Time Stamp, Bid and Ask. Each instance shows price variation for each millisecond.

Table 1 Description of attributes in the dataset

S. no.	Conversion symbol	Timestamp	Bid	Ask
1	AUD/JPY	20170301 00:00:00.021	86.425	86.44
2	AUD/JPY	20170301 00:00:00.165	86.425	86.439
3	AUD/JPY	20170301 00:00:00.171	86.425	86.44
4	AUD/JPY	20170301 00:00:00.232	86.425	86.439
5	AUD/JPY	20170301 00:00:00.328	86.426	86.439
6	AUD/JPY	20170301 00:00:00.339	86.426	86.44
7	AUD/JPY	20170301 00:00:00.377	86.427	86.44
8	AUD/JPY	20170301 00:00:00.486	86.427	86.439
9	AUD/JPY	20170301 00:00:00.487	86.426	86.44
10	AUD/JPY	20170301 00:00:00.543	86.426	86.44
11	AUD/JPY	20170301 00:00:00.661	86.427	86.439
12	AUD/JPY	20170301 00:00:00.708	86.427	86.439
13	AUD/JPY	20170301 00:00:00.776	86.425	86.439
14	AUD/JPY	20170301 00:00:00.895	86.424	86.439
15	AUD/JPY	20170301 00:00:00.953	86.426	86.439
16	AUD/JPY	20170301 00:00:02.302	86.422	86.437
...	...	...	...	...
535415	AUD/JPY	20170301 23:59:59.972	87.398	87.415
535416	AUD/JPY	20170302 00:00:00.078	87.398	87.414
535417	AUD/JPY	20170302 00:00:00.083	87.398	87.413

AUD/JPY	20170301 00:00:00.165	86.425003	86.439003
AUD/JPY	20170301 00:00:00.171	86.425003	86.440002
AUD/JPY	20170301 00:00:00.232	86.425003	86.439003
AUD/JPY	20170301 00:00:00.328	86.426003	86.439003
AUD/JPY	20170301 00:00:00.339	86.426003	86.440002

5.2 Methodology

The events stream [13] generates an infinite sequence of fragments corresponding to real-life events. The update stream produces fragments corresponding to updates to the event context and other related temporal data. By capturing both events and updates as fragments with temporal extents, we provide a unified continuous stream preparation.

5.2.1 Algorithm Creation of Virtual Live Streams Takes Place by the Following Steps

1. Extraction of required data values from the dataset.
2. Prepare the data instances to pass them through a socket channel.
3. Publish the data from a non-i/o blocking server into a socket.
4. Subscribe the socket using a web application and process the data.
5. The processed data is fetched to a client application.

Extraction of required data involves data preprocessing where the raw data are made free from noisy and redundant data. Then the data is further processed by a set of Linux shell scripts where the data will be arranged in a category wise manner. By using a shell script with grep and other commands, the required features were extracted from the overall historical data.

- The required symbol was extracted using a separate script for each folder which results.

  - Masters: Single file of a required feature for each month.
  - Bhavcopy: Single file of a required feature for each day.
  - Trades: Single file of a required feature for each day.
  - Snapshot: 5 files of a required feature for each day.

- Snapshot entries were again processed.

  - Timestamp was added at the beginning of each extracted entry from each discrete timestamp folder. Ex. 110000, 120000, 130000, 140000, 150000.
  - All entries of a day were concatenated into a single file with various timestamps of that day.

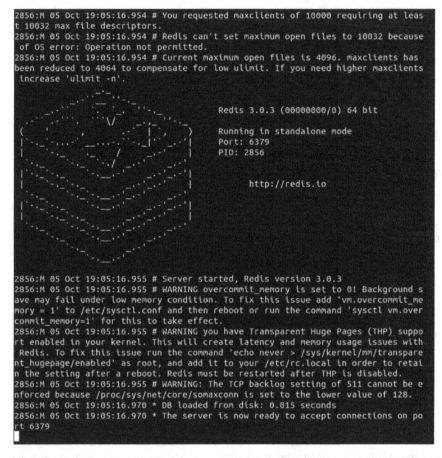

Fig. 2 The Redis server had been initialized and set to on the go with a default port at port number 6379. The Node.js web application connects to Redis server using this port and later a logical socket connection will be established between the two servers. This server plays an important role in our architecture and maintenance of this server involves dealing controlling transparent huge pages

- Finally, left with a single file of extracted entries of order book snapshot for each working day.
- This data, which is feed to sever using Redis–client module will be published into a socket which can be identified by a particular socket identity name (Figs. 2, 3 and 4).
- A web application server in our context a server running with Node.js is configured to receive the data from the socket.
- Node.js server is configured with a Redis.socket.io package which is predefined with basic procedures.

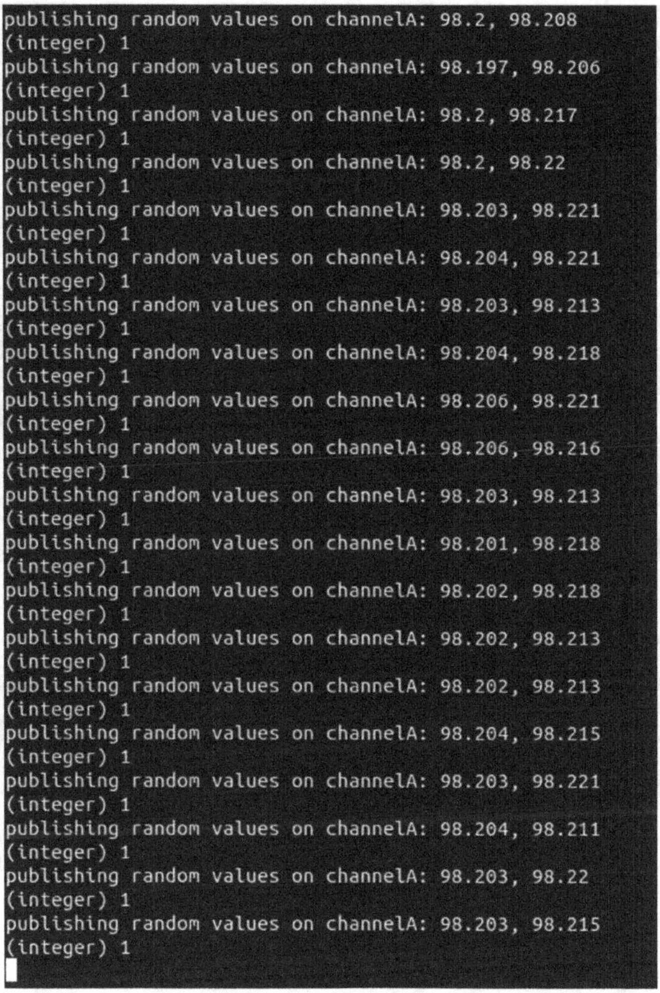

```
publishing random values on channelA: 98.2, 98.208
(integer) 1
publishing random values on channelA: 98.197, 98.206
(integer) 1
publishing random values on channelA: 98.2, 98.217
(integer) 1
publishing random values on channelA: 98.2, 98.22
(integer) 1
publishing random values on channelA: 98.203, 98.221
(integer) 1
publishing random values on channelA: 98.204, 98.221
(integer) 1
publishing random values on channelA: 98.203, 98.213
(integer) 1
publishing random values on channelA: 98.204, 98.218
(integer) 1
publishing random values on channelA: 98.206, 98.221
(integer) 1
publishing random values on channelA: 98.206, 98.216
(integer) 1
publishing random values on channelA: 98.203, 98.213
(integer) 1
publishing random values on channelA: 98.201, 98.218
(integer) 1
publishing random values on channelA: 98.202, 98.218
(integer) 1
publishing random values on channelA: 98.202, 98.213
(integer) 1
publishing random values on channelA: 98.202, 98.213
(integer) 1
publishing random values on channelA: 98.204, 98.215
(integer) 1
publishing random values on channelA: 98.203, 98.221
(integer) 1
publishing random values on channelA: 98.204, 98.211
(integer) 1
publishing random values on channelA: 98.203, 98.22
(integer) 1
publishing random values on channelA: 98.203, 98.215
(integer) 1
```

Fig. 3 The data preprocessed by our shell scripts are fed to sockets using Redis server. The Fig. 3 shows that, the terminal window with outputs of data being sent to the socket. This socket is sent to clients through port number 6379 with identity named channel A

- Socket.io is another package used for establishing a generic networking connection suitable for sockets.
- Express.js is a javascript-based framework that provides backend support to Node.js.

5.3 Results

See Figs. 5, 6, 7, 8 and 9.

```
    info - at webserver, recieved (channel, message): channelA, {"y1":98.199,"y2":98.207}
{ y1: 98.199, y2: 98.207 }
    info  - the y1 & y2 values parsed:
98.199 98.207
    info - at webserver, recieved (channel, message): channelA, {"y1":98.199,"y2":98.207}
{ y1: 98.199, y2: 98.207 }
    info  - the y1 & y2 values parsed:
98.199 98.207
    info - at webserver, recieved (channel, message): channelA, {"y1":98.2,"y2":98.217}
{ y1: 98.2, y2: 98.217 }
    info  - the y1 & y2 values parsed:
98.2 98.217
    info - at webserver, recieved (channel, message): channelA, {"y1":98.2,"y2":98.215}
{ y1: 98.2, y2: 98.215 }
    info  - the y1 & y2 values parsed:
98.2 98.215
    info - at webserver, recieved (channel, message): channelA, {"y1":98.198,"y2":98.216}
{ y1: 98.198, y2: 98.216 }
    info  - the y1 & y2 values parsed:
98.198 98.216
    info - at webserver, recieved (channel, message): channelA, {"y1":98.198,"y2":98.213}
{ y1: 98.198, y2: 98.213 }
    info  - the y1 & y2 values parsed:
98.198 98.213
    info - at webserver, recieved (channel, message): channelA, {"y1":98.196,"y2":98.204}
{ y1: 98.196, y2: 98.204 }
    info  - the y1 & y2 values parsed:
98.196 98.204
    info - at webserver, recieved (channel, message): channelA, {"y1":98.202,"y2":98.215}
{ y1: 98.202, y2: 98.215 }
    info  - the y1 & y2 values parsed:
98.202 98.215
    info - at webserver, recieved (channel, message): channelA, {"y1":98.196,"y2":98.215}
{ y1: 98.196, y2: 98.215 }
    info  - the y1 & y2 values parsed:
98.196 98.215
    info - at webserver, recieved (channel, message): channelA, {"y1":98.201,"y2":98.217}
{ y1: 98.201, y2: 98.217 }
    info  - the y1 & y2 values parsed:
98.201 98.217
```

Fig. 4 The data being received by a web application server and parsing it to the remote client. In our context, web application server that was deployed by us was working on Node.js framework and uses external packages like Express.js and Redis server

Result with On Demand Utilities	
real	9m11.675s
user	2m42.164s
sys	7m15.742s

Fig. 5 The time taken to create the complete live stream from historical data of 1-year, with system resource consumption set to on-demand utilities. The real time taken for this is 9 min 11.675 s, in that usage by system is 7 min 15.742 s and the usage by user is 2 min 42.164 s

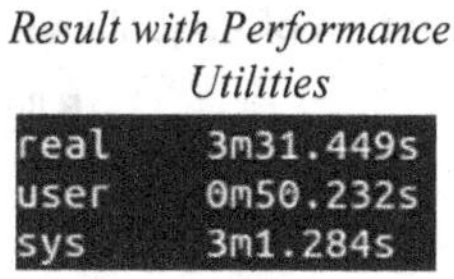

Result with Performance Utilities

```
real    3m31.449s
user    0m50.232s
sys     3m1.284s
```

Fig. 6 The time taken to create the complete live stream from historical data of one year, with system resource consumption set to high performance. The real time taken for this is 3 min 31.449 s, in that, usage by system is 3 min 1.284 s and the usage by user is only 50.232 s

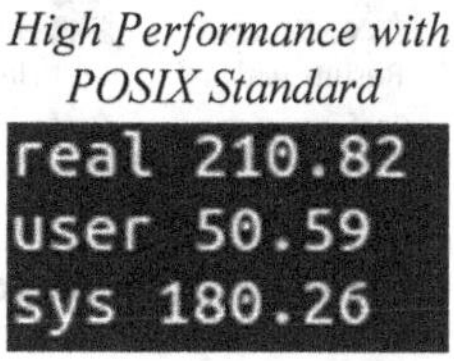

High Performance with POSIX Standard

```
real 210.82
user 50.59
sys 180.26
```

Fig. 7 The time taken to create the complete live stream from historical data of 1-year, with system resource consumption set to high-performance utilities but the results are in Portable Operating System Interface Standard (POSIX) format specifying the POSIX compatibility of project. The real time taken for this is 210.82 s, in that CPU used 180.26 s and the time used by user is only 50.59 s

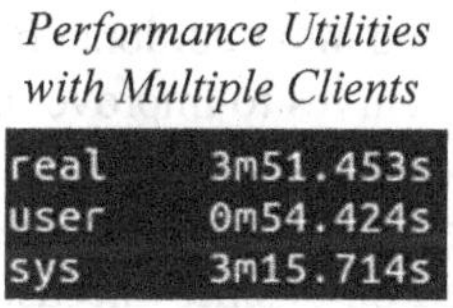

Performance Utilities with Multiple Clients

```
real    3m51.453s
user    0m54.424s
sys     3m15.714s
```

Fig. 8 The time taken to create the complete live stream from historical data of 1-year, with system resource consumption set to high performance utilities, when server serving multiple clients

6 Conclusion

The objective is, to create a virtual live stream using Historical data and was able to create a live stream data with non-i/o blocking strategy. This live stream data can be further contributed to analysis of financial time series data that allows having a complete overview of the characteristics of an asset for various holding times and times of sale. User-based relevance [14] functions allow to emphasize single or multiple regions of interest and to focus on these areas for an advanced analysis.

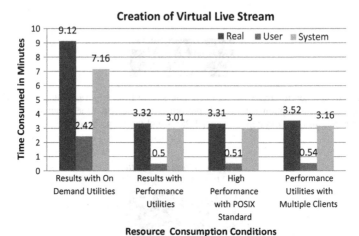

Fig. 9 The time taken to create the complete live stream from historical data of 1-year with system resource consumption set to various conditions listed on X axis. The Y axis represents the time consumed in minutes. The real represents actual time taken for creating the stream, user represents the time used by the user and the system represents the time used by the system

We have shown how our relevance functions that can be used to compute improved performance and risk measures that also take the investors regions of interest into account, and how can be integrated into traditional financial analysis techniques like Dominance Plots and efficiency curves to improve the decision making process the detailed characteristics of assets. Visual Data Analysis [1] had bagged lots of investigation attention nowadays, and an ample range of novel techniques and applications have evolved based on business needs.

Acknowledgements It is our responsibility to thank our Advisor Dr. V. Srikanth, Director, Citi Bank, U.K. for his valuable advice in this area. Thanks to our Technical Supporter Mr. Vamsi Nadella, Grad Student, University of Georgia, Athens, USA. Thanks to Dr. K. Suvarna Vani, Professor, CSE, VRSEC for her moral support and valuable suggestions.

References

1. Hartmut Ziegler, Tilo Nietzschmann, Daniel A. Keim.: Visual Analytics on the Financial Market: Pixel-based Analysis and Comparison of Long-Term Investments: 12th International Conference Information Visualisation, 2008.
2. Erographics Digital Library, http://diglib.eg.org.
3. Allen D. Malony Daniel, A. Reedy: An Integrated Performance Data Collection, Analysis, and Visualization System: NTRS, Center for Supercomputing Research and Development University of Illinois Urbana., 1998.
4. Havre, S., Hetzler, E. G., Nowell L.T.: "ThemeRiver: Visualizing Theme Changes over Time", Proceedings of the Infovis 2000, pp. 115–124.

5. Dwyer T., Gallagher D.R.: Visualising Changes in Fund Manager Holdings in Two and a Half-dimensions. Information Visualization 3, 4 (2004), 227–244.
6. TrueFX by Integral, https://www.truefx.com.
7. Deboeck G. J., Kohonen T. K., Visual Explorations in Finance with Self Organizing Maps. Springer-Verlag New York, Inc., Secaucus, NJ, USA, 1998.
8. David Leinweber: If You Had Everything Computationally, Where Would You Put It, Financially?: Oreially Money Tech Conference, New York, 2008.
9. Vizualization Group, http://www-vis.lbl.gov/.
10. Hochheiser, H.: Interactive Graphical Querying of Time Series and Linear Sequence Data Sets Ph.D. Dissertation, University of Maryland, Dept. of Computer Science, May 2003.
11. The Odysci Academic Search System, academic.odysci.com.
12. The Science Search Engine, http://research.omicsgroup.org.
13. Lepmodas Fegaras.: Data stream management for historical XML data.: International Conference on Management of Data, ACM SIGMOD, 2004.
14. Ziegler, Hartmut, Nietzschmann, Tilo, Keim, Daniel A., Relevance driven visualization of financial applications.

Context-Based Word Sense Disambiguation in Telugu Using the Statistical Techniques

Palanati DurgaPrasad, K. V. N. Sunitha and B. Padmaja Rani

Abstract The statistical technique proposed in this paper assigns a correct sense to the targeted polysemous word which has different meanings in different contexts. The methodology proposed in this paper which solves the well-known AI-Complete problem IS related to natural language processing which is called as Word Sense Disambiguation (WSD). The polysemous word may belong to anyone parts-of-speech assigned by the POS taggers. But there are some words which belong to same parts of speech but their meaning differs based on the context. Currently, the system disambiguates nouns and verbs. The system gives 100% coverage. The proposed method for word sense disambiguation gives best results while translation between different languages. A step forward in this field would have an impact on information extraction applications.

Keywords Context · Cosine similarity · Indowordnet · Word sense disambiguation · Telugu polysemous word

P. DurgaPrasad (✉)
Department of Computer Science and Engineering, UCET, MGU,
Nalgonda, Telangana, India
e-mail: dp.cse5@gmail.com

K. V. N. Sunitha
Department of Computer Science and Engineering, BVRIT,
Hyderabad, Telangana, India
e-mail: k.v.n.sunitha@gmail.com

B. Padmaja Rani
Department of Computer Science and Engineering, JNTUHCEH,
Hyderabad, Telangana, India
e-mail: padmaja_jntuh@jntuh.ac.in

© Springer Nature Singapore Pte Ltd. 2018
V. Bhateja et al. (eds.), *Proceedings of the Second International Conference on Computational Intelligence and Informatics*, Advances in Intelligent Systems and Computing 712, https://doi.org/10.1007/978-981-10-8228-3_25

1 Introduction

1.1 Motivation

The communication among the persons in the natural language contains the chunks/ words which are associated with several meanings based on the co-references and the semantic relations. So, many times there is an ambiguity to determine the underlying meaning of the specific word in the context. Here, is an example in the Telugu language:

1.పత్తి సాగు లాభదాయకం. 2.సమాజంలో మార్పు రావాలంటే నువ్వు ముందుకు సాగు.

Example 1. Telugu sentence with one ambiguous word

In the above two sentences, the word 'సాగు' has two meanings in first sentence it is *CULTIVATION* and in second it is 'PROCEED IN ADVANCE'. These types of problems are considered as WORD SENSE DISAMBIGUATION (WSD) in the field of computational linguistics. Like this when a word is mapped to multiple senses then this type of lexical semantic ambiguity arises. WSD is to define, analyze and identify the relationship between word, meaning and context. The solution for the WSD problem in its way helps to resolve some other difficult problems like common sense reasoning, encyclopedic knowledge and finally natural language understanding. The researches have described this computational problem as an 'AI-complete' problem [1]. This WSD problem plays a vital role in machine translation, information retrieval, content analysis, grammatical analysis, thematic analysis, speech processing and text processing.

1.2 Dealing Word Sense Disambiguation

In general, the WSD requires first to determine the different senses of the targeted word relevant to the context and secondly assigning the word to the appropriate sense. These two things are to be done as a prerequisite for any method to be employed for WSD problem. As an example, let us consider the following sentence:

"Farmer Bill Dies in House"

Example 2. English sentence with ambiguous words linked to each other

Where is the ambiguity? The nouns *Bill* and *House* are ambiguous. In the first sense, *Bill* may be a person's name, and *house* may be a place of residence. In the second sense, *bill* may be proposed legislation and *House* may be the House of Commons. This is an issue of lexis. In addition, the verb *dies* has a literal meaning (the real death of a living organism) and a metaphorical one (the end of something). This is an issue of semantics. Steps to deal with Word Sense Disambiguation may vary depending on the approaches followed whether it is the deep or shallow approach. Word sense disambiguation is achieved using either supervised model which uses annotated corpus or unsupervised models which do not use un-annotated corpus. Consider a Telugu sentence "దేశాభివృద్ధిలో యువకుల *పాత్ర* కీలకం." (The *role* of youth is very important in the development of country). This sentence has the word పాత్ర which has two meanings one is "role" and second is "bowl". So to assign correct meaning one should know the semantic relation between the targeted words పాత్ర and the surrounding words. The context of the sentence must be known in some approaches. However, there are many approaches which are following different strategies to solve WSD. The remainder of this paper is organized as follows. Section 2 contains literature survey followed by Sect. 3 with the proposed method, Sect. 4 contains the evaluation and results, Sect. 5 conclusion and future work.

2 Literature Survey

Word sense disambiguation problem was simply explained by Weaver [2] while using an opaque mask with a hole of one-word size to read the text which yields to the concept of linking context with the word's meaning. Here is the difficulty to obtain the correct meaning of the particular word in the absence of context. Later many efforts have been done on this problem and these efforts lead to different approaches like knowledge-based and machine learning based approaches. Knowledge-based disambiguation uses some restrictions like co-occurrence which is the idea of the information-theoretic measures proposed by Rensik [3] which combines the statistical methods with knowledge-based methods used for selecting conditions. Rensik [3] concluded that even though selectional preference is an important factor in sense disambiguation but practically its application coverage is limited because it is impossible to establish fixed rules for the correct placing of verb and modifiers in a sentence.

Lesk algorithm [4] which assigns a sense to the target word whose gloss shares maximum count of words with the glosses of other words. The experiments were done by the Banerjee et al. [5] using a modified version of basic Lesk algorithm which needs the glosses of words so one finds difficulty to tackle the sense disambiguation if there is no proper resource of glosses. Unlike Lesk algorithm which uses the concept of word sharing among the glosses, there is another approach called semantic similarity [6] applied in query expansion where the words that are related, share a common context and, therefore, the appropriate sense is chosen by those meanings, found within smallest semantic distance. The WSD problem can be solved easily if there are large scale lexical resources and corpora available. Peter F. Brown [7] first described a statistical technique which questions about the context to assign a correct sense to the polysemous word and the techniques are used in machine translation system which decreased the error rate by 13%.

David Yorowsky [8] proposed an unsupervised sense disambiguation algorithm with constraints one sense per discourse and one sense per collocation which shows an accuracy around 95%. Zhizhuo Yang [9] have implemented two novel methods based on Bayesian classifiers, which have used synonym feature and basic features for supervised WSD by substituting synonyms for the context around the targeted polysemous word. Manish [10] developed an overlap based approach for sense disambiguation using the statistical techniques in the absence of morphological variants of the words. In this work, they have assigned a correct sense to a word by measuring the overlap between the content bag and word bag. Khapra [11] have got approximately 75% of F1-score by applying iterative word sense disambiguation algorithm and a modified version of page rank algorithm developed by Rada [12]. Agirre [13] developed an algorithm which performs random walks over the extended WordNet for word sense disambiguation. They have implemented word-to-word heuristic with personalized PageRank on both English and Spanish datasets. Kang [14] have proposed a loopy belief propagation algorithm which uses the pairwise random fields over the Connotation WordNet [15] resulting in 74.83% of accuracy.

3 Proposed Method

In natural language, there may be some polysemous words which have to be assigned correct sense which otherwise leads to ambiguity in the meaning of the sentence. For example, 'నీలం/blue రంగు/colour ఇంటిపై/on house కొబ్బరికాయ/coconut కొట్టు/shop or break/beat'. In this sentence, the polysemous word కొట్టు has different meanings like shop, beat and break which raise the lexical ambiguity. The following section describes the proposed method for word sense disambiguation.

3.1 Methodology

The given problem contains some sentences with polysemous words. The polysemous word is assigned a sense with the help of context, information related to the word and statistical method. The context considered here is the text in which polysemous word presents. Our approach is as follows:

Algorithm: Assigning proper sense for a word

1. Let *w* is a polysemous word with senses *S1, S2,...*so on.
2. Collect surrounding words of *w* call this collection as CONTEXT BAG *CB*
3. For every sense S1, S2,...so on of word *w*

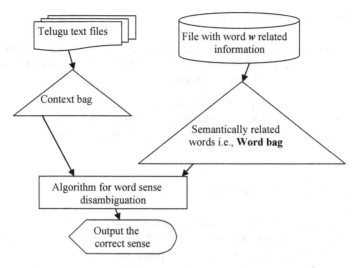

Fig. 1 Process of creating context bag and word bag to measure similarity to assign correct sense

    (i) Form a collection of words using the synonyms, example sentences, hypernyms, meronyms and hyponyms
    (ii) Call this bag as **WORD BAG BW_1, BW_2,...so on**

4. Measure the similarity between the CONTEXT BAG *CB* and WORD BAGS BW_1, BW_2,...*so on*
5. Find the WORD BAG with maximum similarity with CONTEXT BAG and assign that sense for the word *w* (Fig. 1).

3.2   Modules in the System

The system developed using our WSD algorithm has the below modules.

3.2.1   Module 1: Collect the Documents

This module collects the documents containing the polysemous word and the related words (synonyms, hypernyms, hyponyms, meronyms and glosses).

3.2.2 Module 2: Tokenization

The documents collected using the module 1 are sent as input to tokenize. The outputs are used to form CONTEXT BAG and WORD BAG. After tokenization bags will be as shown in the below example,

CONTEXT BAG: CB-['మళ్ళి', 'నీరు', 'చల్లగా', 'ఉంటుంది']
SENSE 1 WORD BAG:BW$_1$—['త్రాగడానికి', 'నీరు', 'ఉంచుతారు', 'తినే', 'పదార్థాలు', etc.]
SENSE 2 WORD BAG:BW$_2$—['పవన', 'చిత్రంలో', 'మంచి', 'పిల్లల', 'ఎదుగుదలకు', etc.]

3.2.3 Module 3: Statistical Method

This module calculates the similarity between the CONTEXT BAG and each WORD BAG using cosine similarity formula. The number generated by the cosine function explained in Eq. 3 indicates the percentage of similarity between the context bag and word bag. The context bag and the word bags are treated as two different vectors and the cosine function mentioned below is applied for measuring the similarity.

Magnitude of context bag,

$$mag(CB) = \sqrt{\sum_{i=0}^{n_{CB}} CB_I^2} \tag{1}$$

n_{CB} = length of context bag

Magnitude of word bag of kth sense,

$$mag(BW_K) = \sqrt{\sum_{i=0}^{n_{BW}} BW_i^2} \tag{2}$$

n_{BW} = length of word bag of kth sense

Similarity between context bag (CB) and word bag (BW),

$$similarity(CB, BW_K) = \frac{\sum_{i=0}^{n} CB_i \times BW_i}{mag(CB) \times mag(BW)} \tag{3}$$

3.2.4 Module 4: Assigning Correct Sense

This module compares the similarity percentage generated by the cosine function explained in Eq. 3 of Sect. 3.2.3. The sense corresponding to the word bag with highest similarity percentage as shown in Fig. 2 is selected as the correct sense of the polysemous word.

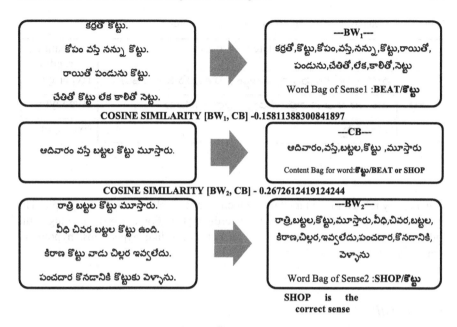

Fig. 2 Process of assigning sense for word 'కొట్టు/SHOP OR BEAT'

4 Evaluation

The text documents which contain Telugu language sentences with polysemous words are used.

4.1 Telugu Text Document

The target word (polysemous) is in the sentence: మట్టి **పాత్రలో** నీరు చల్లగా ఉంటుంది..

The word **పాత్ర** Meaning Sense 1: పదార్థాలు నిల్వ ఉంచు వస్తువు/Dish or basin

Example:1.

త్రాగడానికి పాత్రలో నీరు ఉంచుతారు. 2.తినే పదార్థాలు వంట పాత్రలో పెడతారు.

Meaning Sense2: వేషం/the actions and activities assigned to person

Example: 1.

పవన్ చిత్రంలో మంచి పాత్ర పోషించాడు. 2.పిల్లల ఎదుగుదలకు తల్లి పాత్ర ముఖ్యం.

The above example sentences and the related words are stored in the respective word bags (BW_1, BW_2,…so on) of the polysemous word. The related words are

Table 1 Polysemous word and its semantically related words

Word	Related words	Semantic relation
పాత్ర	మానవ_నిర్మితం ,వస్తువు ,పదార్థం , సొత్తు,అస్తిత్వం ,హాజరు	Hypernym
	మంగళకారుడు , సూత్రధారుడు , ప్రధాన_నటుడు	Hyponymy
	వేషం, గిన్నె	Synonym

semantically related to the target polysemous word. Table 1 shows an example. The related words are collected from Indowordnet [16]. For a word in Telugu, the Indowordnet consists of semantically related words like hypernyms, hyponyms, meronyms, holonyms, antonyms and glosses.

4.2 Results

There are many polysemous words in Telugu but the algorithm is tested for 150 words which include nouns and verbs. For each word minimum of 2 senses and a maximum of 5 senses are considered. The proposed method is tested over 1500 Telugu text documents containing around 80,000 words of different contexts. The metrics for the developed system are discussed below (Fig. 3).

Precision = number of correctly disambiguated words/number of disambiguated words
Recall = number of correctly disambiguated words/number of tested set words
Coverage = number of disambiguated words/number of tested set words.

Our proposed disambiguation algorithm is able to give answers to all polysemous words so the coverage is 100%. As we have considered all the words of the entire text where the target polysemous word is present as context the disambiguation algorithm gives approximately 63% of precision and recall.

Fig. 3 Statistics of disambiguated words

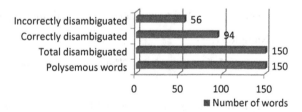

5 Conclusion

The proposed disambiguation algorithm uses different bags which stores all the words which are semantically related to the targeted polysemous word. The percentage of similarity is calculated between the context in which the word is present and each sense of the polysemous word. In this approach, we have two limitations. One is considering all the words of the text in which the target word is present as the context and second is all semantically related words are not considered because of their unavailability.

In our future work, we are planning to change the collection of words for context bag using n-gram-based algorithms. The recall of our disambiguation algorithm can be increased by increasing the size of sense bag so that the algorithm can correctly assign proper sense for the polysemous word.

References

1. Ide, Nancy & Jean Véronis.: Word sense disambiguation: The state of the art. Computational Linguistics, 24(1): pp 1–40, 1998
2. Translation, Warren Weaver 1949
3. Resnik Philip.: Selectional preference and sense disambiguation in Proceedings of ACL Workshop on Tagging Text with Lexical Semantics, Washington, U.S.A., pp 52–57, 1997
4. M. Lesk.: Automatic sense disambiguation using machine readable dictionaries: How to tell a pine cone from an ice cream cone. In: Proceedings of SIGDOC'86, 1986
5. Banerjee. S, Pedersen. T: An adapted Lesk algorithm for word sense disambiguation using WordNet. In: the 3RD International Conference, CICLing, Mexico City, February, 2002
6. Mittal, Amita Jain.: Word sense disambiguation method using semantic similarity measures and owa operator. ICTACT Journal on Soft Computing, Vol. 5 Issue 2, pp 896–904, 2015
7. Brown et al.: Word Sense Disambiguation using Statistical Methods. In: Proceedings of the 29th annual meeting Association for Computational Linguistics, pp 264–270, 1991
8. David Yorowsky.: Unsupervised word sense disambiguation rivaling supervised methods.: www.aclweb.org/anthology/P95-1026
9. Zhizhuo Yang, Heyan Huang.: Chinese Word Sense Disambiguation based on Context Expansion. In: Proceedings of COLING, Mumbai, pp 1401–1408, December 2012
10. Manish Sinha, Mahesh Kumar Reddy, R Pushpak Bhattacharyya, Prabhakar Pandey, Laxmi Kashyap.: Hindi Word Sense Disambiguation. IJCTEE, Volume 2, Issue 2, 2012
11. Mitesh M. Khapra et al.: Projecting Parameters for Multilingual Word Sense Disambiguation. In: Proceedings of the Conference on Empirical Methods in NLP, ACL and AFNLP, pp 459–467, Singapore, 6–7 August 2009
12. Mihalcea Rada.: Large vocabulary unsupervised word sense disambiguation with graph-based algorithms for sequence data labeling. In: Proceedings of the Joint HLT/EM in Natural Language Processing Conference, Vancouver, Canada, pp 411–418, 2005
13. Eneko Agirre, et al.: Random Walks for Knowledge Based Word Sense Disambiguation. Computational Linguistics, Vol 40, Issue 1, pp 57–84, March 2014
14. Jun Seok Kang, et al.: Connotation WordNet: Learning Connotation over the Word Sense Network. In: Proceedings of the 52nd Annual Meeting of the ACL, pp 1544–1554, Baltimore, Maryland, USA, June 23–25 2014

15. Song Feng et al.: Connotation lexicon: A dash of sentiment beneath the surface meaning. In: Proceedings of the 51st Annual Meeting of the ACL, pp 1774–1784, Sofia, Bulgaria, August 4–9 2013
16. Indo Wordnet http://tdil-dc.in/indowordnet/first?langno=15

EEG-Controlled Prosthetic Arm for Micromechanical Tasks

G. Gayathri, Ganesha Udupa, G. J. Nair and S. S. Poorna

Abstract Brain-controlled prosthetics has become one of the significant areas in brain–computer interface (BCI) research. A novel approach is introduced in this paper to extract eyeblink signals from EEG to control a prosthetic arm. The coded eyeblinks are extracted and used as a major task commands for control of prosthetic arm movement. The prosthetic arm is built using 3D printing technology. The major task is converted to micromechanical tasks by the microcontroller. In order to classify the commands, features are extracted in time and spectral domain of the EEG signals using machine learning methods. The two classification techniques used are: Linear Discriminant Analysis (LDA) and K-Nearest Neighbor (KNN). EEG data was obtained from 10 healthy subjects and the performance of the system was evaluated for accuracy, precision, and recall measures. The methods gave accuracy, precision and recall for LDA as 97.7%, 96%, and 95.3% and KNN as 70.7%, 67.3%, and 68% respectively.

Keywords BCI · Coded eyeblinks · EEG signals · Major tasks
LDA · KNN · Micromechanical tasks · 3D-printed prosthetic arm

G. Gayathri (✉) · G. Udupa (✉)
Department of Mechanical Engineering, Amrita School of Engineering, Amrita Vishwa Vidyapeetham, Amrita University, Amritapuri, India
e-mail: siva.gaya3g@gmail.com

G. Udupa
e-mail: ganesh@am.amrita.edu

G. J. Nair · S. S. Poorna
Department of Electronics and Communication Engineering, Amrita School of Engineering, Amrita Vishwa Vidyapeetham, Amrita University, Amritapuri, India

© Springer Nature Singapore Pte Ltd. 2018
V. Bhateja et al. (eds.), *Proceedings of the Second International Conference on Computational Intelligence and Informatics*, Advances in Intelligent Systems and Computing 712, https://doi.org/10.1007/978-981-10-8228-3_26

1 Introduction

Research in brain-controlled interfaces has wide application in medical, navigation, entertainment, education, and other fields. Prosthetics are one of the major tools for amputees and persons with impaired motor disabilities. At present, a functional bionic hand costs between 12,000 USD and 50,000 USD [1]. Due to this, amputees in developing countries find it difficult to afford such devices. The paper presents a novel method of controlling a prosthetic hand using processed electroencephalogram (EEG) signals. In addition to EEG signals, Electromyograms (EMG) and Electrooculograms (EOG) are also used to control the prosthetic hand with reduced functionalities [2, 3]. Neuro-prosthetics, controlled by EEG, can be used for people suffering from severe motor impairments since they can control appliances without motor movements or speech [4, 5]. EEGs are electric potentials, produced by the firing of the neutron in the brain, which results in different states of the brain. These potentials are measured by positioning sensors on the scalp of the subject. The artifacts in EEG are usually high amplitude signals produced by eyeblinks, chewing, muscular movements, external electrical pickups, and epilepsy. Even though eyeblink signals in EEG are removed in most of the bio-signal analysis [6, 7], it can also serve as an efficient tool for control applications [8]. This paper proposes the use of eyeblinks and coded eyeblinks from EEG as a new concept for controlling a prosthetic arm. Each command generated by eyeblink is assigned to a predefined task.

The paper is divided into eight sections. After the introduction, Sect. 2 discusses on the related work. Section 3 deals with the methodology adopted for blink extraction from EEG. Section 4 discusses the experimental setup for feature extraction and classification of objective generation of major commands. Classification algorithms and validation are discussed in Sect. 5. Section 6 explains the translation and implementation of commands into microtask. The last two sections deal with the results and conclusion.

2 Literature Survey

Eyeblinks have distinguished peaks with relatively strong voltages compared to other brain activities. The amplitude and duration of eyeblinks vary based on the movement of eyelids and type of control. The blink can occur under voluntary or involuntary control [5, 9]. Some researchers have proposed that eye artifacts are a viable source of EOG data, which can be measured from EEG, and used for communication and control applications such as wheelchair navigation, drowsiness detection systems, smart appliances, etc. [6, 10, 11, and 12]. The paper of Aydemir et al. [13] gives an approach to classify EOG and EMG signals from EEG. They used the features obtained by the root mean square method, polynomial fitting, and Hjorth descriptors, with 100 trials. Using a k-nearest neighbor (KNN) classifier, they achieved an accuracy of 94%. Gebrehiwot et al. [14] proposed a method to extract blink rate variability. They separated the blinks from EEG data by fast ICA algorithm and thresholding, which gave very high precision (100%) and recall (100%). A robotic arm opening and closing using winks obtained from EEG were done by Santiago et al. [15]. They used spectrogram features and thresholding to distinguish the commands with 95% accuracy. Rani et al. [8] suggested a method using noise filtering, thresholding, and peak detection to detect the eyeblinks for switching home lighting system. In their analysis, they were able to detect eyeblink successfully in 85% of the records. Similar work on using eyeblink-based control using time domain features of EEG can be seen in the works of Poorna et al. [16] and artificial neural network (ANN) to classify the blinks gave promising results.

The literature survey on related work gives an idea of developing a window based system using the secure and distinctive eye signals from EEG signals as a control for the low-cost prosthetic arm. The main difference between the developed system and systems available in the literature is that the subject can be trained using different coded commands for managing the micromechanical tasks. Using some of the above features as well as windowing and coding the commands into micromechanical tasks, an EEG-controlled prosthetic arm is developed at Amrita University. The design, implementation, and validation of the system are presented below.

3 Methodology

The main methodology used is to generate EEG commands to control the prosthetic arm through micromechanical tasks. The main framework includes: (1) Extraction of blinks and commands from EEG signals for the tasks such as: training the subject to code and to control the blinks, acquiring the command signal in the time window, preprocessing the EEG signals, selection of channels, (2) Generation of major task commands using machine learning methods: LDA or KNN using suitable feature extraction techniques. Another simpler method of counting the peaks to generate the command is also explained in this paper. Converting this blink count from any of the above-mentioned methods to pulse-coded commands (PCC). (3) The major task commands are transmitted to the translator for generating microtasks. The micro-controller uses these commands to control the servo motors in the prosthetic arm.

4 Experimental Setup

The experimental setup consists of digital acquisition and Wi-Fi transmission to the command processor, the processing of EEG data for command extraction, the transmission of commands to an Arduino microcontroller command interpreter (CI) and execution of commands as mechanical tasks by the CI.

4.1 Digital Acquisition of EEG Data

A montage of 10–20 standards with 14 electrodes was used to acquire the EEG data. The headset—Emotiv Epoc+ acquires digital EEG data at a 125 sample/sec/channel sampling rate and a 14-bit resolution. Data was sent to a Windows-based laptop over Wi-Fi, where it was decrypted and stored as integer arrays. Twenty untrained individuals were used to collect commands in random order. For ease of remembrance inverted thermometer coded pulses were used as numeric values corresponding to task. It was visually seen that the EEG signals corresponding to blinks appear in eight channels (AF3, F7, F3, FC5, AF4, F8, F4, and FC6). EOG signals are prominent in AF3, AF4, and F3 and F7. However, a weightage scheme described below was used for the final representation of channels.

4.2 Preprocessing of EEG Recording

The samples collected using the EMOTIV Epoc+ was processed and extracted in MATLAB platform. Raw EEG data from the 14 channels is shown in Fig. 1. The data collected contains varies noises. These are preprocessed by removing DC offset, and using filtering and smoothing techniques for each channel. A Butterworth bandpass filter with lower cut off at 4 Hz and upper cut off at 40 Hz was used. The time-varying baseline correction was done by polynomial fitting followed by subtraction. The signals are smoothed and plotted.

4.3 Extraction of Commands from EEG

The peaks were identified in each channel and weights were found. Weights are a nonlinear function of signal/noise (S/N), variance and the mean of peak amplitude and peak width. The weights were normalized to a maximum of 1 and weights less than 0.3 were set to zero. The weighted average of the eight channels gives the averaged pulses. A sample waveform of weighted EEG is shown in Fig. 2a. Channel amplitude below a threshold computed from peak amplitude maximum, mean and variance was also set to zero. The unipolar weighted peaks were further used for command identification. The EEG pulse train used for command extraction is shown in Fig. 2b. In the pulse train, peaks were identified using a time-based window. We can see that small stray peaks in Fig. 2a were removed using suitable threshold and were not detected in Fig. 2b. The number of peaks in each window was converted into pulse coded commands.

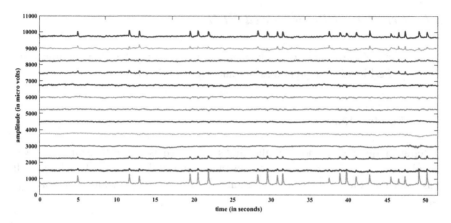

Fig. 1 Raw EEG data from 14 channels

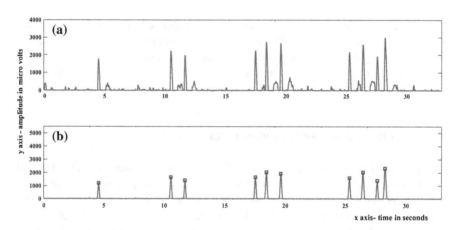

Fig. 2 **a** Weighted average of channels, **b** peak detection

5 Classification of Blinks Using Machine Learning Techniques

The preprocessed EEG signals corresponding to the blinks were then subjected to feature extraction. The features in time as well as spectral domain capable of classifying these blinks were extracted. The features include the spectral energy and the first and second spectral peaks and their corresponding locations in frequency domain and pulse width and number of pulses in the variable window in the time domain.

5.1 Features of Preprocessed EEG

The spectra of the blinks were obtained by applying log fast Fourier transform (FFT) to the blinks after preprocessing and segmenting them.

5.1.1 Spectral Energy

Spectral energy corresponding to the blinks was taken as one of the features. It was calculated as the sum of the squared absolute value of the spectra of each blink. It was seen that the spectral energy feature linearly increases with the number of blinks.

5.1.2 First and Second Spectral Maxima and Locations

The spectral maxima of the log spectrum were extracted using the peak picking method. The spectral maxima give the indication of dominant frequencies present in the eyeblink signals. The peak amplitudes in the magnitude spectrum for commands 1–4 is given in Fig. 3. The first and second spectral maxima in the log magnitude spectrum, as well as the corresponding frequencies locations, are used in this work.

5.1.3 Width and Number of Pulses in the Window

The width of the pulses is also a good indicator of the blink duration. The width of all pulses in each window was extracted using the peak picking algorithm. The count of peaks will give an indicator to the number of blinks. The feature corresponding to the width was calculated by summing up the width of all peaks inside the window. A total of seven feature vectors, one from spectral energy, four from the first and second spectral maxima and locations, two from pulse width and number of pulses were given to the supervised learning methods.

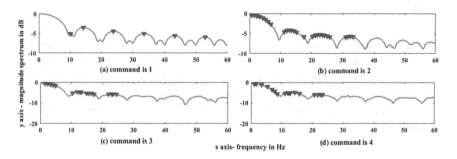

Fig. 3 Figure showing the log magnitude spectra of the commands

5.2 Supervised Learning Methods

Two supervised machine learning algorithms namely Linear Discriminant Analysis (LDA) and K-Nearest Neighbor (KNN) were applied to analyze how much accurately these features could distinguish and classify the blinks. The segregated blink number could be used to generate the control commands for the microcontroller. The performance of the acquired dataset containing the blinks is evaluated below.

5.2.1 Linear Discriminant Analysis (LDA)

LDA is a supervised learning method, which uses the weighted combination of features that separate the classes. The separating plane used in this method will be always linear in nature.

5.2.2 K-Nearest Neighbor (KNN)

KNN uses a clustering method to classify the features based on the highest vote of K-nearest neighbors. In this method, the votes were assigned to these neighbors based on a minimum Euclidean distance measure. 112 trials of eyeblink coded EEG data was recorded, noise eliminated, and features extracted to apply supervised learning. Seventy percent of the data is used for training the algorithms and 30% for testing and evaluating the performance.

5.3 Performance Evaluation of Classifiers

The performance of the classifiers was evaluated using the measures: accuracy, precision, and recall. The calculation of these parameters is given in Table 1. In the Table TP, TN, FP, and FN represent, respectively, the true positive, true negative, and false positive, and false negative elements in classification.

Table 1 Performance measures of LDA and KNN classification (all values in percentage)

Performance measure	Equation	LDA	KNN
Accuracy	$\frac{TP+TN}{TP+TN+FP+FN}$	97.7	70.69
Precision	$\frac{TP}{TP+FP}$	96	67.3
Recall	$\frac{TP}{TP+TN}$	95.24	68

6 Command and Control for Prosthetic Hand

The next sections explain how the major commands—the blinks—were converted into minor commands. The flowchart for command generation is given in Fig. 4. For this only one feature—the number of peaks of the blinks in each window—was used as the command to control the prosthetic arm.

6.1 Translation of Commands and Feature Extraction

The commands extracted from the EEG pulse were transferred to the microcontroller in the prosthetic arm via Bluetooth.

6.2 Arduino NanoV3.0

The microcontroller used in the process is Arduino Nano. The Arduino Nano 3x version used has the ATMEGA328 chip as its core microcontroller. It has 8 analog pins and 14 digital input–output pins of which 6 have pulse width modulated (PWM) output. The Arduino acts as a translator. It breaks the commands into a micromechanical movement that controls servo motors in the prosthetic arm. The commands were sent to Arduino via Bluetooth module placed inside the prosthetic arm.

6.3 Prosthetic Arm

An in-house built prosthetic arm was used as the end application. The design and manufacturing of anthropomorphic prosthetic hand were done using 3D printing

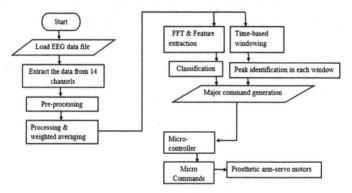

Fig. 4 Command generation

technique. The hand has 5 degrees of freedom with individually actuated fingers. The hand is designed to lift up to 2 kg [2].

7 Results and Discussion

In this work, we developed an EEG-controlled prosthetic arm. For this work coded eyeblinks from the EEG signals were used as the control. A method for improving the SNR of the blinks by averaging the channels was adopted in this work. Peak finding methods and thresholding help to remove the stray peaks and pickups that come in the region of consideration of EEG signals, from where the blinks were extracted. This will improve the robustness and reliability of the control application.

Here the major commands—the EEG signal with the coded eyeblinks were translated into minor commands—the controls by Arduino Nano, which in turn controls the prosthetic arm. The evaluation of commands was done using two methods. The first one is a simple method of counting the number of pulses in the eyeblinks and converting this feature into the controls. The second one uses the time and spectral domain features extracted from the EEG signals to generate the control with the help of supervised machine learning methods. Validation of the second system was done using two algorithms: Linear Discriminant Analysis (LDA) and K-Nearest Neighbor Classifier. The methods gave accuracy, precision, and recall for LDA as 97.7%, 96%, and 95.25% and KNN as 70.69%, 67.3%, and 68% respectively. Figure 5a–c shows the micromechanical tasks—Spherical grasp, Power gasp, Precision grasp—done by the EEG-controlled arm.

(a)   **(b)**   **(c)**

Fig. 5 Micromechanical tasks done by the prosthetic arm **a** spherical grasp, **b** power grasp, **c** precision grasp

8 Conclusion

A reliable prosthetic arm using EEG-based eyeblink commands and Arduino translator has been successfully designed, tested, and implemented. This system is immune to external pickups, artifacts, and unintentional ocular activities. In this paper, we are proposing two methods for generating the commands from the blinks. One is a direct method and the other using feature extraction and supervised learning methods to classify the blinks. The latter one will be superior compared to the simple counting of the blinks, used normally. The validation using these classifiers gave higher performance measures for LDA compared to KNN. We have shown only the preliminary results using LDA and KNN classifiers. The analysis can be extended to other classification methods also. The prosthetic arm is tested successfully to accomplish some of the micromechanical tasks such as precision, power, and spherical grasp through the commands extracted from the EEG signals. This is a real-time device and can thus potentially be used in daily life to support hand activities.

Acknowledgements We would also like to thank our mentors, department, and friends for giving us the courage and encouragement for completing this project.

References

1. http://health.costhelper.com/prosthetic-arms.html.
2. Jeethesh Pai U., Sarath N P., Sidharth R., Anu P Kumar., Pramod S and Ganesha Udupa. Design and manufacture of 3D printed Myoelectric multi-fingered hand for prosthetic application. International Conference on Robotics and Humanitarian Applications, Amritapuri, India. (2016).
3. R. K. Megalingam, Thulasi, A. Asokan, Krishna, R. Raj, Venkata, M. Katta, BV, A. Gupta, and Dutt, T. Uday, "Thought Controlled Wheelchair Using EEG Acquisition Device", Proceedings of 3rd International Conference on Advancements in Electronics and Power Engineering (ICAEPE'2013). Kuala Lumpur, Malaysia, 2013.
4. Abo-Zahhad, M., Ahmed, S.M. and Abbas, S.N., A new EEG acquisition protocol for biometric identification using eye blinking signals. International Journal of Intelligent Systems and Applications, 7(6), p. 48 (2015).
5. Kanoga, S., Nakanishi, M. and Mitsukura, Y., Assessing the effects of voluntary and involuntary eyeblinks in independent components of electroencephalogram. Neurocomputing, 193, pp. 20–32 (2016).
6. Belkacem, A.N., Hirose, H., Yoshimura, N., Shin, D. and Koike, Y., Classification of four eye directions from EEG signals for eye-movement-based communication systems. life, 1, p. 3 (2014).
7. Noureddin, B., Lawrence, P.D. and Birch, G.E., Online removal of eye movement and blink EEG artifacts using a high-speed eye tracker. IEEE Transactions on Biomedical Engineering, 59(8), pp. 2103–2110 (2012).
8. Rani, M.S. and Wahidah, B.T., 2009. Detection of eye blinks from EEG signals for home lighting system activation. In 2009 6th International Symposium on Mechatronics and its Applications, ISMA 2009.

9. Savelainen, A., An introduction to EEG artifacts. Independent research projects in applied mathematics (2010).
10. Fred Achic, Jhon Montero, Christian Penaloza, and Francisco Cuellar "Hybrid BCI System to Operate an Electric Wheelchair and a Robotic Arm for Navigation and Manipulation Tasks". IEEE International Workshop on Advanced Robotics and its Social Impacts (ARSO) Shanghai, China, (2016).
11. Megalingam, R.K., Gupta, B.A., Dutt, T.U. and Sushruth, A.H., August. Switch and thought controlled robotic arm for paralysis patients and arm amputees. In Global Humanitarian Technology Conference: South Asia Satellite (GHTC-SAS), 2013 IEEE (pp. 243–248). IEEE (2013).
12. Megalingam, R.K., Venkata, M.K., Ajithesh Gupta, B.V. and Dutt, T.U., EEG Acquisition Device For A Thought Controlled Robotic Arm. International Journal of Applied Engineering Research, 7(11), p. 2012.
13. Aydemir, O., Pourzare, S. and Kayikcioglu, T., Classifying various EMG and EOG artifacts in EEG signals. Przegląd Elektrotechniczny, 88(11a), pp. 218–222 (2012).
14. Gebrehiwot, T., Paprocki, R., Gradinscak, M. and Lenskiy, A., Extracting Blink Rate Variability from EEG Signals. International Journal of Machine Learning and Computing, 6 (3), p. 191 (2016).
15. Santiago Aguiar, Wilson Y´anez and Diego Ben´ıtez "Low Complexity Approach for Controlling a Robotic Arm Using the Emotiv EPOC Headset" IEEE International Autumn Meeting on Power, Electronics and Computing (ROPEC 2016). Ixtapa, Mexico, (2016).
16. Poorna S. S, PMVD Sai Baba, Lakshmi Ramya Gujjalapudi, Prasanna Poreddy, Aashritha L. S, Renjith S, G J Nair "Classification of EEG based control using ANN and KNN- A Comparison". IEEE International Conference on Computational Intelligence and Computing Research (ICCIC) 15–17th of December, 2016.

Wilcoxon Signed Rank Based Feature Selection for Sentiment Classification

S. Fouzia Sayeedunnisa, Nagaratna P. Hegde
and Khaleel Ur Rahman Khan

Abstract Sentiment analysis process is about gaining insights into the consumer's perception using the inputs like comments and opinions shared over the web platform. Most of the existing sentiment analysis models envisaged the complexities, which is due to high volume of features notified through standard selection/extraction process. In this manuscript, the proposed solution is about using statistical assessment strategies for selecting optimal features under sentiment lexicon context. The proposed solution relies on Wilcoxon signed score for finding significance of feature towards positive and negative sentiments. Concerning to performance analysis of the proposed solution, the experimental study is conducted using benchmark classifiers like SVM, NB and AdaBoost. Results from the experimental study depict that the proposed solution can support in attaining effective classification accuracy levels of 92%, upon using less than 40% of the features too.

Keywords Sentiment analysis · Opinion mining · Sentiment lexicons
Classification · Wilcoxon signed rank

S. F. Sayeedunnisa (✉)
Department of IT, M.J. College of Engineering and Technology, Hyderabad,
Telanagana, India
e-mail: fouzia.qadri@gmail.com

N. P. Hegde
Department of CSE, Vasavi College of Engineering, Hyderabad, Telanagana, India
e-mail: nagaratnaph@gmail.com

K. U. R. Khan
Department of CSE, ACE Engineering College, Hyderabad, Telanagana, India
e-mail: khaleelrkhan@aceec.ac.in

© Springer Nature Singapore Pte Ltd. 2018
V. Bhateja et al. (eds.), *Proceedings of the Second International Conference
on Computational Intelligence and Informatics*, Advances in Intelligent Systems
and Computing 712, https://doi.org/10.1007/978-981-10-8228-3_27

1 Introduction

Key aim of the sentiment analysis is about identifying and extracting opinions, perceptions and sentiments from the vast content generated online. In the exponential growth of internet usage and millions of users posting the content online expressing their views and opinions, such posts and published contents are becoming an integral source of information for the stakeholders of the business. As such, inputs in the form of comments and other such factors if organized properly can provide more insights to the decision-makers over the market trends and the consumer perceptions over a brand, product and the services [1, 2].

The sentiment or opinion analysis approaches vividly relies on information retrieval and natural language processing techniques. Inferring sentiment is one of the significant challenges envisaged in the process of sentiment analysis [3]. The key process of sentiment analysis is carried out in two steps, wherein the first step involves the process of selecting and extracting the features from textual opinions and in second step, the classification of sentiments from samples to multi-classes is carried out.

In the sentiment classification process, the key challenge is about large size dimension, irrelevant and the features that are overlapping [4]. Feature Selection (FS) plays a wide role in the sentiment analysis, and selecting an optimal subset feature based on certain criteria, without making much change to the actual data content, is the crux of process [5].

Numerous techniques are discussed earlier for the feature selection process using the machine learning solutions [6]. It is imperative from [7] that if right kind of feature selection methods were proposed. Such solutions can lead to the elimination of irrelevant features from the vector, thus leading to reduced size of feature vector and increased accuracy of sentiment classification.

FS and extraction techniques are categorized with three key dimensions. First, the techniques are developed for addressing the problem of over-fitting and improving the performance of SA. The second dimension is about emphasis on cost-effective and less time complexity oriented solutions. The third dimension is to gain insights into the basic process for the data generated from sources.

In a study [8], the FS methods classified into three approaches and they are filtering, wrapping and embedded approaches. In the filter model, it offers choice of selecting optimal subset of features using scaling and elimination of low-scoring features. Though the process is scalable to high-dimensional feature space, the challenge is that the interaction with classifier and relationship among the features is ignored in this approach.

In the case of wrapper model, generating and evaluation of different subsets are used as prime aspect for optimal feature selection. Heuristic search methods play a key role in solving the exponential time toward finding a feature subset. The key benefit of wrapper method is that there is a semantic relationship between model selections and subsets of varied search features. Some of the challenges in the approach are about dependency on classification algorithms, over-fitting risks and computational complexities.

Embedded approach handles the search process using combination of model hypothesis and feature subsets space toward a classifier structure. Such process reduces the time complexity than the other approaches and offers the benefit of interaction with classifier models. However, the challenge is about the classifier-dependent selection process [9].

2 Related Work

Feature selection and extraction play a key role in the selection and extraction of highly relevant features, thus leading to collection of optimal subset of features that could use for training the models over the machine learning and pattern recognition solutions [10].

Sentiment analysis is profoundly dependent on three levels of granularities like the document level, sentence level and aspect level [11, 12]. Classification of sentiment based on document-level granularity depends on classifying one whole document into expression in terms of positive or negative sentiment [11, 13].

Research by [14, 15] has adapted sentence-level granularity for determining the sentiment as positive or negative or as neutral, from every sentence level. Such a task often classified as subjectivity classification in the contemporary literature. In a study by [16], the similarity between two types of feature selection methods like the invariant and multivariate is evaluated using data-intensive classification tasks. The study [17] has used the filter methods for feature selection. Five methods are used for feature selections scores, and they are Document Frequency (DF), Information Gain (IG), Chi-Square (CHI), Mutual Information (MI) and Term Frequency (TF). Using the previously mentioned elements, 97% of the low-scoring features are eliminated, and the result has significant improvement in accuracy levels.

In another work by [18], classification accuracy has significantly improved by eliminating low-scoring features, using IG and CHI methods than the other three methods [9].

However, in another work by [19], features are scored based on features distribution using Standard Deviation (SD) of features. Such a hybrid model has been resourceful in solving the problem of low-accuracy sentiment classification in the filtering methods. In addition, the method of higher computation burden in wrapper methods is also addressed, when the wrapper method applied to feature subset is selected in filtering method [20]. A high-dimensional reduction in terms of integration of feature selection and feature extraction methods is reviewed in the study [10].

The model devised in [21] is optimizing the features based on the sentiment polarity such as positive and negative sentiments. In order to optimize the features, this model devised a statistical approach called count-vectorizer. This model is much close to the model that is proposed here in this manuscript. The considerable constraint of the count-vectorizer is that it relies on traditional metrics like term frequency and inverse of document frequency; moreover, the feature selection is performed at document level, which would lead to increase the number of features selected.

It is imperative that majority of the researches relying on machine learning techniques for sentiment classification in a supervised approach adapts Multilayer Perceptron (MLP), NB [22], ME [23], SVM [24], linear discriminant function [25] and AdaBoost [26].

It is imperative from the review of literature that integration of two strategies toward feature selection can be more resourceful. In the proposed study, the focus is on integration of varied feature vectors and feature subsets using term association frequency metric and Wilcoxon signed score [27].

3   Methods and Materials

This section explores metrics used to select optimal features and classifiers used to estimate the significance of the proposed feature selection metric.

3.1   Feature Selection Metrics

This explores the metrics term occurrence, term presence and Wilcoxon signed rank score used for feature selection.

3.1.1   Term Occurrence

Pang et al. [28] originally used the feature occurrence method in sentiment analysis. The method used the term occurrence, i.e., the occurrence that each unigram exists within a document, as the feature values for that document. Therefore, if the word "excellent" appeared in a document ten times, the associated feature would have a value of ten. This approach is adapted to here in this proposal, which is notifying the term occurrence in entire corpus instead of each document.

3.1.2   Term Presence

Pang et al. [28] were first to use feature presence in sentiment analysis. Feature presence is very similar to feature frequency, except that rather than using the frequency of a unigram as its value, we would merely use it to indicate that the sentiment lexicon exists in the opinion document. Multiple occurrences of the same lexicon are ignored, so we get a vector of binary values, with ones for each unique lexicon that occurs in the document and zeros for all lexicons that appear in the corpus but not in the opinion document

3.1.3 Wilcoxon Signed Rank

The Wilcoxon signed rank test [27] is a close sibling of the dependent samples t-test. Because the dependent samples t-tests analyses if the average difference of two repeated measures is zero, it requires metric (interval or ratio) and normally distributed data; the Wilcoxon signed test uses ranked or ordinal data. Thus, it is a common alternative to the dependent samples t-test when its assumptions not met.

3.2 Classifiers

In the process of machine learning and statistics modelling, identifying the classification of a set of relevant categories (including subsets) and right kind of mapping for the new observation is the key process. The whole process relies on efficacy of the data training set comprising relevant observations and its category membership defined. In the machine learning models, instance of supervised learning defined as classification and the unsupervised procedure considered as clustering. The process of clustering is grouping data into various categories based on certain measures of innate similarities or features.

Support Vector Machine (SVM) [24] is a structured supervised machine learning algorithm that can be adapted for classification and regression challenges. However, it is usually used in the classification problems. In the algorithm, every data item is plotted as a point in n-dimensional space (Let n be the number of features comprised.) with value of every coordinate marked as value of each of the features; then, the process of classification is performed by observing the hyperplane, which can differentiate the two classes in an effective manner. Support vectors are simple coordinates of individual observation, and SVM is a frontier that effectively segregates the two classes.

Naive Bayes technique [22] that depends on independence among predictors assumes that no particular feature present in the class is related to any other feature. For instance, apple is considered as fruit, if it is red, round and with a specific measure of diameter. Though such features might rely upon each other and upon existence of other features, certainly all such properties contribute independently toward the probability that the fruit is an apple and such process is known as Naive Bayes model. The NB model is very easy to build and specifically resourceful for handling huge volume datasets. In addition to the simple process, NB is known for its efficacy and performance compared to many of high-end classification methods. Using the predictive probability and prior probability of a class, estimating posterior probability is feasible with Bayes theorem.

AdaBoost [26] is resourceful for boosting the performance of decision trees over the binary classification problems. It is also very effective in boosting the performance of machine-level algorithms, and profoundly in the case of weak learners. Such models attain accuracy with random chance of a classification problem. Decision trees of one level are most suited and common algorithm used with AdaBoost. As the trees are short and comprise one decision for classification, they often have known as decision stumps.

4 Feature Selection

The adopted method to select optimal features is naval and aimed to nullify the impact of the volume of opinion documents given as input for classifier training. The devised feature selection approach is neutral to the 1-gram, 2-gram, tri-gram and n-gram scalable to prune the insignificant features. In the initial level of the approach, the significant sentiment lexicons presented in given opinion documents shall select. Further, the terms appeared in the given documents will be selected as features, which is based on their frequency associated with the selected sentiment lexicons. Further, the significant features among the selected will notify, which is based on the Wilcoxon signed rank obtained for features associated with positive sentiment lexicon and corresponding negative sentiment lexicon. The sentiment lexicons appeared in the given training set will be selected using the feature selection method called term presence. Further, the terms appeared in given documents will be selected as features by their feature association frequency, which is the ratio between the "occurrence of term in association with corresponding sentiment lexicon and the actual term occurrence".

Then, these feature association frequencies are represented as matrix such that each row consists of feature association frequency observed between a sentiment lexicon and all features, and feature association frequency is observed for a feature

	Feature 1	Feature 2	.	.	Feature n
Lexicon 1	Term Association Frequency between lexicon 1 and feature 1	Term Association Frequency between lexicon 1 and feature 2	.	.	Term Association Frequency between lexicon 1 and feature n
Lexicon 2	Term Association Frequency between lexicon 2 and feature 1	Term Association Frequency between lexicon 2 and feature 2	.	.	Term Association Frequency between lexicon 2 and feature n
.	.	.	.	.	.
.	.	.	.	.	.
Lexicon m	Term Association Frequency between lexicon m and feature 1	Term Association Frequency between lexicon m and feature 2	.	.	Term Association Frequency between lexicon m and feature n

Fig. 1 Depiction of feature and term association frequencies as matrix

to all lexicons as a column, which is depicted in Fig. 1. The similar matrix is built for both positive and negative sentiment lexicons.

Further process depicts Wilcoxon signed rank for each feature, which is based on the variance observed between association frequency of the corresponding feature, and positvie sentiment lexicons and association frequency of the corresponding feature and negative sentiment lexicons. These association frequencies can obtain from the matrix projected in Fig. 1.

Further, the optimal features will be selected based on their Wilcoxon signed rank significance.

4.1 Corpus Preprocessing

Let given corpus *CRP* contains opinion documents for training that labelled as positive or negative. The initial step of the preprocess is to build a record set, such that each row represents the bag of words found in each document in given corpus. Then, the non-textual words and stop words will prune from the record set formed. Further, the stemming is applied to the words found in each record. Then, these resultant records partitioned into two sets *PR, NR* that represent the corresponding positive and negative opinions in the given corpus. Upon completion of the corpus processing, sentiment lexicons and term selection process will be initiated on positive and negative record sets *PR, NR*, respectively.

4.2 Sentiment Lexicons Selection

In order to select sentiment lexicons [29], we adapt term presence metric, which is a binary metric that denotes the selected lexicon is presented in respective record set or not. This metric is used to select the positive and negative lexicons presented in respective record sets *PR, NR*. The lexicons selected for positive and negative record sets *PR, NR* will be represented as respective sets *pl, nl*.

4.3 Sentiment Term Selection as Features

The metric term occurrence is used initially that produces term occurrence record set *TC*. Each record in *TC* denotes the term t and its occurrence count $c(t)$, which is the sum of its occurrence in both positive and negative record sets *PR, NR*.

Further, two matrices tlm_{PR}, tlm_{NR} will be formed, such that each row in tlm_{PR} represents the occurrence of each positive lexicon $\{l \exists l \in pl\}$ in association with each term $\{t \exists t \in TC\}$ and each row in tlm_{NR} represents the occurrence of each negative lexicon $\{l \exists l \in nl\}$ occurrence in association with each term $\{t \exists t \in TC\}$. The sizes of these matrices are $|pl| \times |TC|$ and $|nl| \times |TC|$ in respective order of tlm_{PR}, tlm_{NR}.

Finally, the matrices taf_{PR}, taf_{NR} will be formed that are congruent to the matrices tlm_{PR}, tlm_{NR} those representing the feature association frequency, which is the ratio between occurrences of term association with corresponding sentiment lexicon and the term occurrence. Each column $[i,j]$ in taf_{PR} represents the feature association frequency, which is the ratio between the value found at column $[i,j]$ in matrix tlm_{PR} and the occurrence $c(t_j)$ of the term t_j that represents the column j in both matrices tlm_{PR} and taf_{PR}. Similarly, the column values of the matrix taf_{NR} will be assessed with respective to matrix tlm_{NR}.

4.4 Optimal Feature Selection

The optimal features will be selected from taf_{PR}, taf_{NR}, which is based on the Wilcoxon signed rank between the feature association with positive and negative lexicons. In order to this, Wilcoxon signed rank is estimated between the congruent columns represented by a feature in both matrices taf_{PR} and taf_{NR}. If Wilcoxon signed rank is found to be best at given p-value [30], then the respective feature is found to be optimal to positive set if and only if sum of Wilcoxon positive rank sum is greater than the Wilcox negative rank sum, else the feature is optimal to negative set. The Wilcoxon signed rank score is estimated as follows:

Find the difference between respective entries in both columns in sets taf_{PR}, taf_{NR} with respect to feature t, rank the differences in ascending order of their absolute value, and then apply the sign of the difference to the ranks. Afterward, find the sum of negative ranks W_- and positive ranks W_+ from the signed ranks. The sum of absolute values of W_- and W_+ is identically equal to $n(n+1)/2$; here, n is the maximum of the lengths of both columns represented by feature t in both matrices taf_{PR}, taf_{NR}. Further, according to Wilcox signed rank test proof [27], if W_- is best at given p-value threshold, which can be the degree of probability (p-value) [30] in w-table [31] for the signed rank score obtained, then the feature t is optimal to positive set if W_+ is greater than the absolute values of W_-, else the feature t is optimal to negative set. The algorithmic representation of the feature selection process is depicted as follows:

Let *corp* be the corpus of labelled opinion documents

Let *PR* be the empty record set that will be used to store all processed positive labelled opinion documents as records

Let *NR* be the empty record set that will be used to store all processed negative labelled opinion documents as records

Let $LS+$ be the positive sentiment lexicons set [29]

Let $LS-$ be the negative sentiment lexicons set [29]

Let l_+ be the set contains lexicons of $LS+$, those exists in one or more records of *PR*

Let l_- be the set contains lexicons of $LS-$, those exists in one or more records of *NR*

> ➤ //PREPROCESSING// (STEP 1 TO STEP 6)

step 1. $\overset{|corp|}{\underset{i=1}{\forall}} \{d_i \exists d_i \in corp\}$ Begin

step 2. Split d_i in to bag of words $bg(d_i)$

step 3. Prune stop-words and non-textual words from $bg(d_i)$

step 4. $if(d_i$ is positive$)$ then $PR \leftarrow bg(d_i)$

step 5. Else $NR \leftarrow bg(d_i)$

step 6. End //of Step 1

> ➤ // LEXICONS SELECTION BY TERM PRESENCE METRIC// (STEP 7 TO STEP 20)

step 7. $\overset{|LS+|}{\underset{j=1}{\forall}} \{l_j \exists l_j \in LS+\}$ Begin // for each positive lexicon in $LS+$

step 8. $\overset{|PR|}{\underset{i=1}{\forall}} \{r_i \exists r_i \in PR\}$ Begin // for each record in *PR*

step 9. $if(l_j \in r_i)$ Begin

step 10. $l_+ \leftarrow l_j$

step 11. Go to Step 7

step 12. End //of step 9

step 13. Else Go to Step 8
End // of step 8
End // of step 7

step 14. $\overset{|LS-|}{\underset{j=1}{\forall}} \{l_j \exists l_j \in LS-\}$ Begin // for each negative lexicon in $LS-$

step 15. $\overset{|NR|}{\underset{i=1}{\forall}} \{r_i \exists r_i \in NR\}$ Begin // for each record in *NR*

step 16. $if(l_j \in r_i)$ Begin

step 17. $l_- \leftarrow l_j$

step 18. Go to Step 14

step 19. End // of step 16

step 20. Else Go to Step 15
End //of step 15
End // of step 14

> ➤ // FEATURE SELECTION BY TERM OCCURRENCE AND FEATURE ASSOCIATION FREQUENCY // (STEP 21 TO 75)
Let *wl* be the empty set that is used to store unique words exist in both PR and NR record sets.
//Collecting all terms exists // (step 21 to step 34)

step 21. $\overset{|PR|}{\underset{i=1}{\forall}}\{r_i \exists r_i \in PR\}$ Begin // each record of PR

step 22. $\overset{|r_i|}{\underset{j=1}{\forall}}\{w_j \exists w_j \in r_i\}$ Begin // Each word exists in record r_i

step 23. $if(w_j \notin l_+ \&\& w_j \notin wl)$ Begin

// if word w_j is not a positive sentiment lexicon and not exists in word list

step 24. $wl \leftarrow w_j$

step 25. End // of step 23
step 26. End // of step 22
step 27. End //of step 21

step 28. $\overset{|NR|}{\underset{i=1}{\forall}}\{r_i \exists r_i \in NR\}$ Begin // each record of NR

step 29. $\overset{|r_i|}{\underset{j=1}{\forall}}\{w_j \exists w_j \in r_i\}$ Begin // Each word exists in record r_i

step 30. $if(w_j \notin l_- \&\& w_j \notin wl)$ Begin

// if word w_j is not a negative sentiment lexicon and not exists in word list

step 31. $wl \leftarrow w_j$

step 32. End // of step 30
step 33. End // of step 29
step 34. End //of step 28

//Measuring Term occurrence// (step 35 to step 46)

step 35. $TC \leftarrow \phi$ // is an empty map that is using to store a map of each term and its occurrence count

step 36. $\overset{|wl|}{\underset{i=1}{\forall}}\{t_i \exists t_i \in wl\}$ Begin// for each term exists in word list wl

step 37. $c(t_i) = 0$ // term occurrence count $c(t_i)$ of term t_i is initialized to 0

step 38. $\overset{|PR|}{\underset{j=1}{\forall}}\{r_j \exists r_j \in PR\}$ Begin // each record of PR

step 39. $if(r_j \ni t_i)$ $c(t_i) = c(t_i) + 1$ // if record r_j contains the term t_i

step 40. End // of step 38

step 41. $\overset{|NR|}{\underset{j=1}{\forall}}\{r_j \exists r_j \in NR\}$ Begin // each record of NR

step 42. $if(r_j \ni t_i)$ $c(t_i) = c(t_i) + 1$ // if record r_j contains the term t_i

step 43. End // of step 41
step 44. $TC \leftarrow \{t_i, c(t_i)\}$
step 45. End //of step 36

//Finding term and sentiment lexicon association occurrence// (Step 46 to step 67)

step 46. $TAC_{PR} \leftarrow \phi$ // is an empty map set contains the map, where term t and positive sentiment lexicon l association and their co-occurrence in record set PR

step 47. $TAC_{NR} \leftarrow \phi$ // is an empty map set contains the map, where term t and negative sentiment lexicon l association maps to their co-occurrence in record set NR

step 48. $\overset{|wl|}{\underset{i=1}{\forall}}\{t_i \exists t_i \in wl\}$ Begin// for each term exists in word list wl

step 49. $\overset{|l_+|}{\underset{j=1}{\forall}}\{l_j \exists l_j \in l_+\}$ Begin // for each positive lexicon in l_+

step 50. $c(t_i, l_j) = 0$ // the co-occurrence count of the term and lexicon in record set PR

step 51. $\overset{|PR|}{\underset{k=1}{\forall}}\{r_k \exists r_k \in PR\}$ Begin // for each record in PR

step 52. $if(t_i \in r_k \& l_j \in r_k)$ Begin // if both term and lexicon exists in record

step 53. $c(t_i, l_j)++$

step 54. End // of step 52
step 55. End // of step 51
step 56. $TAC_{PR} \leftarrow \{(t_i, l_j), c(t_i, l_j)\}$

step 57. End // of step 49

step 58. $\overset{|L|}{\underset{j=1}{\forall}}\{l_j \exists l_j \in l_-\}$ Begin // for each negative sentiment lexicon in l_-

step 59. $c(t_i, l_j) = 0$ // the co-occurrence count of the term and lexicon in record set NR

step 60. $\overset{|PR|}{\underset{k=1}{\forall}}\{r_k \exists r_k \in PR\}$ Begin // for each record in PR

step 61. $if(t_i \in r_k \& l_j \in r_k)$ Begin // if both term and lexicon exists in record

step 62. $c(t_i, l_j)++$

step 63. End // of step 61
step 64. End // of step 60
step 65. $TAC_{NR} \leftarrow \{(t_i, l_j), c(t_i, l_j)\}$

step 66. End // of step 58
step 67. End // of step 48

//Finding Feature Association Frequency// (Step 68 to Step 75)

step 68. $\begin{aligned} FAF_{PR} &\leftarrow \phi \\ FAF_{NR} &\leftarrow \phi \end{aligned}$

// Empty map sets that are using further to store term and lexicon association and the respective Feature Association Frequency as a map in respective to record sets PR, NR

step 69. $\overset{|wl|}{\underset{i=1}{\forall}}\{t_i \exists t_i \in wl\}$ Begin // for each term exists in word list wl

step 70. $\overset{|L|}{\underset{j=1}{\forall}}\{l_j \exists l_j \in l_+\}$ Begin // for each positive lexicon in l_+

step 71. $FAF_{PR} \leftarrow \left\{ (t_i, l_j), \dfrac{\{c(t_i, l_j) \exists TAC_{PR} \ni c(t_i, l_j)\}}{\{c(t_i) \exists TC_{PR} \ni c(t_i)\}} \right\}$

// ratio between term and positive sentiment lexicon co-occurrence count in record set PR and the respective term occurrence in both PR, NR

step 72. End //of step 70

step 73. $\overset{|L|}{\underset{j=1}{\forall}}\{l_j \exists l_j \in l_-\}$ Begin // for each negative lexicon in l_-

step 74. $FAF_{NR} \leftarrow \left\{ (t_i, l_j), \dfrac{\{c(t_i, l_j) \exists TAC_{NR} \ni c(t_i, l_j)\}}{\{c(t_i) \exists TC_{NR} \ni c(t_i)\}} \right\}$

//ratio between term and negative sentiment lexicon co-occurrence count in record set NR and the respective term occurrence in both PR, NR

End // of step 73
step 75. End //of Step 69

➤ **//FINDING OPTIMAL FEATURES BY WILCOX SIGNED RANK SCORE// (STEP 76 TO STEP 86)** step 76. $\bigvee_{i=1}^{|wl|} \{t_i \exists t_i \in wl\}$ Begin// for each term exists in word list wl .

step 77. $\begin{aligned} P &\leftarrow \phi \\ Q &\leftarrow \phi \end{aligned}$ //The empty sets that are using to store feature association frequency of a

term with respective positive and negative sentiment lexicons

step 78. $\bigvee_{j=1}^{\max(|l_+|,|l_-|)} \{p_j \exists p_j \in l_+, q_j \exists q_j \in l_-\}$ Begin // for each positive and negative lexicon

step 79. $P \leftarrow p_j$ // add p_j to P

step 80. $Q \leftarrow q_j$ //add q_j to Q

step 81. Find W-value (P,Q) (see sec 4.4) // Finding Wilcoxon signed rank score between P,Q .

step 82. If (W-value is not fit probability degree threshold given) Begin

step 83. Discard t_i from the features list.

step 84. End //step of 82

step 85. End //step 78

step 86. End //step 76

5 Experimental Study

At the initial phase of the experimental study, the given datasets are preprocessed and converted to set of records such that each record is of the set of words representing respective opinion document. As a furtherance, proposed feature association frequency and Wilcoxon signed rank score techniques are applied to find the optimal features from the processed dataset.

The implementation of proposed model and other contemporary model [21] considered for performance analysis is carried out using Java 8 running on a computer with i5 processor and 4 GB ram. The classification accuracy assessment is performed by scripts defined in R programming language [32].

The proposed feature selection technique delivers a unique reduced set of features. Then, the resultant optimal feature sets from distinctive datasets tested for accuracy using three different classifiers called SVM, NB and AdaBoost. Similar kind of corroboration is carried out on other technique called count-vectorizer [21], which merely have the similar context of proposal. This model [21] is reducing the features using count-vectorizer technique that is applied on the polarity of the sentiment like positive and negative. In the final step, validation and analysis of the minimized feature sets are performed depending on anatomic relevance. Figure 2 indicates the figurative representation of the approach.

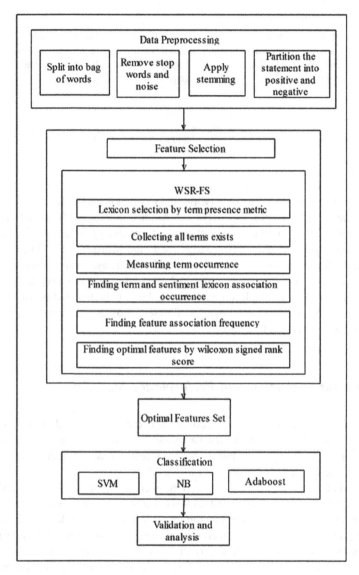

Fig. 2 Schematic diagram of proposed approach

5.1 *Datasets and Statistics*

Twitter datasets for sentiment analysis [33], movie reviews dataset [34] and product reviews dataset [35] were three datasets that are considered for analysis. The movie review dataset is having 27886 reviews; among them, 21933 reviews are considered for experiments and rest are discarded due to the difficulty noticed

Table 1 Data statistics for each dataset

Dataset	Total # of records	Number of records used for training	Number of records used for testing	No # sentiment representative lexicons found in training set
Twitter dataset	29700	20790	8910	39
Movie reviews	21933	15353	6580	37
Product reviews	28310	19817	8493	21

at preprocessing step. The original twitter dataset that is available at [33] is having tweets labelled as positive and negative. Among them, 29700 records retained after preprocessing. The other dataset product reviews of Amazon Instant Video from Amazon product data are provided by Julian McAuley; UCSD [35] that is considered for experiments is having 37126 reviews. Among these, 28310 reviews retained after preprocessing. All the datasets were preprocessed and number of instances available for each of the datasets is depicted in Table 1. The movie reviews and product reviews datasets possess uniform distribution, whereas Twitter dataset is skewed.

5.2 Feature Selection and Performance Statistics

The feature selection techniques Wilcoxon signed rank score and count-vectorizer matrix that are chosen and the list of reduced set of attributes are depicted in Table 2 and Fig. 2. In furtherance, reduced features are validated using three classifiers and the results are depicted in Fig. 3.

Validation is carried out based on rightly classified instances. And from the implementation of Wilcoxon signed rank score and count-vectorizer matrix feature selection techniques, it is evident that the features selected by Wilcoxon signed rank score are almost subset of the features selected by count-vectorizer matrix and marginal difference found to be nearly 40%. Performance dip is observed for NB and for a specific dataset when SVM is a classifier (see Fig. 4) compared to AdaBoost. In overall, the results depict that the feature selection by proposed feature association frequency and Wilcoxon signed rank score retains the overall performance of the classifiers under minimal number of features compared to the count-vectorizer matrix approach.

AdaBoost classifier has performed best for three datasets and resulted in accuracy of above 85%. All the other classifiers too have shown significant improvement in the performance. The classifier NB is resulting above 75%, and SVM is resulting above 80%. Despite that the twitter dataset and the movie review dataset are

Table 2 The statistics of the results obtained from feature selection strategy

Dataset	No of features found in training set	No # optimal features by Wilcoxon signed rank score	No # optimal features by count-vectorizer matrix
Twitter dataset	1985	205	337
Movie reviews	2371	152	263
Product reviews	1411	86	141

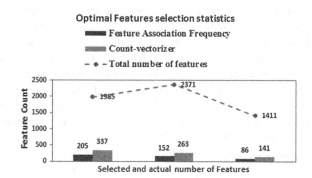

Fig. 3 Optimal features selection statistics

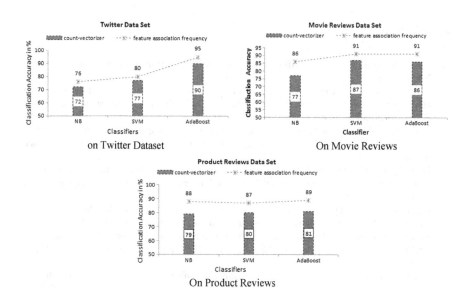

Fig. 4 Classifier accuracy statistics

leveraging maximum performance of 95% and 91%, respectively, it is imperative that actual relevance is of features selected by Wilcoxon signed rank score, thus leading to study of features selected by count-vectorizer matrix for each dataset.

6   Conclusion

Sentiment analysis has become an integral need for the organizations to understand the consumer expectations, buying behaviour and toward analysing the key factors that could influence the decision-making. With the emerging BI trends, sentiment analysis has gained momentum and numerous researches have carried out in the domain. In this paper, the emphasis has been overusing the three classifiers SVM, NB and AdaBoost for feature selection strategies that could make significant impact over the outcome. From the experimental studies that carried over vivid range of datasets, it is imperative that the AdaBoost classifier has outperformed with accuracy levels near around of 90% across all the three datasets. Though optimal features discovered under Wilcoxon signed rank score is lesser than the optimal features discovered using count-vectorizer matrix, however, the accuracy of classification is much higher in the case of Wilcoxon signed rank score analysis. Such an outcome signifies the impact of AdaBoost classifier that outperforms the other two classifiers for features selected by Wilcoxon signed rank score and count-vectorizer matrix.

References

1. Hu, Minqing, and Bing Liu. "Mining opinion features in customer reviews." AAAI. Vol. 4. No. 4. 2004.
2. Liu, Yang, et al. "ARSA: a sentiment-aware model for predicting sales performance using blogs." Proceedings of the 30th annual international ACM SIGIR conference on Research and development in information retrieval. ACM, 2007.
3. Gangemi, Aldo, Valentina Presutti, and Diego Reforgiato Recupero. "Frame-based detection of opinion holders and topics: a model and a tool." IEEE Computational Intelligence Magazine 9.1 (2014): 20–30.
4. Elawady, Rasheed M., Sherif Barakat, and Nora M. Elrashidy. "Different feature selection for sentiment classification." International Journal of Information Science and Intelligent System 3.1 (2014): 137–150.
5. Liu, Huan, and Lei Yu. "Toward integrating feature selection algorithms for classification and clustering." IEEE Transactions on knowledge and data engineering 17.4 (2005): 491–502.
6. Koncz, Peter, and Jan Paralic. "An approach to feature selection for sentiment analysis." Intelligent Engineering Systems (INES), 2011 15th IEEE International Conference on IEEE, 2011.
7. Agarwal, Basant, and Namita Mittal. "Sentiment classification using rough set based hybrid feature selection." Proceedings of the 4th workshop on computational approaches to subjectivity, sentiment and social media analysis (WASSA'13), NAACL-HLT. Atlanta. 2013.

8. Saeys, Yvan, IñakiInza, and Pedro Larrañaga. "A review of feature selection techniques in bioinformatics." bioinformatics 23.19 (2007): 2507–2517.
9. Yousefpour, Alireza, et al. "A comparative study on sentiment analysis." Advances in Environmental Biology (2014): 53–69.
10. Bharti, Kusum Kumari, and Pramod Kumar Singh. "Hybrid dimension reduction by integrating feature selection with feature extraction method for text clustering." Expert Systems with Applications 42.6 (2015): 3105–3114.
11. Liu, Bing. "Sentiment analysis and opinion mining." Synthesis lectures on human language technologies 5.1 (2012): 1–167.
12. Yousefpour, Alireza, Roliana Ibrahim, and Haza Nuzly Abdull Hamed. "A novel feature reduction method in sentiment analysis." International Journal of Innovative Computing 4.1 (2014): 34–40.
13. Taboada, Maite, et al. "Lexicon-based methods for sentiment analysis." Computational linguistics 37.2 (2011): 267–307.
14. McDonald, Ryan, et al. "Structured models for fine-to-coarse sentiment analysis." Annual Meeting-Association for Computational Linguistics. Vol. 45. No. 1. 2007.
15. Nakagawa, Tetsuji, Kentaro Inui, and Sadao Kurohashi. "Dependency tree-based sentiment classification using CRFs with hidden variables." Human Language Technologies: The 2010 Annual Conference of the North American Chapter of the Association for Computational Linguistics. Association for Computational Linguistics, 2010.
16. Dessì, Nicoletta, and Barbara Pes. "Similarity of feature selection methods: An empirical study across data intensive classification tasks." Expert Systems with Applications 42.10 (2015): 4632–4642.
17. Rogati, Monica, and Yiming Yang. "High-performing feature selection for text classification." Proceedings of the eleventh international conference on Information and knowledge management. ACM, 2002.
18. Yang, Yiming, and Jan O. Pedersen. "A comparative study on feature selection in text categorization." Icml. Vol. 97. 1997.
19. Yousefpour, Alireza, et al. "Feature reduction using standard deviation with different subsets selection in sentiment analysis." Asian Conference on Intelligent Information and Database Systems. Springer International Publishing, 2014.
20. Liao, T. Warren. "Feature extraction and selection from acoustic emission signals with an application in grinding wheel condition monitoring." Engineering Applications of Artificial Intelligence 23.1 (2010): 74–84.
21. Tripathy, Abinash, Ankit Agrawal, and Santanu Kumar Rath. "Classification of sentiment reviews using n-gram machine learning approach." Expert Systems with Applications 57 (2016): 117–126.
22. Murphy, Kevin P. "Naive bayes classifiers." University of British Columbia (2006).
23. Speriosu, Michael, et al. "Twitter polarity classification with label propagation over lexical links and the follower graph." Proceedings of the First workshop on Unsupervised Learning in NLP. Association for Computational Linguistics, 2011.
24. Suykens, Johan AK, and Joos Vandewalle. "Least squares support vector machine classifiers." Neural processing letters 9.3 (1999): 293–300.
25. Guyon, Isabelle, et al. "Gene selection for cancer classification using support vector machines." Machine learning 46.1–3 (2002): 389–422.
26. An, Tae-Ki, and Moon-Hyun Kim. "A new diverse AdaBoost classifier." Artificial Intelligence and Computational Intelligence (AICI), 2010 International Conference on. Vol. 1. IEEE, 2010.
27. Rey, Denise, and Markus Neuhäuser. "Wilcoxon-signed-rank test." International encyclopedia of statistical science. Springer Berlin Heidelberg, 2011. 1658–1659.
28. B. Pang, L. Lee, and S. Vaithyanathan. Thumbs up?: sentiment classification using machine learning techniques. In EMNLP '02: Proc. of the ACL-02 conf. on Empirical methods in natural language processing, pages 79–86. ACL, 2002.

29. http://web.stanford.edu/class/cs424p/materials/ling287-handout-09–21-lexicons.pdf.
30. Sahoo, PK, and Riedel T. Mean Value Theorems. Functional Equations. World Scientific, 1998.
31. http://math.ucalgary.ca/files/math/wilcoxon_signed_rank_table.pdf.
32. Ihaka R, Gentleman R. R: a language for data analysis and graphics. Journal of computational and graphical statistics. 1996 Sep 1;5(3):299–314.
33. http://thinknook.com/wp-content/uploads/2012/09/Sentiment-Analysis-Dataset.zip.
34. http://www.cs.cornell.edu/people/pabo/movie-review-data/.
35. http://jmcauley.ucsd.edu/data/amazon/.

A Color Transformation Approach to Retrieve Cloudy Pixels in Daytime Satellite Images

Rachana Gupta, Satyasai Jagannath Nanda and Pradip Panchal

Abstract Accurate cloud detection is a key focus area for several researchers to determine the parameters of earth's energy budget. It is a challenging task in the visible range due to resembling characteristics of thick clouds and snow/ice, difficulty in classifying faded texture soil and seasonal clouds (helpful for weather forecast), combined separation of vegetation and water against clouds. In this paper, a new color transformation approach is developed to classify thick clouds against three natural territories, i.e., water, vegetation, and soil. The proposed approach is implemented in two steps: preprocessing (includes color filter array interpolation method) and detection (a color transformation approach is introduced for classification). Extensive simulation studies are carried out on benchmark images collected from NOAA, VIRR, and MODIS databases. The superior performance of the proposed approach is demonstrated over the official cloud mask of VIRR database.

Keywords Cloud detection · Natural territories · Demosaic
Color filter array · Color transformation approach

1 Introduction

A detailed conception of atmosphere needs thorough information of many different and connected roles of clouds. Clouds have a significant impact on the atmosphere by means of radiative impact, precipitation, and latent heat [1]. For a long time,

R. Gupta (✉) · S. J. Nanda
Department of Electronics and Communication Engineering, Malaviya National
Institute of Technology, Jaipur 302017, Rajasthan, India
e-mail: 2015rec9517@mnit.ac.in

S. J. Nanda
e-mail: nanda.satyasai@gmail.com

P. Panchal
Department of Electronics and Communication Engineering, Charotar University
of Science and Technology, Changa, Gujarat, India
e-mail: pradippanchal.ec@charusat.ac.in

© Springer Nature Singapore Pte Ltd. 2018
V. Bhateja et al. (eds.), *Proceedings of the Second International Conference
on Computational Intelligence and Informatics*, Advances in Intelligent Systems
and Computing 712, https://doi.org/10.1007/978-981-10-8228-3_28

algorithms based on cloud detection were flourished, undertaken and authenticated for satellite images utilizing pixel by pixel processing [2]. Clouds are generally portrayed through lower temperature in the infrared region and brighter in the visible region when contrasted with the various natural territories. It is difficult to differentiate cloud from the underlying surface because of their similarity in reflectance when the underlying earth surface is covered with ice, snow or desert, or coastal region and also when the cloud is thin cirrus, low-level stratus or small cumulus.

The visible and infrared channels give calculable proficiency to cloud detection. Because of lacking in contrast during nighttime, few clouds (i.e., lower level stratus, cumulus, thin clouds, etc.) are hard to distinguish. More troubles are occur when the detection of cloud edges are focused around. Clouds can be detected both over visible as well as an infrared region while only thermal region-based channels are available to design an algorithm to detect cloud during the nighttime [3]. A few advantages of images obtained in infrared region are: to get the information of cloud-top, different ground features and height of different types of cloud. Various algorithms were designed based on radiance and reflectance to discover various high- and mid-level clouds properties which are affected by the polar region [4]. In the polar region, it is found that algorithm based on visible and thermal region is not successful. Also, it is found that because of the similarity of reflectance of snow/ice in the visible band, it can be differentiated against clouds in the infrared region [5].

World Climate Research Programme (WCRP) was first introduced by the International Satellite Cloud Climatology Project (ISCCP) to create new cloud climatology by analyzing radiance database at visible ($0.6 \mu m$) and infrared ($11 \mu m$) channels to minimize faulty cloud detection. Advanced Very High-Resolution Radiometer (AVHRR) used by Apollo performs five threshold tests which are $3.55–3.93 \mu m$, $0.58–0.68 \mu m$, $0.72–1.10 \mu m$, $11.5–12.5 \mu m$, and $10.3–11.3 \mu m$ using visible and infrared channels through which pixel is marked as cloudy that it comes up short any of the five tests [6, 7]. The CASPR system on National Oceanic and Atmospheric Administration (NOAA) used radiance property of AVHRR instruments to detect water clouds, warm clouds, cold clouds, cirrus, and low stratus-thin cirrus. The CO_2 absorption band is used to detect high cloudy region, level and amount of clouds using infrared radiance centered at $15 \mu m$ by using High-resolution Infrared Radiation Sounder (HIRS) [8]. The Geostationary Operational Environmental Satellite (GOES) used an algorithm based on reflectance at 12, 11, 13.3, and $3.7 \mu m$ [9]. Many algorithms were developed for discrimination of clouds based on time-sequence of Sea Surface Temperature (SST) images, HT5, pixels separation, wavelet transform, and numerous others. Inspired by the recent trend literature, a new color transformation approach is proposed in this paper to identify thick clouds against three natural territories i.e. water, vegetation, and soil. The proposed approach is implemented in two steps: preprocessing and detection. The simulation section demonstrates the superior performance of the proposed approach on benchmark images of NOAA, VIRR, and MODIS databases.

2 Proposed Approach: Cloud Detection Algorithm

The proposed algorithm is characterized by two modules: preprocessing and detection. Detector over satellite gives the Bayer image into preprocessing module. The preprocessing module uses demosaicing algorithm and proselyte it into a true color image. In the wake of changing over into true color image, cloud detection algorithm is carried out by detection module (Fig. 1). The target image Fig. 2a was taken on July 22, 2013 by Meteosat-10 showing the development of tropical wave at east–southeast of the Cape Verde Islands. It demonstrates the increment of thunderstorm activity but it will be organized better environment conditions over the next days. Before moving towards the discussion of system modules, the original image (Fig. 2a) is converted to the raw image (Bayer image in RGGB format) in order to generate the query image (Fig. 2b).

2.1 Preprocessing

The preprocessing is done using a new strategy based on Color Filter Array (CFA) interpolation method called demosaicing. It is helpful to reduce the hardware

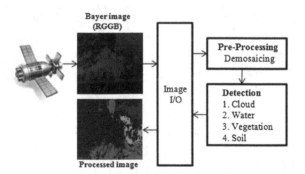

Fig. 1 Flow diagram of the proposed system

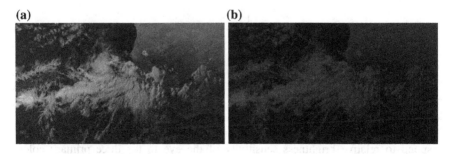

Fig. 2 a Target image from NOAA datasets. **b** Bayer image in "RGGB" format

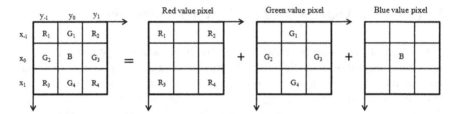

Fig. 3 Blue pixel at the seed pixel location for 3 × 3 matrix

by avoiding the utilization of one sensor per channel for the visible range of electromagnetic spectrum. Preprocessing module is used to reproduce a true color image from the inadequate color samples which are yielded from an image sensor, overlaid with a color filter array (CFA). In this paper, we are going to utilize the bilinear interpolation. If the pixels arrangement is the same as Fig. 3, the bilinear interpolation applied to obtain red pixel value (f_R(x, y)) and green pixel value (f_G (x, y)) with centered blue pixel value (f_B(x, y)), which can be seen from Eqs. (1)–(5). In these equations, x_0 and y_0, x_1 and y_1 and x_{-1} and y_{-1} demonstrate the position of seed pixel, next position from seed pixel and previous position from seed pixel in x- and y-direction, respectively.

$$f_R(x_0, y_0) = \left(\frac{y_1 - y_0}{y_1 - y_{-1}} f(x_0, y_{-1})\right) + \left(\frac{y_0 - y_{-1}}{y_1 - y_{-1}} f(x_0, y_1)\right) \tag{1}$$

$$f_R(x_0, y_0) = \left[\frac{1}{2}\left(\frac{x_1 - x_0}{x_1 - x_{-1}} R_1 + \frac{x_0 - x_{-1}}{x_1 - x_{-1}} R_3\right)\right] + \left[\frac{1}{2}\left(\frac{x_1 - x_0}{x_1 - x_{-1}} R_2 + \frac{x_0 - x_{-1}}{x_1 - x_{-1}} R_4\right)\right] \tag{2}$$

$$f_R(x_0, y_0) = \frac{1}{4}[R_1 + R_2 + R_3 + R_4] \tag{3}$$

$$f_G(x_0, y_0) = \left[\frac{1}{2}\left(\frac{y_1 - y_0}{y_1 - y_{-1}} G_2 + \frac{y_0 - y_{-1}}{y_1 - y_{-1}} G_3\right)\right] + \left[\frac{1}{2}\left(\frac{x_1 - x_0}{x_1 - x_{-1}} G_1 + \frac{x_0 - x_{-1}}{x_1 - x_{-1}} G_3\right)\right] \tag{4}$$

$$f_G(x_0, y_0) = \frac{1}{4}[G_2 + G_3 + G_1 + G_3] \tag{5}$$

2.2 Detection

The luminance signal is formed by adding percentages of VR, VG and VB correspond to relative brightness sensitivity of the eye to the three primary colors which can be seen in Eq. (6) [10]. VR, VG, and VB are red, green, and blue color voltage, respectively.

$$100\% \, V_Y = 30\% \, V_R + 59\% \, V_G + 11\% \, V_B \tag{6}$$

In the same way, different natural territories can be detected and extracted by selecting the vector coefficients correspond to the percentage of R, G and B value of RGB color model through the pixel to pixel processing. The algorithm model described in Fig. 4 demonstrates the differentiation of cloudy pixel against various natural territories. Test on discrimination of presence of natural territories has been chosen to be a mixture of color light fluxes in the ratio as defined in Eqs. (7)–(9) based on the sensitivity of the detector. These percentages correspond to the relative brightness sensitivity of the detector to the three primary colors. Here, m = 3 is considered as the number of primary colors used. ϑ demonstrates the red, green, and blue pixel value at (h, l) location of the image. ξ, ψ and ζ are the vector perimeters of water, vegetation and soil surfaces, respectively. Matrix multiplication of vector

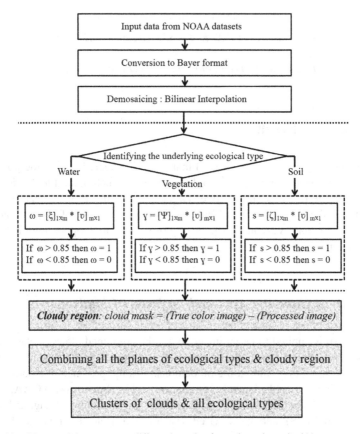

Fig. 4 Algorithm model: process to differentiate cloudy against clear pixel(s)

perimeters and ϑ is done and the result is stored in the seed pixel of ω, γ, and s which demonstrate the discrimination of presence of water, vegetation and soil regions, respectively. ω, γ and s are arranged in each plane to form three planes processed image. This processed image is subtracted from true color image to generate cloud mask image (Eq. 10).

$$\begin{aligned}
\omega(h, l) &= [\xi]_{1 \times m} * [\vartheta]_{m \times 1} \\
[\xi]_{1 \times m} &= [-3.9299 \quad -1.6191 \quad 3.0444]_{1 \times m} \\
[\vartheta]_{m \times 1} &= [R(h, l) \quad G(h, l) \quad B(h, l)]_{1 \times m}^{T} \\
100\% \, \omega(h, l) &= 4 * (-98.2\% \, Red - 40.4\% \, Green + 76.11\% \, Blue)
\end{aligned} \tag{7}$$

$$\begin{aligned}
\gamma(h, l) &= \psi * \vartheta \\
\psi &= [2.8571 \quad -2.6374 \quad 0.8864] \\
100\% \, \gamma(h, l) &= 3 * (95.2\% \, Red - 87.9\% \, Green + 29.54\% \, Blue) \\
s(h, l) &= \zeta * \vartheta
\end{aligned} \tag{8}$$

$$\begin{aligned}
\zeta &= [0.594 \quad -0.38 \quad -0.322] \\
100\% \, s(h, l) &= 59.4\% \, Red - 38\% \, Green - 32.2\% \, Blue
\end{aligned} \tag{9}$$

$$Cloud\,mask = (True\,color\,image) - (Processed\,image) \tag{10}$$

3 Simulation Results

In the target image, it can be observed that cloud is covered over some portion of water, soil, and vegetation region. To differentiate all such natural territories, CTA for individual territory is applied over target image using Eqs. (7)–(9). Figure 5a demonstrates the target image and Fig. 5b–d shows the result of discrimination of water (blue region), vegetation (green region) and soil (copper region) against remaining territories.

3.1 Test on Combine Approach to Detect Clouds

Equation (10) is applied to target image to get the combined effect of classification of clouds and natural territories. The combine approach of CTA can be seen in Fig. 6b. where maroon colored region demonstrates the presence of cloudy region. Now take a look at Table 1 where recognition rate is shown over NOAA, Visible Infrared Imaging Radiometer (VIRR) and MODerate-resolution Imaging Spectroradiometer (MODIS) databases. Used images and Detected images show the

(a) (b)

(c) (d)

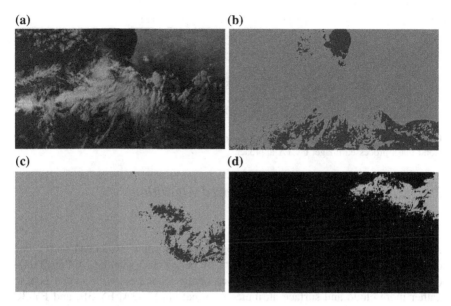

Fig. 5 Cloud detection over individual territory: **a** target image, **b–d** processed image discriminating water, vegetation and soil region, respectively

(a) (b)

Fig. 6 Combine approach. **a** Target image; **b** processed image

Table 1 Recognition rate of CTA method over distinguish databases

Databases	used images	detected images	detection rate (%)
NOAA [11]	5000	4539	90.78
VIRR [12]	10003	8956	89
MODIS [13]	1000	857	85.7

number of images used and number of images with a better result, respectively. Detection rate demonstrates the average rate of the Detected image from the used images which is found approximately more than 85%.

(a)   (b)   (c)

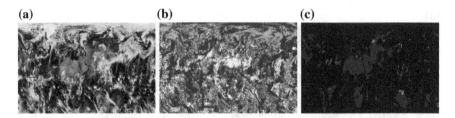

Fig. 7 Illustration of FY-3A/VIRR cloud mask. **a** Target image of VIRR channels; **b** official cloud mask processed image; **c** CTA approach

3.2 Comparing Results of Proposed Algorithm Against Existing Algorithm

As a way to increase the reliability of proposed algorithm associated with cloud detection, it is compared with VIRR data of FengYun-3 satellite. A FengYun-3 (FY-3) is a polar-orbiting satellite of China of the second era with an end goal to gather more cloud and surface attributes information. FY-3A, FY-3B, and FY-3C with VIRR consist of 10-channel VIS/IR multi-reason imaging radiometer (0.43–12.5 μm) and a nadir resolution of 1.1 km. It is investigated that better cloud detection result is obtained using proposed CTA approach (Fig. 7c) as compare to VIRR official cloud mask (Fig. 7b).

4 Conclusions

This article progresses an algorithm centered over vector multiplication and also using RGB color model for daytime cloud detection algorithm. The surface is differentiated into three natural territories with distinguishing cloud location forms. For every natural territory, the vector is characterized to test whether a pixel is sullied by each of them. After recognizing these natural territories, they are removed from the true color image to get the cloud contaminated region. The operational result gives the attractive results over water region and blunder over snow/ice surfaces.

References

1. Jensen, J.R., 2009. Remote sensing of the environment: An earth resource perspective 2/e. Pearson Education India.
2. Weather Prediction: http://www.theweatherprediction.com.
3. He, Q.J.: Night-time cloud detection for FY-3A/VIRR using multispectral thresholds. International journal of remote sensing, 34, 8, 2876–2887 (2013).

4. Long, C.N., Sabburg, J.M., Calbo, J. and Pages, D. Retrieving cloud characteristics from ground-based daytime color all-sky images. Journal of Atmospheric and Oceanic Technology, 23, 5, 633–652 (2006).
5. Rossow, W.B. and Ferrier, J., Evaluation of long-term calibrations of the AVHRR visible radiances. Journal of Atmospheric and Oceanic Technology, 32, 4, 744–766 (2015).
6. Stowe, L.L., McClain, E.P., Carey, R., Pellegrino, P., Gutman, G.G., Davis, P., Long, C. and Hart, S., Global distribution of cloud cover derived from NOAA-AVHRR operational satellite data. Advances in Space Research, 11, 3, 51–54 (1991).
7. He Q.J.: A daytime cloud detection algorithm for FY-3A/VIRR data. International journal of remote sensing, 32, 21, 6811–22 (2011).
8. Menzel, W.P., Frey, R.A., Zhang, H., Wylie, D.P., Moeller, C.C., Holz, R.E., Maddux, B., Baum, B.A., Strabala, K.I. and Gumley, L.E., MODIS global cloud-top pressure and amount estimation: Algorithm description and results. Journal of Applied Meteorology and Climatology, 47, 4, 1175–1198 (2008).
9. Dybbroe, A., Karlsson, K.G. and Thoss, A., NWCSAF AVHRR cloud detection and analysis using dynamic thresholds and radiative transfer modeling. Part I: Algorithm description. Journal of Applied Meteorology, 44, 1, 39–54 (2005).
10. Gulati, R.R., Modern Television Practice Principles, Technology and Servicing. New Age International (2007).
11. NOAA Environmental Visualization Laboratory, Online: http://www.nnvl.noaa.gov/imagegallery.php.
12. VIRR image data sets of FY3C satellite, Online: http://satellite.cma.gov.cn/portalsite/maps/arssmaps.aspx?rasterkey=fy3aglobepage.
13. MODIS: Moderate-resolution imaging spectro-radiometer, Online: http://www.visibleearth.nasa.gov.

Identifying Trustworthy Nodes in an Integrated Internet MANET to Establish a Secure Communication

Rafi U. Zaman and Rafia Sultana

Abstract In MANET, one of the important provocations is to find whether or not a routing message emerges from a trustworthy node. The solution so far to this is to locate a route consisting of trustworthy nodes within the network. The existing approach defines trust concept based on soft security system to eliminate security issues where each node utilizes a trust threshold value. Therefore, a new system is proposed as ITWN (identifying trustworthy nodes) where three trust elements are considered to provide the distinctive features of trust in integrated MANET with collaboration, information, and social networking. These constituents explain the trust capabilities, message integrity, and dynamic social behavior of node. After finding trustworthy nodes, secure route to external network is established through gateway node using appropriate secure route selection value. The simulation results show improvement in performance of the network and minimize routing load and E2E delay within the network.

Keywords Mobile ad hoc networks (MANETs) · Trust management integrated Internet MANET (IIM) · Network security gateway routing

1 Introduction

A mobile ad hoc network (MANET) is defined as a combination of wireless network of moving devices (mobile nodes) created for specific purpose without any fixed support. MANET does not assist any centralized entities and wanted to connect with

R. U. Zaman (✉)
Department of Information and Technology, Muffakham Jah College
of Engineering and Technology, Hyderabad, India
e-mail: rafi.u.zaman@mjcollege.ac.in

R. Sultana
Department of Computer Science and Engineering, Muffakham Jah College
of Engineering and Technology, Hyderabad, India
e-mail: rafiasultana60@yahoo.com

© Springer Nature Singapore Pte Ltd. 2018
V. Bhateja et al. (eds.), *Proceedings of the Second International Conference
on Computational Intelligence and Informatics*, Advances in Intelligent Systems
and Computing 712, https://doi.org/10.1007/978-981-10-8228-3_29

the fixed network such as Internet or LAN to employ the resources presented by them. The interconnection of fixed network with wireless network is known as integrated Internet MANET [1]. The major issues in MANETS are the limited number of resources and applications, limited bandwidth, battery power and wireless coverage, dynamic network topology, and security. Since security is one of the important subjects in MANET, therefore, it is necessary to provide security of network [2].

Trust plays a pivotal role in MANETS in defining the trust values among the mobile nodes [3]. Trust management applies in many situations like authentication, access control, intrusion detection, etc. Trust is based on three things such as trust establishment, trust update, and trust revocation which are the basic factors of trust management [4]. Trust is evaluated based on different metrics in different ways. The various properties of trust are as follows: it is dynamic, subjective, asymmetric, dependent on context, and not inevitably transitive [5]. Trust evaluation includes occurrences, recommendation, and knowledge of trust. It can be either distributed or centralized evaluation. The three phases of trust management are trust propagation, collection, and predictions. The "experience" part of trust is measured by their immediate neighbors and kept upgraded in the trust table. The neighbors also provide "suggestions" another part of trust, and "learning" part of trust is a segment of aggregate trust. The above concepts are discussed in [6–9].

The rest of the paper is organized as follows: Sect. 2 gives the overview of the existing work related to proposed approach; Sect. 3 gives the details of proposed work; Sect. 4 contains the performance evaluation of the proposed system; and Sect. 5 concludes the paper with its future enhancement.

2   Related Work

Cho et al. [10] proposed "Composite trust-based public key management (CTPKM)" in MANET based on threshold where the authors suggested the three trust components to be important parameters in any managing trust solution. To eliminate security affairs within a network, each node makes use of trust threshold value that is collated with trust values. The three trust elements: integrity, competence, and social contact are used to estimate the trust values in ad hoc network. The various limitations observed with the existing systems are as follows: the methods to manage trust in integrated Internet MANET are not defined; it does show how exactly the trust values are calculated by considering only the past experiences of nodes, and there is a high routing load perceived on the node in the ad hoc network.

3   Proposed Work

Security of MANET is one of the major concerns to yield trust among the nodes. Trust management is defined as establishing, updating, or revocation of trust among the nodes [11]. We need to evaluate the trust values of the nodes in order to exchange the information. The source node anxious to discover a secure gateway node connecting to the wired-network transmits solicitation message through neighbor nodes till it reaches gateway. Gateway on receiving request computes trust values and replies with advertisement message. Each intermediate node evaluates trust values and collates with threshold value to identify the most trustworthy nodes, forward the reply further. The source node gets multiple replies and starts computing secure route selection values (SRSVs) to select the secure route with highest SRSV (Fig. 1).

A new system as identifying trustworthy nodes (ITWN) is proposed where three trust ingredients such as trust capabilities, message integrity, and dynamic social behavior of a node are determined in order to produce trust among the nodes. Gateway and intermediate nodes compute trust values based on these three trust elements. The work is implemented in network simulator ns-2.34, and the results obtained are collated with the results of existing systems and WLB-AODV routing protocol.

3.1 Trust Value Calculation

Trust capabilities or competence (T^C) refers to nodes capability to serve the request measured by number of packets dropped to the number of packets forwarded by the nodes.

$$T^C = \frac{\text{Number of Packets Dropped by the node}}{\text{Number of Packets Forwarded by the node}} \tag{1}$$

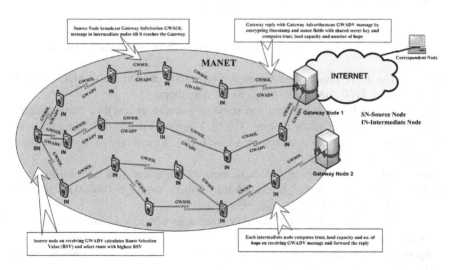

Fig. 1 Architecture of proposed system

Trust message integrity (T^I) refers to nobility of a node in attack deportment measured using the concept of CRC (cyclic redundancy check) defined as the message integrity to check whether the received messages are the actual messages sent by the sender.

$$T^I = \text{CRC calculated for data send equals to CRC calculated for data received} \tag{2}$$

Trust social behavior (T^S) refers to the positive social behavior of a node measured as the number of nodes experienced by another node during a trust update period to the total number of nodes in the network.

$$T^S = \frac{\text{Number of nodes experienced}}{\text{Total number of nodes}} \tag{3}$$

Trust values based on these three components are evaluated and collated with a threshold value to identify the trustworthy nodes. A secured path to gateway is selected based on SRSV calculated from trust values, load capacity, and number of hops of all the nodes in the network. The computation of SRSV is very beneficial in obtaining a secure path to the gateway node. This approach mitigates the security concerns and minimizes the routing overhead and probability of the network. The SRSV is computed as follows:

$$SRSV = T^C + \text{Load Capacity} + \text{Number of Hops.} \tag{4}$$

3.2 Algorithm

Step 1: Create a wired-wireless network of mobile nodes and start gateway discovery.

Step 2: The node broadcast a Gateway-Solicitation message GW-SOL [RREQ, Addr_{SN}, K_{SN}, GW_ADDR, N, t]K_{SN}^{-1} to the neighbors.

Step 3: If neighbor is a gateway and route to gateway exists, then compute trust values based on three trust elements (T^C, T^I, T^S), else if neighbor is not a gateway but route to gateway exists, then rebroadcast GW-SOL message to next node till it reaches the gateway node.

Step 4: Gateway on receiving GW-SOL message computes trust values using trust elements and replies with Gateway-Advertisement message GW-ADV [RREP, Addr_{SN}, GW_{ID}, $(GWADV)S_{SN\text{-}GW}$, N, t]K_{GW}^{-1}.

Step 5: Gateway generates GW-ADV message with Timestamp and Nonce encrypted using a shared secret key.

Step 6: For each intermediate node from gateway to source node, compute trust values, load capacity, and number of hops.

Step 7: Forward GW-ADV to neighbor node.

Step 8: If TV >=Tth, then node is trustworthy and repeat step 6, else if TV < Tth, then node is a malicious node.

Step 9: Source node receives GW-ADV decrypt message with shared secret key and calculates SRSV and select route with highest SRSV value.

Step 10: Source node issues registration request, and gateway on receiving the request verifies it and replies through the established route.

Step 11: Source node is registered with the gateway and connected to fixed network.

4 Simulation Results

4.1 Experimental Setup

The simulation is conducted using the object-oriented network simulator NS2. The simulation parameters used are PDF, NRL, and E2E Delay (Table 1). The AODV+ framework [12] for integrated internet MANET is used for the simulation of the proposed protocol.

Packet Delivery Fraction: It is the ratio of no. of packets delivered to the no. of packets sent within the network.

Normalized Routing Load: It is the ratio of the no. of control packets generated to no. of data packets generated.

End-to-End Delay: It is defined as average delay experienced by the packets in the network (Fig. 2).

Figure 3 shows that performance of proposed protocol is superlative to the existing system and WLB-AODV protocol in terms of packet delivery fraction for all the node speeds. It is observed that the proposed protocol (ITWN) achieves higher packet delivery fraction than the existing protocol.

Figure 4 shows that performance of proposed protocol is superlative to the existing system and WLB-AODV protocol in terms of normalized routing load for all the node speeds. It is observed that the proposed protocol (ITWN) achieves lesser normalized routing load than the existing protocol.

Figure 5 shows that performance of proposed protocol is superlative to the existing system and WLB-AODV protocol in terms of average end-to-end delay for all the node speeds. It is observed that the proposed protocol (ITWN) achieves lesser average end-to-end delay than the existing protocol. The performance gets much better as the number of nodes increases in the proposed approach.

Table 1 Simulation parameters of the proposed protocol

Simulation parameters	Value				
No of mobile nodes	15	25	50	75	100
Topology	800 × 500 m	1000 × 1000 m	1200 × 1200 m	1200 × 1200 m	1200 × 1200 m
No of gateways	2	3	5	5	5
Node radio range	250 m				
Simulation time	900 s				
Traffic sources	5				
Traffic type	CBR				
Mobility model	Random waypoint				
Node speed	1–6 mts/s				
Packet sending rate	5 packets/s				
Destination nodes	2				
Pause time	60 s				
Routing protocol	AODV+				

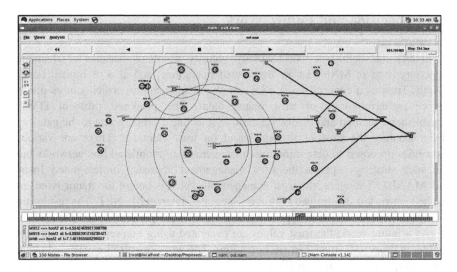

Fig. 2 Simulation scenario of the system with 100 nodes

Fig. 3 Packet delivery fraction versus mobile node speed

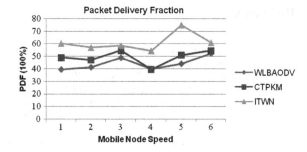

Fig. 4 Normalized routing load versus mobile node speed

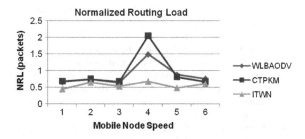

Fig. 5 Average end-to-end delay versus mobile node speed

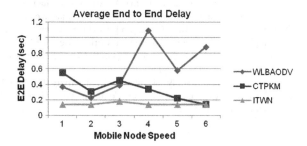

5 Conclusion and Future Work

MANETs are highly at risk to many security problems because of dynamic topology used in MANETs, its distributed operations, and also of limited bandwidth. Trust as a concept has a wide variety of applications, which causes divergence in terminology of trust management. The proposed protocol ITWN establishes a secured and trusted scenario in selecting a proper route to the gateway by evaluating the trust of a node based on trust elements. There are various strategies proposed to determine trust management in mobile ad hoc networks but no such strategy explains the trust management approach in integrated Internet MANET. Therefore, the trust management which is one of the major issues in ad hoc network is determined within an integrated Internet MANET. The results are compared with the existing approaches and show better performance of the system.

As a future work, the trust values are evaluated using non-repudiation concept based on proper key management operations. The work can also be extended using the encryption techniques in IIM.

References

1. Khaleel Ur Rahman Khan, Rafi U Zaman, A. Venugopal Reddy.: Integrating Mobile Ad Hoc Networks and the Internet: challenges and a review of strategies, Communication Systems Software and Middleware and Workshops, COMSWARE 2008. 3rd International Conference on 6–10 Jan., (IEEE CNF) page(s):536–543 (2008).
2. Bin Xie, and Anup Kumar.: A framework for Internet and Ad hoc Network Security, IEEE Symposium on Computers and Communications (ISCC), June (2004).
3. R. Manoharan, S. Mohanalakshmie.: A Trust based gateway selection scheme for integration of MANET with Internet, IEEE-International Conference on Recent Trends in Information Technology, ICRTIT 2011 MIT, Anna University, Chennai. June 3–5 (2011).
4. Vijayan R. and Jeyanthi N.: A Survey of Trust Management in Mobile Ad hocNetworks, IEEE Xplore (2016).
5. M. V. Search Murtuza Jadliwala, Madhusudhanan Chandrasekaran, Shambhu Upadhyaya.: Quantifying Trust in Mobile Ad-Hoc Networks, IEEE Xplore (2005).
6. B.-J. Chang, S.-L. Kuo.: Markov chain trust model for trust-value analysis and key management in distributed multicast MANETs, IEEE Transactions on Vehicular Technology, 58 (5) (2009).
7. J.-H. Cho, A. Swami, I.-R. Chen.: Modeling and analysis of trust management with trust chain optimization in mobile ad hoc networks, Journal of Network and Computer Applications, 35 (3) (2010).
8. M. Anugraha, Dr. S.H. Krishnaveni.: IEEE Xplore document-recent survey on efficient trust management in mobile ad hoc networks, IEEEXplore (2016).
9. Aida Ben Chehida Douss, Ryma Abassi, Sihem Guemara El Fatmi.: A trust-based security environment in MANET: definition and performance evaluation, Science Direct, Paris (2016).
10. Jin Hee cho, Ing-Ray Chen, and Kevin S. Chan.: Trust Threshold based Public Key Management in Mobile Ad Hoc Networks, Elsevier (2016).
11. Jin Hee cho, Ing-Ray Chen, and Kevin S. Chan.: A Survey on Trust Management for Mobile Ad Hoc Networks, Elsevier (2016).
12. Hamidian A.: A Study of Internet Connectivity for Mobile Ad Hoc Networks in NS2, Master's Thesis, Department of Communication Systems, Lund Institute of Technology, Lund University (2003).

Energy-Efficient Routing in MANET Using Load Energy Metric

Rafi U. Zaman and Juvaria Fatima Siddiqui

Abstract A mobile ad hoc network (MANET) is a collection of mobile nodes that are ceaselessly self-designing and work on foundation-less network. A mechanism for broadcasting is flooding where a request is retransmitted by a node at least once, but it is ineffectual in terms of bandwidth and energy. Energy management plays an important role in network. The proposed scheme uses two techniques: First technique is calculating queue length of every node to reduce normalized routing load. The second technique is selecting the route based on a threshold value to maintain efficient transmission to destination. The proposed protocol uses load energy metric algorithm which conserves the energy levels of nodes. The algorithm considers the queue length of every node. The work is implemented using ns2 simulator. The performance evaluation shows a reduction in routing overhead when compared to existing approach and AODV routing protocol.

Keywords MANET · Energy management · Energy efficiency
Queue length · CLEM · Energy level · Normalized routing load

1 Introduction

Wireless mobile nodes are accumulated to form an MANET (mobile ad hoc network) which communicates in a non-framework network without the need of any established architecture or framework [1]. Thus, MANETs are used in a non-infrastructure-based setup.

R. U. Zaman (✉)
Department of Information Technology, Muffakham Jah College of Engineering and Technology, Hyderabad, India
e-mail: rafi.u.zaman@mjcollege.ac.in

J. F. Siddiqui
Department of Computer Science and Engineering, Muffakham Jah College of Engineering and Technology, Hyderabad, India
e-mail: juvariafatima1@gmail.com

© Springer Nature Singapore Pte Ltd. 2018
V. Bhateja et al. (eds.), *Proceedings of the Second International Conference on Computational Intelligence and Informatics*, Advances in Intelligent Systems and Computing 712, https://doi.org/10.1007/978-981-10-8228-3_30

Energy management is an important issue in MANETs, since the nodes have less power to carry out the transmission [2]. Basically, energy management controls three issues [3]: the energy resources are maintained by continuously supervising the battery discharge, tailoring the transmission power and increasing the lifetime of nodes by reorganizing power sources. Efficiently managing the battery, controlling the transmission power and system power management are the three methods to increase the lifetime of node [4–6].

Energy efficiency is measured as the duration of time over which a network can maintain certain performance level [7], that is, network lifetime. If the energy efficiency is high, then more number of packets can be transmitted by the node in every amount of time [8].

The rest of the paper is organized as follows. Section 2 gives an overview of the existing work related to the proposed protocol. Section 3 briefs the proposed load energy metric (LEM) algorithm. Section 4 contains the performance evaluation of proposed protocol in comparison to existing approach and AODV routing protocol. Section 5 concludes the paper with its future scope.

2 Related Work

In [9], the work was carried out using the variable transmission-power levels of node. The nodes are divided based on transmission-power levels into gateways and non-gateways. The different values of transmission-power levels are as follows: high, medium and low. The main idea is to forward packets to nodes which have values different from its own transmission-power level value. The main disadvantages of this approach are, first, as the number of nodes increases, the performance evaluation of nodes decreases in terms of throughput and packet delivery fraction. Second, there is an increase in normalized routing load, which degrades the network performance.

3 Load Energy Metric Algorithm

Nodes energy level plays a very important role in determining the route to the destination. So, the proposed scheme uses two techniques: First technique calculates the queue length of every node which helps to calculate the remaining energy of node reducing routing load and energy level of nodes, and the second technique sets an optimal threshold value to select the best route among the available routes, to maintain efficient transmission of packets among all nodes. The proposed scheme reduces the normalized routing load significantly.

According to the algorithm, the energy levels of individual nodes are calculated at RREQ and stored in their packets. The main idea is to identify the most reliable nodes among the available nodes and select the best path to the destination based on the optimal threshold value. Using the queue length, the nodes whose queue is full is eliminated from the transmission, and only the nodes which are reliable are selected.

Two main goals are achieved when taking the nodes energy level into account while transmitting the RREQ packet: since only some nodes are used while performing retransmission, the energy is conserved and control overhead and congestion are reduced as they degrade network performance. This scheme also decreases unwanted excess retransmission, decreases packet loss and packet drop ratio and expands the reachability of data packet.

3.1 Update Queue Length at Every Node

The main idea in using the queue length for all nodes is to select the nodes which are available for transmission and exclude the nodes whose queue is full. Whenever a request is received by a node, the nodes update the queue to indicate that the node is performing a transmission. If the queue is full, it means that the node is busy performing other transmissions, and another node is selected.

When a mobile node receives an RREQ, it updates its own queue with the RREQ and calculates the CLEM metric. After calculating the CLEM metric, it forwards the request to another node. If the mobile nodes queue is full, it does not consider the node for transmission as it is an unreliable node. In the same way, the queue length is updated at every node, and the CLEM metric is calculated with respect to the queue length until a destination is reached.

The destination on receiving the request broadcasts a reply on the reliable paths by including the CLEM metric value calculated at every node. The idea of using load balanced approach is to negate using heavily loaded paths and use only lightly loaded paths for transmitting data.

3.2 Calculate Energy Level of Nodes

The energy model represents the level of energy in a mobile node. The energy model in a node has initial energy value which is the level of energy the node has at the beginning of the simulation. Here, it is the battery level of the node. It also has a given energy usage for every packet that it transmits and receives which is the load at a node. The idea of using a load balanced approach is to negate using heavily loaded paths and use only lightly loaded paths for transmitting data. The protocol uses lesser loaded paths which are also energy efficient irrespective of their lengths. Every node calculates energy considering the load and battery level of the node. The load is calculated with respect to the queue length. The load energy metric is calculated as

$$\text{Load Energy Metric (LEM)} = \frac{\text{Load}}{\text{Battery Level (BL)}} \qquad (1)$$

The LEM metric value is calculated at every node on receiving the request. The load is calculated with respect to the queue which is updated at every node, as explained in Sect. 3.1. The CLEM is calculated by taking the LEM value and number of hops available.

$$\text{Cumulative Load Energy Metric (CLEM)} = \frac{\text{Load}}{\text{Number of hops}} \qquad (2)$$

The CLEM is calculated at every node and stored in the RREQ packet by adding the previous nodes CLEM with the current nodes CLEM value.

$$\text{ALEM} = \text{previous nodes CLEM} + \text{current nodes CLEM} \qquad (3)$$

The destination updates the ALEM value in the reply packet and forwards the packet to the source on the paths reliable paths available (Fig. 1).

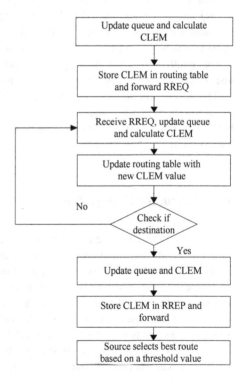

Fig. 1 Working of proposed protocol based on Load Energy Metric

3.3 Select the Reliable Route

The source on receiving the reply packet selects the most reliable route among the available routes. The route is selected based on an optimal threshold value, which should be less than the ALEM value. The route selected should be sufficient for performing the transmission. The main idea in using the threshold value is to select the reliable route among the available routes. The proposed protocol identifies the unreliable nodes using the energy level values calculated for each node.

Algorithm: Load Energy Metric (LEM)

Step 1: The routing starts by source sending a request to neighbouring nodes by updating its queue and calculating the energy. The energy of node is stored in routing table.

Step 2: The energy is calculated as follows:

$$\text{Load Energy Metric (LEM)} = \frac{\text{Load}}{\text{Battery Level (BL)}} \qquad (4)$$

$$\text{Cumulative Load Energy Metric (CLEM)} = \frac{\text{Load}}{\text{Number of hops}} \qquad (5)$$

Step 3: Nodes receiving RREQ update queue and calculate the LEM value as step 2 and update the routing table with their values as follows:

$$\text{ALEM} = \text{previous nodes CLEM} + \text{current nodes CLEM} \qquad (6)$$

Step 4: Intermediate nodes perform the operation of steps 2 and 3 until a destination node is reached.

Step 5: When the RREQ packet reaches the destination, the destination also performs the same steps 2 and 3 and stores the calculated energy value in reply and send the reply on the routes from the request is received.

Step 6: The source on receiving the reply selects the best route on a threshold value such that the received energy is greater than the threshold.

4 Simulation Parameters and Performance Evaluation

The simulation is carried out using ns-2.34 framework. The simulation parameters are shown in Table 1. Different speeds (ranging from 1 to 6 m/s) are used to show the node mobility using a random waypoint model which shows no effect on performance. The simulation was carried out on different nodes (15, 25, 50, 75, 100)

Table 1 Simulation parameters of the proposed protocol

Parameters	Values
Number of mobile nodes	15, 25, 50, 75, 100
Topology size	800 × 500
Transmission range	250 m
Packet size	512 bytes
Bandwidth	2 Mb/s
Simulation time	900 s
Pause time	100 s
Node speed	1–6 m/s
Traffic type	CBR
Number of connections	8

over a fixed area of 800 m × 500 m. The simulation is 900 s. The following metrics were used to evaluate proposed protocol in MANET.

Routing Overhead: It is the total number of packets transmitted for route discovery and maintenance needed to deliver the data packet.

Packet Delivery Ratio: It is the ratio of packets received by the destination to the packets originated by the source.

Throughput: It is the ratio of the amount of data that is received by the destination to the simulation time (Fig. 2).

The following figures show the performance evaluation of proposed protocol with number of nodes taken as 100. Each data point is an average of five simulation runs on the graph.

Figure 3 shows the routing load incurred by proposed protocol (EERP) with respect to existing approach and AODV routing protocol. It shows that the

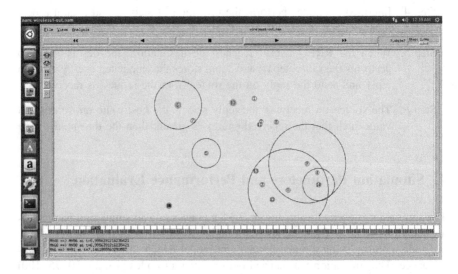

Fig. 2 Simulation of the proposed protocol in a scenario of 100 nodes

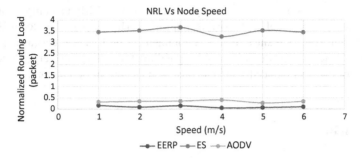

Fig. 3 Normalized routing load versus mobile speed

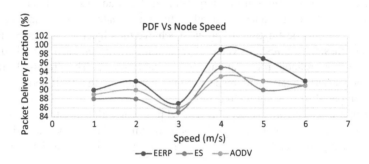

Fig. 4 Packet delivery fraction (PDF) versus mobile speed

proposed protocol performs better and significantly reduces the routing load. The proposed protocol maintains low load by negating unwanted retransmission, thus reducing the normalized routing load.

Figure 4 shows the PDF incurred by EERP in comparison with AODV routing protocol and existing approach. The goal of PDF is to deliver the packets to nodes and is one of the significant metrics. The proposed protocol significantly improves the PDF, as the approach ease more congestion in the network.

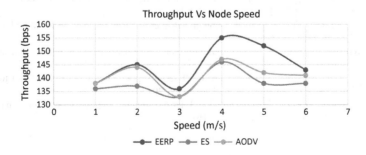

Fig. 5 Throughput versus mobile speed

Figure 5 shows the data throughput incurred by EERP in comparison with AODV routing protocol and existing approach. The proposed protocol demonstrates better execution since there is a confinement on RREQ broadcasting, hence decreasing the measure of unwanted signalling for route discovery. Hence, the routing overhead is reduced that can cause congestion in the network.

5 Conclusion

In this paper, a new scheme based on load energy metric algorithm is proposed for energy efficient routing of data packets in MANET. The proposed protocol identifies the nodes that drain out of energy during data transmission. Energy level of each node is calculated, and nodes select the route that is least loaded for transmitting data to the destination. In order to achieve the objective of energy efficient data transmission, a threshold value of remaining energy on each node is used to select the most reliable route among the available routes. The proposed protocol shows better performance with respect to QoS parameters like PDR and throughput when compared to existing approaches.

The performance comparison of proposed protocol with respect to existing work is being undertaken. In future work, more optimization will be carried out for reducing congestion and energy consumption by using different techniques.

References

1. Shuo Ding: "A survey on integrating MANETs with the internet: Challenges and designs," Australia: University of South Australia (2008).
2. C. Siva Ram Murthy and B. S. Manoj, Ad Hoc Wireless Networks: Energy Management in ad hoc wireless networks, India: Pearson Education, (2005), 607–659.
3. Mohammad A. Mikki: "Energy Efficient location aided routing protocol for wireless MANETs," Gaza, Palestine: IUG (2009).
4. K. Sumathia & A. Priyadharshinib: "Energy optimization in MANETs using on-demand routing protocol," Coimbatore: India (2015).
5. Laura Marie Feeney: "An Energy consumption model for performance analysis of routing protocols for MANET," Kista: Sweden (2001).
6. M. Tamilarasai & T. G. Palanivelu: "Integrated Energy-aware mechanism for MANETs using on-demand routing," (2008).
7. G. Ravi & K. R. Kashwan: "A new routing protocol for Energy efficient mobile application for ad hoc network," Tamil Nadu: India (2015).
8. S. Sridhar, R. Baskaran, P. Chandrasekar: "Energy supported AODV (EN-AODV) for QOS routing in MANET", Chennai: Anna University, 2013.
9. Atif Alghamdi, Peter J B King et al., "Energy-Efficient adaptive forwarding scheme for MANETS," Heriot-Watt University (2016).

Design Optimization of Robotic Gripper Links Using Accelerated Particle Swarm Optimization Technique

Golak Bihari Mahanta, B. B. V. L. Deepak, B. B. Biswal,
Amruta Rout and G. Bala Murali

Abstract Robotic gripper plays a key player in the industrial robotics application such as pick and place, assembly, etc., and it is necessary to design and develop the best possible robotic gripper which needs a lot of mathematical formulation as well as analysis. In this paper, an optimized design of the robotic gripper is obtained using the accelerated PSO. The objective function of the robotic gripper developed is a complex, constraint optimization problem. For solving the problem, the developed two objective functions are used by the APSO with seven constraints. Seven decision variables are chosen to develop the objective function for the robotic gripper, and using APSO algorithm, the optimized best dimensions for the gripper configuration are obtained.

Keywords APSO · Multi-objective · Gripper force

1 Introduction

In this twenty-first century competitive environment for mass production and quality products, industrial robots are playing a significant role. With the advancement of technologies, various methods are used to control the gripper. For performing the grasping and manipulation tasks such as pick place, the role of a robot gripper mechanism is crucial. Robotic grippers are used in the simple, precision, high accuracy, and repetitive work like in automobile industries, aircraft, and

G. B. Mahanta (✉) · B. B. V. L.Deepak · B. B. Biswal · A. Rout · G. Bala Murali
Department of Industrial Design, National Institute of Technology, Rourkela 769008, India
e-mail: golakmahanta@gmail.com

© Springer Nature Singapore Pte Ltd. 2018
V. Bhateja et al. (eds.), *Proceedings of the Second International Conference on Computational Intelligence and Informatics*, Advances in Intelligent Systems and Computing 712, https://doi.org/10.1007/978-981-10-8228-3_31

337

micro/nano-fabrication industries. Industrial robots are mainly having manipulator to interact with the environment, energy source and to control the robot computer controller. End effector or robotic gripper is a part of the manipulator which uses to hold the objects. Various types of gripper designed to handle the object and operated using electrical, pneumatic types of energy source. Grippers are designed considering the physical constraints and a different design for various processes.

Cutkosky [1] carried out a study on the human grasp choice according to the grasp taxonomy and proposed various types of the model and design the grasp. Ceccarelli et al. [2] used Cartesian coordinates methodology for the dimensional synthesis of gripper mechanisms. Chen [3] summarized the various gripping mechanism of the industrial robots. Osyczka [4] proposed a method to solve the robot gripper design problem by converting it into a multi-objective optimization problem. He considered various configurations and obtained the optimized designed parameters for the robotic gripper. Osyczka et al. [5] formulated a multicriteria optimization problem of the robotic gripper in which by using min-max and genetic algorithm, the design parameters were obtained. Osyczka et al. [6] solved the nonlinear multicriteria optimization problem such as interval objective function, beam design problem, and robot gripper design using GA. Krenich [7] solved the design optimization problem of a robotic gripper using an evolutionary algorithm and weighted min-max approach. Abu-Zitar et al. [8] proposed evolutionary programming method to solve the nonlinear frictional gripper problem in robotic grasping considering three-finger contact. Osyczka et al. [9] addressed various types of multicriteria design optimization problem such as gripper design, multiple clutch break design, and shaft design using evolutionary algorithms. Abu Zitar et al. [10] proposed a novel method using ant colony optimization to find the optimum grasping force on a rigid body. They considered three-finger contact and solved the nonlinear problem. Saravanan et al. [11] used an intelligent technique such as genetic algorithm, sorting algorithm, differential evolution, etc., to find the best dimensions for the gripper, i.e., optimum geometric dimensions of a robotic gripper which is used to the manufacturing of the gripper. They considered three configurations for the robotic gripper. They formulated five different objective functions and various constrained condition and solved the design optimization problem by using multi-objective genetic algorithm (MOGA), elitist non-dominated sorting genetic algorithm (NSGA-II), and multi-objective differential evolution (MODE). Zaki et al. [12] designed an intelligent gripper using ANFIS and solved the developed criteria. Datta et al. [13] proposed an objective function used to obtain the optimized geometric configuration of the robotic gripper, which minimizes the difference between the maximum and minimum gripping force. They solve the problem using NSGA-II. Rao et al. [14] formulated three robot gripper configurations [11] and solved them by using TLBO and compared the results. They show that their algorithm performs better to get the results as shown in Fig. 1.

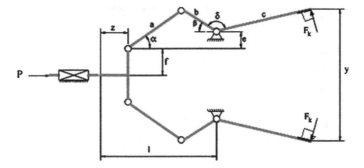

Fig. 1 Sketch of robot gripper configuration [11]

2 Gripper Configuration Design

The robot configurations considered in this paper for the analysis is originally formulated by Osyczka [5], and detailed description of the same is mentioned in the following text. The developed multi-objective function or fitness function uses seven decision variables such as $X = [a, b, c, e, f, l, \delta]^T$. The decision variables $a, b, c, e, f,$ and l are the dimensions of the gripper, and δ is the joint angle between elements b and c of the gripper. The actuating force used to operate the gripper is P, and the angle of the link b w.r.t. horizontal reference line is β and angle of the link a w.r.t. horizontal reference line is α. For holding the object with the help of the robotic gripper, it is necessary to give some gripping force (F_k) to hold the object without slipping. All the links are connected with each other as shown in Fig. 2. Figure 3 shows the geometrical constraints of the gripper. The mathematical model for the robotic configuration for Fig. 2 is presented as follows:

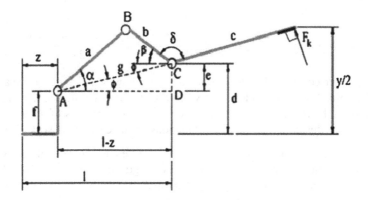

Fig. 2 Geometrical dependencies of the gripper [11]

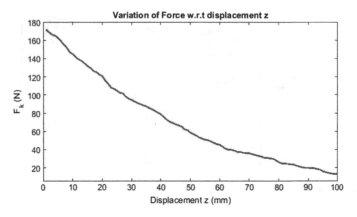

Fig. 3 Variation in the force w.r.t. displacement z

$$g = \sqrt{(l-z)^2 + e^2},$$
$$b^2 = a^2 + g^2 - 2 \cdot a \cdot g \cdot \cos(\alpha - \varphi),$$
$$\alpha = arccos\left(\frac{a^2 + g^2 - b^2}{2 \cdot a \cdot g}\right) + \varphi,$$
$$a^2 = b^2 + g^2 - 2 \cdot b \cdot g \cdot \cos(\beta + \varphi), \qquad (1)$$
$$\beta = arccos\left(\frac{b^2 + g^2 - a^2}{2 \cdot b \cdot g}\right) - \varphi,$$
$$\varphi = arctan\left(\frac{e}{(l-z)}\right).$$

The force generated in the gripper is calculated by obtaining the free body diagram of the systems.

$$R \cdot b \cdot sin(\alpha + \beta) = F_k \cdot c, \quad R = \frac{P}{2 \cdot cos\alpha}, \quad F_k = \frac{P \cdot b \cdot sin(\alpha + \beta)}{2 \cdot c \, cos\alpha} \qquad (2)$$

In the above equation, R is the reaction developed on the link a due to the applied actuating force P from left side as shown in Fig. 1. By using the above correlation derived from the robotic gripper mechanism, objective functions of the system are formulated. In this case, two objective functions are taken from Osyczka [5].

1. The first objective function is coined as "the difference between the maximum and minimum gripping forces developed when grasping the rigid objects for the assumed range of gripper ends displacement."

$$f_1(x) = |max\,F(x, z) - \min F(x, z)| \tag{3}$$

The minimization of above objective is necessary to ensure that there is less variation in the gripping force during the entire range of operation of the gripper

2. The second objective function is "the force transmission ratio which is the ratio between the applied actuating force P and the resulting minimum gripping force at the tip of link c."

$$f_2(x) = \frac{P}{min_z F_k(x, z)} \tag{4}$$

The objective functions formulated above consist of the decision variable of the robotic gripper, i.e., link length and the joint angles ($X = [a, b, c, e, f, l, \delta]^T$) and on the displacement z which is the displacement of the actuator. The minimum value for the actuator displacement is 0 when it is in initial position and maximum value at $Zmax$. The corresponding force variation with respect to the displacement is shown in Fig. 3, which shows that maximum force when $z = 0$ and minimum when $z = Z_{max}$.

By considering all criteria, the objective function is developed as follows:

$$f_1(x) = F_x(x, 0) - F_x(x, Z_{max}) \quad f_2(x) = \frac{P}{F(x, Z_{max})}. \tag{5}$$

The above said objective function having the constraints equation is defined as follows:

$$\begin{aligned}
&g1(x) = Ymin - y(x, Zmax) \geq 0, \quad g5(x) = (a + b)^2 - l^2 - e^2 \geq 0, \\
&g2(x) = y(x, Zmax) \geq 0, \quad g6(x) = (l - Zmax)^2 + (a - e)^2 - b^2 \geq 0, \\
&g3(x) = y(x, 0) - Ymax \geq 0, \quad g7(x) = l - Zmax \geq 0, \\
&g4(x) = YG - y(x, 0) \geq 0, \quad g8(x) = minz\,Fk(x, z) - FG \geq 0.
\end{aligned} \tag{6}$$

The range of the seven decision variables by using which the objective function is formulated is as follows:$10 \leq a \leq 250$, $10 \leq b \leq 250$, $100 \leq c \leq 300$, $0 \leq e \leq 50$, $10 \leq f \leq 250$, $100 \leq l \leq 300$, $1.0 \leq \delta \leq 3.14$

The geometric parameters used in the constraints equation for the objective function are as follows: Y_{min} is the minimal dimension of the object to be grasped $= 50$ mm; YG is the maximum gripper ends displacement $= 150$ mm; Y_{max} is the maximum size of the object to be grasped $= 100$ mm; Z_{max} is the maximum displacement of the actuator $= 100\,mm$; and the actuating force of the gripper is $P = 100$ N.

3 Accelerated Particle Swarm Optimization

Particle swarm optimization (PSO) uses current *global best* and the *individual best* to get the result whereas accelerated particle swarm optimization (APSO) uses only the *global best*. As a result, APSO is simpler than that of the PSO and, at the same time, is quite powerful. So, the velocity of each particle in APSO algorithm is calculated by the more straightforward formula as compared to PSO

$$v_i^{t+1} = v_i^t + \alpha\left(\epsilon - \frac{1}{2}\right) + \beta\left(g^* - x_i^t\right), \tag{7}$$

where ε is a random variable having range from 0 to 1.

g^* is the global best, v_i^t represents the velocity of ith particle for tth iteration, x_i^t is the position of ith particle for tth iteration, and α, β are the acceleration coefficients. The following formula can obtain the updated position:

$$x_i^{t+1} = x_i^t + v_i^{t+1}. \tag{8}$$

For achieving better results, it is necessary to tune the various parameters of the algorithms. However, it is observed that for accelerated PSO, various parameters gave best results such as $\alpha \approx 0.1 - 0.4$ and $\beta \approx 0.1 - 0.7$.

4 Proposed Methodology

The methodology to use accelerated PSO to obtain the optimal value of the geometric parameters of the robotic gripper is illustrated in a flowchart as shown in Fig. 4.

After obtaining the objective function, APSO algorithm is used to obtain the best geometric configuration for the robotic gripper. All the necessary parameters are initialized (α, β, number of population, number of iteration, etc.). Evaluate the fitness for each population using Eqs. (1–6). Then record the best global solution and update the velocity and position of the particle using Eqs. (7) and (8). Then, check whether maximum iteration achieved or best solution found. Either of the above achieved, then displaced the output as the global best which is the best solution for the decision variables.

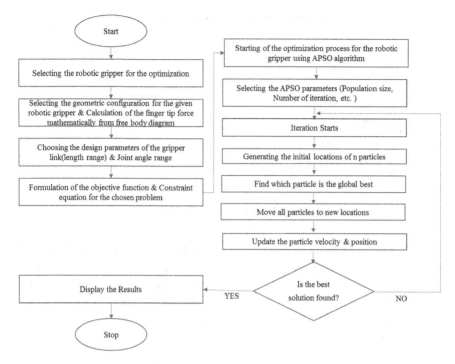

Fig. 4 Proposed methodology using APSO

5 Result and Discussions

For checking the effectiveness of the APSO algorithm, a number of simulations are conducted. For the first case, the parameters used for the simulation are as follows: population size = 50 and maximum number of generations = 150 [14]. All the link lengths of the grippers are in mm, and the angle between the links a and b is in radian. For simulation, the methodology proposed here using APSO to solve for the robotic configuration was implemented using MATLAB 2014a on an Intel i7 2600, CPU 3.40 GHz personal computer with 4 GB of RAM. After successfully implementing the APSO algorithm to solve the gripper optimization problem, the variation in the obtained output is shown in the Fig. 4.

Initially, there is a rapid variation in the decision variables. But around iteration number 60, all the variables are converged to a fixed value giving the best results as shown in Fig. 5. As the parameters of the APSO are crucial to get the best optimized results, it is necessary to tune the parameters properly. Hence, in this case, we varied the population size and iteration number keeping the value of α and β constant, and recorded the change in the decision variable and force value in Tables 1 and 2. The objective functions f_1 and f_2 are in N.

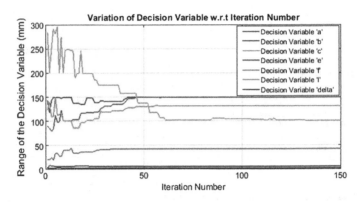

Fig. 5 Variation in the decision variable w.r.t. iteration number

Table 1 Optimal obtained parameters (Population = 50, P = iteration number, $\alpha = 0.2, \beta = 0.5$)

P	f_1	f_2	a	b	c	e	f	l	δ
150	168.8	3.04	149.9	149.8	131.3	5.6	42.8	101.4	2.1
160	169.8	3.4	149.8	149.8	130.3	6.6	44.8	106.9	3.1
170	178.4	3.9	148.5	150.0	135.3	5.9	42.4	102.7	2.4

Table 2 Optimal obtained parameters (Max iteration = 150, P = Population size, $\alpha = 0.2, \beta = 0.5$)

P	f_1	f_2	a	b	c	e	f	l	δ
50	168.8	3.04	149.9	149.8	131.3	5.6	42.8	101.4	2.1
60	168.4	4.04	147.6	150.0	135.3	6.6	40.4	103.3	3.0
70	166.3	4.4	148.7	145.8	133.3	5.9	43.6	100.4	3.1

6 Conclusion

In the present work, APSO algorithm used to solve for the multi-objective design optimization of robot grippers and performances is recorded. From the simulation results, it was found that the objective function f_1 is more, compared to objective function f_2. A thorough study on the influence of the output results was conducted by changing the APSO parameters. It has been observed that there is a very slight change in the results by varying the iteration number and population size. The obtained set of optimal dimensions of the robotic gripper can be used to develop the real-time gripper.

References

1. Cutkosky, Mark R. "On grasp choice, grasp models, and the design of hands for manufacturing tasks." IEEE Transactions on Robotics and automation 5.3 (1989): 269–279.
2. Ceccarelli, M., J. Cuadrado, and D. Dopico. "An optimum synthesis for gripping mechanisms by using natural coordinates." Proceedings of the Institution of Mechanical Engineers, Part C: Journal of Mechanical Engineering Science 216.6 (2002): 643–653.
3. Chen, Fan Yu. "Gripping mechanisms for industrial robots: an overview." Mechanism and Machine Theory 17.5 (1982): 299–311.
4. Osyczka, Andrzej. "Evolutionary algorithms for single and multicriteria design optimization." (2002).
5. Krenich, Stanislaw, and Andrzej Osyczka. "Optimization of Robot Gripper Parameters Using Genetic Algorithms." Romansy 13. Springer Vienna, 2000. 139–146.
6. Andrzej, Osyczka, and Krenich Stanislaw. "A new constraint tournament selection method for multicriteria optimization using genetic algorithm." Evolutionary Computation, 2000. Proceedings of the 2000 Congress on. Vol. 1. IEEE, 2000.
7. Krenich, Stanisław. "Multicriteria design optimization of robot gripper mechanisms." IUTAM Symposium on Evolutionary Methods in Mechanics. Springer Netherlands, 2004.
8. Abu-Zitar, R., and AM Al-Fahed Nuseirat. "An evolutionary based method for solving the nonlinear gripper problem." International Journal of Applied Science and Engineering 2.3 (2004): 211–221.
9. Osyczka, Andrzej, and Stanislaw Krenich. "Some methods for multicriteria design optimization using evolutionary algorithms." Journal of Theoretical and Applied Mechanics 42.3 (2004): 565–584.
10. Abu Zitar, R. A. "Optimum gripper using ant colony intelligence." Industrial Robot: An International Journal 32.1 (2005): 17–23.
11. Saravanan, R., et al. "Evolutionary multi criteria design optimization of robot grippers." Applied Soft Computing 9.1 (2009): 159–172.
12. Zaki, A. M., et al. "Design and implementation of efficient intelligent robotic gripper." Modelling, Identification and Control (ICMIC), the 2010 International Conference on. IEEE, 2010.
13. Datta, Rituparna, and Kalyanmoy Deb. "Multi-objective design and analysis of robot gripper configurations using an evolutionary-classical approach." Proceedings of the 13th annual conference on Genetic and evolutionary computation. ACM, 2011.
14. Rao, R. Venkata, and Gajanan Waghmare. "Design optimization of robot grippers using teaching-learning-based optimization algorithm." Advanced Robotics 29.6 (2015): 431–447.

Cognitive Decision Support System for the Prioritization of Functional and Non-functional Requirements of Mobile Applications

P. Saratha and G. V. Uma

Abstract Requirements based on goal-oriented approach provide different possibilities and alternatives. Quality criteria are measured for each goal of the system, and its possible quality characteristics are identified. The functional requirements pertaining to each quality characteristics are needed to be prioritized for development. Formal decision-making relieves us from manual and time-consuming tasks. The decision-making based on criterion yields best results. The decision-making is the cognitive process and done based on experience, knowledge, and intuition. This paper discusses cognitive decision-based prioritization along with Analytic Hierarchy Process (AHP) and Fuzzy AHP for functional and non-functional requirements. The experimental evaluation has been done for the mobile-based applications.

Keywords Decision-making · Software requirements · AHP
Fuzzy AHP

1 Introduction

Software development activities are executed based on the size of the project. The size is measured using the number of requirements. Software systems are considering every requirement to ensure the success criteria of software development. Decision-making is the challenging task for stakeholders to prioritize the requirements. Conflicts among stakeholders create uncertainty and indistinct values for selection. It is essential to identify the important requirements to deliver immediately due to time and resource constraints. The requirements with the highest priority are

P. Saratha (✉) · G. V. Uma
Department of Information Science and Technology, College of Engineering,
Anna University, Chennai, India
e-mail: psaratha1104@auist.net

G. V. Uma
e-mail: gvuma@annauniv.edu

© Springer Nature Singapore Pte Ltd. 2018
V. Bhateja et al. (eds.), *Proceedings of the Second International Conference on Computational Intelligence and Informatics*, Advances in Intelligent Systems and Computing 712, https://doi.org/10.1007/978-981-10-8228-3_32

347

selected for immediate release. The important requirements must be prioritized based on defined criteria. Requirements prioritization has been identified as the important activity of requirements engineering.

The stakeholder's needs and preferences are different in nature [1]. The different needs and alternatives are considered during requirements elaboration process. The alternatives must be evaluated using qualitative analysis. The importance of the requirements based on stakeholder's opinion is considered and analyzed. The score is assigned for each criterion which involves opinion. This paper discusses the AHP and Fuzzy AHP techniques for ranking analysis. This paper is containing the following sections. Section 2 illustrates the related studies for prioritization. Section 3 indicates the proposed tactics for processing the prioritization. Sections 4 and 5 show the experimental evaluations and results. The conclusion of the proposed work and its future enhancement are discussed in Sect. 6.

2 Related Work

There are several factors associated with the prioritization techniques. It includes the business context, time to market, stakeholders, preferences, risk associated with the project, functional and non-functional constraints, cost, and time [1]. Business analyst or the requirements engineer meets the number of challenges during prioritization. The requirements prioritization techniques were proposed by a number of researchers. The techniques are based on different underlying concepts. Each technique produces different priority results on context basis and is shown in Table 1.

Table 1 Different prioritization approaches and their techniques

Name of the approach	Name of the author	Prioritization technique
Value-based fuzzy logic	Ramzan et al. [2]	Value-based fuzzy
AHP-based approach	Sadiq et al. [3]	AHP
AHP-based approach	Dabbagh and Lee [4]	AHP
Quality criteria-based approach	Otero et al. [5]	Quality attribute measurement
Case-based ranking approach	Perini et al. [6]	Pairwise comparison and machine learning
HAM-based approach	Dabbagh et al. [7]	Pairwise comparison
IPA-based approach	Dabbagh and Lee [7]	Integrated prioritization

3 Proposed Work

In this section, the new approach called Cognitive Decision-Making for Order Preference (CDMOP) has been proposed for the process of requirements prioritization and is shown in Fig. 1.

3.1 Identification of Quality Attributes

Software quality and software requirements are interrelated concepts. The software system behavior is determined by the functional requirements and its constraints accumulated by the quality characteristics. The quality characteristics are also called as Non-functional Requirements (NFR). The system goal or the stakeholders' objectives are used as input to derive the list of stakeholder concerns [8]. After the concerns are analyzed, the list of quality characteristics is derived.

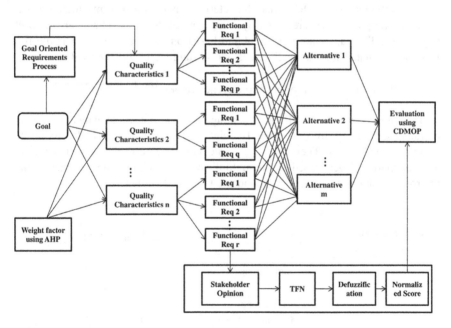

Fig. 1 Proposed prioritization approach for CDMOP

3.2 Prioritization Using AHP

The identified quality characteristics are prioritized using AHP. AHP method has been introduced by Saaty [9] during the year 1980, for decision-making problem based on multiple criteria. It is also used to determine the priority among multiple alternatives for the criteria.

3.2.1 Basic Principles of AHP

AHP is represented by the matrix which expresses the relative importance of the set of criteria. The importance is measured using scale values and represented as numbers. The AHP decision model is created for the factors impacting the decisions. The factors are taken as x- and y-axis elements in the matrix. The pairwise comparisons between the factors are considered, and it depends on the stakeholders, context of the project, and importance. The relative importance is evaluated between each factor and represented by numerical scale [7] values as shown in Table 2.

The preference for each quality characteristics is obtained from different stakeholders. The preferences are given by comparing each quality characteristics with others [10]. The values are numerical and based on the Saaty rating. The nth root value is calculated for each cell values and is summed up for the total.

$$\text{Nth root value} = (\text{product of preference values})^{1/\text{total number of preferences}}$$

The eigenvector is calculated by taking the nth root value which should be divided by the total nth root value. The eigenvector elements are added to get the normalized value of 1. Then, the λ_{max} is calculated to receive the consistency ratio. The calculation is performed by the multiplication of each preference value against the row eigenvector for all values, and the total is identified.

Table 2 Saaty rating scale

Option	Numerical value(s)
Equal	1
Little strong	3
Strong	5
Very strong	7
Extremely strong	9
Intermediate values	2, 4, 6, 8
Second alternatives	Reciprocals

3.3   Identification of Functional Requirements

The decomposition is performed, and the leaf level requirements are elaborated until it is assigned to an agent. The importance of functional requirement is facilitated with non-functional quality rankings. The priority based on AHP is linked as a weight factor to rank the functional requirements. The Fuzzy AHP method is applied for the evaluation of the priority for the functional requirements, and it is represented by single-valued numbers.

3.4   Stakeholder Opinion and Triangular Fuzzy Number (TFN)

The subjective opinions obtained from the stakeholders are measured and aggregated using TFN. TFN is a triplet (Lw, Md, Hi) where Md represents the modal value, Lw represents the low (minimum) value, and Hi represents the high (maximum) value.

The scale values are mentioned for each requirement in Table 3. The TFN values must be calculated into quantifiable values for defuzzification process. It is represented by Eq. (1):

$$DF^a(X_i) = a\,fn_R\left(y_j\right) + (1-a)fn_L\left(y_j\right),\tag{1}$$

where $x_i = TFN_i$ represents the triangular fuzzy number, a specifies the preference value, and the ranges are between 0 and 1. When $a = 1$, the equation provides the optimistic view such as

$$DF^1\left(y_j\right) = a\,fn_R\left(y_j\right).\tag{2}$$

Table 3 Scale and its numerical values

Scale	Actual numbers
Very high importance (VHI)	1
High importance (HI)	0.75
Low importance (LI)	0.5
Very low importance (VLI)	0.25
No importance (NI)	0.001

When $a = 0$, the equation provides the pessimistic view such as

$$DF^0\left(y_j\right) = fn_L\left(y_j\right),\tag{3}$$

$$fn_{Lw}\left(y_j\right) = Lw_j + \left(Md_j - Lw_j\right)b,\tag{4}$$

$$fn_R\left(y_j\right) = Hi_j + \left(Md_j - Hi_j\right)b.\tag{5}$$

3.5 Defuzzification and Normalization

Defuzzification is calculated using Eqs. (6) and (7):

$$DF^a\left(y_j\right) = \frac{1}{2}a\left(Md_j + Hi_j\right) + \frac{1}{2}(1 - a)(Lw_j + Md_j),\tag{6}$$

$$DF^a\left(y_j\right) = \frac{1}{2}\left[\left(a\,Hi_j\right) + Md_j + (1 - a)(Lw_j)\right].\tag{7}$$

Finally, the normalized score is evaluated using Eq. (8)

$$NDF_j = DF_j / \sum_{j=1}^{m} DF_j.\tag{8}$$

The priorities are decided based on the normalized score value for each requirement.

3.6 Alternatives and Evaluation Using CDMOP

The decision matrix is created for "n" criteria and "m" alternatives using CDMOP approach. The criteria have been decided based on the environmental context parameters as mentioned in Table 4. It assigns the normalized matrix with positive values from 0.1 to 1, and values which are nearest to 1 get the highest order.

Table 4 Alternatives for prioritization

Alternatives based on contextual parameters

Budget	Time	Performance	Installation	Technology	Maintenance	Privacy	User type	Revenue	UIPatterns
1	0.8	0.8	0.7	1	0.5	0.8	0.7	0.9	0.7

4 Case Studies

In this section, this paper discussing the experience in applying the proposed approach to the online food ordering system through Mobile APP case study. The non-functional requirements for each system have been derived from the project goal and represented in Table 5. The preference matrix, the eigenvectors, and the consistency index have been evaluated and represented in Table 6.

1. The weight factors are assigned for each quality criteria as represented in Table 6.
2. Stakeholder opinions are obtained, and the relevant scale values are mentioned in Table 7.
3. TFN is calculated based on Eqs. (4) and (5).
4. Defuzzification is calculated based on Eq. (7).
5. Normalized score is calculated using Eq. (8).
6. Based on the alternatives and its weight scores mentioned in Table 4 and normalized score value in Table 8, the CDMOP score is calculated. The calculation is based on the multiplication of each normalized score with all the parameter weight.

Table 5 Project goal and quality criteria for mobile applications

System goal	Objectives	User concerns
Food delivery in 45 min	Enable easy order placement	Placing order with promptness and secured payment in minimum time
		Placing the order at all time with easy options
		Protect the information from failures and errors and maintain the privacy
	Reduce idle time for order processing	Allocation of orders for preparation and reducing the time
		Ability to handle volume of customers' order
	Remove bottlenecks for order delivery	Delivery location identification and removing the bottleneck
		Provision to track the customers' order

Table 6 Priority and consistency index using AHP for food ordering system

Online food ordering system using APP

	Efficiency	Security	Reliability	Usability	Portability	Nth root of product values	Eigenvector	Priority	λ_{max}
Efficiency	1	½	1/3	1	9	1.084	0.181	4	1.047
Security	2	1	1	1/3	7	1.363	0.228	3	1.3
Reliability	3	1	1	1	3	1.552	0.260	2	1.445
Usability	1	3	1	1	5	1.719	0.288	1	1.623
Portability	1/9	1/7	1/3	1/5	1	0.251	0.042	5	0.212
						5.969	0.999		

Table 7 Fuzzy AHP for online food ordering system using APP

Weight factor	4	3	2	1	5
FR/NFR	Efficiency	Security	Reliability	Usability	Portability
Order placement	HI (0.75)	LI (0.5)	HI (0.75)	VHI (1)	HI (0.75)
Order dashboard	VHI (1)	LI (0.5)	HI (0.75)	HI (0.75)	LI (0.5)
Order customization	HI (0.75)	LI (0.5)	HI (0.75)	VHI (1)	LI (0.5)
Cart summary	HI (0.75)	LI (0.5)	HI (0.75)	VHI (1)	LI (0.5)
Personalized order	VHI (1)	LI (0.5)	LI (0.5)	HI (0.75)	LI (0.5)
OTP authentication	HI (0.75)	VHI (1)	HI (0.75)	LI (0.5)	VLI (0.25)
Delivery notification	HI (0.75)	HI (0.75)	VHI (1)	LI (0.5)	NI (0.001)
Payment recovery	HI (0.75)	HI (0.75)	VHI (1)	LI (0.5)	VLI (0.25)
Order allocation	VHI (1)	LI (0.5)	HI (0.75)	LI (0.5)	LI (0.5)
User authorization	HI (0.75)	VHI (1)	HI (0.75)	LI (0.5)	VLI (0.25)
Order preparation	HI (0.75)	VLI (0.25)	LI (0.5)	VHI (1)	HI (0.75)

Table 8 CDMOP evaluation for online food ordering system using APP

FR/NFR	TFN	Defuzzification	Normalized scores	CDMOP scores
Order placement	(0.5, 0.75, 1)	0.75	0.1	0.79
Order dashboard	(0.5, 0.7, 1)	0.725	0.097	0.7663
Order customization	(0.5, 0.7, 1)	0.725	0.097	0.7663
Cart summary	(0.5, 0.7, 1)	0.725	0.097	0.7663
Personalized order	(0.5, 0.65, 1)	0.7	0.094	0.7426
OTP authentication	(0.25, 0.65, 1)	0.638	0.085	0.6715
Delivery notification	(0.001, 0.6, 1)	0.550	0.074	0.5846
Payment recovery	(0.25, 0.65, 1)	0.638	0.085	0.6715
Order allocation	(0.5, 0.65, 1)	0.7	0.094	0.7426
User authorization	(0.25, 0.65, 1)	0.638	0.085	0.6715
Order preparation	(0.5, 0.6, 1)	0.675	0.090	0.711

5 Results and Discussion

This paper discussed the CDMOP evaluation with AHP and Fuzzy AHP to deal with the functional and non-functional requirements prioritization, respectively. The technique improves the decision-making and produces the best accuracy for prioritizing the requirements. The techniques were applied for three different online applications such as Paytm, Online Banking, and Desktop application. The evaluation was done based on two important properties: ease of use and accuracy in prioritization. Multiple alternatives are evaluated and considered for preference based on the applications. The further examination will be continued with the different set of applications with different techniques.

6 Conclusions and Future Work

Requirements prioritization is performed as an important activity during elaboration stage of requirements engineering. This paper used the prioritization of goal-oriented functional and non-functional prioritization. The involvement of multiple stakeholders creates conflicts and the issue must be solved. Multiple decision-making has used widely by most of the systems for ranking selections. The prioritization resolves those conflicts, and the proper decisions lead to time-consuming and cost-effectiveness. This approach for any software system leads to focus on the important requirements during the early phase of life cycle. It also drives the architectural phase to design with the proper guidelines. In future, the CDMOP will be evaluated with fuzzy-based alternatives to get more accuracy in prioritization.

References

1. Leffingwell, D., Widrig, D.: Managing Software Requirements: A Unified Approach. Addison-Wesley (2003).
2. Ramzan, M., Jaffar, M.A., Iqbal, M.A.: Value based fuzzy requirement prioritization and its evaluation framework. In 4th IEEE International Conference on Innovative Computing, Information and Control, pp. 1464–1468 (2009).
3. Sadiq, M., Shahid, M., Ahmad, S.: Adding threat during software requirements elicitation and prioritization. International Journal of Computer Applications, vol 1(9), pp 50–54 (2010).
4. Tahriri, F., Dabbagh, M., Ale Ebrahim, N.: Supplier assessment and selection using fuzzy analytic hierarchy process in a steel manufacturing company. Journal of Scientific Research & Reports, vol 3(10), pp 1319–1338 (2014).
5. Otero, C.E., Dell, E., Qureshi, A., et al.: A quality-based requirement prioritization framework using binary inputs. In 4th Asia International Conference on Mathematical/Analytical Modeling and Computer Simulation, pp 187–192 (2010).
6. Perini, A., Ricca, F., Susi, A.: Tool-supported requirements prioritization: comparing the AHP and CBRank methods, Information and Software Technology, vol 51(6), pp 1021–1032 (2009).
7. Dabbagh, M., Lee, S.P.: An approach for integrating the prioritization of functional and nonfunctional requirements. The Scientific World Journal, 13 pages, Article ID 737626 (2014).

8. Lapouchnian, A. Goal-Oriented Requirement Engineering. An Overview of the Current Research, Technical Report, University of Toronto, Toronto (2005).
9. Saaty, T.L.: Decision making with the analytic hierarchy process. International Journal of Services Sciences, vol $1(1)$, pp 83–98 (2008).
10. Dabbagh, M., Lee, S.P.: A consistent approach for prioritizing system quality attributes. In 14th ACIS International Conference on Software Engineering, Artificial Intelligence, Networking and Parallel/Distributed Computing, pp 317–322 (2013).

Audio CAPTCHA Techniques: A Review

Sushama Kulkarni and Hanumant Fadewar

Abstract Audio CAPTCHA is a reverse Turing test to discriminate the user as human or bot using audio as a medium. As W3C guidelines mandate the need of accessibility of web for people with disabilities like visual or motor impairment, audio CAPTCHA is an essential form of CAPTCHA for web security. Very less work has been done in this area as compared to other types of CAPTCHAs like text, image, or video. Hence, an evaluation of current scenario is important in enhancement of audio CAPTCHA. This paper is an attempt to understand existing work and its accessibility in the arena of audio CAPTCHA. It also explores the obstacles in use of audio CAPTCHA.

Keywords Audio CAPTCHA · HCI · Web security · Human interactive proof (HIP) · Automatic speech recognition (ASR)

1 Introduction

CAPTCHA (Completely Automatic Public Turing Test to Tell Computer and Human Apart) is a reverse Turing test to distinguish between human and bot users. A CAPTCHA test commonly presents a visual interface challenge to be solved by the user, which is not suitable for people with visual or motor disabilities. Sometimes visual CAPTCHA is not legible for a normal person as well. The format of the CAPTCHA must be easy for human users and difficult for bots. Ironically, unapproachability of CAPTCHA has been a major hindrance to popularity of CAPTCHAs [1]. Several studies have assessed the usability of text, image, and video-based CAPTCHA [2–7]. Although the development of CAPTCHAs based on

S. Kulkarni (✉) · H. Fadewar
School of Computational Sciences, S. R. T. M. University, Nanded, India
e-mail: sushama.s.kulkarni@gmail.com

H. Fadewar
e-mail: fadewar_hsf@yahoo.com

© Springer Nature Singapore Pte Ltd. 2018
V. Bhateja et al. (eds.), *Proceedings of the Second International Conference on Computational Intelligence and Informatics*, Advances in Intelligent Systems and Computing 712, https://doi.org/10.1007/978-981-10-8228-3_33

text, image, and video has been seen remarkable efforts, the field of audio CAPTCHA has been very less explored.

Audio CAPTCHA is usually designed to aid visually impaired users [8]. It requires the user to listen to the audio file and provide their answer. But the audio quality is distorted so that it can thwart the automated speech recognition software. Audio CAPTCHA can be broadly classified as follows:

- Speech CAPTCHA—It uses spoken words, letters, or numbers in audio format to design CAPTCHA challenge.
- Acoustic CAPTCHA—It utilizes acoustic event sounds to design the CAPTCHA.

2 Related Work

Most of the researchers have utilized the better capability of identifying speech or acoustic event sounds in human being as compared to bots.

2.1 Speech CAPTCHA

Speech CAPTCHA designs rely on superior human ability to recognize letters, numbers, or phrases among the background noises like chatter, music, or acoustic sounds. The speech is generated either by human speaker or using the speech synthesizer.

Speech-Based Reverse Turing Test. Speech-based reverse Turing test used superior capabilities of human in recognition of distorted speech as compared to ASR techniques [8]. It analyzed 18 different sets of distortions to exhibit possible variety of ways to make the recognition problem hard for machines. It proved the applicability of difference between patterns of errors made by human and ASR techniques [8].

Voice CAPTCHA for Internet Telephony. This CAPTCHA was proposed for the Internet telephony. As telephony-based CAPTCHA has no auxiliary interface to provide detail information about CAPTCHA, the voice CAPTCHA presented five random digits in audio format as a CAPTCHA challenge [9]. The user has to input the same sequence of digits via numeric keypad in order to solve the CAPTCHA. This CAPTCHA was implemented in the form of Skype plugin [9]. An audio CAPTCHA was introduced for Voice over IP (VoIP) telephony to combat SPam Over Internet Telephony (SPIT) bots [10]. It recommended vocabulary, background noise, time, and audio production as important attributes to design an efficient audio CAPTCHA for normal and visually impaired humans and prevent SPIT bots. The last stage of security testing of this audio CAPTCHA for VoIP telephony achieved a promising score of less than 2% success for bots [10].

Secure Audio CAPTCHA/No Nonsense CAPTCHA. This CAPTCHA focused on preserving intelligibility by presenting a test which requires the user to identify meaningful words from an audio [11]. The audio file contained undistorted meaningful words along with randomly inserted nonsense speech sound. It relied upon the better language understanding and speech recognition ability of humans as compared to bots. It achieved four times higher average human success rate in comparison with the automated speech recognizer [11, 12].

Auditory CAPTCHA. It generates audio signals by superimposing pair of words on each other [12]. Since there is no distinct separation within a pair of successive words, word isolation becomes virtually impossible for bots attacking this CAPTCHA. But superimposition of pair of words is limited to some degree by adding a signal-dependent time delay at the beginning of two adjacent words [12]. This way it helps the human user to recognize the designated sequence order of words from the audio signal. But it fills all speech pauses with a multitalker babble noise so as to thwart word segmentation attack by a bot. This CAPTCHA has accomplished 2.8 times higher average human success rate as compared to the ASR attacks [12].

C-instance. In this experiment, a C-instance (CAPTCHA instance) was generated using CART (Classification and Regression Tree) machine learning technique [13]. This audio CAPTCHA was synthesized through a diphone speech synthesizer. Text-to-Speech synthesizer generated foreground utterance in the form of a digit sequence. A background noise of some English words in the same voice was added to create a C-instance. Foreground and background utterances were mixed by overlapping them [13]. After testing the C-instances on human and ASR, the recognition accuracy percentage and comparison of human and ASR performance were done to generate the experience dataset. CART technique was then applied to use this experience dataset for discarding the C-instances that are likely to be easy to solve for a bot. Authors identified 3 dB SNR (Signal-to-Noise Ratio) to be the most favorable for humans.

2.2   Acoustic CAPTCHA

Acoustic CAPTCHA employs human auditory perception for identification of human and bots apart.

Picture/Sound Form of CAPTCHA/HIPUU (Human Interaction Proof, Universally Usable). It was proposed for the users with or without visual impairment [14]. It presented an image and the sound effect suitable to hint the image from 15 possible combinations. The user was asked to watch the image or listen to the audio effect and choose his/her answer from the dropdown list of possible answers. The database used to create this CAPTCHA was of substantially small size [14].

SoundsRight CAPTCHA. Later an improved version of HIPUU series of CAPTCHA was proposed, namely, SoundsRight CAPTCHA [15]. It was more secure in comparison with HIPUU. SoundsRight CAPTCHA asked the user to identify a specific sound from a series of 10 sounds played through the computer's audio system. User selected the specific sound by pressing the space bar. Usability testing exhibited 90% of success rate for blind users using SoundsRight CAPTCHA [15].

Nonspeech Audio CAPTCHA. This CAPTCHA utilized acoustic sound events mixed with an environmental scene, achieving independence of linguistic abilities [16]. It achieved 85% of relative improvement in human success rate as compared to existing audio CAPTCHAs. Only 21% of attacks got success in breaking this CAPTCHA, which is a better score than the existing audio CAPTCHA security levels [16].

3 Attacks on Audio CAPTCHA

Visual impairment in a person makes it difficult to perform any online transaction independently and securely. Thus, audio CAPTCHAs which are intended for blind users have greater responsibility in terms of security. But in line with all other types of CAPTCHAs, audio CAPTCHAs have been attacked in several ways.

AdaBoost, SVM, and k-NN techniques were used to break Google audio CAPTCHA, Digg CAPTCHA, and an older version of reCAPTCHA's audio CAPTCHA. SVM technique yielded a 92% of pass rate using the "one mistake" passing conditions and a 67% of exact solution match rate for Google audio CAPTCHA. SVM achieved 96% of "one mistake" pass rate and 71% of exact solution match rate for Digg CAPTCHA. For an older version of reCAPTCHA's audio CAPTCHA, SVM technique produced 58% of best pass rate for "one mistake" passing conditions and 45% of pass rate for the exact solution match [17].

Decaptcha tool was developed to break e-Bay audio CAPTCHA [18]. Authors attacked e-Bay audio CAPTCHA using three strategies. A scraper tool was developed to create CAPTCHA corpus database of 26,000 audio CAPTCHAs for the purpose of analysis. A state-of-art speech recognition system, Sphinx, was evaluated for CAPTCHA breaking accuracy. It was noted that Sphinx was not efficient enough to break e-Bay audio CAPTCHA. Thus, authors developed Decaptcha tool which was capable of breaking 75% of e-Bay audio CAPTCHAs. With a medium sized IP pool, Decaptcha was able to register over 75,000 fake accounts on e-Bay per day. Decaptcha used a Discrete Fourier Transform (DFT) to the audio file and then isolated the voice energy spikes. It applied a supervised learning algorithm to build the model based on the isolated voice energy spikes. This model was further used to recognize digits in the e-Bay audio CAPTCHA [18].

An audio reCAPTCHA solver was developed to break Google's continuous audio reCAPTCHA. It applied Hidden Markov Models (HMMs) for speech recognition. Audio reCAPTCHA solver used three steps, namely, cluster segmentation, spectral feature extraction, and cluster labeling with the help of HMM-based ASR. The accuracy of the solver was evaluated using five metrics, namely, off-by-one accuracy, strict accuracy, per-cluster accuracy, per-digit accuracy, and n-segment accuracy. Closed test and open test methods were used to evaluate the performance of the solver. The closed test was performed for fivefold cross-validation. Audio reCAPTCHA solver achieved 52% of accuracy for the current version of audio reCAPTCHA. Authors suggested that increasing uncertainty in the number of target voices and adopting proper semantic noise can improve the security of CAPTCHAs [19].

A Decaptcha was developed to attack the noise-based noncontinuous audio CAPTCHAs. This Decaptcha consisted stages, namely, segmentation stage that extracts spoken digits, a representation scheme for the extracted digits, and a classification stage that recognizes each digit. Segmentation stage was unsupervised while the classification stage was supervised one. It applied Regularized Least Squares Classification (RLSC) algorithm for classification. Scraped CAPTCHAs were labeled using Amazon Mechanical Turk. It attained 49% of success for solving Microsoft's audio CAPTCHAs and 45% of success for solving Yahoo's audio CAPTCHAs. Training cost was also low because it required only 300 labeled CAPTCHAs and approximately 20 min of training time to defeat the hardest schemes [20]. reCAPTCHA was attacked using HMM-based speech recognition system [21]. It achieved 62.8% of success rate. Authors had proposed a new acoustic CAPTCHA scheme, which included overlapping speakers and artificial room reverberation effects to combat bot attacks. This new acoustic CAPTCHA was then attacked using ASR system. It achieved only 5.33% of success rate for attacks on the acoustic CAPTCHA proposed by the authors. These attacks on audio CAPTCHAs depict the advancement of bots against security measures.

Table 1 depicts a comparison of attacks on various audio CAPTCHAs.

4 Techniques to Prevent Attacks on Audio CAPTCHA

Audio CAPTCHAs are considered to be a soft target as compared to the other form of CAPTCHAs. Thus, audio CAPTCHAs implement some of the following defensive strategies in order to prevent bot attacks:

- Overlap of target voices [19],
- Random number of target voices in a cluster [19],
- Addition of stationary noise [19],
- Filtering of high-frequency features [19],
- Use of vocabulary, background noise, time, and audio production quality [10],
- Random insertion of nonsense speech sound in meaningful words [11],
- Better language understanding of human being [12],
- Superimposing pair of words along with signal-dependent time delay [12],

Table 1 Comparison of attacks on audio CAPTCHAs

CAPTCHA	Method of attack	Success rate
reCAPTCHA (older version)	SVM [17]	58% (with "one mistake" passing conditions)
		45% (exact solution match)
Google audio CAPTCHA	SVM [17]	92% (with "one mistake" passing conditions)
		67% (exact solution match)
Digg	SVM [17]	96% (with "one mistake" passing conditions)
		71% (exact solution match)
e-Bay audio CAPTCHA	DFT and supervised learning algorithm [18]	75%
Google's continuous audio reCAPTCHA	Cluster segmentation, spectral feature extraction, and cluster labeling with the help of HMM [19]	52%
Microsoft's noncontinuous audio CAPTCHA	Unsupervised segmentation, labeling using Amazon mechanical turk, regularized least squares classification (RLSC) [20]	49%
Yahoo's noncontinuous audio CAPTCHA	Unsupervised segmentation, labeling using Amazon mechanical turk, regularized least squares classification (RLSC) [20]	45%
reCAPTCHA (current version 2014)	HMM-based automatic speech recognition [21]	62.8%
Acoustic CAPTCHA	HMM-based automatic speech recognition [21]	5.33%

- Use of machine learning technique like CART (Classification and Regression Tree) to synthesize background and foreground utterances [13], and
- Use of acoustic sound events mixed with an environmental scene [16].

5   Accessibility

Audio CAPTCHAs are perceived to be the possible solution for the users with visual impairment. But many of the existing audio CAPTCHAs have exhibited lesser human success rate for normal humans as well. Thus, accessibility of audio CAPTCHA needs close review. A study was performed to analyze the accessibility barriers, and possible solutions were proposed based on types of disabilities [22]. It emphasized the need of implementing the W3C Web Content Accessibility Guidelines 2.0 for the ease of people with various kinds of impairments. A field study was conducted to measure the amount of inconvenience telephony-based CAPTCHA causes to users, and how various features of the CAPTCHA, such as duration and size, influence usability of telephony-based CAPTCHA [23]. It suggested some guidelines to improve existing CAPTCHAs for use in telephony systems. It recommended some design conventions like one-time instruction, loss/error tolerance, availability of feedback, and verbal responses instead of key press [23]. A study conducted with 89 blind users evaluated the usability of audio CAPTCHA interface [24]. They identified major obstacles in use of audio CAPTCHA, which are as follows:

- The speech representations are not clearer; background noise is louder and disturbing.
- Contextual hints are not provided.
- Interfaces of audio CAPTCHA are clumsy and sometimes confusing.
- Navigation elements in the interface are not optimized for nonvisual use.

Although the audio CAPTCHAs are designed to enhance the accessibility, many of them are not suitable for people with various types of disabilities. We have evaluated some of the available audio CAPTCHAs with an assumption that the web page containing the audio CAPTCHA fulfills WCAG 2.0. Table 2 shows the accessibility evaluation of audio CAPTCHAs discussed in Sect. 2. It clearly shows that the current audio CAPTCHAs are not suitable for the people with auditory disabilities, cognitive disabilities, and the combination of disabilities except the visual and motor disability combination. People with motor disabilities have high dependency on WCAG 2.0 Compliance and AT (Assistive Technology).

Speech-based reverse Turing test [8] uses distorted speech as CAPTCHA, which is not suitable for people with auditory and cognitive disabilities. Voice CAPTCHA for Internet telephony [9, 10] utilizes digits, vocabulary, background noise, time, and audio production for CAPTCHA designing, which causes difficulties for people with auditory and cognitive disabilities. Secure audio CAPTCHA/no nonsense CAPTCHA [11] uses better language understanding and speech recognition ability of human being for CAPTCHA designing which is an obstacle for people with auditory and cognitive disabilities. Auditory CAPTCHA [12] relies on superimposition of pair of words and a multitalker babble noise for CAPTCHA designing which is an accessibility barrier for people with auditory and cognitive disabilities. C-instance [13] combined a foreground utterance in the form of a digit sequence and a background noise of some English words in the same voice which is an impediment for people with auditory and cognitive disabilities.

Picture/sound form of CAPTCHA/HIPUU [14] used image or audio effect to hint the answer of CAPTCHA which is a barrier for people with cognitive disabilities and combination of visual and auditory disabilities both. SoundsRight CAPTCHA [15] used a series of 10 sounds played through the computer's audio system to generate a CAPTCHA challenge, which causes difficulties for people with auditory and cognitive disabilities. Although nonspeech audio CAPTCHA [16] achieved independence of linguistic abilities by combining acoustic sound events with an environmental scene, it still remained a barrier for people with auditory and cognitive disabilities.

Table 2 Accessibility evaluation of audio CAPTCHAs discussed in Sect. 2

Audio CAPTCHA	Disabilities									
	Visual	Auditory	Motor	Cognitive	Visual and auditory	Visual and motor	Visual and cognitive	Auditory and motor	Auditory and cognitive	Motor and cognitive
Speech-based reverse Turing test [8]	✓	X	✓[a]	X	X	✓[a]	X	X	X	X
Voice CAPTCHA for Internet telephony [9, 10]	✓	X	✓[a]	X	X	✓[a]	X	X	X	X
Secure audio CAPTCHA/no nonsense CAPTCHA [11]	✓	X	✓[a]	X	X	✓[a]	X	X	X	X
Auditory CAPTCHA [12]	✓	X	✓[a]	X	X	✓[a]	X	X	X	X
C-instance [13]	✓	X	✓[a]	X	X	✓[a]	X	X	X	X
Picture/sound form of CAPTCHA/HIPUU [14]	✓	✓	✓[a]	X	X	✓[a]	X	✓	X	X
SoundsRight CAPTCHA [15]	✓	X	✓[a]	X	X	✓[a]	X	X	X	X
Nonspeech audio CAPTCHA [16]	✓[a]	X	✓[a]	X	X	✓[a]	X	X	X	X

[a]High interdependence with the—WCAG 2.0 Compliance—and support with keyboard and AT (screen reader)

6 Conclusion

Audio CAPTCHA is an important part of accessibility initiatives for visually impaired users. Hence, it needs careful design and deployment strategies in order to provide convenience for human users especially the blind users. The aim of this paper was to provide a survey of existing audio CAPTCHA techniques. We have summarized some prominent audio CAPTCHA techniques and their accessibility for the people with various disabilities. It highlighted the accessibility barriers faced by common and blind users, emphasizing the need of efficient use of W3C WCAG 2.0 guidelines. We have also outlined the bot attack methods applied on current audio CAPTCHAs and their success rate, providing a pointer for future development of secure audio CAPTCHAs. With the ever improving state-of-art ASR systems threatening the security of audio CAPTCHAs, current audio CAPTCHAs need to improvise on the security aspect so as to provide robust strength against bot attacks. Encouraging the use of cognitive skills of human users can be the key to more secure and user-friendly audio CAPTCHAs for visually impaired users.

References

1. World Wide Web Consortium (W3C). Inaccessibility of CAPTCHA (2007), http://www.w3.org/TR/turingtest/ (accessed: March 20, 2017).
2. Baird, H.S., Bentley, J.L.: Implicit CAPTCHAs. In: Proceedings of Document Recognition and Retrieval XII (IS&T/SPIE Electronic Imaging), San Jose, CA, January 2005, vol. 5676, pp. 191–196 (2005).
3. Ahmad, A.S.E., Yan, J., Ng, W.-Y.: CAPTCHA design: Color, usability, and security. IEEE Internet Computing 16(2), 44–51 (2012).
4. Chellapilla K., Larson K., Simard P.Y., Czerwinski M. (2005) Building Segmentation Based Human-Friendly Human Interaction Proofs (HIPs). In: Baird H.S., Lopresti D.P. (eds) Human Interactive Proofs. Lecture Notes in Computer Science, vol. 3517. Springer, Berlin, Heidelberg.
5. Chew, M., Baird, H.S.: BaffleText: a Human Interactive Proof. In: Proceedings of 10th IS&T/SPIE Document Recognition & Retrieval Conf., Santa Clara, CA, January 22 (2003).
6. Kluever KA, Zanibbi R (2009) Balancing usability and security in a video CAPTCHA. In: ACM symposium on usable privacy and security, Article 14, ACM Press, 11 p.
7. Shirali-Shahreza, M., Shirali-Shahreza, S.: Advanced collage captcha. In: Proceedings of IEEE ITNG (2008).
8. Kochanski, G., Lopresti, D., Shih, C.: Reverse Turing Test Using Speech. In: Proceedings of ICSLP 2002, pp. 1357–1360. Causal Productions Pty Ltd. (2002).
9. Markkola, A., Lindqvist, J.: Accessible Voice CAPTCHAs for Internet Telephony. In: The Symposium on Accessible Privacy and Security (SOAPS 2008) (2008).
10. Bigham, J.P., Cavender, A.C.: Evaluating existing audio captchas and an interface optimized for non-visual use. In: CHI 2009, pp. 1829–1838. ACM (2009).
11. Meutzner H., Gupta S., Kolossa D.: Constructing Secure Audio CAPTCHAs by Exploiting Differences between Humans and Machine. In: CHI 2015, pp. 2335–2338. ACM (2015).
12. Meutzner, H., Gupta, S., Nguyen V-H., Holz T., Kolossa D.: Toward Improved Audio CAPTCHAs Based on Auditory Perception and Language Understanding. In: ACM

S. Kulkarni and H. Fadewar

Transactions on Privacy and Security (TOPS) (formerly known as TISSEC), vol. 19, no. 4, pp. 10:1–10:31 (2016).

13. Chan T.Y.: Using a Text-to-Speech Synthesizer to Generate a Reverse Turing Test. In: Proceedings of 15th IEEE ICTAI, pp. 226–232 (2003).

14. Holman, J., Lazar, J., Feng, J.H., D'Arcy, J.: Developing usable captchas for blind users. In: Assets 2007, pp. 245–246. ACM (2007).

15. Lazar J., Feng J., Brooks T., Melamed G., Wentz B., Holman J., Olalere A., Ekedebe N.: The SoundsRight CAPTCHA: An Improved Approach to Audio Human Interaction Proofs for Blind Users. In: CHI 2012, pp. 2267–2276. ACM (2012).

16. Meutzner H., Kolossa D.: A Non-speech Audio CAPTCHA Based on Acoustic Event Detection and Classification. In Proceedings of EUSIPCO, Budapest, Hungary (2016).

17. Tam J., Simsa J., Hyde S., Von H.L.: Breaking audio captchas. Advances in Neural Information Processing Systems, vol. 1, no. 4, pp. 1 (2008).

18. Bursztein E., Bethard S.: Decaptcha: Breaking 75% of eBay audio CAPTCHAs. In Proceedings of 3rd USENIX conference on Offensive technologies, vol. 1, no. 8, pp. 8 (2009).

19. Sano S., Otsuka T., Okuno H.G. (2013) Solving Google's Continuous Audio CAPTCHA with HMM-Based Automatic Speech Recognition. In: Sakiyama K., Terada M. (eds) Advances in Information and Computer Security. IWSEC 2013. Lecture Notes in Computer Science, vol 8231. Springer, Berlin, Heidelberg.

20. Bursztein, E., Beauxis, R., Paskov, H., Perito, D., Fabry, C., Mitchell, J.: The failure of noise-based non-continuous audio CAPTCHAs. In: IEEE Symposium on Security and Privacy, pp. 19–31. IEEE (2011).

21. Meutzner H., Nguyen V.-H., Holz T., Kolossa D.: Using Automatic Speech Recognition for Attacking Acoustic CAPTCHAs: The Trade-off between Usability and Security. In Proceedings of ACSAC, pp. 276–285. ACM (2014).

22. Moreno L., González M., Martínez P.: CAPTCHA and accessibility. Is this the best we can do? In Proceedings of WEBIST, vol. 2, pp. 115–122 (2014).

23. Sachdeva N., Saxena N., Kumaraguru P. (2015) On the Viability of CAPTCHAs for use in Telephony Systems: A Usability Field Study. In: Desmedt Y. (eds) Information Security. Lecture Notes in Computer Science, vol 7807. Springer, Cham.

24. Bigham, J.P., Cavender, A.C.: Evaluating existing audio CAPTCHAs and an interface optimized for non-visual use. In: Proceedings of the SIGCHI Conference on Human Factors in Computing Systems. ACM (2009).

An Efficient Cache Refreshing Policy to Improve QoS in MANET Through RAMP

A. Vijay Vasanth, K. Venkatachalapathy, T. P. Latchoumi,
Latha Parthiban, T. Sowmia and V. OhmPrakash

Abstract Cache routing is an important phenomenon practiced in mobile ad hoc networks (MANET) to overcome the unnecessary process of route discovery, and overall, this route discovery process serves as a routing overhead. The concept of route cache is basically manipulated with the dynamic source routing (DSR) protocols with less percentage of QoS constraints. In this paper, a cache refreshing policy (CRP) is proposed to increase the overall QoS factors in MANET using reputation-aware multi-hop routing protocol (RAMP). RAMP basically provides secure and trusted access between nodes but fails in maintaining efficiency in QoS factors. In this paper, a challenging technique has been proposed to deal RAMP to work out with CRP to improve QoS.

A. Vijay Vasanth · K. Venkatachalapathy
Department of Computer Science and Engineering, Annamalai University,
Tamil Nadu, India
e-mail: omsumeethaa@rediffmail.com

A. Vijay Vasanth (✉)
Department of Computer science and Engineering, Vel Tech University,
Tamil Nadu, India
e-mail: a.vijayvasanth@gmail.com

T. P. Latchoumi
Department of Computer Science and Engineering, Vignan's University,
Vadlamudi, Andhra Pradesh, India
e-mail: lax.senthil@gmail.com

L. Parthiban
Department of Computer Science, Pondicherry University, Puducherry, India
e-mail: lathaparthiban@yahoo.com

T. Sowmia
Department of Electronics and Communication Engineering,
Christ College of Engineering and Technology, Puducherry, India
e-mail: sowmiavasanth@gmail.com

V. OhmPrakash
Electronics and Communication Engineering, Panimalar College of Engineering,
Chennai, India
e-mail: prakashohm96@gmail.com

© Springer Nature Singapore Pte Ltd. 2018
V. Bhateja et al. (eds.), *Proceedings of the Second International Conference on Computational Intelligence and Informatics*, Advances in Intelligent Systems and Computing 712, https://doi.org/10.1007/978-981-10-8228-3_34

Keywords DRS · RAMP · CRP · Cache routing · Route cache
MANET · QoS

1 Introduction

The cache routing concept is primarily a fast and simple hash lookup table which stores the routing information in a forwarding node (intermediate node or forwarding node between the source and destination node) in a mobile network environment.

The routing information is maintained as cache in the forwarding node. The forwarding node holds the information and passes it to the neighboring nodes when it requires [1]. The use of cache routing principle in MANET had more advantages [2, 3], i.e., any node can access the information in the cache rather than depending on the source nodes, and the time taken for the route discovery is almost reduced. This results in an overall reduction of routing overhead.

The cache routing concept deals with positive signs of development in routing overhead [4, 5] and also contains certain drawbacks, which are listed as follows:

1. Broken link condition—When a link between any two nodes is broken, the route cache will intimately only to the source node, and it does not possess any signal to intimate the broken link to the connected nodes.
2. No expiry—The route cache does not possess any order or priority in maintaining the routing information. All the entries in the lookup table will be saved as information.
3. Renew stale routes—There may also exist a condition to add stale routes, i.e., if a stale route is determined, it will be removed immediately. Since the broken information is intimated only to the source node alone, the other intermediate nodes may again add the same stale routes to the cache which is being removed in the near futures.
4. Broken link notification—There will be a delay of broadcasting of information about broken links to all nodes. Due to delay in broadcasting, there may be a chance of creation of stale nodes.
5. Negative cache—This technique may be better in determining the broken link but to notify the broken link, it consumes bandwidth and energy.
6. Time-based route expiry—In this technique, a time-out is maintained for each route, but still there exists a problem in accessing the routes in case if the cache route information accessed in multiple paths in sequence. The first node will utilize the cache information, and the successor nodes may not go for route cache access if the routing information has reached the time-out strategy.

The role of protocols plays a greater impact on improving QoS in MANET [6]. The reactive protocols are dynamic in nature and it is very easier to obtain higher range of QoS primitives. The DSR protocol plays a vital role in determining the QoS in terms of metrics such as energy, packet delivery ratio, end-to-end delay, routing overhead, and packet drop [7, 8].

Through DSR, the specific values of QoS are delivered, but the values responded at the average rate [9]. To improve the metric values of QoS, many versions and updates have been focused on DSR protocols, and one among them is [10] RAMP which shows high potential for security and maintains the trust value of each node [11]. When compared with all successive versions of DSR, only RAMP shows similar values (average QoS metrics) equivalent to DSR [12]. The RAMP can be improved by working out with a cache refreshing policy which is further worked out in this paper. By working out RAMP with the cache refreshing policy, RAMP reflects the overall improvement in all aspects of data [13].

This paper organized as follows: Sect. 2 gives the preliminaries of DSR, characteristics of RAMP, and cache optimization techniques. Section 3 describes the proposed methodology. In Sect. 4, we describe simulation environment, and in Sect. 5, we describe simulation parameters. Section 6 shows the results and discussions. Finally, in Sect. 7, a conclusion is made.

2 Preliminaries

2.1 Overview of DSR

The most familiar protocol where the cache routing process should be carried out is DSR. The DSR is a reactive protocol in MANET which works dynamically on existing conditions. When a particular node in a network needs to communicate with the other node. The searching process is initiated in the local cache during the absence of information followed by setting up a control message (RREQ) and broadcasting the same to all nodes. If the destination node exists, it replies to the source node in the form of a route reply (RREP), and at the later stage, the communication starts. In case if the source node receives a nil response, again, it adds the identity of destination node in the control message by modifying it and rebroadcast it on the same network [2, 14].

2.2 Characterization of RAMP

RAMP is a routing protocol basically derived from DSR. It is mostly used to perform routing in a mobile ad hoc network and for the purpose of congestion control in TCP for reputation/trust management [15, 16]. RAMP coordinates with all the nodes connected in the mobile ad hoc network by calculating the trust value of each node. RAMP uses the additive increase multiplicative decrease (AIMD) algorithm to examine the behavior of each node and its reliability [17]. RAMP performs best in monitoring nodes based on its behavior. RAMP shows high reliability toward security, reputation on nodes, node cooperation, and selfish behavior of an individual node, but shows less tolerant toward QoS when comparing with DSR.

2.3   Role of Cache Optimization Techniques

The nodes in MANET are infrastructure less and there should be cooperative communication between the nodes and with an assumption that nodes are connected with each other [18, 19]. When the source and the destination nodes are not in the communication range, an intermediate node can receive the data and forward it toward the destination node which is commonly termed as route caching. That way it is essentially a multi-hop network where each node acts as a transmitter and receiver.

It is ascertained that various aspects of caching in the MANET and the advantages of caching optimize the data availability in MANET besides improving throughput and decreasing computational overhead [4, 20]. This paper proposes a new caching mechanism that improves efficiency in data dissemination and network performance. The basic cache optimization mechanisms are as follows:

(i)   **Role of Cache Routing**—The caching mechanism avoids unnecessary route discovery process and causes performance improvement in MANET [21]. This scheme mostly works in such a way that only stale routes are removed from the cache while valid routes are never removed.

(ii)   **Client Cache for Consistency and Cache Sharing Interface**—The client cache is the temporary memory that helps in the presence of network disconnection.

(iii)   **Impact of Cache Time-out and Backup Routes**—These deals with the impact of backup routes and the cache time-out on reliable source routing. The research revealed that in the case of high mobility environment, time-out mechanism has its impact on performance while the backup routes have its impact on the robustness of mobility.

(iv)   **Caching Alternatives and DB Caching in MANETs**—It is achieved by caching queries and responses. The mobile nodes are granted access to the database through cache nodes (CNs), query directories (QD) and backup for Query Directories (BQDs). This results in improving the status of mobile nodes.

(v)   **Transparent Cache-Based Mechanism for MANET**—It caches data and data paths that are repeatedly used in order to reduce the communication overhead. The mechanism reduces the consumption of bandwidth besides making it an energy-efficient approach. It regulates cache capacity from time to time so as to make it more efficient. It achieves better routing performance.

3 Proposed Methodology

This paper focuses on cache refreshing policies which are a transparent cache-based approach where the caching information is renewed frequently in dynamic time [1]. The cache updating sequences use distributed route cache update algorithm. In earlier approach, only the trusted node values are determined, and the link is established and maintained. But in MRAMP by following distributed cache replacement algorithm, the source node broadcasts the route error information of size 60 bytes to all its neighbors. Hence, all neighbors replace the stale route in their cache and maintain the trusted value of existing nodes.

Figure 1 shows the process of cache update process sequence in the modified RAMP. Experimental evaluation is done using the network simulator (NS2) of monarch group. The proposed approach improves the performance up to 10–30% using different QoS parameters like packet delivery ratio (PDR), end-to-end delay, packet drop, and energy consumption. The main challenge in routing protocol is how effectively they handle when the topology changes.

When broken link information is detected, RERR message is propagated and route cache is updated only the nodes involved in the routing path. New protocol MRAMP updates the cache using distributed route cache update algorithm in a distributed manner of 60 bytes for all the nodes present in the neighbors. The nodes present in the routing path receives RERR messages, whereas the nodes which are not in the routing path lacks in hearing RERR messages. To avoid the problem of lacking in hearing RERR messages, an explicit approach of RERR message is made and broadcasted to all reachable nodes in the selected topology and the route cache date is updated in all nodes by means of distribution pattern. Such approach improves the QoS of network using different parameters. For route cache updates, the following algorithms are used:

3.1 Algorithm 1: Add Route

Step: 1. The node adds a route for data packet for a fresh communication.
Step: 2. If the node is the destination, then it stores the source node and sets data packet (DP) to 0 as the route is not used.

Fig. 1 Block diagram of proposed system

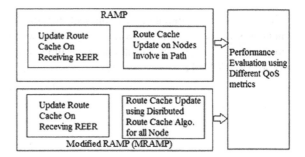

Step: 3. If it is not an intermediate node, then it checks the cache.

> Step: 3a.
> Carried out by DP in 0 creates a Reply Record in which neighbor should learn downstream link.
> Step: 3b.
> If the route is available, then an entry will be added to the Replay Record Field.

Step: 4. If source route exists in cache, then DP is 1.
Step: 5. If the route does not exist, then it is the destination node and DP is 1.
Step: 6. If the route does not exist, then the node is intermediate node and increments DP by 1.
Step: 7. If not having full path, then it creates a cache table entry and sets DP to 1.

3.2   Algorithm 2: Find Route

Step: 1. Cache is null.
Step: 2. For each entry, path is stored in the cache table.
Step: 3. If node finds route, then it adds an entry to Replay Record, including neighbors to which ROUTE REPLAY is sent.
Step: 4. If the node is source node and finds route, then DP is incremented by 1.
Step: 5. If the derived route is sub route, then an entry is added in the cache table and DP is incremented by 1.
Step: 6. If a data packet can be recovered, then it is unnecessary to be included in the cache table.

Cache update algorithms are work using above two algorithms in a distributed manner and send REER notification of 60 bytes to all the nodes present in the topology.

4   Simulation Environment

Table 1 illustrates simulation environment use for route cache update. The experiment was carried by the network simulator software (NS2), and the aspects are plotted in Table 1 which constitutes the stimulation environment for improving RAMP to MRAMP.

Table 1 Simulation environment

Parameters	Values
Area	500 m x 500 m
No. of node	10, 20, 30, 40, 0, 60, 70, 80, 90, 100
Pause time	3, 6, 9, 12, 15
Mac type	802.11
Interface type	Phy/WirelessPhy
Queue length	50 packets
Antenna type	Omni antenna
Propagation type	TwoRayGround
Mobility model	Random waypoint
Application agent	CBR
Transport agent	UDP
Routing protocol	UDSR
Simulation time	50 s

5 Simulations

The new approach is measured using different QoS parameters as mention below.

5.1 Packet Delivery Ratio (PDR)

PDR is the amount of packet, successfully reached to the destination.

$$PDF = \frac{P_{nr}}{P_{ns}} \qquad (1)$$

where

Pnr Number of packets received, and
Pns Number of packets send.

5.2 End-to-End Delay

It is the ratio of time difference between numbers of packet send and received over the total time required to reach the destination. If delay reduces, then the performance of the network gives better output.

$$D_y = \frac{1}{N} \sum_{}^{J} [PR_J - PS_J] \qquad (2)$$

where

D_y Delay,
PR_j Number of packets received at the jth time,
PS_j Number of packets send at jth time, and
N Number of node.

5.3 Overhead

The number of routing packets generated in the network is routing overhead. If routing overhead is high, then the performance of protocol is reduced, which affects the QoS in MANET.

5.4 Packet Drop

Packet Drop rate are termed as the number of packets that have not reached the destination end.

5.5 Energy Consumption

It is the average energy consumed by each node. When numbers of control message flooded are high, then more energy is consumed by each node.

$$AEC = \sum_{i=1}^{n} \frac{E_{ci}}{N} \qquad (3)$$

AEC Average energy consumption,
E_c Energy consumed by each node, and
N Number of nodes.

6 Results and Discussions

The abovementioned QoS parameters are considered for evaluating the performance of the new approach for updating the cache.

6.1 PDR

Figure 2 shows the improvement for PDR between RAMP and MRAMP.

6.2 End-to-End Delay

Figure 3 shows the reduction of delay between RAMP and MRAMP.

6.3 Overhead

Figure 4 shows the reduction of overhead between RAMP and MRAMP.

6.4 Packet Drop

Figure 5 shows the packet drop ratio which reduces from 5 to 10%.

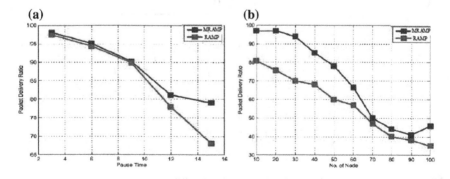

Fig. 2 **a** PDR versus pause time and **b** PDR versus no. of nodes

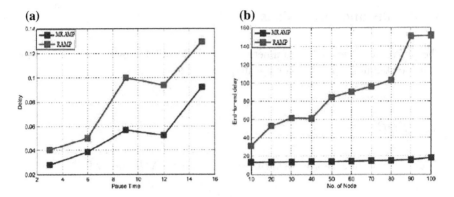

Fig. 3 a End-to-end delay versus pause time and **b** end-to-end delay versus no. of nodes

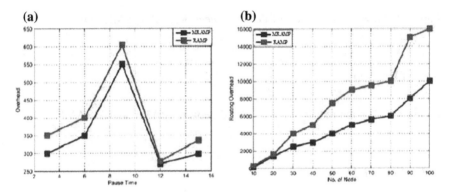

Fig. 4 a Overhead versus pause time and **b** overhead versus no. of nodes

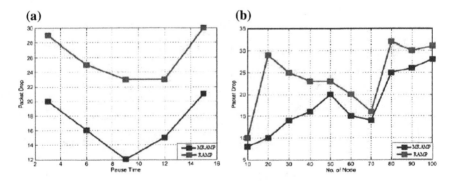

Fig. 5 a Overhead versus pause time and **b** overhead versus no. of nodes

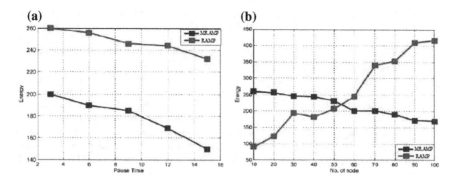

Fig 6 **a** Energy Consumption vs. Pause Time. **b** Energy Consumption vs. No. of Nodes

6.5 *Energy Consumption*

Figure 6 shows a saving of energy to sufficient levels.

7 Conclusion

Thus, from the results and discussions, the MRAMP shows an increase in performance of routing in terms of QoS metrics. The improvement shows both least group of nodes and average group of nodes. The MRAMP apart from recording trusted values of node also shows a feasible improvement through the cache refreshing policy, and it proves that cache optimization mechanisms play a vital role in improving the QoS through the dynamic protocols of the mobile ad hoc networks.

Further experiments may be carried out to observe the role of cache optimization mechanisms in the high density of ad hoc networks.

1. Smith, T.F., Waterman, M.S.: Identification of Common Molecular Subsequences. J. Mol. Biol. 147, 195–197 (1981).
2. May, P., Ehrlich, H.C., Steinke, T.: ZIB Structure Prediction Pipeline: Composing a Complex Biological Workflow through Web Services. In: Nagel, W.E., Walter, W.V., Lehner, W. (eds.) Euro-Par 2006. LNCS, vol. 4128, pp. 1148–1158. Springer, Heidelberg (2006).

References

1. Spanakis Emmanouil and Apostolos Traganitis,: "Applying the Time-To-Live Parameter in On Demand Route Caching in MANETs", ICTACT, February (2009).

2. Narinderjeet Kaur, Maninder Singh,: Caching Strategies in MANET Routing Protocols, International Journal of Scientific and Research Publications, Volume 2, Issue 9, September (2012).
3. Vivek Arya and Charu,: A survey of Enhanced Routing protocols for MANETs, International Journal on Adhoc Networking Systems, Vol. 3, issue No. 3, July (2013).
4. V.V. Mandhare, R.C. Thool,: Improving QoS of Mobile Ad-hoc Network using Cache Update Scheme in Dynamic Source Routing Protocol, 7th International Conference on Communication, Computing and Virtualization, pp. 692–699, Procedia Computer Science 79, (2016).
5. Ying-Hong Wang, Jenhui Chen, Chih-Feng Chao and Chien-Min Lee,: A Transparent Cache-based Mechanism for Mobile Ad Hoc Networks, Proceedings of the Third International Conference on Information Technology and Applications (ICITA '05), IEEE (2005).
6. Yaoda Liu Shengming Jiang, Yuming Jiang, Dajiang He,: An Adaptive Link Caching Scheme for On-Demand Routing in MANETs, IEEE (2003).
7. Mostafa Dehghan, Anand Seetharam, Ting He, Theodoros Salonidis, Jim Kurose, and Don Towsley,: Optimal Caching and Routing in Hybrid Networks, IEEE Military Communications Conference, (2014).
8. Wenbo Zhu, Xinming Z hang, Yongzhen Liu, Nana Li,: Improve Preemptive Routing Performance in Mobile Ad hoc Networks with Cache-enabled Method, National Natural Science Foundation (2008).
9. Vineet Joshi, Xuefu Zhu, and Qing-An Zeng,: Caching-based Multipath Routing Protocol, Proceedings of International Conference on Computational Science and Engineering (2009).
10. P. Sateesh Kumar and S. Ramachandram,: The Performance Evaluation of Cached Genetic Zone Routing Protocol for MANETs, IEEE (2008).
11. CH. V. Raghavendran, G. Naga Satish, P. SureshVarma and K.N.S.L. Kumar,: Challenges and Advances in QoS Routing Protocols for Mobile Ad Hoc Networks, International Journal of Advanced Research in Computer Science and Software Engineering, Volume 3, Issue 8, August (2013).
12. Hao-jun Li, Fei-yue Qiu, Yu-jun Liu,: Research on Mechanism Optimization of ZRP Cache Information Processing in Mobile Ad Hoc Network, IEEE (2007).
13. Sanjeev Gangwar and Dr. Krishan Kumar,: Mobile Ad Hoc Networks: A detailed Survey of QoS Routing Protocols, International Journal of Distributed and Parallel Systems (IJDPS) Vol. 2, No. 6, November (2011).
14. Chao-Tsun Chang,: Hash caching mechanism in source-based routing for wireless ad hoc networks, Journal of Network and Computer Applications, October (2011).
15. G. Santhi and Alamelu Nachiappan,: A survey of QOS routing protocols for Mobile Ad Hoc Networks, International journal of computer science & information Technology (IJCSIT) Vol. 2, No. 4, August (2010).
16. Mayur Bhalia,: Analysis of MANET Characteristics, Applications and its routing challenges, International Journal of Engineering Research and General Science, Volume 3, Issue 4, Part-2, July–August, (2015).
17. Anuj Rana, Sandeep Gupta,: Review on MANETs Characteristics, Challenges, Application and Security Attacks, International Journal of Science and Research, Volume 4 Issue 2, February (2015).
18. Song Guo and Oliver Yang,: Effects of Backup Routes and Cache Timeout Mechanism on Reliable Source Routing in Mobile Ad-hoc Networks, IEEE (2005).
19. N. Ashokraj, C. Arun and K. Murugan,: Route Cache optimization mechanism using smart packets for on-demand routing protocol in MANET, Proceedings of International Conference on Information Technology, IEEE (2003).
20. Fenglien Lee, Carl T. Swanson and Jigang Liu,: "Efficient On-Demand Cache Routing for Mobile Ad Hoc Networks", IEEE (2009).

21. Weibo Chen, Kun Yang and Achilleas Achilleos,: RZRP: A Pure Reactive Zone-based Routing Protocol with Location-based Predictive Caching Scheme for Wireless Mobile Ad Hoc Networks, IEEE (2008).
22. Roberto Beraldi and Roberto Baldoni,: Caching Scheme for Routing in Mobile Ad Hoc Networks and Its Application to ZRP, IEEE transactions on computers, Vol. 52, No. 8, August (2003).

A Novel Parity Preserving Reversible Binary-to-BCD Code Converter with Testability of Building Blocks in Quantum Circuit

Neeraj Kumar Misra, Bibhash Sen, Subodh Wairya and Bandan Bhoi

Abstract The reversible logic circuit is popular due to its quantum gates involved where quantum gates are reversible and noted down feature of no information loss. In this paper, parity preserving reversible binary-to-BCD code converter is designed, and effect of reversible metrics is analyzed such as gate count, ancilla input, garbage output, and quantum cost. This design can build blocks of basic existing parity preserving reversible gates. The building blocks of the code converter reversible circuit constructed on Toffoli gate based as well as elemental gate based such as CNOT, C-V, and C-V+ gates. In addition, qubit transition analysis of the quantum circuit in the regime of quantum computing has been presented. The heuristic approach has been developed in quantum circuit construction and the optimized quantum cost for the circuit of binary-to-BCD code converter. Logic functions validate the development of quantum circuit. Moving the testability aim are figured in the quantum logic circuit testing such as single missing gate and single missing control point fault.

Keywords Reversible computation · Quantum circuit · Quantum Toffoli gate Testability · Code converter

N. K. Misra (✉) · S. Wairya
Department of Electronics and Communication Engineering, Bharat Institute of Engineering and Technology, Hyderabad 501510, India
e-mail: neeraj.mishra@ietlucknow.ac.in; neeraj.mishra3@gmail.com

B. Sen
Department of Computer Science and Technology, National Institute of Engineering and Technology, Durgapur, India

B. Bhoi
Department of Electronics & Telecommunication, Veer Surendra Sai University of Technology, Burla 768018, India

© Springer Nature Singapore Pte Ltd. 2018
V. Bhateja et al. (eds.), *Proceedings of the Second International Conference on Computational Intelligence and Informatics*, Advances in Intelligent Systems and Computing 712, https://doi.org/10.1007/978-981-10-8228-3_35

1 Introduction

Over the last few decades, the reversible circuit has gained interest due to their low energy dissipation [1]. The reversible circuit properties of recovered inputs and no loss of information are a unique feature to look upon. In fact, the amount of heat dissipation is related to information loss [2]. The noted down features of reversible computing are recovered input states from output states, controllable inputs and outputs, and zero energy dissipation [3]. The reversible technology in quantum computation is expanded and becomes one of the prominent technologies in the new era. Toffoli gates have found extensive building blocks in quantum circuit construction. Increasing demand of the complex reversible circuit, it is a major challenge to optimize the reversible parameters. To design a compact reversible design, Toffoli gate blocks can be preferred [4]. The effort has been put to increase the performance of the reversible circuit by optimizing the ancilla input, gate count, garbage output, and quantum cost [5].

In this paper, we have proposed a parity preserving reversible binary-to-BCD code converter. The quantum equivalent circuit is built from a standard library of reversible gates such as NCT (NOT, CNOT, and Toffoli gates), MCT (multiple control Toffoli gate), and NCV (NOT, CNOT, and the square root of NOT) [6]. Further, the high-level block of binary-to-BCD code converter to the respective quantum equivalent circuit is framed successfully. The constructed circuit is optimized regarding gate count, ancilla input, and quantum cost. The noted work of the proposed circuit is the fault testing of the quantum circuit under a single missing gate and single missing control point. To the best of our literature review on reversible binary-to-BCD code converter circuit with the testability of quantum circuit is not covered elsewhere.

The major pillars of this workaround quantum reversible code converter with testability are highlighted as follows:

- In this work, a heuristic approach is used to synthesize the quantum circuit of binary-to-BCD code converter. This approach depicts a promise to synthesize the Toffoli gate-based quantum circuit and synthesis result, such as quantum cost at a faster time in few seconds.
- A testability of quantum circuit is achieved by a single missing gate and single control point-based fault detection. The target quantum reversible gates such as FRG, F2G, NFT, RDC, and DFG gates are included with testability aspects.

This connection of paper is organized as follows. First, the essential background of reversible and quantum computing is given. A term of the testing method base of the quantum circuit is also covered as well. In Sect. 3, the previous work constraints are provided. In Sect. 4, the synthesis of parity preserving reversible binary-to-BCD code converter is presented. Section 5 provides an individual building block testability methodology. The conclusion is discussed in Sect. 6.

2 Background of Reversible and Quantum Computing

We have discussed some terminologies such as reversible computing, quantum computing, existing reversible gates implementation in Toffoli gate based, and elemental quantum gates in this section.

2.1 Reversible and Quantum Computing

Definition 1 All reversible gate Boolean functions maintain the Bijective mapping (onto and one-to-one).

Proof The reversible gate executes Bijective mapping in which a unique input vector is mapped to unique output vectors and therefore an equal number of input and output. Let us assume that reversible gate does not fulfill the criteria of Bijective mapping, i.e., input is not equal to the output. The output is not recovered the input and loss the reversibility. Hence, all reversible gates must hold the Bijective mapping [7]. The feature of quantum gates is inherently reversible [8]. The quantum circuit construction with minimal quantum cost by using given Boolean satisfiability (known as SAT) can be considered more effective in quantum computing regime [4].

Multiple control Toffoli (MCT): The involvement of cascade MCT gates makes reversible circuit compact. In MCT, gates are represented with $\oplus$ which is the control gate and black dots means the control point. An n-input Toffoli gate is shown as Y_n (C, t) that map the reversible structure pattern as $(p_{i1}, p_{i2}, \ldots\ldots p_{ik})$ to $(p_{i1}, p_{i2}, \ldots\ldots p_{k-1}, p_k \oplus p_{i1} p_{i2} \ldots\ldots\ldots p_{k-1} p_{k+1} \ldots p_{ij})$, where $C = (p_{i1}, p_{i2}, \ldots\ldots p_{ik})$, $t = \{p_{i1}\}$, C is denoted as control point, and t is donated as control gate [9, 11]. CCNOT (TG) gate has three inputs and outputs. First two inputs (a, b) are the control-bit, and the third input c is the target-bit (Fig. 1a). If the first control-bit a is zero, it maps input CNOT (FG) gate (Fig. 1b). Again, if the first and second control-bit $a = b=0$, it maps input NOT gate (Fig. 1c). In n-input Toffoli gate, first $(n - 1)$ input is noted as control-bit, and n-input is target-bit (Fig. 1d).

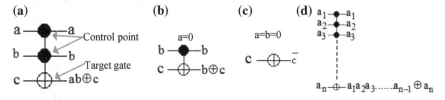

Fig. 1 Multiple control Toffoli structures of **a** CCNOT, **b** CNOT, and **c** NOT **d** MCT

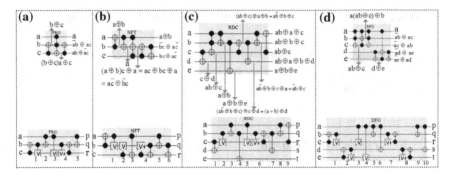

Fig. 2 Reversible and quantum circuit of **a** FRG, **b** NFT, **c** RDC, and **d** DFG

2.2 Existing Parity Preserving Reversible Gates

Reversible decoder gate (RDC), and NFT gates: A 5×5 reversible decoder (RDC) conserves parity at both inputs and outputs also known as parity preserving reversible RDC gate [10]. The logical functions can be expressed as $P = ab \oplus a \oplus c, q = ab \oplus b \oplus c, r = ab \oplus c, s = ab \oplus a \oplus b \oplus d, t = a \oplus b \oplus e$. Figure 2a–c presents the Toffoli gate block and the quantum equivalent of FRG and RDC gates, respectively.

The NFT gate executes the following three output functions: $p = a \oplus b, Q = a\bar{c} \oplus \bar{b}c, r = a\bar{c} \oplus bc$, where a, b, and c are the inputs. Toffoli gate block and elemental quantum gate-based structure of NFT have been shown in Fig. 2b.

Double Fredkin gate (DFG): The functional relationship between inputs and outputs of DRG can be connected as $p = a, q = \bar{a}b \oplus ac, r = \bar{a}c \oplus ab, s = \bar{a}d \oplus ae, t = \bar{a}e \oplus ad$. Figure 2d presents the Toffoli gate block and the quantum equivalent of DFG gate. In all the quantum circuits in this work, we consider black dot as the control point and the $\oplus$ (presented by exclusive OR) control gate.

3 Previous Work

In this section, we have reviewed the existing design of reversible binary-to-BCD code converter. Researchers have worked on the various areas of reversible binary-to-BCD code converters such as synthesis and optimization [12–15]. No any previous research work on binary-to-BCD code converter such as testing of the quantum logic circuit. However, the existing reversible code converter circuit still suffers from difficulties such as more number of GC, GO, CI, and QC. Lower values of reversible parameters are always preferred for circuit synthesis [16, 17]. In existing work presented in [12] is reversible approach but not parity preserving binary-to-BCD code converter. This design noted down parameters such as 17 GC,

12 GO, 13 CI, and 78 QC. Another existing work in [13] constructs the reversible binary-to-BCD converter that takes GC of 7, GO of 2, and QC of 28. But this circuit utilizes TG, which is non-parity preserving reversible gate. However existing methodology is based on the quantum reversible circuit of binary-to-BCD converter but not target the testing of quantum circuit. But here, lower value of gate count is synthesized, and hence, the proposed binary-to-BCD code converter does not suffer from the pitfalls discussed above of reversible design.

4 Synthesis of Proposed Binary-to-BCD Code Converter

The proposed binary-to-BCD code converter contains eight parity preserving reversible gates the details of the utilizing gate such as one RDC, one DFG, one F2G, two FRG, and three NFT; it has 19 inputs and 19 outputs.

Algorithm 1: Binary-to-BCD code converter

Input: $I = (a, b, c, d)$ in binary, **Output:** $Y = (Y_0, Y_1, Y_2, Y_3)$ in BCD.
1: **For** i=0 to n-1 **do**
2: **If** i=1 **then**
      $(c, d, 0, 1, 0) \rightarrow$ 1-RDC   // Assign input to RDC gate
      1-RDC$\leftarrow (a'_{1\text{-RDC}}, b'_{1\text{-RDC}}, c'_{1\text{-RDC}})$ // Three intermediate output
      End if, Else
3:    $(b, d, c, 0, 0) \rightarrow$ 2-DFG   // Assign input to 2-DFG
      2-DFG$\leftarrow (a'_{2\text{-DFG}}, c'_{2\text{-DFG}}, e'_{2\text{-DFG}})$ // Three intermediate output to 2-DFG
      End if, Else
4: **If** i=3 **then**
      $(a'_{2\text{-DFG}}, 1, 0) \rightarrow$ 3-F2G   // Assign input to 3-F2G
      $(b'_{3\text{-F2G}}) \leftarrow$ 3-F2G   // One intermediate output of 3-F2G
      End if, Else
5: $(b'_{1\text{-RDC}}, 0, b'_{3\text{-F2G}}) \rightarrow$ 4-F2G   // Assign input to 4-F2G
      $(Y_4) \leftarrow$ 4-F2G   // Catch the one target output
6: **If** i=5 **then**
      $(c'_{2\text{-DFG}}, 0, c) \rightarrow$ 5-FRG   // Assign input to 5-FRG
      $(a'_{5\text{-FRG}}, b'_{5\text{-FRG}}) \leftarrow$ 5-FRG   // Two intermediate output of 5-FRG
End if, Else
      $(a'_{5\text{-FRG}}, 0, c'_{1\text{-RDC}}) \rightarrow$ 6-NFT // Assign input to 6-NFT
      $(Y_3) \leftarrow$ 6-NFT   // Catch the one target output
6: **If** i=7 **then**
      $(e'_{2\text{-DFG}}, 0, a'_{1\text{-RDC}}) \rightarrow$ 7-NFT // Assign input to 7-NFT
      $(Y_2) \leftarrow$ 7-NFT   // Catch the one target output
End if, Else
      $(b'_{5\text{-FRG}}, 0, a'_{6\text{-NFT}}) \rightarrow$ 8-NFT // Assign input to 8-NFT
      $(Y_1) \leftarrow$ 8-NFT   // Catch the one target output
      1-RDC$\leftarrow (g_1, g_2)$, 2-DFG$\leftarrow (g_3, g_4)$, 3-F2G$\leftarrow (g_5, g_6)$, 4-F2G$\leftarrow (g_7, g_8)$, 5-FRG$\leftarrow (g_9)$,
      6-NFT$\leftarrow (g_{10})$, 7-NFT$\leftarrow (g_{11}, g_{12})$, 8-NFT$\leftarrow (g_{13}, g_{14})$ // Remaining as garbage output
5: **End if, end if, end for,**
6: **Return**(Y_i), **End;**

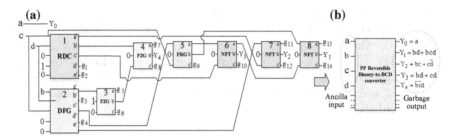

Fig. 3 The proposed binary-to-BCD code converter: **a** functional diagram and **b** cell diagram

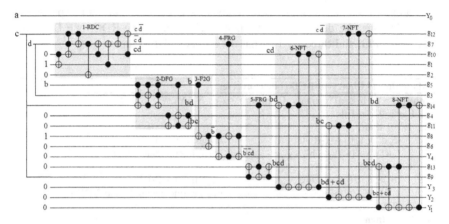

Fig. 4 The proposed quantum representation of the binary-to-BCD code converter

The inputs comprise four input information (a, b, c, d) in binary and 15 ancilla inputs which are specified as constant 1 and 0. It produces four outputs such as $Y_0 = a$, $Y_1 = bcd + bd$, $Y_2 = bc + c\bar{d}$, $Y_3 = bd + cd$, and $Y_4 = \bar{b}\bar{c}d$. This circuit garbage output is denoted by g. The functional diagram and quantum equivalent of binary-to-BCD are presented in Figs. 3 and 4. In the quantum circuit construction, we have used exact synthesis technique such as Boolean satisfactory (known as SAT) to acquire compact circuit for reversible functions [18]. In this work, the quantum circuit of proposed binary-to-BCD is achieved with Toffoli gate based. However, the limitation of page dimensions is that we are not able to show the quantum circuit of binary-to-BCD converter by elemental quantum gate based. The QC of binary-to-BCD converter is 46. Algorithm 1 is shown for the synthesis of this circuit. The QC (binary-to-BCD) is presented as $1QC_{RDC} + 1\ QC_{DEG} + 2QC_{F2G} + 1QC_{FRG} + 3QC_{NFT} = 9 + 10 + 2 \times 2 + 5 + 3 \times 6 = 46$.

5 Individual Building Block Testability Methodology

In the quantum circuit, the testing is essentially used to confirm a validate output. For quantum circuit embedded to Toffoli gate, various faults were reviewed in state-of-the-art [18]. In this paper, we have covered the single missing gate type fault (SMGF) and single missing control type fault (SMCF). To test the fault in the quantum circuit, two points should be considered: (i) if control gate is faulty, then the information is unchanged in a quantum wire. (ii) In case of the CNOT gate, if a control point is faulty, then the control gate changes the information in the quantum wire [18]. Thus, these points must be remembered that the fault testing at the quantum circuit. Therefore, we present the fault pattern structure of building blocks used in binary-to-BCD. The building blocks are used in binary-to-BCD such as F2G, FRG, NFT, RDC, and DFG. The procedure for fault patterns is generated in the following description: First chose any test input vectors and found the target output vectors. In the quantum circuit, faults are SMGF and SMCF. In SMGF change the bit when the control gate is faulty as in the case of CNOT quantum logic structure. In SMCF change the bit when the control point is faulty in the case of CNOT quantum logic structure. The fault patterns of NFT, FRG, DFG, and RDC gates are shown in Figs. 5, 6, 7, 8 and 9, respectively. The fault pattern lookup table for NFT, FRG, DFG, and RDC gates is presented in Tables 1, 2, 3 and 4. By examining the fault testing in the quantum circuit, it can be expected that control point and control gate have played an important role in the synthesis of qubit transition in the quantum circuit. The key feature of this work is integrated such as synthesis, optimisation, and testing for a design flow of binary-to-BCD code converter.

According to Table 5, the proposed circuit provides a better result over the counterpart circuits. The preliminary study reveals the significant improvement of reversible parameters. The notations used in Table 5 are as follows: PP: Parity preserving, QE: Quantum equivalent, GC: Gate count, CI: Constant input, GO: Garbage output, QC: Quantum cost, -: Not mentioned, Y: Yes, and N: No.

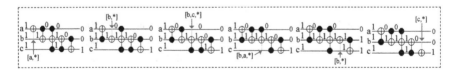

Fig. 5 Illustration of SMCF for NFT gate

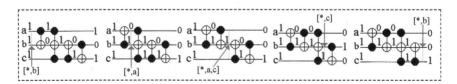

Fig. 6 Illustration of SMGF for NFT gate

(a) **(b)**

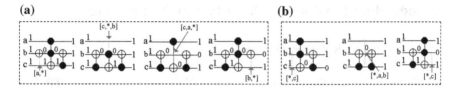

Fig. 7 Illustration of fault testing in FRG **a** SMCF and **b** SMGF

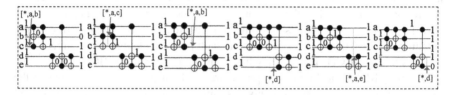

Fig. 8 Illustration of SMGF for DFG gate

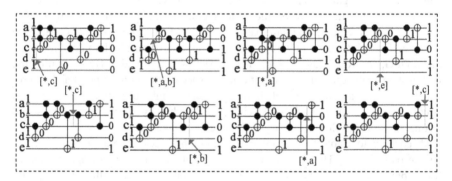

Fig. 9 SMGF faults pattern in RDC gate

Table 1 Table for NFT gate with capture of both faulty and fault-free outputs

Design	SMCF			
	Marking method	Test in	Target out	Fault pattern
Figure 5	[a,*]	111	001	001
	[b,*]	111	001	001
	[b,c,*]	111	001	001
	[b,a,*]	111	001	001
	[b,*]	111	001	011 (Faulty)
	[c,*]	111	001	001
	SMGF			
Figure 6	[*,b]	111	001	101 (Faulty)
	[*,a]	111	001	001
	[*,a,c]	111	001	001
	[*,c]	111	001	010 (Faulty)
	[*,b]	111	001	001

Table 2 Table for FRG gate with the capture of both faulty and fault-free outputs

Design	SMCF			
	Marking method	Test in	Target out	Fault pattern
Figure 7	[a,*]	111	111	111
	[c,*,b]	111	111	111
	[c,a,*]	111	111	101 (Faulty)
	[b,*]	111	111	101 (Faulty)
	SMGF			
	[*,c]	111	111	110 (Faulty)
	[*,a,b]	111	111	111
	[*,c]	111	111	101 (Faulty)

Table 3 Table for DFG gate with the capture of both faulty and fault-free outputs

Design	SMGF			
	Marking method	Test in	Target out	Fault pattern
Figure 8	[*,a,b]	11111	11111	10111 (Faulty)
	[*,a,c]	11111	11111	11111
	[*,a,b]	11111	11111	11011 (Faulty)
	[*,b]	11111	11111	11101 (Faulty)
	[*,a,c]	11111	11111	11111
	[*,d]	11111	11111	11110 (Faulty)

Table 4 Table for RDC gate with the capture of both faulty and fault-free outputs

Design	SMGF			
	Marking method	Test in	Target out	Fault pattern
Figure 9	[*,c]	11111	11001	10000 (Faulty)
	[*,a,b]	11111	11001	00111 (Faulty)
	[*,a]	11111	11001	10010 (Faulty)
	[*,e]	11111	11001	11011 (Faulty)
	[*,c]	11111	11001	11001
	[*,b]	11111	11001	11001
	[*,a]	11111	11001	10001 (Faulty)
	[*,c]	11111	11001	11001 (Faulty)

Table 5 Comparison with counterpart designs and novel binary-to-BCD converter

Designs	PP	QE	GC	CI	GO	QC
Novel	Y	Y	8	12	14	46
[12]	N	N	17	13	12	78
[13]	N	N	7	–	–	28
[14]	N	N	15	21	20	–
[15]	N	N	49	61	64	–

6 Conclusions

In this paper, a quantum reversible circuit of binary-to-BCD code converter and the testability of building blocks used in the circuit are proposed. The heuristic approach is adopted for quantum circuit construction and gives better results with optimizing quantum cost. The compact quantum circuit design is achieved when Toffoli gate is embedded in the quantum circuit. The proposed circuit offers to reduce the gate count, ancilla input, and quantum cost. In addition, we have illustrated the individual quantum gates testability used in the schematic of binary-to-BCD code converter by single missing control point fault and single missing gate fault for it. An interesting future work would be to extend the data bits in reversible binary-to-BCD code converter and fewer reversible parameters.

References

1. De Vos, A.: Reversible computing: fundamentals, quantum computing, and applications. John Wiley & Sons, pages 261 (2011).
2. Bennett, C.H.: Logical reversibility of computation. IBM Journal of Research and Development, 17(6), 525–532 (1973).
3. Misra, N.K., Sen, B. and Wairya, S.: Towards designing efficient reversible binary code converters and a dual-rail checker for emerging nanocircuits. Journal of Computational Electronics, 16(2), 442–458 (2017).
4. Maslov, D. and Dueck, G.W.: Improved quantum cost for n-bit Toffoli gates. Electronics Letters, 39(25), 1790–1791 (2003).
5. Misra, N.K., Wairya, S. and Singh, V.K.: Optimized Approach for Reversible Code Converters Using Quantum Dot Cellular Automata. In Proceedings of the 4th International Conference on Frontiers in Intelligent Computing: Theory and Applications (FICTA) Springer India, 367–378 (2016).
6. Sasanian, Z., Wille, R. and Miller, D.M.: Clarification on the Mapping of Reversible Circuits to the NCV-v1 Library. arXiv preprint arXiv:1309.1419, (2013).
7. Misra, N.K., Wairya, S. and Sen, B.: Design of conservative, reversible sequential logic for cost efficient emerging nano circuits with enhanced testability. Ain Shams Engineering Journal. (2017).
8. Misra, N.K., Sen, B., Wairya, S. and Bhoi, B.: Testable Novel Parity-Preserving Reversible Gate and Low-Cost Quantum Decoder Design in 1D Molecular-QCA. Journal of Circuits, Systems and Computers, 26(09), p. 1750145 (2017).
9. Deb, A., Das, D.K., Rahaman, H., Wille, R., Drechsler, R. and Bhattacharya, B.B.: Reversible Synthesis of Symmetric Functions with a Simple Regular Structure and Easy Testability. ACM Journal on Emerging Technologies in Computing Systems (JETC), 12(4), pages. 34 (2016).
10. Rahman, M.R.: Cost Efficient Fault Tolerant Decoder in Reversible Logic Synthesis. International Journal of Computer Applications, 108(2), 7–12 (2014).
11. Sen, B., Dutta, M., Some, S. and Sikdar, B.K.: Realizing reversible computing in QCA framework resulting in efficient design of testable ALU. ACM Journal on Emerging Technologies in Computing Systems (JETC), 11(3), pages. 30 (2014).
12. Gandhi, M. and Devishree, J.: Design of Reversible Code Converters for Quantum Computer based Systems. International Journal of Computer Applications, 3, (2013).

13. Haghparast, M. and Shams, M.: Optimized nanometric fault tolerant reversible bcd adder. Research Journal of Applied Sciences, Engineering and Technology, 4(9), 1067–1072 (2012).
14. Murugesan. P. and Keppanagounder.: An improved design of reversible binary to binary coded decimal converter for binary coded decimal multiplication. American Journal of Applied Science, 11(1), 69–73 (2014).
15. Rajmohan, V. Ranganathan and Rajmohan, M.: A reversible design of BCD multiplier. Journal of Computer, vol. 2, 112–117 (2010).
16. Misra, N.K., Wairya, S. and Singh, V.K.: Approach to design a high performance fault-tolerant reversible ALU. International Journal of Circuits and Architecture Design, 2(1), 83–103 (2016).
17. Misra, N.K., Sen, B. and Wairya, S.: Designing conservative reversible n-bit binary comparator for emerging quantum-dot cellular automata nano circuits. Journal of Nanoengineering and Nanomanufacturing, 6(3), 201–216 (2016).
18. Nagamani, A.N., Ashwin, S., Abhishek, B. and Agrawal, V.K.: An Exact approach for Complete Test Set Generation of Toffoli-Fredkin-Peres based Reversible Circuits. Journal of Electronic Testing, 32(2), 175–196 (2016).

Differentiated WRED Algorithm for Wireless Mesh Networks

B. Nandini, Suresh Pabboju and G. Narasimha

Abstract As wireless multimedia communication for next-generation broadband wireless Internet access deploying WMN, one emerging challenge is meeting the Quality of Service (QoS) requirements with efficient resource utilization. Different QoS guarantees are required for real-time and non-real-time traffics such as delay constraints for voice traffic and throughput for delay insensitive data applications. Quality of services experienced by the users in wireless mesh networks, i.e., end-to-end delays, bandwidth utilization, and packet losses, is heavily dependent upon the effective congestion avoidance techniques employed in the wireless nodes. Congestion can be detected by computing the average queue size and either dropping the packet or marking special bits in the packet header when average queue buffer usage crosses the threshold. Several algorithms and packet dropping probabilities are discussed in Floyd and Jacobson (IEEE/ACM Transactions on Networking, 1(4), 397–413, 1993, [1]), Chaegwon Lim et al. (A Weighted RED for Alleviating Starvation Problem in Wireless Mesh Networks, IEEE, 2008, [2]). In this paper, we discuss the existing average queue length computation and packet drop methods employed and propose improvements to consider dynamic buffer allocation for Wireless Random Early Detection (WRED) algorithm.

Keywords Quality of service · Wireless mesh networks · Random early detection · WRED

B. Nandini (✉)
Department of CSE, Telangana University, Nizamabad, India
e-mail: cnuvnk@gmail.com

S. Pabboju
Department of CSE, CBIT, Hyderabad, India
e-mail: plpsuresh@gmail.com

G. Narasimha
Department of CSE, JNTUHCE, Jagityal, Karimnagar, India
e-mail: narsimha06@gmail.com

© Springer Nature Singapore Pte Ltd. 2018
V. Bhateja et al. (eds.), *Proceedings of the Second International Conference on Computational Intelligence and Informatics*, Advances in Intelligent Systems and Computing 712, https://doi.org/10.1007/978-981-10-8228-3_36

1 Introduction

As the demand for broadband Internet access is increasing, wireless local area networks play an important role in providing high-rate data services, which is further aided by emerging new technologies like Multiple Input Multiple Output (MIMO) at low cost. Wireless mesh networks are emerging as future wireless broadband technology as a form of wireless ad hoc networks [3]. Wireless mesh networks consist of routers without cabling between them like traditional Wireless Local Area Networks (WLAN) access points, Wireless Mesh Networks (WMNs) can be, spread across large areas, by carrying data over multiple hops using intermediate nodes such as wireless routers. These intermediate nodes can boost signal strength and make forwarding decisions based on the networking protocols like Internet Protocol (IP).

WMNs are organized in three-tier architecture [3, 4] as shown in Fig. 1, consisting of wireline gateways, mesh routers, and mesh clients. A mesh client is connected to one or two mesh routers, which are in static locations comprising wireless mesh backbone. Wireless mesh backbone provides relay services to mesh clients, cellular networks, and WLAN networks and provides loose coupling for interworking of heterogeneous wireless access networks [5]. Wireline gateways connect wireless mesh backbone to the Internet backbone. With extensive deployments of WMNs, many new challenges are emerging such as fading mitigation, routing with quality of service (QoS) requirements, call admission control, link-layer resource allocation, seamless roaming, and network security.

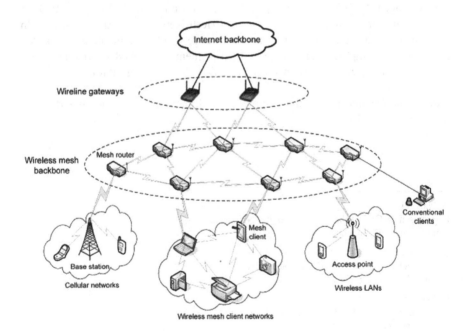

Fig. 1 Wireless mesh network illustration

As wireless multimedia communication for next-generation broadband wireless Internet access deploying Wireless Mesh Network (WMN), one emerging challenge is meeting the QoS requirements with efficient resource utilization. In shared wireless medium, simultaneous transmission by multiple nodes results in collisions causing the retransmission of the packets consuming the packet data rates. Different QoS guarantees are required for real-time and non-real-time traffics such as delay constraints for voice traffic and throughput for delay insensitive data applications. Due to contention between mesh clients and mesh router for the available limited radio bandwidth, Medium Access Control (MAC) has become essential to coordinate the transmission of data packet to/from the nodes in efficient and effective ways.

So the minimum requirements of MAC protocols are to be adaptive to various types of traffic meeting the performance metrics such as delay, throughput, fairness, robustness to link vulnerability, and multimedia support [6, 7]. In this paper, we overview congestion control mechanism already in place and propose improvements.

2 Related Work

Guaranteed QoS requirements in a WMNs node are met by proper allocation of bandwidth and memory space for buffering in a node [8, 9], and various types of traffic are differentiated based on the service requirements by distributing them into different packet queues each having a different priority [10]. We investigate queue management policies when packet queues congested and mechanisms proposed to detect early congestion based on the queue buffer size. The Random Early Detection (RED) algorithm proposed in [1] provides an effective mechanism for congestion avoidance at the mesh nodes or gateways. RED algorithm drops an incoming packet with some probability if the average queue size is I in between predetermined minimum and maximum thresholds, and if the average queue size is above the maximum threshold, then all of the incoming packets are dropped. The packet losses result in traffic control at the source for application like Transmission Control Packets (TCP) if congetion notification mechanism like ECN (explicit congetion notification alogorithm) is used, and for application like User Datagram Protocol (UDP), RED algorithm provides fairness when multiple traffic streams are competing for resources. While the RED algorithm proposes effective mechanisms for congestion avoidance, the calculation of average queue size and drop probabilities are complex and take a significant amount of processing resources at the WMN nodes. In [2], a new method called weighted RED (WRED) is proposed to evaluate drop probabilities which takes number of hops a packet traversed. When the network is congested, packet from the farthest source node gets higher priority than the packet generated at the local router. In this paper, we propose a new mechanism to mark packets with loss priority so that packets will get differentiated queue threshold levels and propose real-time evaluation mechanism for queue thresholds.

3 Proposed Improvements

Data packets are distributed across multiple queues in the node, when there is congestion in the node, packet will be buffered in memory waiting to be served, once the other queues are served according to the scheduling algorithm [11, 12]. An example of scheduler is shown in Fig. 2 for distributed deficit round robin [13]. Each queue will buffer the packet until the scheduler serves it, as soon as it is served, the buffer will be made empty and ready to be occupied by the next outgoing packet. If the cumulative outgoing traffic rate is more than the mesh node can send out, the packets will be dropped once the allocated buffer per queue filled up.

Traffic classifiers [4] are defined to distribute the incoming packets to various queues, wherein ToS bits in the IP header are used to classify the packets. We propose to associate each incoming packet with particular ToS bits with loss priority. Loss priority is the measure of susceptibility of the packet to be dropped. Loss priorities of low, medium, and high can be defined for the packets to be classified.

Low—loss priority is low, which means that the packets are less susceptible to be dropped.
High—loss priority is high, which means that the packets are more susceptible to be dropped.
Medium—loss priority is medium, which means that the packet's susceptibility of being dropped falls between that of low and high loss priorities.

Loss priorities are user configurable, which gives the user more flexibility to drop traffic stream in case of network congestion. When multiple traffic streams are

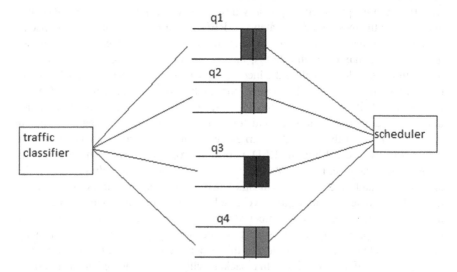

Fig. 2 Illustration of scheduler serving the packet queues

available with different ToS bits classified to the same queue, loss priorities can be used to apply different drop probabilities and queue fill levels. Min and max threshold values used in the RED algorithm proposed in [1] are static while average queue size is computed at the time of packet arrival with fixed queue size. For better buffer resource utilization, a new buffering mechanism is proposed in [14] where each queue will have two components in its buffer resources: one is a fixed number of buffers which are dedicated to that particular queue, and the second component is shared buffers to which all of the queues in the gateway router have equal access to when required, and it is shown in Fig. 3. The shared buffers and the buffer access algorithm ensure better resource utilization and fairness among competing queues. The number of buffers being used by a queue at any time depends on the other competing queues, which makes total queue buffer usage fluctuate from time to time. In this case, usage of fixed min and max buffer thresholds results in incorrect behavior of the RED algorithm. To evaluate the real-time buffer usage of a queue, we propose the following buffer evaluation for a queue where the queue size is composite of fixed dedicated buffers and shared buffers:

$$\text{Avg\_queue\_size} = (1 - \alpha) * \text{avg\_queue\_size} + \alpha * \text{sample\_queue\_size} \qquad (1)$$

avg_queue_size is weighted average of the sample_queue_size. Equation (1) puts more weight on the recent sample_queue_sizes, so that avg_queue_size reflects the current congestion in the network. α is chosen such that recent queue sizes get more weight while the old queue sizes get less weightage. Above avg_queue_size is used to determine min and max rather than to mark the packet to be dropped as given in [1].

Min and max buffer fill levels can be specified per queue, as percentage of the avg_queue_size for each of the loss priorities independently so that packet with low loss priority can occupy more buffers before being dropped or marked for congestion marking of the packet. High loss-priority packet can be assigned with small percent of avg_queue_size so that the packet can be dropped or marked with congestion happened bit in the IP header.

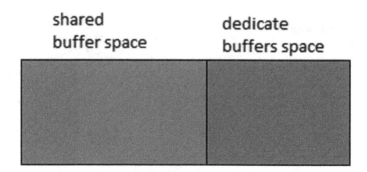

Fig. 3 Buffer memory division

Instead of using only ToS bits of the IP header, various values of the packet fields can be utilized to decide loss priorities. By utilizing the multiple packet header options, greater flexibility in achieving resource allocation for bandwidth and buffers can be achieved. Some of the values from packet fields are listed below.

Layer 4:

- TCP/UDP destination port,
- TCP/UDP source port, and
- TCP flags.

Layer 3:

- Destination/source IP addresses,
- TTL,
- ToS,
- IP options,
- Is-fragment bit, and
- IP protocol type.

Layer 2:

- Source/destination MAC addresses.

An example for the WRED packet drop profile is shown in Figs. 4 and 5 for a three-node mesh network. Node n3 is the gateway node where congestion will happen for the traffic streams from WMNs nodes n1 and n2. Figure 5 shows a graph of WRED for average queue fill levels versus packet drop probability with min and max buffer threshold of 30 and 50% with drop probability of 80%. Till buffer fill level reaches min of 30% of total buffer by the queue, no packet will be dropped; once minimum fill level is reached, packets begin to drop. The percentage of dropped packet increases linearly as the queue fill level increases. When the fill level reaches the drop end point of 50%, drop probability is increased to 80%; once the fill levels exceed 50%, all packets are dropped with drop probability of 100% until queue fill level drops below 50%.

Fig. 4 Three-node wireless mesh network

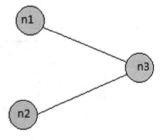

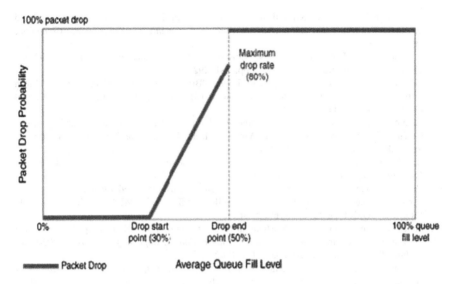

Fig. 5 WRED packet drop profile

4 Conclusion

In this paper, we have proposed improvements to the WRED algorithm, which differentiates the packets based on loss priorities. User has the flexibility to configure classifiers based on the ToS filed of the IP header so that buffer access and congestion notification can be controlled effectively on per packet basis. The average queue size evaluation is modified to take into account the dynamic buffer usage of the queue. To give more flexibility to differentiate incoming packets, various packet values are listed to assign bandwidth and buffer allocation per queue.

References

1. S. Floyd and V. Jacobson, "Random early detection gateways for congestion avoidance," In: *IEEE/ACM Transactions on Networking*, vol. 1, no. 4, pp. 397–413, August 1993.
2. Chaegwon Lim, Chong-ho Choi, Hyuk Lim, "A Weighted RED for Alleviating Starvation Problem in Wireless Mesh Networks" **978-1-4244-2413-9/08 - 2008 IEEE**.
3. Akyildiz IF, Wang X, Wang W. Wireless mesh networks: A survey. *Computer Networks* 2005; **47**(4): 445–487.
4. Jiang H, Zhuang W, Shen X, Abdrabou A, Wang P. Differentiated services for wireless mesh backbone. In: *IEEE Communications Magazine*; **44**(7): 113–119.
5. Song W, Jiang H, Zhuang W, Shen X. Resource management for QoS support in cellular/ WLAN interworking. In: *IEEE Network* 2005; **19**(5): 12–18.
6. Chandra A, Gummalla ACV, Limb JO. Wireless medium access control protocols. In: *IEEE Communications Surveys and Tutorials* 2000; **3**(2): 2–15.

7. Issariyakul T, Hossain E, Kim DI. Medium access control protocols for wireless mobile ad hoc networks: issues and approaches. In: *Wireless Communications and Mobile Computing* 2003; **3**(8):935–958.
8. G. Appenzeller, I. Keslassy, and N. McKeown, "Sizing router buffers". In: SIGCOMM Comput. Commun. Rev. 34, 4 (2004), pp. 281–292.
9. K. Jamshaid, B. Shihada, L. Xia, P. Levis, "Buffer Sizing in 802.11 Wireless Mesh Networks", Mobile Adhoc and Sensor Systems (MASS). In: 2011 IEEE 8th International Conference on, (2011), pp. 272–281.
10. RumipambaZambrano R.; Vázquez-Rodas A.; de la Cruz Llopis L.J.;SanvicenteGargallo E. Dynamic Buffer Size Allocation in Wireless Mesh Networks for Non-Elastic Traffic. In: RevistaPolitécnica-Febrero 2015, Vol. 35, No. 1.
11. Banchs A, P´erez X. Distributed weighted fair queuing in 802.11 wireless LAN. In: *Proceedings of IEEE ICC* 2002; 3121–3127.
12. Ho Ting Cheng, Hai Jiang and Weihua Zhuang. Distributed medium access control for wireless mesh networks. In: WIRELESS COMMUNICATIONS AND MOBILE COMPUTING. 2006; **6**:845–864.
13. Pattara-AukomW, Banerjee S, Krishnamurthy P. Starvation prevention and quality of service in wireless LANs. In: *Proceedings of 5th International Symposium on Wireless Personal Multimedia Communications* 2002; 1078–1082.
14. B. Nandini "On demand buffer allocation in wireless mesh networks", In: Proceedings of National conference on Current Research Advances in Computer Science 2017; ISBN: 978-93-5230-183-6.

Robust Estimation of Brain Functional Connectivity from Functional Magnetic Resonance Imaging Using Power, Cross-Correlation and Cross-Coherence

Nivedita Daimiwal, Betty Martin, M. Sundararajan
and Revati Shriram

Abstract Functional Magnetic Resonance Imaging (fMRI) is a non-invasive method for investigating the structure and function of the brain. Using fMRI, brain functions and areas responsible for the particular activities are investigated. The objective of the image processing methods using fMRI is to investigate the functional connectivity. To localize mental functions of specific brain regions and to identify the brain regions, those are activated simultaneously. Correlation and cross-coherence of the time series of the pixels are used for the detection of functional connectivity in fMRI images for the different motor movements (upper and lower limb movement and finger tapping action). The methodology was applied to three groups (six subjects) consisting aged between 10 and 75 years: (1) Normal and healthy subject performing finger tapping actions, (2) brain tumour patient performing lower limb movement (LL), and (3) brain tumour patient performing upper limb movements (UL). The threshold applied for the cross-correlation is 5000. Similarly, the threshold applied for cross-coherence and power parameters is in the range of (0.6–0.9). The algorithm implemented is found to be non-destructive, and there is no loss of temporal or spatial data. The result shows that for the normal subject, functionally connected pixels are more as compared to the brain tumour patients.

Keywords Cross-correlation · Cross-coherence · Discrete cosine transform
fMRI · Functional connectivity · Time series

N. Daimiwal (✉) · B. Martin
Sathyabama University, Chennai, India
e-mail: nivedita.daimiwal@gmail.com

M. Sundararajan
Bharat University, Chennai, India

N. Daimiwal · R. Shriram
Cummins College of Engineering for Women, Pune, India

B. Martin
Sastra University, Thanjavur, India

© Springer Nature Singapore Pte Ltd. 2018
V. Bhateja et al. (eds.), *Proceedings of the Second International Conference on Computational Intelligence and Informatics*, Advances in Intelligent Systems and Computing 712, https://doi.org/10.1007/978-981-10-8228-3_37

1 Introduction

Functional Magnetic Resonance Imaging (fMRI) is a non-invasive and powerful technique that characterizes brain functions under various tasks. To study the human brain functions, fMRI has become a predominant modality [1]. To understand the neural mechanism of the human brain, a concurrent EEG-fMRI setup can provide better functional information. Concurrent EEG-fMRI recording combines the information acquired through EEG and fMRI. For high spatial and temporal resolution, the two modalities can be merged together [2].

fMRI-based brain mapping methods used to identify active or functionally connected brain regions rely on statistical methods such as a hypothesis test. A pixel is identified as 'active' or 'connected', if its task-related signal exceeds the threshold [3]. Brain mapping can be done by providing different stimuli and detecting the brain regions activated in response to the stimulus. The functional connectivity is one method of characterizing the relationship between the time series of two or more regions. This method refers to correlations in activity between spatially remote regions of the brain. First, it could reflect the direct influence of one region on another (direct influence). Second, it could reflect the influence of another region that is mediated by a third region (indirect influence). Third, it could reflect a common input to both regions (shared influence). The different ways in which correlated activity between two regions A and B can arise by either direct influence, indirect influence, or shared influence are shown in Fig. 1.

The correlations in activity in different regions of the brain to detect functional connectivity can be done between subjects and also within the subject. Correlation between subjects is the simplest approach to detect connectivity. Some estimate of the activation is extracted for a particular region, and those values are entered into a whole brain. If the two regions are functionally connected, they show similar levels of activity across subjects. Correlation within subjects involves computing functional connectivity by finding out the correlation between the time series of different regions. First, the time series from seed pixel or seed region is extracted and then used to compute correlation with all the pixels across the brain [4]. Effective connectivity reflects the direct influence of one region on another.

Fig. 1 Various ways of correlating activity between regions A and B [4]

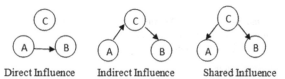

Direct Influence   Indirect Influence   Shared Influence

2 Materials and Methods

The fMRI analysis is done in order to detect seed pixel-based brain connectivity. This method for analysing functional connectivity shows how spatially distant brain regions work and interact with each other to create mental functions. Parameters employed in functional connectivity detection are period, power, mean, cross-correlation and cross-coherence. Activated pixels are detected using period and power by applying periodicity transform [5].

2.1 fMRI Time Series

In each cycle of image acquisition, recording of images was done during ON state and OFF state: M_{ON} and M_{OFF}, respectively. In the experiment, $M_{ON} = M_{OFF}$, which results in a square wave stimulus. The image sequence is $I(k_1, k_2, l)$, where k_1, k_2 are spatial dimensions, and l is the discrete time index. The fMRI time series has both periodic and non-periodic components. Non-periodic components form the noise in the fMRI time series data.

The time series of fMRI is represented by

$$TS(k, l) = \pi(TS(k, l), P_T) + \sum_i \pi(TS(k, l), Pp_i) + n(l), \tag{1}$$

where T is the stimulus period, $\pi(TS(k, l), P_T)$ is the projection of time series (TS) onto the T periodic basis elements, and Pp_i is the periodicity subspaces. The periodicities other than the stimulus period T and n is the non-periodic noisy component. The time series is applied to periodicity transform to extract period and power of the pixel time series. Activated pixels are extracted by applying manual power threshold in the range (0–1) and the stimulus period of 20. Periodicity transform is applied for all intracranial pixel time series for period and power extraction [5–7].

2.2 Cross-Coherence

Wiener (1949) introduced the concept of coherence of time series. Rosenberg invented its applicability to functional imaging data [8, 9]. Coherence is defined in the frequency domain. Coherence is a technique for investigation of functional connectivity across brain regions [10–12]. Coherence measures lag between two different (time courses) regions. If a time series of two regions is similar, the coherence will be high.

Coherence $Coh_{ij}(\lambda)$ between any two different time courses (i, j) at frequency λ is defined as [13]

$$Coh_{ij} = \left|R_{ij}(\lambda)\right|^2 = \frac{\left|f_{ij}(\lambda)\right|^2}{f_{ii}(\lambda)f_{jj}(\lambda)}. \tag{2}$$

2.3 Cross-Correlation

Correlation is defined in the time domain. Zero-order correlation is used to measure the simultaneous linear coupling relationship between two time series. Positively correlated time courses imply that the two regions are active (average or less) at the same times. Negatively correlated time courses show that one region is active, and the other is inactive. fMRI zero-order correlation can be used to measure interregional relationships [14]. Interregional relationships can be measured using lagged cross-correlation. Lagged correlation measures the delayed or lagged linear relationships between the time courses of different regions [15].

Cross-correlation of any two different time series (i, j), at lag h, $\rho_{ij}(h)$ is given by [13]

$$\rho_{ij}(h) = \frac{cov_{ij}(t, t+h)}{\sqrt{var_i(t)var_j(t+h)}}. \tag{3}$$

2.4 High-Pass Filter

Noise arises from physical sources, and it is sometimes due to as scanner drift. It also arises from physiological sources, from residual movement effects and their interaction with the static magnetic field. When the subject is performing a task, signal components are added to this noise. The most obvious characteristic of noise in BOLD fMRI data is the presence of low-frequency drift. The low-frequency trend in a fMRI time series is observed in the time domain and by the power spectrum in the Fourier domain.

The filters used in imaging are designed to allow particular portions (or bands) of the frequency spectrum to pass through while removing others. Removing low-frequency drift is to apply a high-pass filter. The high-pass filtering is implemented using Discrete Cosine Transform (DCT) basis set [1, 16].

Mathematically, for time points $t = 1, \ldots , N$, the discrete cosine set functions are

Fig. 2 Power spectrum of original time series

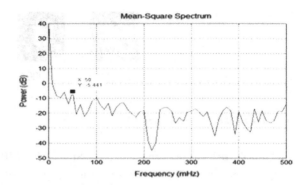

$$f_r(t) = \sqrt{2/N}\left(\cos(r\pi\frac{t}{N})\right). \tag{4}$$

The integer index r ranges from 1 to user-specified maximum R.

Original time series fit and DCT of time series in both the cases the fit is very similar. The power spectrum is acquired by taking the Fourier transform of the time series. The X-axis of the plot refers to different frequencies, whereas the Y-axis refers to the power or strength of that frequency.

The power spectrum of the original time series is shown in Fig. 2. The original time series exhibited a low-frequency noise. The spike in the power spectrum at 50 MHz corresponds to the frequency of the task (once every 20 s) in this experiment.

Figure 3 shows time series in which low-frequency components are removed using high-pass filtering (DCT).

Fig. 3 Power spectrum of time series after high-pass filtering

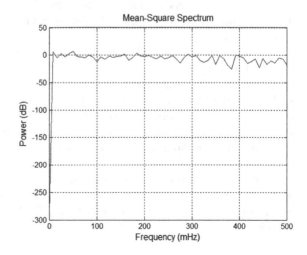

Fig. 4 fMRI images

500001   500002   500003   500004

3 fMRI Data Acquisition

Functional MRI was performed on a GE Signa 1.5 T MRI system (Global Hospital, India) for six brain tumour patients; three patients were performing lower limb movement (LL), two patients while performing upper limb movement (UL), and one normal subject was made to perform finger tapping action. While recording the data set, during activation or ON state, the subject performs movements. During OFF state or rest, no activity is performed. The stimulus period is 20. The data set was in DICOM (Digital Image Communication in Medicine) format. T2* weighted images with BOLD contrast were acquired using an Echo-Planar Imaging (EPI) sequence of a repetition time (TR) of 3000 ms, with an echo time of 60 ms, and in-plane resolution of 64 * 64 pixels [17]. EPI of a sequence that were recorded with respect to time are shown in Fig. 4 [18, 19]. For each motor movement, 512 images were recorded.

4 Signal Processing

For fMRI data, the signal processing steps are explained below and described in Fig. 5:

- The analysis is restricted to the intracranial region only.
- Mask is used to separate the intracranial pixels.
- The average of all the images of the time series belonging to a particular slice is calculated.
- The valley point is selected from histogram of average image, to detect pixels in intracranial region.
- Select time series within intra-cranial region.
- Periodicity transform is applied to a pixel time series.
- Using periodicity transform period, power is calculated for all time series.
- If the periodicity matches with the stimulus period and if the signal power greater than threshold (manually in the range of 0–1). That pixel is declared as activated pixel.
- Activated seed pixel is selected manually.
- Seed pixel time series is considered as reference time series.
- The high-pass filtering is implemented using DCT to remove the low-frequency noise.

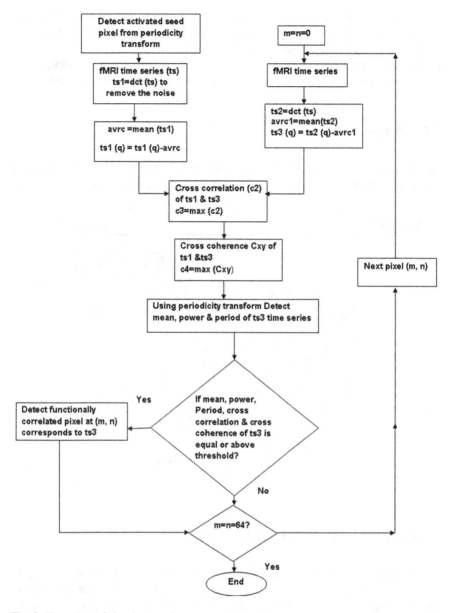

Fig. 5 Flow chart of signal processing

- The cross-correlation and cross-coherence between reference time series and pixel time series are measured. Cross-coherence is used to detect the time series of pixel which is in phase with the seed pixel time series.

- If the period of time series is same that of a stimulus and the signal power, cross-correlation and cross-coherence are greater than or equal to the thresholds then that pixel is functionally connected pixel.
- For all the intra-cranial pixel time series, the above procedure is repeated.
- The seed pixel and functionally connected pixel is marked in white colour.

4.1 Program Code

Matlab code for the cross-correlation and cross-coherence of time series and functional connectivity is shown in Algorithms 1 and 2, respectively.

Algorithm 1	Algorithm 2:

```
% Extract the pixel time series

for q=1:120
        ts(q)= x(38,33,q);
        ts1(q)= dct(ts(q));
end
        avro=mean(ts1);
        ts1(q)=ts1(q)-avro;
        for m=1:64
        for n=1:64
            if mask (m,n)==100
                for q=1:120
                ts(q)= x(m,n,q);
                ts2(q)=dct(ts(q));
            end
        avrc1=mean(ts2);
        % Subtract mean from the time series
            for q=1:120
            ts3(q)=ts2(q)-avrc1;
            end

        c2=xcorr(ts1,ts3);
        c3=max(c2)
        Cxy = mscohere(ts1,ts3)
        c4=max(Cxy)
```

```
% Call the M-BEST Algorithm, which inputs the
        % pixel time series and the desired periodicity
        [per,pow,bas]=mbest(ts3,1,20);
    % If the period is 20 and power is greater than threshold then
    % pixel is declared activated
    % Threshold has to be selected by the user manually between 0 to 1,
        % here the threshold set is 0.98.
if (per == 20 & pow > 0.65 & c3>5000 &c4>0.6& avrc1>=89.8)
    IN=600 * pow;
    uu(m,n)=IN;
    a=m;
    b=n;
else
    uu(m,n)=0;
end
            end
        end
    end
        % Plot the Detected Activation pattern
        uu(38,33)=IN;
figure(3);
image(uu);
colormap('gray');
title('The Activation pattern');
figure(4);
image(double(A)/5+ uu)
colormap('gray');
title('The Activation pattern superimposed onto the anatomic image');
```

5   Results on fMRI Data Set

The threshold applied for cross-correlation is 5000. The threshold for determination of functional connectivity using cross-coherence and power is in the range (0.6–0.9). The analysis shows the functionally connected pixels for three different motor activities like finger tapping, lower limb movement and upper limb movement. A computationally efficient algorithm is presented, which shows functional connectivity in different parts of the brain based on the time series analysis. The algorithm compared time series of seed pixel with the rest of the pixel time series. Figure 6 shows the time series of seed and functionally connected pixel and their cross-correlation.

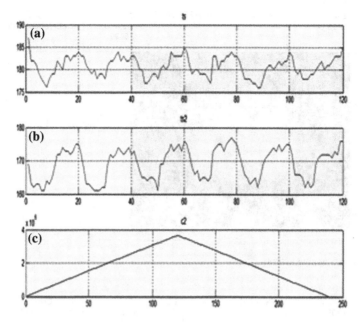

Fig. 6 a Time series of seed pixel, **b** time series of functionally connected pixel, and **c** cross-correlation between seed and functionally connected pixel

For functionally connected pixels, it has been observed that the cross-correlation is positive and maximum, and the cross-coherence is observed to be high. Functionally connected pixels show the pixels that are simultaneously activated. Cross-coherence between seed pixel and functionally connected pixel is shown in Fig. 7. The result shows that, for normal subject performing finger tapping activity, the functionally connected pixels are more as compared to brain tumour patients performing the lower and upper limb movements. Seed pixel and functionally connected pixels are marked with white colour.

Fig. 7 Cross-coherence between seed and functionally connected pixel

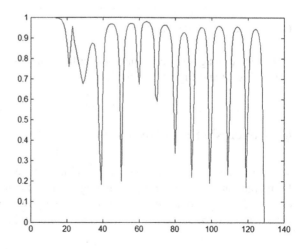

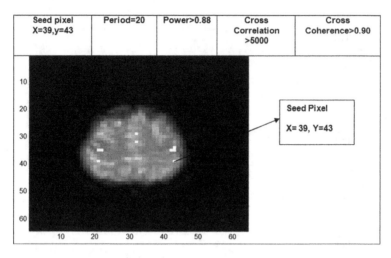

Fig. 8 Finger tapping activity

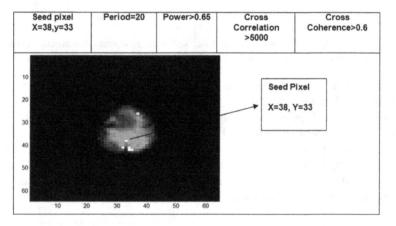

Fig. 9 Upper limb movement

Resultant images finger tapping, upper limb and lower limb activity are shown in Fig. 8, Fig. 9 and Fig. 10, respectively.

6 Conclusion and Discussion

The approach for functional connectivity using time series and periodicity transform eliminates the need for pre-filtering the fMRI data. The periodicity transform can detect the periodicities and power in the noisy waveform [19]. The stimulus period needs to be set as per the user's requirement. No loss of temporal or spatial

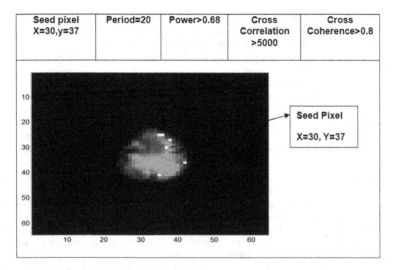

Seed pixel X=30,y=37	Period=20	Power>0.68	Cross Correlation >5000	Cross Coherence>0.8

Fig. 10 Lower limb movement

data is found while using this algorithm. Periodicity transform for the signal and image processing methods is extended in this paper for measurement of functional connectivity. The system identifies the brain regions that are simultaneously activated by using cross-correlation and cross-coherence. This is analogous to the 'phase' of the fMRI time series [20]. Alternative approaches for functional connectivity analysis using fMRI can be achieved by dynamic causal modelling, Granger Causality Mapping (GCM), Independent Component Analysis (ICA), Partial Least Squares (PLS), Conventional Correlation Analysis and Principal Component Analysis (PCA) [21–23].

References

1. Frackowiak RS, Friston KJ, Frith CD and Zeki S. Human brain function. 2nd ed. USA: Academic Press, 2004.
2. Ahmad R, Malik A. Optimization and Development of Concurrent EEG-FMRI data acquisition setup for understanding Neutral mechanisms of Brain. In: IEEE 2015, Instrumentation and Measurement technology Conference. 11–14 May 2015; Giacomo Matteotti, Pisa, Italy: IEEE. pp. 476–481.
3. Song X, Chen N. A unified Machine learning method for task related and resting state fMRI data Analysis. In: IEEE 2014, Engineering and Medicine and Biology Society (EMBC); 36th Annual International Conference of the IEEE. 26–30 Aug. 2014; Chicago, Illinois, USA: IEEE. pp. 6426–6429.
4. Poldrack RA, Mumford JA, Nichols TE. Handbook of Functional MRI Data Analysis. USA: Cambridge University press, 2011.

5. Deshmukh A, Shivhare V, Gadre V. Functional MRI activation signal detection using the periodicity transform. In: IEEE 2004, International Conference on Signal and Communication (SPCOM); 11–14 Dec. 2004; Bangalor, India: IEEE. pp. 121–125.
6. Friston KJ, Jezzard P, Turner R. Analysis of Functional MRI time-series. Human Brain Mapping 1994; 2: 69–78.
7. Daimiwal N, Sundharajan M, Shriram R. Application of fMRI for Brain Mapping. (IJCIS) International Journal of computer science and information Security 2012; 10: 23–27.
8. Wiener N. Extrapolation, Interpolation and Smoothing of Stationary Time Series: With Engineering Applications. Las Vegas, NV, USA: MIT Press, 1949.
9. Rosenberg JR, Amjad AM, Breeze P. The Fourier approach to the identification of functional coupling between neuronal spike trains. Progress in Biophysics and Molecular Biology 1989; 53: 1–31.
10. Leopold DA, Murayama Y, Logothetis NK. Very slow activity fluctuations in monkey visual cortex: Implications for functional brain imaging. Cerebral Cortex 2003; 13: 422–433.
11. Sun FT, Miller LM, D'Esposito M. Measuring interregional functional connectivity using coherence and partial coherence analyses of fMRI data. Neuroimage 2004; 21: 647–658.
12. Salvador R, Suckling J, Schwarzbauer C. Undirected graphs of frequency-dependent functional connectivity in whole brain networks. Philosophical Transactions of the Royal Society B: Biological Sciences 2005; 360: 937–946.
13. Zhou D. Functional connectivity analysis of fMRI time-series data. B.S. in Statistics, Beijing Normal University, China, 2005. M.A. in Statistics, University of Pittsburgh, 2007.
14. Biswal B, Yetkin FZ, Haughton VM. Functional connectivity in the motor cortex of resting human brain using echo-planar MRI. Magnetic Resonance in Medicine 1995; 34: 537–541.
15. Siegle GJ, Thompson W, Carter C, Steinhauer SR. Increased amygdala and decreased dorsolateral prefrontal BOLD responses in unipolar depression: Related and independent feature. Biological Psychiatry 2007; 61: 198–209.
16. Friston KJ, Ashburner JT, Kiebel SJ, Nichols TE. Statistical parametric mapping the analysis of functional brain images. San Diego, CA, USA: Academic Press, 2007.
17. Stehling MK, Turner R. Echo-planar Imaging: magnetic Resonance imaging in a fraction of a second. Science; 254: 44–49.
18. Deshmukh A, Shivhare V, Parihar R, Gadre VM. Periodicity Analysis of fMRI data in the Wavelet Domain. In: 2005 National Communications Conference; 28–30 January 2005; IIT, Kharagpur, India: pp. 460–464.
19. Sethares WA, Staley TW. Periodicity transforms. IEEE Trans. on Signal processing 1999; 47: 2953–2964.
20. Deshmukh A, Gadre VM. Functional magnetic Resonance Imaging. Novel transform method. Novel transform method. New Delhi, India: Narosa Publishing, 2008.
21. Friston KJ, Harrison L. Dynamic causal modelling. NeuroImage 2003; 19: 1273–1302.
22. Roebroeck A, Formisano E, Goebel R. Mapping directed influence over the brain using Granger causality and fMRI. NeuroImage 2005; 25: 230–242.
23. Calhoun VD, Adali T, Pearison GD. Spatial and temporal independent component analysis of functional MRI data con-training a pair of task-related waveforms. Human Brain Mapping 2001; 13: 43–53.

Statistical Analysis of Derivatives of Cranial Photoplethysmogram in Young Adults

Revati Shriram, Betty Martin, M. Sundhararajan
and Nivedita Daimiwal

Abstract Every day risk of cardiovascular diseases is increasing in young adults. Now researchers are working on study related to a single bio-signal for prediction of maximum physiological parameters. One of such a bio-signal is photoplethysmogram (PPG). Non-invasive measurement of blood volume change is carried out by using PPG. PPG captured from a cranial site is known as cranial photoplethysmogram (CPPG). Most of the time various bio-signals acquired from the brain are used to study only the brain-related disorders. Near-infrared spectroscopy-based sensor used to record CPPG from frontal region can be used to predict heart rate, oxygen saturation, blood pressure, cardiac output and respiration rate. This paper explains the design specifications of sensor used and study of various time and amplitude indices of differentiated CPPG signal. Authors have studied two levels of differentiation of CPPG obtained by applying MATLAB-based algorithm. Features obtained from differentiated CPPG signal are compared with the standard available values to check the feasibility of this brain signal in the prediction of vascular health. The study was carried out on 19 healthy subjects aged between 19 and 30 years. Our results showed that the optical brain signal used to study hemodynamic changes in brain can also be used for the prediction of vascular health.

Keywords Acceleration photoplethysmogram · Cranial photoplethysmogram Near infrared spectroscopy · Optical sensor · Vascular health and velocity photoplethysmogram

R. Shriram (✉)
Department of ECE, Sathyabama University, Chennai, India
e-mail: revatishriram@yahoo.com

R. Shriram · N. Daimiwal
Cummins College of Engineering for Women, Pune, India

M. Sundhararajan
Bharath University, Chennai, India

B. Martin
SASTRA University, Thanjavur, India

© Springer Nature Singapore Pte Ltd. 2018
V. Bhateja et al. (eds.), *Proceedings of the Second International Conference on Computational Intelligence and Informatics*, Advances in Intelligent Systems and Computing 712, https://doi.org/10.1007/978-981-10-8228-3_38

415

1 Introduction

Non-invasive brain imaging techniques are becoming famous these days, because of its simplicity, easy analysis and shorter estimation time. Various non-invasive neuroimaging techniques currently available are single photon emission computed tomography (SPECT), positron emission tomography (PET), functional magnetic resonance imaging (fMRI) and electroencephalography (EEG). Among these techniques, magnetic and electrical imaging techniques are the fastest. Each technique has its advantages and disadvantages. With each heartbeat, oxygenated blood is pumped to the peripheral site and deoxygenated blood is drawn back to the heart and lungs. Near-infrared (NIR) light-based neuroimaging technique having good temporal resolution is used to record the change in blood volume at frontal lobe of a subject. This acquired brain signal is known as cranial photoplethysmogram (CPPG). Wave nature of PPG signal captured from any measurement site on the human body is almost same; only reduction in the amplitude can be observed at the measurement site having higher tissue thickness [1].

Ageing and the association of various diseases affect all the body organs more or less. In healthy subjects, resistance to the arterial wall is less and the capacitance is high. With increase in age or onset of arteriosclerosis, the resistance of the arterial wall to blood pressure increases and the capacitance decreases reducing the amplitude of pulsating CPPG signal [2]. This paper explains various time- and amplitude-related indices of the CPPG signal and statistical analysis of the same.

1.1 *Beer–Lambert's Law*

Optical PPG acquisition is based on Beer–Lambert's law of spectroscopy. It states that the light travelling through a uniform medium with absorptive substance reduces exponentially with the optical path length through the medium and absorptive coefficient. During the acquisition of PPG, light travels through various mediums such as skin, bone, tissue, skin pigmentation, venous blood and arterial blood

$$I(z) = I0 \exp(-\mu a z), \tag{1}$$

where $I(z)$ is the attenuated intensity as a function of the distance z in the tissue, $I0$ is the incident intensity and μa is the optical absorption coefficient at the wavelength of interest [1].

1.2 *Sensor Specifications*

PPG can be acquired by using transmission type of sensor or reflection type of sensor. Transmission type of sensor has limited application as performance of such

Table 1 Sensor specifications [1]

Category	Specifications
Sensor	
Type	Reflection type of sensor
Source	860 nm (5 mm LED)
Detector	OPT 101 (Si Burr brown diode)
Sensor casing	Black polyurethane
Optode distance	1.5 cm
Supply	10 V DC-signal conditioning Ckt; 5 V, 2 kHz AC to source
Application	
Measurement site	Anywhere on the human body from head to toe

(a)   **(b)**

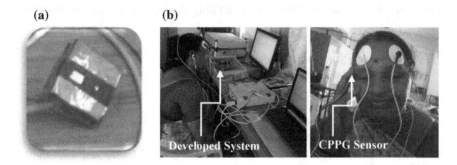

Fig. 1 **a** CPPG sensor and **b** system setup to acquire CPPG signal

a sensor is totally dependent on the tissue thickness at the measurement site. Signal acquisition using reflection type of sensor is independent of the tissue thickness at the measurement site. The penetration depth of light is dependent on its wavelength and infrared light reaches deeper tissues than visible light. At cranial region, bone/ tissue thickness is higher, so infrared light source of wavelength 860 nm is used by the authors to capture CPPG. OPT 101 Silicon burr brown diode is used as a detector in the developed prototype. Table 1 shows the specifications of the developed prototype.

Figure 1 shows the sensor developed by the authors and the system setup to capture CPPG signal.

2 Methods

Pressure wave propagation is divided into two phenomena related to heart activity. Every time when heart contracts blood is ejected for circulation in the form of a pulse wave. This pulse wave propagates through whole systemic circulation and reaches the peripheral site. The propagation velocity of the pulse wave depends on various factors such as intravascular pressure, compliance, diameter, elasticity and thickness of arteries [2]. These parameters can change with various diseases or age of the subject.

Forehead PPG signal is recorded by using IR LED-based reflective type of sensor developed by the authors. All the subjects were in seated position with eyes closed condition while capturing the CPPG from prefrontal area. The recorded CPPG is first digitally filtered (to avoid the amplification of noise present in the measured signal) before the next operation of derivative. The first derivative of the CPPG signal is known as velocity photoplethysmogram (VPG), whereas the second derivative of the CPPG signal is known as acceleration photoplethysmogram (APG) [3, 4]. Filtering and first and second derivatives of CPPG are obtained by applying MATLAB-based algorithm on 19 CPPG signals captured from healthy subjects aged between 19 and 30 years. Two-level differentiation (high pass filtering) of CPPG signal is carried out to amplify the high-frequency components.

2.1 Velocity Photoplethysmogram (VPG)

It is the indicator of velocity of blood at a given site.

- **Crest Time (CT)**: It is the time measured from base of the CPPG waveform to its first positive maxima, i.e. peak.
- **Delta T (ΔT)**: It is the time between the systolic peak of CPPG to diastolic peak. CT and ΔT are the best features for classification of cardiovascular disease using VPG [5].

2.2 Acceleration Photoplethysmogram (APG)

APG-related studies are more commonly carried out than the VPG studies [4]. It is the indicator of acceleration of blood at a given site. Second derivative curve parameters are interconnected with state of cardiovascular system parameters such as elasticity modulus and vascular compliance [5–7]. Following are the five extrema related to APG which are widely studied [8–10].

- **'a', 'b'and 'c'**: 'a' is first positive extrema and is known as early systolic positive wave. 'b' is first negative extrema and is known as early systolic negative wave. 'c' is second positive extrema and is known as late systolic reincreasing wave.
- **'d' and 'e'**: 'd' is second negative extrema and is known as redecreasing wave. 'e' is third positive extrema and it represents dicrotic notch (Fig. 2).

The height of each extrema is measured from the baseline. Points above the baseline are treated as positive extrema, whereas points below the baseline are

Fig. 2 a Finger PPG and **b** VPG of PPG signal [5]

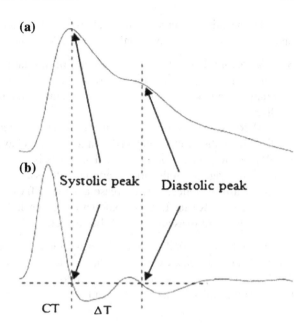

Fig. 3 a Finger PPG and **b** APG of PPG signal [5]

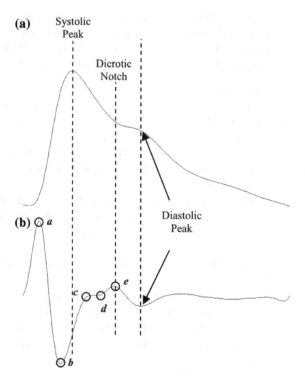

treated as negative extrema. Figure 3 shows the APG and various extrema indicated on the APG waveform. Commonly studied APG-related features are as follows [5]:

- **'b/a' Ratio**: This ratio is the reflection of arterial stiffness. It is a very useful non-invasive index of athrosclerosis. This ratio is directly correlated to the Framington risk score, which is used to estimate the risk of cardiovascular heart diseases.
- **'c/a' Ratio**: This ratio is also the reflection of arterial stiffness. It is used to distinguish the healthy subjects from the subjects having blood pressure history.
- **'d/a' Ratio**: Evaluation of vasoactive agents is carried out with the help of this ratio. It is again a reflection of arterial stiffness.
- **'e/a' Ratio**: It reflects the decrease arterial stiffness.
- **'(b-c-d-e)/a' Ratio**: It is also known as ageing index (AGI), and it is used to check the cardiovascular age of the subject.

'b/a' and '(b-c-d-e)/a' ratio increases with age of a subject, whereas 'c/a', 'd/a' and 'e/a' ratio decreases with age. The standard values of these ratios from finger APG for healthy young adults are as follows: b/a: -0.9 (SD = 0.09), c/a: 0.2 (SD = 0.09), d/a: -0.1 (SD = 0.1) and e/a: 0.1 (SD = 0.1) [7].

2.3 Statistical Parameters

VPG- and APG-related quantitative data are analysed by carrying out statistical analysis summarized by mean, standard deviation and Pearson product. Linear regression fit is applied to the various features of VPG and APG to find out the equation and R^2-value for the given data by using Excel 2016.

- **Mean**: Addition of all the data points divided by the total number of data points gives the mean value for that data set. It is the average value of the data set.
- **Standard Deviation**: It is one of the very important and commonly studied statistical feature. It is the measure of dispersion or variation from the average or mean value of the data set. High standard deviation shows that the data points are widely spread out, while lower standard deviation shows that the data points are less dispersed.
- **Pearson Product**: Pearson product returns the moment correlation coefficient 'r', which ranges from -1 to 1. It is one of the normalized measures of coupling. $+1$ indicate positive correlation, zero indicate no correlation and -1 indicate negative correlation.
- **Linear Regression Fit**: It calculates an equation which minimizes the distance between the fitted line and the set of data points. R^2 is the statistical measure of how well the model can predict the data, and the value is always between the '0' and '1'. Higher value of R^2 closer to '1' shows that the predicting model is the better fit for the data studied. As the order of polynomial fit increases the R^2 value also increases, decreasing the variation between the predicting model and the data set studied.

3 Results

Analysis of VPG and APG of the CPPG captured from frontal lobe for the 19 subjects is carried out by the authors. Time indices like CT and ΔT were measured based on VPG, while amplitude indices like 'a', 'b', 'c', 'd', 'e' and related ratios were calculated based on APG. Pearson correlation product was calculated between the various measured extrema of APG and is as shown in Table 2. Table 3 shows the mean and standard deviation of first derivative-related features such as crest time and delta T.

'a', 'c' and 'e' are positive extrema so the Pearson product of correlation between these three extrema is positive, whereas 'b' and 'd' are the negative extrema so the Pearson product between 'a' and 'b' 'd' is negative. Linear regression fit was checked for all the measured time and amplitude indices. Figure 4 and 5 show the linear fit, R^2 value, mean and standard deviation for CT, ΔT, 'b/a', 'c/a', 'd/a', 'e/a' and '(b-c-d-e)/a' ratio, respectively. Table 4 shows the correlation coefficient R^2 for the regression fit of various VPG and APG features.

Table 2 Pearson product correlation feature of APG features

APG features	Pearson product	Correlation
a & b	−0.912962	Highly negative
a & c	0.8979923	Medium-high positive
a & d	−0.908638	Highly negative
a & e	0.8531958	Medium-high positive

Table 3 Mean and standard deviation of VPG and APG features

VPG and APG features	Mean ± standard deviation
Crest time	49.05 ± 8.28
Delta T	70 ± 8.0
b/a ratio	−0.4911 ± 0.1787
c/a ratio	0.3269 ± 0.1441
d/a ratio	−0.3191 ± 0.1048
e/a ratio	0.1488 ± 0.0796
(b-c-d-e)/a ratio	−0.6478 ± 0.2233

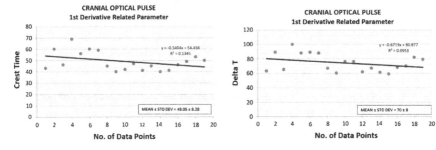

Fig. 4 Linear fit for first derivative (velocity photoplethysmogram) based features such as crest time and delta T. For crest time, $-0.5404X + 54.456$ is the equation of linear fit, $R^2 = 0.1345$. For delta T, $-0.6719X + 80.877$ is the equation of linear fit, $R^2 = 0.0953$

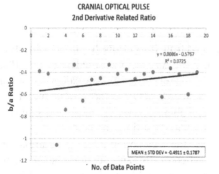

Linear Fit for b/a ratio of APG: Equation of Linear Fit is y=0.0086X-0.5767, R^2=0.0725

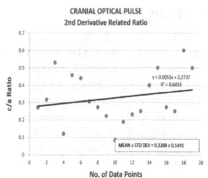

Linear Fit for c/a ratio of APG: Equation of Linear Fit is y=0.0053X+0.2737, R^2= 0.0433

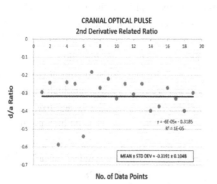

Linear Fit for d/a ratio of APG: Equation of Linear Fit is y=-6E-05X-0.3185, R^2=1E-05

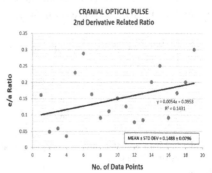

Linear Fit for e/a ratio of APG: Equation of Linear Fit is y=0.0054X+0.0953, R^2=0.1431

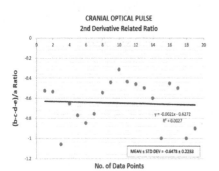

Linear Fit for (b-c-d-e)/a ratio of APG: Equation of Linear Fit is y=-0.0021X-0.6272, R^2=0.0027

Fig. 5 Linear fit for various velocity photoplethysmogram and acceleration photoplethysmogram features

Table 4 Correlation coefficient R^2 for VPG and APG features

VPG and APG features	R^2	Correlation
Crest time	0.1345	Weak correlation
Delta T	0.0953	Weak correlation
b/a ratio	0.0727	Weak correlation
c/a ratio	0.0433	Weak correlation
d/a ratio	1E-05	Very weak correlation
e/a ratio	0.1431	Weak correlation
(b-c-d-e)/a ratio	0.0027	Very weak correlation

4 Discussion

NIRS-based brain imaging technique is relatively new and has great potential in various daily/routine bio-signal monitoring [11]. By detecting the absorption of NIR light cerebral blood flow changes are detected from cortical brain regions. This technique offers good temporal and spatial resolution. Acquisition of PPG is easy as compared to the acquisition of electroencephalogram (EEG) brain signal or electrocardiogram (ECG) heart signal. Though it is much difficult to estimate the stiffness of blood vessels in vascular bed at the cranial site. Brain-related various parameters such as activity-related blood flow changes and coherence can be found out from the CPPG. This acquired signal is not only useful for study of brain but heart-related various parameters such as heart rate (HR), respiratory rate (RR), cardiac output, blood oxygen saturation (SpO$_2$), blood pressure (BP), haemoglobin (Hb) and vascular health can be predicted from an acquired CPPG brain signal. Study related to CPPG signal can be carried out in depth for accurate interpretation of head–heart coupling. NIRS technique-based CPPG is a promising technique in early diagnosis of vascular diseases (atherosclerosis) and can be used as a cheaper screening technique. Atherosclerosis means clogging of arterial anywhere in the body. The relationship between cardiovascular risk factors and the second derivative-related indices has been analysed widely. Fourth-order differentiator is also used by the researchers. Order of differentiator used decides the suppression of high-frequency components. Various levels or order of differentiator can be implemented by using only hardware components [10, 12–14].

5 Conclusion

Arterial stiffness measurement can be carried out non-invasively by using PPG signal. PPG signal varies in amplitude, shape and upstroke time with respect to the measurement site [15]. Work was carried out by the authors on the various time indices and amplitude indices of CPPG signal. Age is an important factor for change in contour of the PPG/CPPG signal. As the age increases, the VPG-related parameter, ΔT, decreases. As age increases, arterial elasticity decreases (artery stiffness increases) and in turn arterial compliance also reduces. This leads to increase in the pulse wave velocity (PWV) and affects the haemodynamics [12]. The result shows a strong correlation between the APG-related amplitude indices. Various APG-related ratios obtained from CPPG signal for healthy young adults are almost in line with the standard values of finger APG. Linear regression fit applied to all the VPG- and APG-related features showed lower R^2 value (closer the value of R^2 to 1, then better the fit and vice a versa). This shows that the studied data set is related to the biological human parameters (which are highly non-linear), so higher order polynomial fit would be the best predicting model than the linear regression fit. NIRS-based optical brain imaging system to capture CPPG is safe (radiations free), compact, inexpensive and easy to use. It provides high temporal resolution and is used safely to study the cortical blood volume changes in infants as well as adults.

References

1. Revati Shriram, M. Sundhararajan & Nivedita Daimiwal, "Cranial PPG Brain Signal based Cardiovascular Parameter Estimation", XII Control Instrumentation System Conference (CISCON), 2nd–4th November 2015, Manipal, India.
2. Stefan Borik, Ivo Cap, "Measurement and Analysis Possibilities of Pulse Wave Signals", Biomedical Engineering Volume: 11, Number: 6, 2013, December.
3. Pilt. K, Ferenets. R, Meigas. K, Lindberg. L, Temitski. K, Viigimaa. M, "New Photoplethysmographic Signal Analysis Algorithm for Arterial Stiffness Estimation", Hindawi Publishing Corporation, The Scientific World Journal, Vol. 2013, pp: 1–9, 2013.
4. Qawqzeh. Y, Uldis. R, Alharbi. M, "Photoplethysmogram Second Derivative Review: Analysis and Applications", Scientific Research and Essays, Vol. 10, No. 21, pp: 633–639, 2015.
5. Mohamed Elgendi, "On the Analysis of Fingertip Photoplethysmogram Signals", Current Cardiology Reviews, 2012, 8, 14–25.
6. R. Mohamad Rozi, Sahnius Usman, M.A. Mohd Ali, M.B.I Reaz, "Second Derivatives of Photoplethysmography (PPG) for Estimating Vascular Aging of Atherosclerotic Patients", 2012 IEEE EMBS International Conference on Biomedical Engineering and Sciences, Langkawi, 17th–19th December 2012.
7. Dae-Geun Jang, Jang-Ho Park, Seung-Hun Park, Minsoo Hahn, "A Morphological Approach to Calculation of the Second Derivative of Photoplethysmography", 10th International Conference on Signal Processing (ICSP), 24th–28th October 2010, Beijing, China.
8. Wei. C, "Developing an Effective Arterial Stiffness Monitoring System Using the Spring Constant Method and Photoplethysmography", IEEE Transactions on Biomedical Engineering, Vol. 60, No. 1, pp: 151–154, 2013.
9. Jaafar. N, Sidek. K, Azam. S, "Acceleration Plethysmogram Based Biometric Identification", International Conference on BioSignal Analysis, Processing and Systems (ICBAPS), 26–28 May, Kuala Lumpur, Malaysia, pp: 16–21, 2015.
10. Jae A, "Wave Detection in Acceleration Plethysmogram", Healthc Inform Res., Vol. 21, No. 2, pp: 111–117, 2015.
11. Jeong, J. Finkelstein, "Applicability of the Second Derivative Photoplethysmogram for Non-invasive Blood Pressure Estimation during Exercise", 2013 Pan American Health Care Exchanges (PAHCE).

12. Sahnius bt Usman, Mohd Alauddin bin Mohd Ali, Md. Mamun Bin Ibne Reaz, Kalaivani Chellapan, "Second Derivative of Photoplethysmogram in Estimating Vascular Aging among Diabetic Patients" International Conference on Technical Postgraduates (TECHPOS), 14th–15th December 2009, Kuala Lumpur, Malaysia.

13. Y. K. Qawqzeh, M. B. I. Reaz, M. A. M. Ali, "The analysis of PPG contour in the assessment of atherosclerosis for erectile dysfunction subjects", WSEAS TRANSACTIONS on BIOLOGY and BIOMEDICIN, ISSN: 1109–9518, Issue 4, Volume 7, October 2010.

14. Pilt. K, Meigas. K, Temitski. K, Viigimaa. M, "Second derivative analysis of forehead photoplethysmographic signal in healthy volunteers and diabetes patients", World Congress on Medical Physics and Biomedical Engineering, IFMBE Proceedings, Vol. 39, pp: 410–413, 2012.

15. Revati Shriram, Betty Martin, M. Sundhararajan and Nivedita Daimiwal, "Effect of source wavelength on second derivative of finger photoplethysmogram in healthy young female volunteers" Biomed Res- India 2016 Special Issue, Special Section: Health Science and Bio Convergence Technology, pp: S454–S45, ISSN 0970–938X, 2016.

Fuzzy Decision-Based Reliable Link Prediction Routing in MANET Using Belief Propagation

K. G. Preetha and A. Unnikrishnan

Abstract Mobile ad hoc networks (MANET) are well suited, when the application demands quick network configuration and faces difficulty to set up the infrastructure for communication. Node mobility is not predictable in MANET, so that the topology keeps changing dynamically, thereby making the decision of routes also dynamic. In this paper, an attempt is made to predict the longevity of the links by utilizing fuzzy decision process and belief propagation. Initially, the link is selected based on the signal strength while the final choice accommodates the node energy also. Two methods are introduced (i) link prediction routing (LPR) and (ii) reliable link prediction routing (RLPR) to discover most reliable link in the context of arbitrarily node and link failure. At each node, the decision to propagate the belief is the consequence of the fuzzy decision taken on the signal strength and node energy. Two methods are compared with other on-demand algorithms AODV and DSR in respect of various performance parameters for different simulation timings, based on the average values generated out of 1000 independent runs.

Keywords MANET · Fuzzy set · Knowledge base · Decision-making
Belief propagation

K. G. Preetha (✉)
Cochin University of Science & Technology, Kochi, India
e-mail: preethamanish@gmail.com

K. G. Preetha · A. Unnikrishnan
Rajagiri School of Engineering & Technology, Kochi, India
e-mail: unnikrishnan_a@live.com

A. Unnikrishnan
DRDO, Kochi, India

© Springer Nature Singapore Pte Ltd. 2018 427
V. Bhateja et al. (eds.), *Proceedings of the Second International Conference on Computational Intelligence and Informatics*, Advances in Intelligent Systems and Computing 712, https://doi.org/10.1007/978-981-10-8228-3_39

1 Introduction

MANETs are always plagued by unstable routes. The frequent movement of nodes and the deteriorating energy levels in nodes undermine the sustained availability of routes. Literature is abuzz with an opulent collection of solutions and techniques to tackle the problems in routing [1]. Routing algorithms in MANETs are generally of two types—proactive and reactive. Simulation studies reveal that the reactive routing algorithms are more effective for MANETs than the proactive algorithms [2, 3]. Among the many reactive algorithms, ad hoc on demand distance vector (AODV) and dynamic source routing (DSR) [1] have attracted the attention of the researchers on account of the simplicity of the approach and the commendable performance. The high mobility of the nodes in network leads to frequent disruption in route [4], when the intermediate node informs the incidents via special packets to the source. In case of AODV and DSR, the routes have to be established all over again. The re-establishment of route is achieved by flooding with a large number of control packets, which results in substantial overhead. Many methods have been reported to alleviate the overhead in route establishment [5, 6].

A method based on random walk has been demonstrated in [7], for establishing connectivity by accumulating the probability of choosing one of the links emanating from a node. Following this idea, a score is generated based on the degree of connectivity and signal strength in the proposed link prediction routing (LPR). The path having the highest accumulated score is declared as the route from source to destination. The idea of LPR is extended to reliable link prediction routing (RLPR), when the signal strength and residual energy of the nodes decide the generation of the score. Selection of the route is carried out in the same way in both the cases. The inclusion of the node life time along with the link life time in the prediction of the route enhances the robustness of the route selection process.

The rest of the paper is organized as follows. Section 2 discusses the related research in the field of belief propagation and link prediction. Proposed algorithms LPR and RPLR are presented in Sect. 3. Results and discussion are given in Sect. 4 and a conclusion in Sect. 5.

2 Related Research

The literature review reveals that many methods have been developed for predicting the good neighbours from a given node [8]. A Gauss-Markov random process presented in [9] estimated the mobility of a node based on speed and direction. The method does not consider life time of a node. Historical-based prediction is carried out in [10]. Storing historical data for large network is challenging. Belief propagation [11] is a message passing technique and recursively update the computations locally based on the message received from the neighbouring nodes. In [12], a load balancing algorithm is proposed using belief propagation. Improving PDR or throughput is out of the scope of the algorithm. A distributed algorithm is proposed in [13] which address the cell association and subcarrier allocation. The algorithm does not predict the future life of nodes and links in the network.

Gaussian sum cubature belief propagation (GSCBP) proposed in [14] uses the Gaussian sum to represent the messages propagated through neighbouring vehicles, which provides accuracy in communication. This method tries to increase the communication accuracy but no attempt is made to predict the life of the link. Paper [15] utilizes the information kept in the cache of each base station (BS) which transmits the cached information to neighbouring BS. The routes are established by the BS combining the cached information through collaborative action. As claimed by authors the collaborative collection can lead to the reduction of the average delay and network overhead in route establishment. However, the restrictions on the memory to cache large amount of data from neighbouring BS often reduce the effectiveness of the approach.

3   Link Prediction with Proposed Approach

The discussions in the previous sections lead to the conclusion that the frequent breakage of links in MANET necessitates regular establishment of route. Link break occurs because of the node failure or the link failure. The present work addresses the two failures as mentioned above and new prediction strategy is evolved to find out the good quality neighbours. The neighbouring nodes finally contribute to the routes by propagating the belief from each node. The route finding process proposed here arrives at the final route by assessing the belief content received at the destination. While the belief is generated from the reliable links from a node in the case of LPR, the RLPR incorporates the node life time also for improving the robustness of the route established. The LPR estimates the link reliability based on the signal strength between the two neighbouring nodes. The signal strength is normalized across all the branches emanating from a given node which generates the transition probability p_{ij} for propagating the belief from a given node i. Referring to Fig. 1

$$p_{ij} = \frac{S_{ij}}{\sum_{j=1}^{n} S_{ij}}, \text{ where n is the degree of the } i\text{th node.} \qquad (1)$$

Each node i receives the belief from the previous node. Node i multiplies the belief b_i with p_{ij} and forwards to node j. For the source node belief is 1. Finally, the path with the highest accumulated belief is selected by the destination node.

In the second method, reliable link prediction routing (RLPR) proposed here, the node and link life time with respect to the node energy (X1) and signal strength (X2) are both considered. A fuzzy function on the node energy and signal strength, which are treated as fuzzy linguistic variables, is computed to establish the node

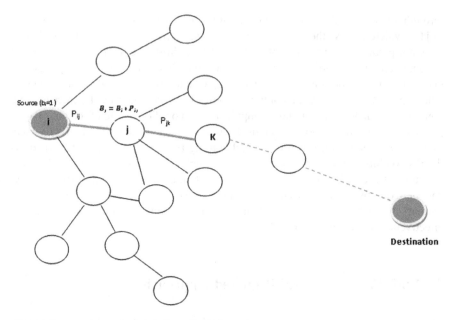

Fig. 1 Computation and propagation of belief

transition probability. The signal strength and node energy are fuzzy field using the fuzzy membership function given by $\mu(x) = e^{-(x-c_i)^2/2}$, for the ith fuzzy set and x could represent either node energy or signal strength.

The support for the fuzzy sets is defined as {0 to 1} and i = 1 to 5 defines the five fuzzy sets on the variable x as shown below:

Very low	Low	Medium	High	Very high

Thus, $c_i \in \{0.05, 0.25, 0.5, 0.75, 1\}$.

For a given node i, the transition probability p_{ij} is computed using a fuzzy function evaluated on the basis of the rule base given in Table 1.

A typical rule could be formulated.

If X1 is low and X2 is very low, then consequent is very low.

Table 1 Rule base

Signal strength/residual energy	Very low	Low	Medium	High	Very high
Very low (V.L)	V.L	V.L	V.L	V.L	V.L
Low (L)	V.L	L	L	L	L
Medium (M)	V.L	L	M	M	M
High (H)	V.L	L	M	H	V.H
Very high (V.H)	V.L	L	H	V.H	V.H

To decide the transition from node i to node j, the belief is accumulated across all the routes and compute the fuzzy decision.

$$q_{ij} = \frac{\sum_{l=1}^{5} \sum_{m=1}^{5} \mu_l(x1)\mu_m(x2)c_{lm}}{\sum_{l=1}^{5} \sum_{m=1}^{5} \mu_l(x1)\mu_m(x2)} \qquad (2)$$

The transition probability p_{ij} is given by

$$p_{ij} = \frac{q_{ij}}{\sum_{j=1}^{n} q_{ij}}, \text{ where } n \text{ is the degree of the } i\text{th node.} \qquad (3)$$

Destination node selects the route with the highest value of belief in the same way as discussed in the case of LPR.

4   Results and Discussions

LPR and RLPR are implemented in NS2 simulator with varying simulation time. The performance of the proposed work is evaluated based on the parameters PDR, network overhead, packet drop and end-to-end delay. Simulation scenario is described in Table 2. The simulation of the network scenario is executed for the

Table 2 Simulation parameters

Parameters	Options/value chosen
Routing protocols	AODV, DSR
Mac protocols	IEEE 802.15.4
Number of nodes	50
Simulation area	500 × 500
Packet size	512 bytes
Simulation time	100, 200, 300, 400 s

Fig. 2 Comparison of AODV, DSR with LPR and RLPR in terms of PDR. The PDR is the average of the result of 1000 independent runs

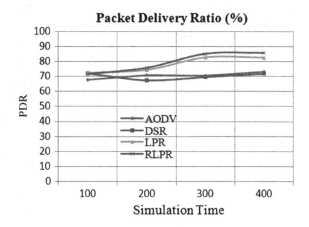

Fig. 3 Comparison of
AODV, DSR with LPR and
RLPR in terms of packet drop

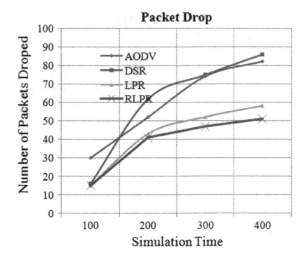

Fig. 4 Comparison of
AODV, DSR with LPR and
RLPR in terms of routing
overhead

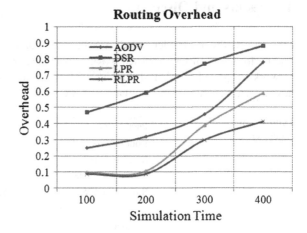

Fig. 5 Comparison of
AODV, DSR with LPR and
RLPR in terms of end-to-end
delay

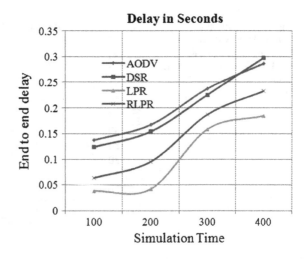

Table 3 Mean and variance of 1000 independent runs. The relatively high value of PDR shows the improvement in performance of LPR and RLPR, while the low value of variance indicates the trustworthiness

Protocol	PDR	Simulation time			
		100	200	300	400
AODV	Mean	67.540	70.758	70.393	73.134
	Variance	8.329	6.010	7.472	6.47
DSR	Mean	72.072	67.344	69.583	71.864
	Variance	7.513	7.456	8.23	8.029
LPR	**Mean**	**72.23**	**74.16**	**82.76**	**82.88**
	Variance	**7.319**	**6.441**	**5.954**	**5.992**
RLPR	**Mean**	**71.69**	**75.846**	**85.149**	**85.759**
	Variance	**8.583**	**5.669**	**4.137**	**4.17**

durations, viz. 100, 200, 300, and 400 s. Figures 2, 3, 4 and 5 clearly bring out the better performance of LPR and RLPR in comparison with AODV and DSR. For evaluating the PDR, 1000 independent runs were carried out for each duration and the average of each is calculated, along with its variance for all the four methods. Results are listed in Table 3, which clearly show that the mean value of the PDR of both LPR and RLPR is better than AODV and DSR. The lower value of the variance also underscores the confidence of the proposed algorithms in respect of PDR. Figure 2 illustrates the improvement in the PDR. Similarly, packet drop and network overhead are also seen to be significantly lower in LPR and RLPR as against the other two methods, which is evident from Figs. 3 and 4. While Fig. 5 portrays the comparison of LPR, RLPR, AODV and DSR based on end-to-end delay, it is observed that both the LPR and RLPR have less delay than AODV and DSR. But the delay in RLPR is slightly more than the LPR. The path established by RLPR may not be the shortest and hence could face increased delay, though RLPR achieves improved robustness in terms of consistently high values of PDR and lesser number of packet drops.

5   Conclusions

The paper has proposed and evaluated two on-demand routing algorithms for MANET. The main objective of the methods proposed, viz. LPR and RLPR, is to select the route that could be stable from source to destination. It is shown that the routing will be most successful by selecting those nodes having the best neighbouring node for data forwarding. Both the methods establishing the route by propagating the belief from the source to the destination nodes. The first method LPR is formulated based on the signal strength only to calculate the transition probability of a given node. The best path is selected based on the belief accumulated at the destination node. The second method RLPR strengthens the path identification by using the node energy, in addition to signal strength. The transition probability is calculated on the basis of a fuzzy decision function, extracted from a rule base with signal strength and residual energy as antecedence. Simulation studies carried out on 1000 independent run for different durations from 100 to 400 s and reveal that the LPR and RLPR perform much better than AODV and DSR. The delay is observed to be little higher for the RLPR in comparison with LPR possibly because of landing in a longer path in the pursuit to achieve a higher robustness, manifest in terms of large PDR. The choice between LPR and RLPR is decided by the application which weighs fast delivery against robustness.

References

1. George Aggelou, "Mobile Adhoc Networks", Tata McGraw-Hill, 2009 ISBN-13:978-0-07-067748-7.
2. Salim El Khediri; Nejah Nasri; Awatef Benfradj; Abdennaceur Kachouri; Anne Wei, "Routing protocols in MANET: Performance comparison of AODV, DSR and DSDV protocols using NS2", International Symposium on Networks, Computers and Communications, 2014.
3. Tejashree S Khanvilkar; K P Patil, "Performance evaluation and comparison of routing protocols in MANETs", Fourth International Conference on Computing, Communications and Networking Technologies (ICCCNT), 2013.
4. Afrah Daas; Khulood Mofleh; Elham Jabr; Sofian Hamad, "Comparison between AODV and DSDV routing protocols in mobile Ad-hoc Network (MANET)", 5th National Symposium on Information Technology: Towards New Smart World (NSITNSW), 2015.
5. Krishna Mahajan; Devesh Malik; M. A. Rizvi; D. Singh Karaulia, "Event Driven Dynamic Path Optimization for AODV in MANET", International Conference on Computational Intelligence and Communication Networks, 2014.
6. Ranjan Kumar; Mansi Gupta, "Route stability and energy aware based AODV in MANET", International Conference on High Performance Computing and Applications (ICHPCA), 2014.
7. Weiping Liu and Linyuan LU, Link Prediction Based on Local Random Walk, Europhysics Letters Association.
8. MD. Osman Gani, Hasan Sarwar and Chowdhury Mofizur Rahman, "Predication of state of wireless network using Markov and Hidden Markov Model", Journal of Networks, Vol. 4, No. 10, December 2009.
9. Javad Akbari Torkestani, "Mobility prediction in mobile wireless networks", Journal of Network and Computer Applications, 35 (2012) 1633–1645, Elsevier.
10. Hui Xu, J.J. Garcia-Luna-Aceves, "Neighbourhood tracking for mobile ad hoc networks", Computer Networks, 53 (2009) 1683–1696.
11. E. B. Sudderth, A. T. Ihler, M. Isard, W. T. Freeman, and A. S. Willsky, "Nonparametric belief propagation", Communications of ACM, vol. 53, no. 10, pp. 95–103, 2010.
12. Youjia Chen, Jun Li, He (Henry) Chen, Zihuai Lin, Guoqiang Maozx, Jianyong Cai, "A Belief Propagation Approach for Distributed User Association in Heterogeneous Networks", IEEE 25th International Symposium on Personal, Indoor and Mobile Radio Communications, 2014.
13. CHEN Siyi, XING Chengwen, FEI Zesong, "Distributed Resource Allocation in Ultra-Dense Networks via Belief Propagation", China Communications, November 2015.
14. Wenyun Gao, Xi Chen, Menglu Wan and Daichen Zhang, "Gaussian Sum Cubature Belief Propagation for Distributed Vehicular Network Navigation", IEEE, 2016.
15. Juan Liu, Bo Bai, Jun Zhang, and Khaled B. Letaief, "Content Caching at the Wireless Network Edge: A Distributed Algorithm via Belief Propagation", ICC 2016 - Mobile and Wireless Networking Symposium, IEEE, 2016.

Adaptive PET/CT Fusion Using Empirical Wavelet Transform

R. Barani and M. Sumathi

Abstract Empirical Wavelet Transform (EWT) is an adaptive signal decomposition technique in which the wavelet basis is constructed based on the information contained in the signal instead of a fixed basis as in standard Wavelet Transform (WT). Its adaptive nature enables EWT in many image processing applications like image denoising, image compression, etc. In this paper, a new adaptive image fusion algorithm is proposed for combining CT and PET images using EWT. EWT first decomposes both the images into approximate and detailed components using the adaptive filters that are constructed according to the content of an image by estimating the frequency boundaries. Then, the corresponding approximate and detailed components of CT and PET images are combined by using appropriate fusion rules. An adaptive EWT image fusion, a newly proposed method, is compared with standard WT fusion using the image quality metrics, image fusion metrics and error metrics. The quantitative analysis proved that the newly proposed method results in better quality than the standard WT method.

Keywords PET/CT Fusion · Medical image fusion · Adaptive image fusion
EWT fusion · Quality metrics · Empirical Wavelet Transform

1 Introduction

In medical scenario, multimodality imaging has a significant consign in many clinical applications like cancer screening, staging, response monitoring, neurological assessment, drug development, etc. Multimodality imaging is the technology that efficiently combines morphological images with functional images either acquired at

R. Barani (✉)
V.V.Vanniaperumal College for Women, Virudhunagar 626001, Tamil Nadu, India
e-mail: baranisrinivasan1994@gmail.com

M. Sumathi
Sri Meenakshi Government Arts College for Women, Madurai 625021, Tamil Nadu, India
e-mail: sumathivasagam@gmail.com

© Springer Nature Singapore Pte Ltd. 2018
V. Bhateja et al. (eds.), *Proceedings of the Second International Conference on Computational Intelligence and Informatics*, Advances in Intelligent Systems and Computing 712, https://doi.org/10.1007/978-981-10-8228-3_40

different times or simultaneously for a better understanding of disease and physiological condition during preclinical and clinical activities. The primary imaging modalities like CT, MRI, PET, SPECT, ultrasound and X-Ray provides either the anatomical information or the functional information. But to diagnose and localize the problems correctly and accurately, the physician needs both the information in a single image. So, the need arises to combine the anatomical imaging like CT and MRI with any of functional imaging PET and SPECT [1].

Image fusion is the technique of achieving a superlative image by gaining the information from two or more images of same or different modality. Recently, image fusion is attaining its importance in many applications, since it results in a unique image which is more valuable than any of the source images in terms of wealthy content, minimized redundancies and artifacts. Image fusion operation can be performed either in the spatial domain or in the transformed domain of an image [2]. In image fusion algorithms like simple and weighted average, maximum/minimum, median and rank, PCA, ICA, CCA and other HIS-based algorithms [3], the fusion operation is based on spatial-related information like pixel intensity value, mean, standard deviation and other statistical-related information of images. These methods do not consider the spectral information of an image. But most of the salient image features are spectral-related information. Whereas image fusion in transform domain considers spectral-related information which is represented as multi-scales or multi-resolutions [4]. At the early stage of multi-scale image fusion, pyramid representation was adopted for image representation. Later, the wavelets have proved its ability in getting more directional information and well performed than pyramid-based methods. Recently, several extensions such as Ridgelet, Ripplet, Curvelet and Contourlet are introduced for multi-scale image representation and also for image fusion.

Though multi-scale representation-based image fusion methods like standard WTs are superior than spatial fusion methods, they are not adaptive to the content of an image. Because they use fixed basis functions for decompositions. But biomedical images and signals normally have nonlinear and non-stationary stuff that cannot be processed using these rigid methods. So, an adaptive image representation where the filters are generated according to the information of an image is required. Huang et al. proposed a new adaptive representation called Empirical Mode Decomposition (EMD) [5]. Its basic idea is to represent a signal as collection of principal modes present in the signal spectrum. But the main drawback of this method is that there is no mathematical background and is too sensitive to any noise existing in the signal. Later, Jerome Gilles introduced a different approach known as empirical wavelet transform [6] based on EMD and WT. EWT is an adaptive transformation method, which constructs its wavelet basis based on the characteristics of a signal. The content-adaptive filters are more useful in many medical image processing applications like automated diagnosis of glaucoma in digital fundus images [7], EEG seizure detection [8], automatic detection and classification of heart sounds and murmurs [9].

This paper proposes a new adaptive image fusion algorithm using EWT to incorporate the functional information from PET image into the anatomical information of CT image and to provide a standalone ideal image for better diagnosis. To perform image fusion, the automatically constructed adaptive filters by EWT are applied on the images to decompose into approximate and detailed coefficients. Then, the corresponding coefficients of both the images are combined using the simple average and maximum selection rule. At the end, the resultant fused image is obtained by inverse EWT. The proposed method is compared with normal WT fusion method using the image quality metrics: Spatial Frequency (SF) and Shannon Entropy (ENT), image fusion metrics: Fusion Factor (FF), Fusion Measure ($Q^{AB/F}$), Fusion Quality Index (Q(A, B, F)), Mutual Information (MI), and Structural Similarity Index Metric (SSIM) and error metrics: Root Mean Square Error (RMSE) and Percentage Fit Error (PFE).

2 EWT: An Overview

The basic idea of EWT is to construct a set of empirical wavelets which are adaptive to the content of processing signal and then to decompose the signal using the constructed empirical wavelets in the conceptual framework of standard wavelet theory.

2.1 Empirical Wavelets

Empirical Wavelets (EW) [6] are the adaptive band-pass filters constructed over each node in the spectrum of a signal. For example, consider the Fourier spectrum with the interval $[0, \pi]$ which is divided into N continuous segments. Let ω_n be the limits between each segment, where $\omega_0 = 0$ and $\omega_N = \pi$. There is a transition phase of width around each ω_n. Each segment in the spectrum can be obtained as the difference between two limits as $|\omega_{n-1} - \omega_n|$. If the segment in the spectrum is specified by Λ_n, the entire Fourier spectrum can be obtained as the union of all the segments and is given by $\bigcup_{n=1}^{N} \Lambda_n$. The partitioning of Fourier spectrum into N segments is shown in Fig. 1.

Then, EW can be defined as band-pass filters on each Λ_n, based on the idea of Littlewood–Paley and Mayers wavelets [10]. The empirical scaling function $\hat{\phi}_n(\omega)$

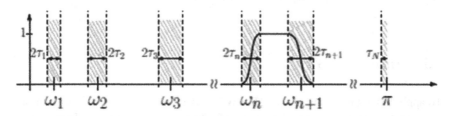

Fig. 1 Partitioning of a Fourier spectrum

and empirical wavelets function $\hat{\psi}_n(\omega)$ are expressed by (1) and (2), respectively.

$$
\hat{\phi}_n(\omega) = \begin{cases} 1 & \text{if } |\omega| \le (1-\gamma)\omega_n, \\ \cos[\frac{\pi}{2}\beta(\frac{1}{2\gamma\omega_n}(|\omega|-(1-\gamma)\omega_n)] & \text{if } (1-\gamma)\omega_n \le |\omega| \le (1+\gamma)\omega_n, \\ 0 & \text{otherwise.} \end{cases} \tag{1}
$$

and

$$
\hat{\psi}_n(\omega) = \begin{cases} 1 & \text{if } |\omega| \le (1-\gamma)\omega_n, \\ \cos[\frac{\pi}{2}\beta(\frac{1}{2\gamma\omega_{n+1}}(|\omega|-(1-\gamma)\omega_{n+1})] & \text{if } (1-\gamma)\omega_{n+1} \le |\omega_n| \le (1+\gamma)\omega_{n+1}, \\ \sin[\frac{\pi}{2}\beta(\frac{1}{2\gamma\omega_{n+1}}(|\omega|-(1-\gamma)\omega_n)] & \text{if } (1-\gamma)\omega_n \le |\omega| \le (1+\gamma)\omega_n, \\ 0 & \text{otherwise.} \end{cases} \tag{2}
$$

where ω_n is the segment limit and $0 < \gamma < 1$.

2.2 Empirical Wavelet Transform

The EWT, $w_f^z(n,t)$ [6] can be defined as similar to standard WT such that the detail coefficients are obtained by the inner product of processing signal with the empirical wavelet $\hat{\psi}_n(\omega)$ as shown in (3).

$$
w_f^z(n,t) = <f, \psi_n> = (\hat{f}(\omega)\overline{\hat{\psi}_n(\omega)})^\vee \tag{3}
$$

and the approximation coefficients can be obtained by the inner product of processing signal with the scaling function $\hat{\phi}_n(\omega)$ as in (4).

$$
w_f^z(0,t) = <f, \phi_1> = (\hat{f}(\omega)\overline{\hat{\phi}_1(\omega)})^\vee, \tag{4}
$$

where $\hat{\phi}_n(\omega)$ and $\hat{\psi}_n(\omega)$ are defined in Eq. (1) and Eq. (2), respectively. The reconstruction of the processing signal is obtained by the inverse EWT and is given by (5).

$$
f(t) = (\hat{w}_f^\epsilon(0,\omega)\hat{\phi}_1(w) + \sum_{n=1}^{N} \hat{w}_f^\epsilon(n,\omega)\hat{\psi}_n(w))^\vee \tag{5}
$$

2.3 Tensor 2D EWT

To apply EWT for an image, first 1D EWT is extended to 2D EWT using a tensor approach [6, 11]. According to this approach, the filter bank containing the set of row

Table 1 Algorithm for 2D tensor empirical wavelet transform

1	*Input the image I of size $r \times c$.*
2	*For each row of I compute 1D FFT, $F_{1,x}(I)(i, \omega_1)$;* *then calculate the average $\tilde{F}_{row}(\omega_1)$.*
3	*For each column of I compute 1D FFT, $F_{1,y}(I)(j, \omega_2)$;* *then calculate the average $\tilde{F}_{col}(\omega_2)$.*
4	*Perform the boundaries detection on $\tilde{F}_{row}(\omega_1)$ to get the set of Fourier boundaries Ω_{row} and obtain the row filters β_{row}.*
5	*Perform the boundaries detection on $\tilde{F}_{col}(\omega_2)$ to get the set of Fourier boundaries Ω_{col} and obtain the column filters β_{col}.*
6	*Filter the input image I along the rows with β_{row} and obtain m output images.*
7	*Filter the input image I along the column with β_{col} and obtain n output images.*
8	*Output the row and column filters and the output images.*

filters and column filters is obtained initially based on the content of an image and then applied on an image to obtain the multi-scale representation. The adaptive filter bank is obtained using the following steps: first apply 1D FFT on each row and obtain the average row spectrum. Then, detect the Fourier boundaries using the detection technique [12] on the average spectrum and obtain the row filter bank. Repeat the same procedure for columns and obtain the column filter bank. Apply now, row and column filter bank on the image to achieve multi-scale decomposition containing one approximate component and many detailed components. On considering I as the image with r rows and c columns, the pseudocode for 2D Tensor EWT algorithm for multi-scale decomposition is given in Table 1.

3 Image Fusion with EWT

To perform image fusion using EWT, initially PET and CT images are decomposed into a set of band-pass filtered components using 2D tensor EWT as mentioned in Table 1. Assume that the 2D tensor EWT algorithm constructs $m1$ row filters and $n1$ column filters for CT image and $m2$ row filters and $n2$ column filters for PET image. Usually, if the filter bank has m row filters and n column filters, after decomposition, it results in $m \times n$ components (one approximate component and mn-1 detailed components). The approximate component contains the blurred version of base information of an image and detailed component provides the important features like edges and corners that exist in an image. The corresponding approximate and the detailed components of CT and PET images are combined using the appropriate fusion rule.

Table 2 Algorithm for adaptive image fusion using empirical wavelet transform

1 Input the source images I_1 and I_2 of size $r \times c$.

2 Obtain the row and column filters and decomposed components for image I_1
   using the steps 2 through 7 in Table 1. Let them be
   Row Filters: $\{rf_1^{(1)}, rf_2^{(1)} \ldots, rf_{m1}^{(1)}\}$, Column Filters: $\{cf_1^{(1)}, cf_2^{(1)} \ldots, cf_{n1}^{(1)}\}$
   and the decomposed components: $\{C_{I_1}^{(1,1)}, C_{I_1}^{(1,2)} \ldots, C_{I_1}^{(m1,n1)}\}$

3 Similarly, obtain the row and column filters and decomposed components
   for image I_2 using the steps 2 through 7 in Table 1. Let them be
   Row Filters: $\{rf_1^{(2)}, rf_2^{(2)} \ldots, rf_{m2}^{(2)}\}$, Column Filters: $\{cf_1^{(2)}, cf_2^{(2)} \ldots, cf_{n2}^{(2)}\}$
   and the decomposed components: $\{C_{I_2}^{(1,1)}, C_{I_2}^{(1,2)} \ldots, C_{I_2}^{(m2,n2)}\}$

4 Combine the approximate coefficients $C_{I_1}^{(1,1)}$ and $C_{I_2}^{(1,1)}$ using adaptive-weighted
   average rule to obtain $C_{I_f}^{(1,1)}$.

5 Combine the detailed coefficients $C_{I_1}^{(1,2)} \ldots C_{I_1}^{(m1,n1)}$ of image I_1 and
   $C_{I_2}^{(1,2)} \ldots C_{I_2}^{(m2,n2)}$ of image I_2 using maximum selection rule to obtain the
   combined detailed coefficients $C_{I_f}^{(1,2)} \ldots C_{I_f}^{(m,n)}$.

6 Construct the combined row and column filters $\{rf_1^{(f)}, rf_2^{(f)} \ldots, rf_m^{(f)}\}$
   and $\{cf_1^{(1)}, cf_2^{(1)} \ldots, cf_{n2}^{(1)}\}$, from row and column filters of I_1 and I_2
   using the average and maximum selection rule.

7 Obtain the fused image I_f by inverse 2D EWT of combined approximate.
   and detailed components using the combined filters.

And also, the corresponding row and column filters of CT and PET images need to be combined using the appropriate fusion rule for reconstruction.

In this paper, adaptive-weighted average rule for combining approximate components and maximum selection rule for combining detailed components has been adopted. Then, obtain the row and column filters of combined component by applying the average and maximum selection rule on row and column filters of two source images. The resultant fused image is obtained using the inverse 2D tensor EWT on the combined approximation and detailed components using the combined row and column filters. The obtained fused image contains the base information and also the salient features from both the images. The obtained fused image is more informative and visible for easy diagnosis. The pseudocode for an adaptive image fusion algorithm using EWT is given in Table 2.

4 Experimental Data and Analysis

The proposed adaptive medical image fusion algorithm using EWT is tested on Rider Lung PET/CT data available in The Cancer Imaging Archive (TCIA) and the real data. In this paper, the source and fused images of two real data sets are presented.

The data set1 is an axial view of CT and PET images of a neck showing necrotic mass lesion of 39 years patient. Data set2 is an axial view of CT and PET images of a neck showing a mass lesion in apicoposterior segment of right upper lobe. Both the data sets are in jpg format. Since the quality and content of output image is more important, the resultant image of the proposed method is compared with standard WT fusion method using the non-reference image quality metrics: SF, ENT, image fusion metrics: FF, Q(AB/F), Q(A, B, F), MI, SSIM and error metrics: RMSE, PFE [13].

5 Results and Discussion

5.1 Data Set1

Figure 2a–d shows the source and fused images of data set1 and of size 336×272. Figure 2a is a CT image which contains the information about bony structures; Fig. 2b is a PET image showing the metabolic activity of necrotic mass lesion in the base of tongue which is not captured in CT image; Fig. 2c and 2d are the fused images of WT and EWT fusion methods, respectively. The fused image from the proposed method extracts the important features from both images and also it retains the contrast and clarity compared to WT method.

In data set1, the 2D tensor EWT detects 4 row and 5 column boundaries for CT image which results in 5 row and 6 column filters. The convolution of these adaptive filters with the CT image results in 5×6 decomposed components. For PET image, 2D tensorEWT detects nine rowboundaries and eight column boundaries that results in ten row filters and nine column filters. The convolution of these adaptive row and column filters with the PET image results in 10×9 decomposed components. For both images, the component at position $(1, 1)$ is an approximate component and the remaining are detailed components.

During the fusion process, the approximate coefficients are combined using simple average rule. The detailed coefficients, row and column filters are combined using the maximum selection rule. The detailed components from the first five rows and six

(a)   **(b)**   **(c)**   **(d)**

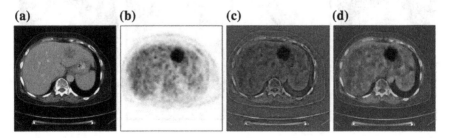

Fig. 2 Source and fused images of data set1, **a** CT image, **b** PET image, **c** WT method and **d** EWT method

Table 3 Quantitative analysis of data set1

Quality metrics	SF	ENT	FF	$Q^{AB/F}$	Q(A, B, F)	MI	SSIM	RMSE	PFE
WT method	0.036	0.817	1.581	0.590	0.473	−0.187	0.966	0.396	57.60
EWT method	0.054	0.850	2.762	0.616	0.954	1.975	0.984	0.349	46.3

columns of CT and PET images were combined using the maximum selection rule and the detailed components from the remaining rows and columns of PET image are just copied for the resultant image. Similarly, five row filters and six column filters are combined using the maximum selection rule, and the remaining row and column filters of PET image are just copied. Once the combined approximate and detailed coefficients, combined row and column filters are formed, the resultant fused image is reconstructed using inverse 2D Tensor EWT transform.

The quantitative analysis of proposed method and WT method is given in Table 3. Higher values of SF and ENT for the proposed method depict that the fused image from proposed method is better in terms of contrast, sharpness and edge clarity than WT method and also the values of FF, $Q^{AB/F}$, and Q(A, B, F), MI and SSIM clarify that the proposed method extracts important features from the source images effectively than the WT method. The error metric also proved that the new method outperforms the WT method by reducing the occurrence of error.

5.2 Data Set2

Figure 3a–d shows the source and fused images of data set2 and of size 624×528. Figure 3a is CT image which contains the information about hard tissues like bone;

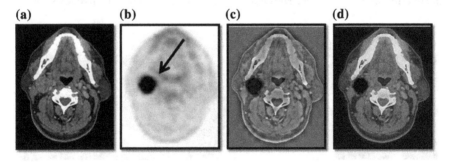

(a)   **(b)**   **(c)**   **(d)**

Fig. 3 Source and fused images of data set2, **a** CT image, **b** PET image, **c** WT method and **d** EWT method

Table 4 Quantitative analysis of data Set2

Quality metrics	SF	ENT	FF	$Q^{AB/F}$	Q(A, B, F)	MI	SSIM	RMSE	PFE
WT method	0.049	0.817	1.582	0.616	1.954	−0.187	0.974	0.3499	57.62
EWT method	0.056	0.965	3.488	0.626	2.385	0.247	0.988	0.3463	40.47

Fig. 3b is a PET image showing the metabolic activity of necrotic mass lesion (as pointed in Fig. 3b) in apicoposterior segment of right upper lobe which is not shown in CT image; Fig. 3c and 3d are the fused images of WT and EWT fusion methods, respectively. The resultant fused image of proposed method shows that the salient features from both images are extracted and combined effectively and also retaining the contrast compared to WT method.

In data set2, the 2D tensor EWT detects 9 row and 16 column boundaries for CT image which results in 10 row and 17 column filters. The convolution of these adaptive filters with the CT image results in 10×17 decomposed components. For PET image, 2D tensor EWT detects five row boundaries and six column boundaries that results in six row filters and seven column filters. The convolution of these row and column filters with the PET image results in 6×7 decomposed components. For both images, the component at position (1, 1) is an approximate component, and the remaining are detailed components.

During the fusion process, the approximate coefficients are combined using simple average rule. The detailed coefficients, row and column filters are combined using the maximum selection rule. The detailed components from the first six rows and seven columns of CT and PET images were combined using the maximum selection rule and the detailed components from the remaining rows and columns of CT image are just copied for the resultant image. Similarly, six row filters and seven column filters are combined using the maximum selection rule, and the remaining row and column filters of PET image are just copied. Once the combined approximate and detailed coefficients, combined row and column filters are formed, the resultant fused image is reconstructed using inverse 2D tensor EWT transform.

The quantitative analysis of proposed method and WT method is given in Table 4. Higher values of SF and ENT are achieved for the proposed method which depicts that the fused image from proposed method is better in terms of contrast, sharpness and edge clarity than WT method and also the values of FF, $Q^{AB/F}$, and Q(A, B, F), MI and SSIM clarify that the proposed method extracts the important features from the source images effectively than the WT method. The error metric also proved that the new method outperforms the WT method by reducing the error rate.

6 Conclusion

This paper successfully implements an adaptive image fusion method for merging of PET and CT images using 2D Tensor EWT. The proposed adaptive medical image fusion method based on EWT is tested on Rider Lung PET/CT data available in The Cancer Imaging Archive (TCIA) and the real data. The objective analysis depicts that the resultant fused image reveals anatomical information with good contrast and sharpness including the functional information at the maximum extent. The subjective analysis of resultant images proved the effectiveness of EWT for combing salient features from different images for future purpose. Even though it shows excellency, the problem occurs while combining the detailed components, since the number of detailed components of CT and PET images differs. And also, the adaptive filter banks that are automatically constructed are different and not same in number. When the filters are averaged for reconstruction, it degrades the performance of EWT. These problems are to be resolved, if the total benefit of EWT has to be gained in image fusion applications. Though it has some disadvantages, its adaptive nature brings its use in many of the image processing domains.

References

1. Constantinos S. Pattichis, Marios S. Pattichis, Evangelia Micheli-Tzanakou: Medical imaging fusion applications: An overview, Thirty-Fifth Asilomar Conf. on Signals, Systems and Computers, 2001; 2: 1263–1267.
2. C. Pohl, J. L. Van Genderen: Review article Multisensor image fusion in remote sensing: Concepts, methods and applications, International Journal of Remote Sensing, 1998; 19(5): 823–854.
3. H.B. Mitchell: Image Fusion Theories, Techniques and Applications, Springer, 2010.
4. Hui Li, B. S. Manjunath. Sanjit K. Mitra, Multi-Sensor Image Fusion using the Wavelet Transform, Graphical Models and Image Processing, 1995; 57(3): 235–245.
5. N.E. Huang and Z. Shen and S.R. Long and M.C. Wu and H.H. Shih and Q. Zheng and N-C. Yen and C.C. Tung and H.H. Liu, The empirical mode decomposition and the Hilbert spectrum for nonlinear and nonstationary time series analysis, Proc. Royal Society London A Mathematical, Physical and Engineering Sciences., 1998; 454(1971): 903–995.
6. Jerome Gilles, Empirical Wavelet Transform, IEEE Transactions on Image Processing, 2013; 61(16): 3999–4010.
7. Maheshwari, Shishir and Pachori, Ram Bilas and Acharya, U Rajendra: Automated diagnosis of glaucoma using empirical wavelet transform and correntropy features extracted from fundus images, IEEE Journal of Biomedical and Health Informatics, 2017, 21(3): 803–813.
8. Bhattacharyya, Abhijit and Pachori, Ram Bilas: A Multivariate Approach for Patient Specific EEG Seizure Detection using Empirical Wavelet Transform, IEEE Transactions on Biomedical Engineering, 2017.
9. Varghees, V Nivitha and Ramachandran, KI: Effective Heart Sound Segmentation and Murmur Classification Using Empirical Wavelet Transform and Instantaneous Phase for Electronic Stethoscope, IEEE Sensors Journal, 2017, 17(12): 3861–3872.
10. I. Daubechies, Ten Lectures on Wavelets, Society for Industrial and Applied Mathematics, CBMS-NSF Regional Conference Series in Applied Mathematics, 1992.
11. Jerome Gilles and Giang Tran, Stanley Osher, 2D Empirical Transforms: Wavelets, Ridgelets and Curvelets revisited, SIAM Journal on Imaging Sciences, 2014; 7(1): 157–186.

12. Scott Shaobing Chen, David L. Donoho, Michael A. Saunders: Atomic Decomposition by Basis Pursuit, SIAM Review, Society for Industrial and Applied Mathematics, 2001; 43(1): 129–159.
13. Jagalingam Pa, Arkal Vittal Hegdeb: A Review of Quality Metrics for Fused Image, Aquatic Procedia, 2015; 4: 133–142.

Sudoku Game Solving Approach Through Parallel Processing

Rahul Saxena, Monika Jain and Syed Mohammad Yaqub

Abstract Sudoku is one of the most popular puzzle games of all time. Today, Sudoku is based on simple rules of placing the numbers from 1 to 9 in the empty cells of the Sudoku board. After solving the Sudoku, each row, column, and mini grid should have only one occurrence of each digit between 1 and 9. There are various approaches introduced to solve the Sudoku game till now. The solutions in the state of art are computationally complex as the algorithmic complexity for the solution falls under the class of NP-complete. The paper discusses various solutions for the problem along with their pros and cons. The paper also discusses the performance of a serial version solution with respect to execution time taken to solve Sudoku. With certain modifications and assumptions to the serial version code, we have evaluated the possible solution and bottleneck in the parallel approach. The results have been evaluated for the parallelized algorithm on a normal user machine first, and then on a supercomputing solution box PARA-MSHAVAK. Finally, the paper is concluded with the analysis of all the three solution aspects proposed.

Keywords Cells · Mini grid · Board · Elimination · Lone ranger
Twins · Triplets

R. Saxena (✉) · M. Jain · S. M. Yaqub
Manipal University, Jaipur, India
e-mail: rahulsaxena0812@gmail.com

M. Jain
e-mail: monikalnct@gmail.com

S. M. Yaqub
e-mail: syed5yaqub@gmail.com

© Springer Nature Singapore Pte Ltd. 2018
V. Bhateja et al. (eds.), *Proceedings of the Second International Conference on Computational Intelligence and Informatics*, Advances in Intelligent Systems and Computing 712, https://doi.org/10.1007/978-981-10-8228-3_41

1 Introduction

Sudoku was first published in the year 1984 in the Dell Magazine by the name "Number Place". Sudoku is a short form of Su-ji wa dokushin ni kagiru (in Japanese) which means "the numbers must be single" in each row, column, and mini grid [1]. Sudoku consists of mini grid, cells, grids, and the board. Different mini grids form different region of the board. Sudoku is well known for its simple rules, which makes it interesting and simple. Various versions of Sudoku puzzle are possible for the players to enjoy and enhance their skills of analyzing. Sudoku is most famous because of the simple rules that it has and in spite of being a game of numbers, one does not need to know any mathematics or calculation to get the

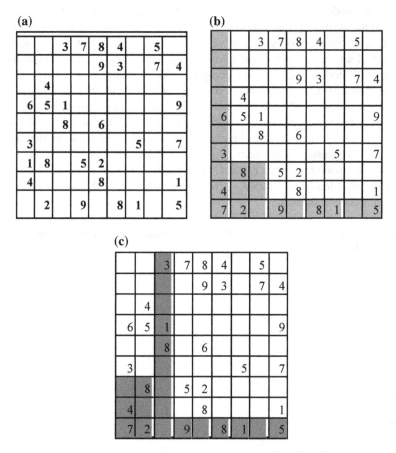

Fig. 1 Showing the description of resolving the Sudoku game: **a** initial board, **b** step1, and **c** step2. The colored regions specify which numbers are considered to calculate the cell that is filled with red (step 1) and green (step 2)

solution of the puzzle, which helps different players to enjoy it. Sudoku board can be of different sizes like $4 \times 4, 6 \times 6, 9 \times 9, 12 \times 12$, and 16×16 with different mini grid sizes and the range of numbers that can be filled depending on the board size. The default Sudoku board consists of filled cells and empty cells, and one has to determine what values need to be filled in the empty cells by analyzing the row, column, and mini grid to which the empty cell belongs. Different Sudoku have different levels of difficulty. The game which has very less number of empty cells is considered easy, with medium number of empty cells it is considered medium level, and with most of the cells being empty it is considered difficult. Sudoku is considered to be an NP-complete problem that is "Non-deterministic Polynomial Time". There are different ways to solve the Sudoku problem like brute force, elimination, assigning values to the cell, and backtracking when no possible move is left. A method to solve the Sudoku game by logic and analyzing the row, columns, and mini grids. This is the default 9×9 board with 3×3 cells (Fig. 1).

2   Existing Solutions for Solving Sudoku Serially

The researchers using algorithmic approaches and heuristics have worked upon game playing and solution strategies intensively. Sudoku has been one such game which has been targeted in this direction with numerous approaches and interestingly various alternatives apart from the traditional graph searching techniques has been suggested in the state of art due to the NP-complete nature of the problem. The paper here discusses different proposed solutions for the problem with backtracking and rule-based methods such as hidden single, lone ranger, twin, triplet, locked candidate, etc. [2]. Further, an attempt has been made to generate a parallel Sudoku solver. The experimental results will help us understand the computation time and which method is best.

Brute Force with Backtracking: Brute force as the name suggests is inserting the numbers in the vacant unit and check [1]. This method is purely based on individuals analyzing and placing values which the players think to be appropriate. If the brute force is safe which means if it yields the **correct** result, so we move on to the next empty cell and repeat the brute force. If the brute force **fails**, then we have to backtrack the entries we made to find out the correct placing of the number, and this is done until the Sudoku board does not have any empty cells. This algorithm has its own advantage of finding the solution no matter how hard the puzzle is as it tries all permutation and combination at every empty cell of the Sudoku board. But the limitation to this algorithm is the time needed to find the solution depends on the number of empty cells in the Sudoku board.

2.1 Rule-Based Methods

Rule-based methods are the methods used to fill the empty cells based on different rules which will eventually leave very less empty cells than the input Sudoku board, and some of the rules are discussed below [3].

Naked Singles: In this method, the empty cells are filled that can have only one value first decreases the number of possible permutation and combination in the empty cells that are in the same row, column, and mini grid [4]. Sometimes, filling the naked singles will eliminate a choice in another empty cell because the number is already used as a naked single in either the same row, column, or mini grid.

Hidden Singles: In some cases, there exist multiple values that can be filled in an empty cell but the extra values do not make the exact solution visible, which again is found out by analyzing the row, column, and mini grid, and the values that cannot be filled are eliminated [5].

Naked Pair: Sometimes, there are two common possible values in an empty cell, and it is repeated in two other empty cells of the same row, column, or mini grid [6]. The rule is to remove this common pair from all other empty cells, and then, we can apply hidden singles or naked singles to fill an empty cell, and hence reducing the combination to the value that needs to be filled in that empty cell.

Hidden Pairs: Hidden pairs can be identified if there exist two cells in a mini grid with the same pair of value that can be filled in that empty cell, then the remaining possible values are discarded from the cells. This method can be used to find hidden triplets or hidden quads only difference being that there should be three or four common values among the empty cells possible values [7].

Naked Triple: There are three common likely values in three empty cells of the same row, column, or mini grid. The value that is common in these empty cells is removed from all other empty cells [8].

3 Algorithm for Serial Code

Step 1: Initialization: input Sudoku board.

Step 2: While (empty_cell_exists) repeat steps 3 to 7: Sudoku board is still unsolved.

Step 3: Find_empty_cell in the Sudoku board: the cells where value needs to be placed.

Step 4: Place a number between 1 to grid size in the empty cell and check if is valid or not: checking all values by placing it in the empty cell to find the correct value.

Step 5: If the placed number is not the same row go to step 6 else go to step 8: the values should be unique in the same row.

Step 6: If the placed number is not the same column go to step 7 else go to step 8: the values should be unique in the same column.

Step 7: If the placed number is not the same mini grid go to step 3 else go to step 8: the values should be unique in the same row.

Step 8: Go to step 4: placing another value in the empty cell and again check if it is valid or not.

Step 9: Print feasible solution of the Sudoku board.

4 Existing Solution for Execution Through Parallel Algorithm

By creating multiple threads in OpenMP, we parallelize the rules that are applied serially in the previous section so that they can be executed parallely. The main idea is to first apply elimination that means to find the empty cells that can have only one value. Lone ranger is found that means finding the hidden single and then finding twins and triplets then applying depth-first search to find the cell with least number of possible value and then applying brute force on it. All these methods are applied PARALLELY on each mini grid. We choose the mini grid in such a way that the rows and columns are not scanned redundantly. The rules followed in executing the Sudoku solver parallel are discussed below:

Elimination: This is same as naked singles where we find the empty cells that can have only one possible value to decrease the number of possibility of the other empty cells in the same row, column, and mini grid.

Lone Ranger: Lone ranger means there exists multiple possible value in an empty cell but on analyzing, we can conclude that only one value can go into that empty cell due to the values present in the partially filled Sudoku board.

Finding Twins: This rule is to find two common values in two cells that can be filled in those two empty cells, and hence, we have to remove that pair from all the possible values of empty cells in the same row, column, and mini grid.

Finding Triplets: This rule is to find three common values that can be filled in the three empty cells in the same row, column, or mini grid. After the triplets are found out, we remove these common values from all the other empty cells in the same row, column, or mini grid.

4.1 Parallel Algorithm

Step 1: Initialization: input Sudoku board.

Step 2: If (empty_cell_exists) repeat steps 3 to 4: Sudoku board is still unsolved.

Step 3: Parallelized elimination process on each mini grid: the empty cells where only one value can be placed are found out in each grid, and the values are placed in those empty cells and this is done parallely on each mini grid.

Step 4: Parallelized lone ranger process on each mini grid: the empty cells where multiple value can be placed are found out in each grid, and one value that actually can be placed is found by eliminating all the other values by analyzing the Sudoku board and this is done parallely on each mini grid.

Step 5: End if.

Step 6: While (empty_cell_exists) repeat step 7 to 8: remaining empty cells are filled.

Step 7: Apply graph traversal algorithm (depth-first search) on all empty cells and repeat step 8. Remaining empty cells are filled one by one on the basis of number of values that can be filled in that empty cell. The empty cell with minimum number of values it can take if first filled by placing value between 1 to grid size.

Step 8: Place a number between 1 to grid size in the empty cell and check if is valid or not repeat steps 9 to 11: checking all values by placing it in the empty cell to find the correct value.

Step 9: If the placed number is not the same row go to step 10 else go to step 12: the values should be unique in the same row.

Step 10: If the placed number is not the same column go to step 11 else go to step 12. The values should be unique in the same column.

Step 11: If the placed number is not the same mini grid go to step 7 else go to step 12: the values should be unique in the same row.

Step 12: Go to step 8: placing another value in the empty cell and again check if it is valid or not.

Step 13: Print feasible solution of the Sudoku board.

Step 14: End.

Table 1 Specification of PARAMSHAVAK

Features	Specification
RAM	64 GB ECC DDR3 1866 MHz RAM in balanced configuration
Operating system	CentOS 6.5
Processor	Dual socket Intel Xeon E5-2600 v2 series/ with 10 cores each with minimum 2.2 GHz clock speed with minimum specification 2006_rate od 580
Accelerator	(i) Intel Xeon Phi 3120A/3120P card (max 2). **Or**
	(ii) NVidia K20/K40 (max 2) series
Number of cores	12 hardware cores with 2 hyper threading on each core which makes 12 × 2 = 24 cores
Hard drive	4 × 2 TB SATA/NL-SAS disks. Total usable space 5.456 TB. 1.455 TB for System, applications and 4.0 TB for home user

PARAMSHAVAK: PARAMSHAVAK is a "Supercomputing Solution in a Box", perfect match to the growing high-end computational needs for Scientific, Engineering and Academic Excel. It provides solution to some of the gigantic computational problems in quick time and at an affordable price. It is an initiative to experience HPC (high-performance computing) power in a true sense. PARAMSHAVAK is installed at Manipal University Jaipur. Some specifications of PRAMASHAVAK are given in Table 1.

5 Experimental Results

The following experimental results have been evaluated and furnished on a single core, quad core, and a 24-core machine, respectively. The grid sizes have been considered as 4 × 4 and 9 × 9 for three different difficulty levels, and the execution time of the program has been observed to degrade with the increase in the number of cores owing to computational bottlenecks for the parallel code (Table 2).

Table 2 Time taken by serial and parallel C code by different number of cores with different grid sizes

Grid size	Difficulty level of Sudoku	Serial C code 1 core	Parallel C code (OpenMP: 4 cores)	PARAMSHAVAK (Serial: 24 cores)
4 × 4	Easy	0m0.002s	0m0.008s	0m0.001s
	Medium	0m0.003s	0m0.012s	0m0.001s
	Hard	0m0.007s	0m0.010s	0m0.001s
9 × 9	Easy	0m0.003s	0m0.009s	0m0.001s
	Medium	0m0.005s	0m0.009s	0m0.001s
	Hard	0m0.004s	0m0.055s	0m0.002s

6 Conclusion and Future Scope

The paper presents a performance analysis of serial and parallel C code for solving the Sudoku maze. The serial code employs the brute force approach for filling the board along with the use of rule-based methods to solve the partially filled Sudoku board. The parallel version of the code first deals with the elimination of naked singles, the hidden pairs, and the triplets, and then the depth-first search is used to find the cell that will be filled first. The time taken by both the codes for solving different Sudoku puzzles with different grid sizes and different levels of difficulty has been evaluated and reported above. Experimental results clearly show that although various tasks are done in parallel on each mini grid, the time taken by the machine to produce a solution for Sudoku puzzle is not as optimal as in case of serial code. The running algorithm at the backend for generating the possible solution involves depth-first search with backtracking. The tree-based depth-first search approach is hard to parallelize as stated in [9–11] as the nature of traversing the solution which is in our case is Sudoku maze state require the threads to wait at the root node. This implies that the children or the possible states of Sudoku cannot be explored simultaneously. The serial code guarantees to find the solution to the Sudoku board if a solution exists because of elimination of empty cells by rule-based methods and then applying brute force to the remaining cells. The serial Sudoku solver is able to solve different levels of difficult puzzle with different grid sizes in the very short span of time.

However, due to combinatorial nature [12, 13] of the problem, it is interesting to investigate it with some heuristics or alternatives to the serial approach as the serial code will significantly slow down as the size of the Sudoku board increases or may not be able to produce the solution due to computational limitations of the machine. In the above study, 4×4 and 9×9 boards have been taken into consideration for which the computational complexity stands out to be nearly $4 \times 4!$ and $9 \times 9!$ So for an $n \times n$ board, the computational complexity will be of order of $n \times n!$ for the serial code. Thus, an attempt was made to break down this computational complexity among the cores but due to heavy internal dependency of the approach, it becomes hard to find distributed regions. It is possible to explore the solution through combinatorial heuristics with some a priori knowledge of the system as discussed in [14]. But still, the algorithmic complexity will remain significantly on the higher end which can then be tackled using the parallel processing power of the modern systems [15–17].

References

1. Abhishek Majumder, Abhay Kumar, Nilu Das, Nilotpal Chakraborty: The Game of Sudoku-Advanced Backtrack Approach. International Journal of Computer Science and Network Security, Vol. 10 No. 8, 255, August (2010).
2. Arnab K. Maji, Sudipta Roy, Rajat K. Pal: A Novel Algorithmic approach for solving Sudoku puzzle in Guessed Free Manner. European Academic Research, Vol. I, Issue 6/ September (2013).

3. Taruna Kumari, Preeti Yadav, Lavina, "Study Of Brute Force and Heuristic Approach to solve sodoku", International Journal of Emerging Trends & Technology in Computer Science (IJETTCS), Volume 4, Issue 5(2), September - October 2015, pp. 052–055, ISSN 2278-6856.
4. Sudoku solving techniques, http://www.su-doku.net/tech.php.
5. The Mathematics of Sudoku, Tom Davis, September 13, 2012.
6. Sudoku Generation Using Human Logic Methods, Halliday rutter. February 17, 2008.
7. Hidden Pairs, Hidden Triples, Hidden Quads Strategy, http://www.thonky.com/sudoku/hidden-pairs-triples-quads.
8. Naked Triple, http://www.sudoku9981.com/sudoku-solving/naked-triple.php.
9. R. Saxena, M. Jain, D.P. Sharma, A. Mundra, "A review of load flow and network reconfiguration techniques with their enhancement for radial distribution network", 2016 Fourth International Conference on Parallel, Distributed and Grid Computing (PDGC), Waknaghat, 2016, pp. 569–574.
10. Dehne F, Yogaratnam K. Exploring the limits of gpus with parallel graph algorithms. arXiv preprint arXiv:1002.4482. 2010 Feb 24.
11. Saxena, R., Jain, M., & Sharma, D. P. (2018). GPU-Based Parallelization of Topological Sorting. In Proceedings of First International Conference on Smart System, Innovations and Computing (pp. 411–421). Springer, Singapore.
12. Hromkovič J. Algorithmics for hard problems: introduction to combinatorial optimization, randomization, approximation, and heuristics. Springer Science & Business Media; 2013 Mar 1.
13. Rahul Saxena, Monika Jain, Siddharth Bhadri, Suyash Khemka. (2017). Parallelizing GA based heuristic approach for TSP using CUDA and OPenMP International Conference on Advances in Computing, Communications and Informatics (ICACCI) (pp. 1934–1940), Volume 6, IEEE.
14. Bello I, Pham H, Le QV, Norouzi M, Bengio S. Neural Combinatorial Optimization with Reinforcement Learning. arXiv preprint arXiv:1611.09940. 2016 Nov 29.
15. Kirk DB, Wen-Mei WH. Programming massively parallel processors: a hands-on approach. Morgan Kaufmann; 2016 Dec 20.
16. Podobas, A. and Brorsson, M., 2015. From software to parallel hardware through the OpenMP programming model.
17. Saxena, R., Jain, M., Singh, D. and Kushwah, A., 2017, October. An enhanced parallel version of RSA public key crypto based algorithm using openMP. In Proceedings of the 10th International Conference on Security of Information and Networks (pp. 37–42). ACM.
18. PARAM SHAVAK-Supercomputing Solution in a Box, http://cdac.in/index.aspx?id=hpc_ss_Param_shavak.

VANET: Security Attacks, Solution and Simulation

Monika Jain and Rahul Saxena

Abstract Vehicular ad hoc network is introduced through MANET in the year of 2000. They are intelligent system which is implemented in the vehicles to provide the luxury, comfort, and security to the peoples. At present, the use of vehicles is growing per year as per user demand. Various researches have been done in VANET through different perspectives, and security issues in VANET are studied and resolved. This paper analyzes the various security issues and their proposed solutions, and the comparative study of this solution is done. Also, the performance of these security solutions is analyzed.

Keywords Scenario · Ad hoc networks · Simulation

1 Introduction

Wireless networks are increasing day by day; there is not a single person who is not influenced by wireless network. Wireless ad hoc network is a kind of network where each node acts as a router. VANET is a special wireless ad hoc network in which the vehicles act as a node and connect to each other to share the information like routing info, weather condition, road condition, collision information, etc.

Every node in the network is armed with WAVE protocol is known as on-board unit (OBU) and fixed part is called road side unit (RSU). Broadly, there are three types of communication in VANET: node to node communication, node to infrastructure, and road side to road side communication. When the network nodes (vehicles) can connect to other network node popularly known as vehicle to vehicle communication and the network nodes (vehicles) communicate to RSU is called vehicle to infrastructure communication. Figure 1 describes the VANET scenario.

M. Jain (✉) · R. Saxena
Manipal University Jaipur, Jaipur, India
e-mail: monikalnct@gmail.com

R. Saxena
e-mail: rahulsaxena0812@gmail.com

© Springer Nature Singapore Pte Ltd. 2018
V. Bhateja et al. (eds.), *Proceedings of the Second International Conference on Computational Intelligence and Informatics*, Advances in Intelligent Systems and Computing 712, https://doi.org/10.1007/978-981-10-8228-3_42

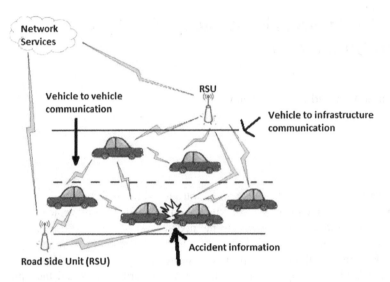

Fig. 1 VANET scenario

The reason behind the popularity of VANET is its characteristics. First, its dynamic nature due to their highly mobile in nature; becuause of this the node predictability is also difficult. Second, the topology of the network is always varying because of the nodes moving at various speeds. Third, sufficient power storage in VANET, since the nodes in VANET is vehicles there is no issue of power consumption. Fourth, the network node shares and accepts the information in a given network, and repeated exchange of routing information is required in VANET [1]. Figure 2 describes the applications of VANET.

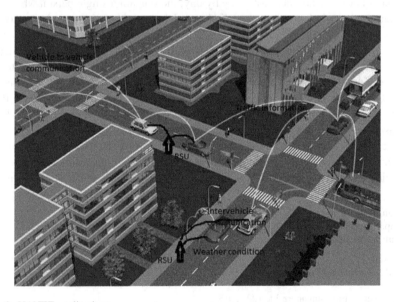

Fig. 2 VANET applications

Some applications of VANET are its safety and comfort application, for example, download the digital map location, safety monitoring, pay toll taxes, detection of surrounding condition, driver support, information of fuel usage, etc. [2]. Security can be a major concern in VANET, and the reason behind security check in VANET is its frequently change of information with other vehicles. If some vehicle is exchanging a malicious information, the whole network can be interrupted and information can be corrupted; this can lead to a serious danger of lives [3].

2 Requirement of Security

Although VANET provides numerous characteristics and benefits for the users, it does not have enough security features implemented at present, and it is vulnerable to the attacks. According to some investigation, it has been found that 60% of collisions can be evaded if the driver gets the notice half a second earlier to the collision [4]. VANET is the answer to that, besides the information shared through network is real. The objective of VANET is to provide the safety and preventing the accidents or loss by informing the driver timely about other vehicles information. The only problem is attacks. Figure 3 describes the block diagram of various types of attacks.

2.1 Routing

These kinds of attacks may attacks on the invincibility of network layer. The malicious node drops the packets or disrupts the network [5].

a. **Black hole attack**

A vehicle claims to have a prime route and attracts other nodes to send their basic information. This could cause the falling of the packet without sending to others [6] as shown in Fig. 4a.

Fig. 3 Block diagram showing the different types of attacks in VANET

Fig. 4 a Black hole attack.
b Gray hole attack. **c** Sybil
attack

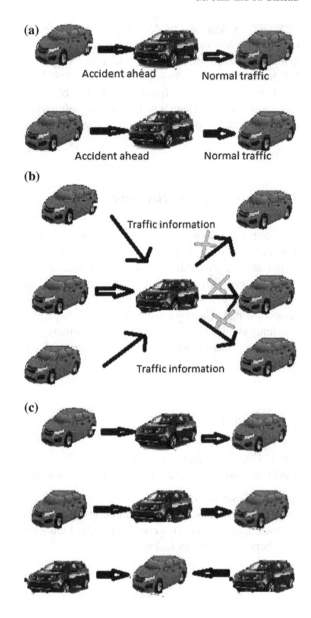

b. **Gray hole attack**

In this attack, the evil node fakes in the region by accepting the information and after receiving the information, the malicious node drops the packets [7] as shown in Fig. 4b.

Fig. 5 Integrity attacks

Normal traffic   Accident ahead

c. **Sybil attack**

It can happen when the node creates a greater number of fake names and behaves like it is larger than hundred nodes to circulate information regarding the traffic and suggests them to follow alternate routes as shown in Fig. 4c.

2.2 Integrity

The message integrity should be preserved in such a way so that attackers cannot alter the information, message contents to be trusted [8]. Alteration attacks come under integrity class. Figure 5 describes the integrity attacks.

2.3 Confidentiality

The messages transmitted between nodes should be prevented from other nodes to read them. The messages should be encrypted such that the eavesdropping of malicious nodes cannot happen, and confidentiality is maintained. Eavesdropping is the common attack on confidentiality. Figure 6 describes the confidentiality attacks.

2.4 Availability

The nodes in VANET available should be all time; this makes the network vulnerable to attacks. The DoS attack, black hole attack, jamming attack, and spamming come under this category.

a. **Denial of service attack**

It is the dangerous one, it mainly attacks availability of network and causes life-threatening effect on vehicles drivers, and this attack is very dangerous even for a small instant of time [9]. Figure 7a describes the denial of service attack.

b. **Jamming**

Jamming attack deliberately transmits radio signals to disturbed the communications by decreasing the signal-to-noise ratio [10]. Figure 7b specifies the jamming attack.

Fig. 6 Confidentiality attacks

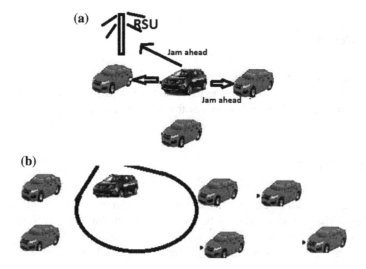

Fig. 7 **a** Denial of service attack. **b** Jamming attack

c. **Spamming**

The messages are transferred so frequently that it can cause the transmission inactivity. It is distributed network, so it is very hard to control [11].

3   Categories of Attackers

Insider

The attacker present in the network and have access and understanding of the network. This attacker is dangerous because it has all the detailed information about the network and can cause damage to it.

Outsider

This kind of attacker is not a participant in the network. However, it tries to damage the network by attacking its network protocols. However, its range is limited in comparison to insider.

- **Active**

  Active attacker generates the malicious packets to disturb the network.
- **Passive**

  This kind of attacker only listen/receive the packets in between the nodes. It does not generate packets.
- **Malicious**

  It aims to harm the nodes and destroy the network without personal intention.
- **Rational**

  It specifically aims to achieve some profit from attack.

4   Security Solution for VANET Attacks

a. **Routing Attack Solutions: SEAD**

   SEAD is the solution for routing attacks in VANET. SEAD, i.e., secure ad hoc routing protocol is based on DSDV protocol. The attacks in VANET cause the consumption of network bandwidth excessively; SEAD is the solution for these kinds of attacks. It uses the one-way based hash function by selecting the random value through the node. After that, the list of values is computed as below: t0, t1, t2… tn, where t0 = x and ti = T (ti − 1) for $0 < I\_n$. For verification, a node with the authenticated value hi can authenticate hi-4 by computing $T(T(T(T(ti − 4))))$ [12].

b. **Eavesdropping Solution**

   When the various vehicles are in a network and communicating with each other, registered vehicles should receive the data transferred between the vehicles only, so that we can prevent the eavesdropping from occurring.

c. **Alteration Attack Solution**

   An illegal manipulation or changes must be detected in the region so that the message between sender and receiver should not be changed.

d. **Denial of Service Attack Solution**

   This attack can be prevented through on-board unit installed in the vehicle nodes. DoS attack consumes the network bandwidth of the system and jams the network so that other nodes cannot transfer the packet. The solution is based on two approaches rate decreasing algorithm and state transition algorithm [12].

e. **Sybil Attack Solution**

There are various solutions introduced till now for Sybil attack like time series clustering algorithm and public authentication key. In the time series approach, the time stamp is noted when the messages are sent to vehicles and between the vehicles. Sybil attack can be identified if many traffic messages contain very similar series of time stamps [13]. In public authentication key, when the sender wants to send the message, it will send the authentication message and encryption key. If this information is validated by the road side unit database, the message is authenticated; otherwise, it can be an attack [14].

5   Simulation Results of VANET

Various number of vehicles is created in network connection has been made through NS2 simulation tool. The AODV algorithm is implemented between the vehicles and road side unit. Table 1 shows the parameters which we have created 3 nodes in first simulation and 12 nodes in second simulation from NS2 as shown in Figs. 8 and 9.

The steps followed for creation of VANET are as follows:

1. Set up the network simulator.
2. Setting the properties of wireless ad hoc network-VANET.
3. Create the topology.
4. Set the number of nodes.
5. Set the road side unit.
6. Set the position of nodes.
7. Set up the timing of start and sink node.

Conclusion and Future Scope

The paper described the overview of VANET and discussed the overview of security attacks in VANET. The different algorithms are also discussed for these security attacks. The simlation has demonstrated on different scenarios of VANET is done along with its parameters and algorithm is demonstrated. In future, we are aiming to evaluate the degradation in performance of VANET scenario with security attacks and how we can improve the performance through these solutions.

Table 1 VANET parameters set in NS2

Parameters	Values
Channel	Wireless channel
MAC	802_11
Packet size	50
Routing protocol	AODV

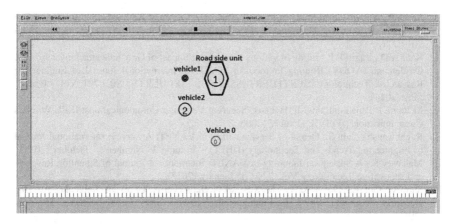

Fig. 8 The simulation of VANET

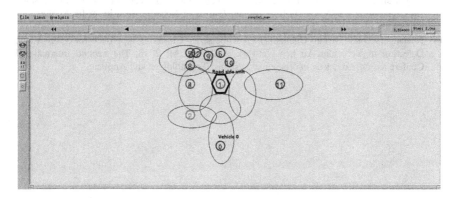

Fig. 9 The simulation of VANET and its range of network

References

1. B. Ducourthial, Y. Khaled, and M. Shawky: Conditional transmissions: Performance study of a new communication strategy in VANET, IEEE Trans. Veh. Technol., vol. 56, no. 6, pp. 3348–3357 (2007).
2. Vishal Kumar1, Shailendra Mishra1, Narottam Chand2: Applications of VANETs: Present & Future, Communications and Network, 2013, 5, 12–15, https://doi.org/10.4236/cn.2013. 51b004, February (2013).
3. M Raya, P Papadimitratos, JP Hubaux: Securing Vehicular Communications, IEEE Wireless Communications, Vol 13, October (2006).
4. Raya M.: The Security of Vehicular Ad Hoc Network: SASN'05, Alexandria, Virginia, USA, pp. 11–21, Nov 7 (2005).
5. Raw R.S., Kumar M., Singh N.: Security challenges, issues and their solutions for VANET, International Journal of Network Security & Its Applications (IJNSA): Vol. 5, No. 5, September (2013).

6. Sirola P.: An Analytical Study of Routing Attacks in Vehicular Ad-hoc Networks (VANETs): International Journal of Computer Science Engineering, ISSN: 2319-7323 Vol. 3 No. 04 Jul 2014 217.
7. Kaur A.D., Ragiri P.R.: Study of various Attacks and Impact of Gray hole attack over Ad-Hoc On demand (AODV) Routing Protocol in MANETS: International Journal of Engineering Research & Technology (IJERT) IJERT ISSN: 2278-0181 IJERTV3IS050721, Vol. 3 Issue 5, May (2014).
8. M Raya, P Papadimitratos, JP Hubaux: Securing Vehicular Communications: IEEE Wireless Communications, Vol 13, October (2006).
9. Raghuwanshi., Jain S.: Denial of Service Attack in VANET: A Survey: International Journal of Engineering Trends and Technology (IJETT) – Volume 28 Number 1 - October (2015).
10. Malebary S.: A Survey on Jamming in VANET: International Journal of Scientific Research and Innovative Technology Vol. 2 No. 1; January (2015).
11. Sari A., Onursal O., Akkaya M.: Review of the Security Issues in Vehicular Ad Hoc Networks (VANET): Int. J. Communications, Network and System Sciences, 552–566, (2015).
12. Adil Mudasir Malla, Ravi Kant Sahu: Security Attacks with an Effective Solution for DOS Attacks in VANET, International Journal of Computer Applications (0975 – 8887).
13. Mina Rahbari1 and Mohammad Ali Jabreil Jamali: Efficient Detection Of Sybil Attack Based On Cryptography In VANET, International Journal of Network Security & Its Applications (IJNSA), Vol. 3, No. 6, November 2011.
14. Neelanjana Dutta, Sriram Chellappan: A Time-series Clustering Approach for Sybil Attack Detection in Vehicular Ad hoc Network, VEHICULAR 2013, The Second International Conference on Advances in Vehicular Systems, Technologies and Applicat.

A Novel Approach for Efficient Bandwidth Utilization in Transport Layer Protocols

Sakshi Garg, Purushottam Sharma and Varsha Singh

Abstract The Internet has in a flash developed into an incomprehensible worldwide system in the developing innovation. TCP/IP Protocol Suite is the fundamental necessity for nowadays Internet. Web use keeps on expanding exponentially. The TCP/IP suite has many plan shortcomings so far as transfer speed, bandwidth utilization, and congestion are concerned. Some of these are protocol outline shortcomings, though rest is deformities in the product that executes the protocols. The real accentuation is on protocol-level issues, as opposed to execution defects. The paper discusses about the packet transmission issues identified with the connection-less and connection-oriented protocols in the transport layer. Subsequently, an approach is proposed for the overcoming of the weaknesses and expanding transmission capacity and decreasing congestion in the network and succeeds in giving the proposed solution complexity as O(n) which is best, that can be practically achieved.

Keywords Internet · TCP/IP protocol · Security · Congestion
Bandwidth utilization · Transmission capacity · Packet retransmission
Connection less · Connection-oriented

S. Garg (✉) · P. Sharma · V. Singh
Amity School of Engineering and Technology, Amity University,
Noida, Uttar Pradesh, India
e-mail: sakshijyotigarg@gmail.com

P. Sharma
e-mail: psharma5@amity.edu

V. Singh
e-mail: varsha.singh502@gmail.com

© Springer Nature Singapore Pte Ltd. 2018
V. Bhateja et al. (eds.), *Proceedings of the Second International Conference
on Computational Intelligence and Informatics*, Advances in Intelligent Systems
and Computing 712, https://doi.org/10.1007/978-981-10-8228-3_43

467

1 Introduction

The Internet Protocol suite [1] is the theoretical model and set of interchanges protocols utilized on the Internet and comparative PC systems. It is generally known as TCP/IP on the grounds that the first protocols in the suite are the Transmission Control Protocol (TCP) [2, 3] and the Internet Protocol (IP). It is sometimes known as the Department of Defense (DoD) model, in light of the fact that the improvement of the systems administration model was supported by DARPA, an organization of the United States Department of Defense [4]. The transport layer [5] in the TCP/IP suite [6] is situated between the application layer and the network layer. It gives administrations to the application layer and gets administrations from the network layer. It gives a procedure-to-process correspondence between two application layers, one at the nearby host and the other at the remote host [7, 8].

Transport layer has different protocols like Sliding Window Scheme [9], Stop-and-Wait Protocol, Go-Back-N, Selective Repeat and ARQ Technique for the above Protocols. These protocols lay controls for the exchange of bundle from sender to beneficiary. Each is a change of the other, however, no protocol clarifies the situation of retransmissions of the packets. Though there exists procedures like timeout based transactions, ARQ Technique and others but each of them deals with the accentuation on the transmission of packets not the retransmission of the similar packet. Such packets either get dropped or they are retransmitted over and over which increment clog at the network and furthermore squander the transmission capacity on the off chance that the parcel needs to hold up until Time to Live. Thus, in the paper a novel way to deal with this hitch in the network is proposed [10].

The principle commitment of this paper is to propose a contemporary approach for the transmission of packets from the sender to the back-to-back recipient which will aid in proficient bandwidth utilization and falling congestion [11, 12] in the system brought about due to the loss of packets and retransmissions. The paper is organized as follows. Section 2 holds the literature review. Section 3 exhibits a relative investigation of different transport layer protocols considering in space of both, connection-less and connection-oriented protocols took after by their working. Section 4 expresses the research gap, proposed solution to it and its preferences. In Sect. 6, results are outfitted for the same. Section 7 contains the conclusions on the exhaustive investigation of the transport layer protocols, the issues distinguished, proposed arrangement, and derivation from it. Ultimately, Sect. 8 furnishes the future scope and bearings.

2   Literature Review

Espina et al., in his paper described the development and the essential usefulness of the TCP\IP protocol. They attempted to uncover the reasons why the main data networks expected to advance to wind up what we know these days as the internet. At long last, a future view is point by point of what the current reviews and tried advances convey as the best answer for bolster the developing requests of the clients and technological upgrades [1]. Jacobson, Van, et al., revised the transport protocol and set forward RTP, the continuous transport protocol that gave end-to-end network transport capacities reasonable for applications transmitting ongoing data, for example, sound, video or recreation data, over multicast or unicast network administrations. The change he proposed was to the scalable timer algorithm for computing when to send RTCP packets keeping in mind the end goal to limit transmission in abundance of the planned rate, when numerous members join a session at the same time in the prior RFC 1889 [2].

Karnati Hemanth et al., gave the TCP/IP suite which has many outline shortcomings as far as security and protection are concerned. In his paper, he concentrated primarily on protocol-level issues, instead of execution defects. In his paper, he examined the security issues identified with the portion of the protocols in the TCP/IP suite [3]. Iren et al., reviewed Transport layer protocols accommodate end-to-end correspondence between at least two hosts. This paper introduced an instructional exercise on transport layer ideas and phrasing, and an overview of transport layer administrations and protocols. The administration and protocol elements of twelve of the most imperative protocols were condensed in his paper [5].

Randall et al., analyzed that a few applications as of now require more noteworthy usefulness than what either TCP or UDP brings to the table, and future applications may require considerably more. Like TCP, STCP offers an indicate point, connection-oriented, dependable conveyance transport benefit for applications imparted over an IP network. Perceiving that different applications could utilize a portion of the new protocol's capacities for call control motioning in voice-over (VoIP) networks, the IETF now holds onto SCTP as a broadly useful transport-layer protocol, joining TCP and UDP over the IP layer [6]. S. M. Bellovin, evaluated the TCP/IP protocol suite, which is broadly utilized today, was produced under the sponsorship of the Department of Defense. In spite of that, there are various genuine security flaws intrinsic in the protocols, regardless of the rightness of any executions. He depicted an assortment of attacks in light of these flaws, including sequence number spoofing, routing attacks, source address spoofing, and verification attacks and furthermore introduced protections against these attacks, and closed with a talk of wide range resistances, for example, encryption [7].

Balwinder Kaur et al., explained the data can get lost, reordered, or copied because of the nearness of switches and support space over the inconsistent divert in the traditional networks. The sliding window protocol will recognize and rectify blunder if they got data have enough repetitive bits or rehash a retransmission of data. The paper demonstrated the working of this duplex protocol of data link network [9]. Prabhaker Mateti, explained the TCP/IP suite has many outline shortcomings so far as security and protection are concerned. Some of those are protocol plan shortcomings, while the rest are imperfections in the product that executes the protocols. In his paper, he depicted these issues from a useful point of view [10].

Purvang Dalal et al., conferred that the Transmission Control Protocol (TCP), an imperative transport layer correspondence protocol, is ordinarily tuned to perform well in customary wired networks, where Bit Error Rate (BER) is low and congestion is the essential driver of packet misfortune. He portrayed the issue by the conduct of wireless links and the parts of TCP operation that influence execution lastly, his report investigated a sorted examination of various existing arrangements similarly, as it is hard to make a "one size fits all" TCP for wireless networks [13]. M. Anand Kumar et al., examined that the network and Internet applications are developing quickly in the current past and the current security system was not satisfactory for today's applications. Since there is no protection for the application layer of the network model. He proposed security engineering for the TCP/IP Protocol Suite talked over Internet use keeps on expanding exponentially. So network security turns into a developing issue. Also, he introduced another design for TCP/IP protocol suite which ensures security to application layer utilizing a protocol Application Layer Security Protocol (ALSP) [14].

3   Various Transport Protocols

The TCP/IP protocol utilizes a transport layer protocol [15] that is either an adjustment or a mix of some of these protocols.

3.1   Sliding Window

Since the arrangement numbers utilize modulo 2m, a circle can speak to the grouping numbers from 0 to 2m − 1. The buffer is taken as an arrangement of slices, called the sliding window that involves some portion of the circle at any instance. At the sender site, when a packet is sent, the comparing slice is stamped. At the point when the slices are marked, it implies that the buffer is full and no further messages can be acknowledged from the application layer. At the point when an acknowledgment arrives, the comparing slice is unmarked. In the event that some successive slices from the earliest starting point of the window are unmarked, the window slides over the scope of the comparing arrangement numbers to permit all the more free slices toward the finish of the window. The succession numbers are in modulo 16 (m = 4) and the measure of the window is 7.

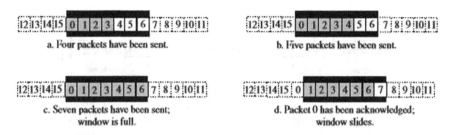

Fig. 1 Sliding window in linear format (*Source* [4])

Take note of that the sliding window is only a reflection: the genuine circumstance utilizes PC factors to hold the grouping quantities of the following packet to be sent and the last packet sent [4, 9] (Fig. 1).

3.2 Simple Protocol

Our first protocol is a basic connection-less protocol with neither stream nor mistake control. We accept that the recipient can quickly deal with any packet it gets. As it were, the collector can never be overpowered with approaching packets. Figure 4 demonstrates the format for this protocol (Fig. 2).

The transport layer at the sender gets a message from its application layer, makes a packet out of it, and sends the packet. The transport layer at the collector gets a packet from its network layer, removes the message from the packet, and conveys the message to its application layer. The transport layers of the sender and collector give transmission administrations to their application layers [2, 4].

3.3 Stop-and-Wait Protocol

Our second protocol is a connection-oriented protocol [11, 12] called the Stop-and-Wait protocol, which utilizes both stream and mistake control. Both the sender and the recipient utilize a sliding window of size 1. The sender sends one packet at any given moment and sits tight for an acknowledgment before sending

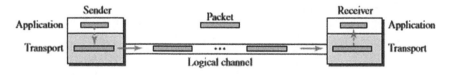

Fig. 2 Simple protocol (*Source* [4])

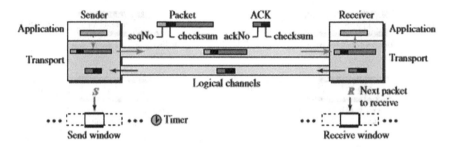

Fig. 3 Stop and wait protocol (*Source* [4])

the following one. To distinguish undermined packets, we have to add a checksum to every data packet. At the point when a packet reaches the collector site, it is checked. In the event that its checksum is off base, the packet is debased and noiselessly disposed of. The hush of the recipient is a flag for the sender that a packet was either ruined or lost. Each time the sender sends a packet, it begins a clock, if an acknowledgment arrives before the clock terminates, the clock is halted and the sender sends the following packet (on the off chance that it has one to send). On the off chance that the clock terminates, the sender resends the past packet, expecting that the packet was either lost or adulterated. This implies the sender needs to keep a duplicate of the packet until its acknowledgment arrives [4] (Fig. 3).

3.4 Go-Back-N Protocol (GBN)

To enhance the effectiveness of transmission [6, 14, 15] (to fill the pipe), numerous packets must be experiencing significant change while the sender is sitting tight for affirmation. As it were, we have to give more than one packet a chance to be extraordinary to keep the channel occupied while the sender is sitting tight for acknowledgment. In this segment, we examine one protocol that can accomplish this objective; in the following segment, we talk about a moment. The first is gotten back to Go N (GBN) (the sound of the name will turn out to be clear later). The way to Go-back-N is that we can send a few packets before accepting affirmations, yet the beneficiary can just buffer one packet. We keep a duplicate of the sent packets until the acknowledgments arrive, however, few data packets and acknowledgments can be in the channel in the meantime (Fig. 4).

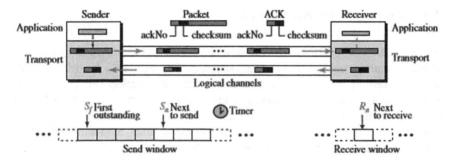

Fig. 4 Go-Back-N protocol (GBN) (*Source* [4])

3.5 Selective Repeat

The Go-Back-N protocol improves the procedure at the recipient. The beneficiary monitors just a single variable, and there is no compelling reason to buffer out-of-request packets; they are essentially disposed of. Be that as it may, this protocol is wasteful if the hidden network protocol loses a ton of packets. Each time a solitary packet is lost or adulterated, the sender resends every single extraordinary packet, despite the fact that some of these packets may have been gotten sheltered and sound however out of request. On the off chance that the network layer is losing numerous packets in view of congestion in the network, the resending of these extraordinary packets aggravates the congestion, and in the end, more packets are lost. This has a torrential slide impact that may bring about the collapse of the network. Another protocol, called the Selective Repeat (SR) protocol, has been concocted, which, as the name infers, resends just particular packets, those that are really lost [4] (Fig. 5).

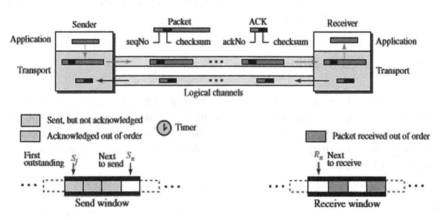

Fig. 5 Selective Repeat protocol (*Source* [4])

3.6 Automatic Repeat Request

Automatic Repeat reQuest (ARQ), otherwise called Automatic Repeat Query, is a mistake control strategy for data transmission that utilizes acknowledgements (messages sent by the beneficiary demonstrating that it has effectively gotten a data casing or packet) and timeouts (indicated time frames permitted to slip by before an acknowledgement is received) to accomplish solid data transmission over untrustworthy service. ARQ protocols can be classified into Stop-and-wait ARQ, Go-Back-N ARQ, and Selective Repeat ARQ/Selective Reject [7, 10].

4 Proposed Solution

The problem statement and solution to the problem identified can be given as follows.

4.1 Research Gap

(a) Factors after various literature surveys and findings it is seen that there is an anomaly in the transport layer protocols that although the protocols are efficient enough to send the packet from source to destination and each protocol has the revised features of the previous ones but none of the protocols takes into account the time being wasted to retransmit the packet that could not be sent due to network congestion, or loss of response, loss of acknowledgement, etc.
(b) It is necessary to know the time spent (rather the no. of the times the packet is retransmitted) to send the same packet again and again since it is wasting the bandwidth and creating congestion over the network by transmitting the same packet over and over again. Consequently, there exists no such framework that suggests any such policy.

4.2 Proposed Solution

Many researchers have progressed with plentiful techniques and algorithms to examine the security issues and other transmission related schemes. The emphasis here is on scrutinizing the gap analyzed during the literature reviews in the networking domain, which can be further explored for efficient retransmission of packets and effective utilization of bandwidth (Fig. 6).

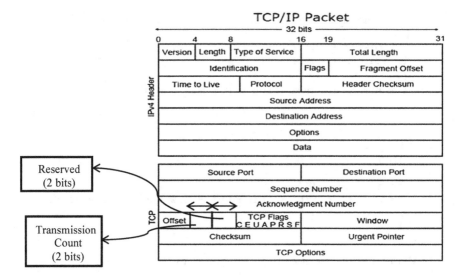

Fig. 6 Modified TCP/IP packet

5   Algorithm

Transmission Count

```
if (int m==0)
{
 trans_count=0;
}
else
{
while (m<15)
{
trans_count++;
}
delay();
}
```

Clock Delay

```
void delay(unsigned int result)
{
clock_t goal = result + clock();
while (goal > clock());
}
```

Random_Number Generator

```
int random_number (int min_num, int max_num)
{
int result = 0, low_num = 0,    high_num= 0;
if (min_num < max_num)
{
low_num = min_num; high_num = max_num + 1;
}
else
{
low_num = max_num + 1;
high_num = min_num;
}
srand (time(NULL));
 result = (rand() % (high_num - low_num)) + low_num;
return result;
}
```

//*Min to max range can be taken between (0 to 10)*//

5.1   Complexity of the Algorithm: O(n)

Since there can be n-number of such packets that have to be retransmitted over the network and each packet takes O(1) time complexity where m is the concerned packet to be retransmitted so, the overall complexity comes out to be O(n). This is the least time complexity one can achieve for this function for n-number of the packets to be retransmitted.

5.2   Justification of the Study

The solution proposed is such that it is taking 2 bits from the RESERVED field of the TCP/IP frame format and introducing an additional field of 2-bits called as transmission count.

This transmission count will keep an entry of the no. of times the packet is retransmitted that is not received at destination due to network reasons like congestion, loss of response, loss of acknowledgment etc.

If the no. of transmissions for the retransmitted packet exceeds 4, the packet shall wait for some random time that can be given by some Probability (p).

This will greatly reduce the traffic at the network by decreasing the no. of retransmission packets at the network and therefore, help in reducing network congestion.

Since definite and less attempts have to be made at a time to send the retransmission packet; this will be a great aid in reducing the wastage of the bandwidth.

6   Results and Discussions

6.1   Increased Bandwidth Utilization

Each time a packet is transmitted from source to goal, in existing cases the packet gets dropped if there should be an occurrence of network congestion and keeps up no mean numerous transmissions of a similar packet in the event that it cannot be conveyed at the primary example, thus data transfer capacity is squandered in it.

Utilizing the proposed calculation, since number of retransmission for the rehashed packets named as transmission count are fixed, so once that count is over then the packet will be transmitted simply after a random time with probability p. In this manner, in the network that packet will be sent simply after the irregular time with likelihood p which will diminish activity at the network, thus more data transfer capacity will be accessible for more packets to be transmitted. Consequently, it will give increased bandwidth utilization.

6.2   Less Network Congestion

In the network that packet will be sent simply after the arbitrary time with likelihood p after it has striven for one total round of Transmission Count, which will decrease activity at the network, thus will prompt less congestion in the network.

6.3   More Robust

Robust means the ability to withstand failures. Failures may occur due to network congestion which shall be recovered using the proposed model.

6.4 Reliable

When the packet cannot be sent at the first instance, it will be tried to send multiple times without keeping a count of number of retransmissions happens to send that packet and also in case of much network congestion the packet will be dropped which means some valuable information may be lost in the existing scenario. But the catch in the proposed approach is that since it is keeping a track of how many times the packet is retransmitted, assume that the packet is not transmitted in the first round of transmission count, then it will be transmitted after a random time with probability p which means the packet will not be dropped, it will be tried after that random time, that is no information will be lost, hence it will be more reliable.

6.5 Higher Efficiency

The overall complexity of the pseudocode comes out to be $O(n)$ which is very less for n-number of packets to be the retransmitted as compared to the existing scenario. So, it has higher efficiency.

7 Conclusion

The current system takes after transmission of packets from source to goal where the packet gets dropped if there should arise an occurrence of network congestion and maintains no count for multiple transmissions of the same packet if it cannot be delivered at the first instance. While in the proposed approach, the packet will be sent simply after the arbitrary time with likelihood p which will decrease movement at the network, thus more transmission capacity will be accessible for more packets to be transmitted, less network congestion, more reliable, more robust, and higher effectiveness. Additionally, the general intricacy of the pseudocode is $O(n)$, which is very less for n-number of packets to be retransmitted.

8 Future Scope and Directions

The future work of the proposed model is the simulation of the proposed algorithm using NS-2 tool.

References

1. Espina, David, and Dariusz Baha. "The present and the future of TCP/IP."
2. Jacobson, Van, et al. "RTP: A transport protocol for real-time applications." (2003).
3. Karnati Hemanth et al, "Security Problems and Their Defenses in TCP/IP Protocol Suite", International Journal of Scientific and Research Publications, Volume 2, Issue 12, December 2012, ISSN 2250-3153.
4. Forouzan, A. Behrouz. Data communications & networking (sie). Tata McGraw-Hill Education, 2006.
5. Iren, Sami, Paul D. Amer, and Phillip T. Conrad. "The transport layer: tutorial and survey." ACM Computing Surveys (CSUR) 31.4 (1999): 360–404.
6. Stewart, Randall, and Christopher Metz. "SCTP: new transport protocol for TCP/IP." IEEE Internet Computing 5.6 (2001): 64–69.
7. Bellovin, Steven M. "A look back at" security problems in the tcp/ip protocol suite." Computer Security Applications Conference, 2004. 20th Annual. IEEE, 2004.
8. Yongguang Zhang, Malibu, C.A. "A multilayer IP security protocol for TCP Performance enhancement in wireless networks", IEEE Journal on Selected areas in communication, 22(4), 767–776 (2004).
9. Kaur, Balwinder, et al. "Importance Of Sliding Window Protocoll." International Journal of Research In Engineering And Technology, EISSN: 2319-1163| PISSN: 2321-7308 (2013).
10. Mateti, Prabhaker. "Security issues in the TCP/IP suite." Security in Distributed and Networking Systems", World Scientific Pub Co Inc (2007): 3–30.
11. Sahu, Yaminee, and Sumit Sar. "Congestion Control Analysis in Network: A Literature Survey." (2016).
12. Chaudhary, Pooja, and Sachin Kumar, "A Review of Comparative Analysis of TCP Variants for Congestion Control in Network", International Journal of Computer Applications 160.8 (2017).
13. Dalal, Purvang, and K. S. Dasgupta. "TCP performance issues and related enhancement schemes over wireless network environment." International Journal 2.4 (2012).
14. Kumar, M. Anand, and S. Karthikeyan. "An Enhanced Security for TCP/IP Protocol Suite." Journal of Computer Science and Mobile Computing, Vol.2 Issue. 11, November-2013, pg. 331–338.
15. Abdelsalam, Ahmed, et al. "TCP Wave: A new reliable transport approach for future internet." Computer Networks 112 (2017): 122–143.

Functional Link Artificial Neural Network-Based Equalizer Trained by Variable Step Size Firefly Algorithm for Channel Equalization

Archana Sarangi, Shubhendu Kumar Sarangi,
Madhurima Mukherjee and Siba Prasada Panigrahi

Abstract In this work, FLANN structure is presented which can be utilized to construct nonlinear channel equalizer. This network has a modest structure in which nonlinearity is instigated by the functional expansion of input pattern by trigonometric and Chebyshev polynomials. This work also defines evolutionary approaches coined as firefly algorithm (FFA) along with modified variable step size firefly algorithm for resolving channel equalization complixeties using artificial neural network. This paper recapitulates techniques with simulated results acquired for given channel with certain noise conditions and justify the efficacy of proposed FLANN-based channel equalizer using VSFFA over FFA and PSO in terms of MSE curves and BER plots.

Keywords Equalization · Trigonometric functional link ANN
Chebyshev functional link ANN · Particle swarm optimization
Firefly algorithm · Variable step size firefly algorithm

1 Introduction

One of the typical signal processing hitches is distortion of the transmitted signal by channel before reaching the receiver. Due to this effect of distortion, overlapping of symbols takes place, which is coined as intersymbol interference. For time-varying channels, equalization schemes must be adaptive [1]. Because of nonlinear potentiality, artificial neural networks turned out to be a potent gadget for composite applications, i.e. functional approximation, system identification, etc. [2]. Functional link ANN initially projected by Pao which can be utilized for function approximation and pattern organization in simplicity and minor computational

A. Sarangi (✉) · S. K. Sarangi · M. Mukherjee
ITER, Siksha 'O' Anusandhan University, Bhubaneswar, India
e-mail: archanasarangi24@gmail.com

S. P. Panigrahi
Department of Electrical Engineering, VSSUT, Burla, India

© Springer Nature Singapore Pte Ltd. 2018
V. Bhateja et al. (eds.), *Proceedings of the Second International Conference on Computational Intelligence and Informatics*, Advances in Intelligent Systems and Computing 712, https://doi.org/10.1007/978-981-10-8228-3_44

intricacy than MLP network. This network gives worldwide approximation competency and quicker convergence. This work implies on FLANN with trigonometric and Chebyshev polynomial expansion in which evolutionary learning algorithms are applied. Gradually weights are updated with diverse adaptive biological-inspired algorithms, i.e. firefly algorithm and its modified versions which when applied to nonlinear channels [3].

This paper is arranged as follows: Sect. 1 labels concise introduction. Section 2 defines channel equalization concept. Section 3 describes multilayer perceptron network. Sections 4 and 5 define trigonometric and Chebyshev expansion-based FLANN. Section 6 discusses variants of evolutionary algorithms. Finally, paper finishes with simulated results along with conclusion and references in further sections.

2   Equalization

In telecommunication, equalization is a setback of distortion sustained by a signal transmitted through medium [1]. Equalizers are utilized to extract frequency response. In communication medium, acknowledged data is identical to conveyed but not for real communication, where signal misrepresentations take place.

Channel equalization is an imperative subsystem in the receiver. Equalization is in practice utilized to eradicate intersymbol interference fashioned due to the restricted bandwidth of broadcast medium. When the channel is band limited, symbols conveyed through will be disseminated. This preceding symbol tends to obstruct with next symbols. Also, the multipath response in wireless infrastructures grounds interference at the receiver. ISI can be considered by Finite Impulse Response filter. At receiver, distortion must be recompensed to reorganize conveyed symbols (Fig. 1).

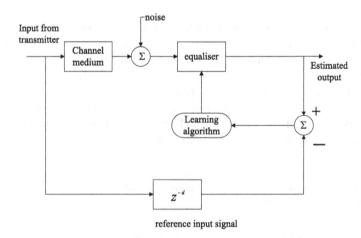

Fig. 1 Channel equalization in digital communication system [1]

3 Multilayer Perceptron Network-Based Equalizer

In MLP, input signal directed through the network on layer-by-layer basis in forward direction. The network is properly trained in supervised custom with a standard algorithm called error backpropagation algorithm. This equalizer consists of the individual level of input with hidden levels set and associated individual output level. Nodes are interrelated by weights. Signal established consecutively is validated to transmit through hidden levels up to node in output level [1]. The outcome of each node is calculated as a weighted summation of nodes outcome in preceding stage and exaggerated by an activation function. Weights are updated by mean square error (MSE) descents beneath the chosen threshold level or when extreme repetitions are proficient. The error signal is utilized to modernize weights and thresholds of hidden levels as well as outcome level.

4 Trigonometric Functional Link ANN

MLP, multilayered network with backpropagation algorithm is quite complex due to which FLANN comes into play. In FLANN, input pattern is functionally boosted by a set of linearly liberated functions preferred from an orthonormal basis set. These functional values are multiplied with equivalent masses and sum to yield output. This output is linked with desired signal and error is fed to adaptive learning algorithm to update masses. Pao has discovered learning and oversimplification features of haphazard vector FLANN and associated with MLP structure skilled by propagation algorithm by captivating limited functional approximation complications. Mostly, two types of functional expanders are used, i.e. trigonometric expansion and Chebyshev expansion [3].

Features of each input x are extracted using trigonometric expansion shown as

$$\text{i.e.} [x1 \cos(\pi x1) \; \sin(\pi x1) \ldots x2 \; \cos(\pi x2) \quad \sin(\pi x2) \ldots x1x2]^{\mathrm{T}}. \tag{1}$$

Let $\mathbf{x}$ be the input course of size $N \times 1$ which embodies N components; kth component is set by

$$x(k) = x_k, \, 1 \le k \le N \tag{2}$$

Functional expansion of component x_k by power series expansion is approved by means of equation set in,

$$s_i = \begin{cases} x_k & \text{for } i = 1 \\ x_k^i & \text{for } i = 2, 3, 4, \ldots, M \end{cases} \tag{3}$$

For trigonometric expansion (Fig. 2),

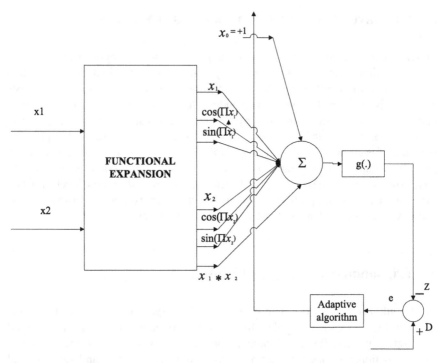

Fig. 2 A FLANN structure with trigonometric expansion [1]

$$s_i = \begin{cases} x_k & \text{for } i = 1 \\ \sin(l\varPi\, x_k) & \text{for } i = 2, 4, \ldots, M \\ \cos(l\varPi\, x_k) & \text{for } i = 3, 5, \ldots, M+1 \end{cases} \tag{4}$$

Elongated components of input vector $\mathbf{E}$, is signified by $\mathbf{S}$ of size $N \times (M + 1)$. Bias input is unity. So additional unity assessment is expanded with $\mathbf{S}$ matrix and dimension of $\mathbf{S}$ matrix leads to $N \times Q$. Single-layer neural network is reflected as a substitute methodology to overwhelmed problems that are linked with multilayer neural network [2, 3].

5 Chebyshev Functional Link ANN

Chebyshev polynomials are a set of orthogonal polynomials demarcated as a solution to Chebyshev differential equation. Pros over trigonometric FLANN is that Chebyshev polynomials are numerically much proficient than trigonometric polynomials to magnify input space. Chebyshev artificial neural network (ChNN) is comparable to FLANN. Only dissimilarity being that, in FLANN input signal is

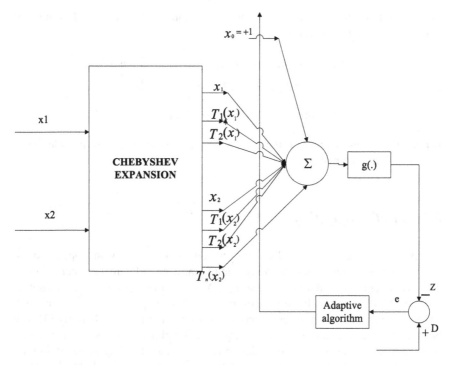

Fig. 3 A FLANN structure with Chebyshev expansion [3]

lengthened to higher dimension using functional expansion. Chebyshev polynomials engendered using recursive formula is specified as (Fig. 3)

$$S_{n+1} = 2x\, S_n(x) - S_{n-1}(x).$$ (5)

Some Chebyshev polynomials are specified as

$$S_0(x) = 1, \quad S_1(x) = x, \quad S_2(x) = 2\,x^2 - 1, \quad S_3(x) = 4\,x^3 - 3x.$$ (6)

The weighted aggregate of components of heightened input is conceded through a hyperbolic tangent nonlinear function to yield desired output.

6 Computational Techniques

6.1 Particle Swarm Optimization (PSO)

PSO is a standard algorithm of population, where a lot of particles fly in the region of search space to determine the superior result. So, particles, regard their individual finest solutions as well as utmost brilliant solution till established. PSO was numerically modeled as follows [4]:

$$velc(t+1) = w \cdot velc(t) + p_1 \cdot rand \cdot (pbest(t) - posc(t)) \\ + p_2 \cdot rand \cdot (gbest(t) - posc(t)) \tag{7}$$

$$posc(t+1) = posc(t) + velc(t+1) \tag{8}$$

where t is existing number of iterations, velc(t) is velocity vector of particles, posc (t) is position vector of particles.

6.2 Firefly Algorithm (FFA)

Fireflies flickering is an incredible observation in summer sky in tropical zones. The flashes are crucial for enchanting mating partners and demand potential prey. The fundamental procedure of FFA are: (1) A firefly might be enchanted to another in spite of their neighborhood (2) For any two flickering fireflies, lighter one will enchant brighter one. But in case of nonexistence then, it moves arbitrarily [5]. The brightness of fireflies can be related to objective function estimation in an inversely proportional manner when global minima is considered. The brightness of a firefly at definite position x is preferred as $G(x)$. For normal cases, light absorption coefficient γ, light intensity G fluctuates for parameter of distance r as

$$G = G_0 e^{-\gamma r^2} \tag{9}$$

where G_0 is original radiant intensity at r = 0.

Distance connecting any two fireflies i and j can be specified as

$$r_{ij} = \|x_i - x_j\| = \sqrt{\sum_{k=1}^{D} (x_{i,k} - x_{j,k})^2} \tag{10}$$

The progress of ith firefly is occupied to one more striking jth firefly and is represented by

$$x_i = x_i + \beta_0 e^{-\gamma r_{ij}^2} (x_j - x_i) + \alpha \left(rand - \frac{1}{2} \right). \tag{11}$$

6.3 Variable Step Size Firefly Algorithm (VSFFA)

Basic firefly uses an invariable step size α. But the investigation of existent solution area requires a big step size, but generally not supportive for convergence to the globally most favourable solution. If step size value is small, the result is dissimilar [6]. Taking this, dynamic modifying model of step size α was selected and can be described as

$$\alpha(t) = 0.4/(1 + \exp(0.015*(t - \max generation)/3)) \tag{12}$$

where t is actual iteration number, Max generation is maximal number of iterations.

7 Simulation Results

An innovative methodology of training trigonometric and Chebyshev FLANN structures using evolutionary algorithms were implemented. A nonlinear channel was taken for testing algorithm's proficiency. The total population was taken as 30. The noise of 20 dB was introduced into the channel. Parameters for PSO are p1 = p2 = 1 and for firefly algorithm are $\alpha = 0.2$, $\beta = 1$, $\gamma = 1$ and maximum iteration was set to 300.

Channel measured for equalization has transfer functions, i.e.

$$H(z) = 0.2600 + 0.9300z^{-1} + 0.2600z^{-2} \tag{13}$$

Figure 4 shows the joint plot of 3 algorithms, for the given channel. The MSE value for PSO is 5.089e-005 with 150 no. of iterations whereas value of MSE for FFA is 3.223e-005 with 99 no. of estimation iterations. But VSFFA showed the best reduction in MSE with 1.379e-005 for 287 no. of iterations.

BER values with SNR = 20 dB for PSO, FFA, and VSFFA trained trigonometric FLANN equalizer at 10 dB are, $10^{-1.002}$, $10^{-1.148}$, $10^{-1.24}$ and at 20 dB are $10^{-1.624}$, $10^{-2.173}$, $10^{-2.552}$ respectively. Hence, we noticed that at 10 dB, VSFFA-trained FLANN equalizer illustrates better BER performance than that of other algorithm's trained equalizer. Similarly, when we compare it at the SNR (dB) = 20, VSFFA-trained FLANN equalizer illustrates better BER performance than other two. When BER values for both dBs are compared, VSFFA-trained equalizer (at SNR = 20) outperforms VSFFA-trained equalizer (at SNR = 10) (Fig. 5).

Figure 6 shows the joint plot of 3 algorithms, for given channel of Chebyshev FLANN. MSE for PSO algorithm is 9.173e-006 with 290 no. of iterations. But MSE value for FFA is 4.816e-006 with 178 no. of iterations. As compared to PSO and FFA, VSFFA algorithm shown a better performance with a MSE of 2.1e-007 for 62 no. of iterations.

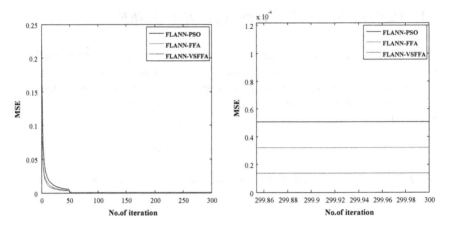

Fig. 4 Convergence curve of trigonometric FLANN

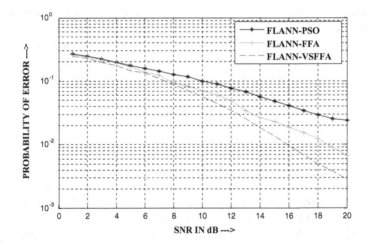

Fig. 5 BER plot for trigonometric FLANN equalizer

BER values with SNR = 20 dB for PSO, FFA, and VSFFA-trained Chebyshev FLANN equalizer at 10 dB are, $10^{-1.001}$, $10^{-1.162}$, $10^{-1.29}$, and at 20 dB are $10^{-1.655}$, $10^{-2.23}$, $10^{-2.8}$ respectively. Hence, we noticed that at 10 dB, VSFFA-trained FLANN equalizer illustrates better BER performance than that of other algorithm's trained equalizer. Similarly, when we compare it at SNR(dB) = 20, VSFFA-trained FLANN equalizer illustrates better BER performance than other two. When BER values for both dB's are compared, VSFFA-trained equalizer (at SNR = 20) outperforms VSFFA-trained equalizer (at SNR = 10) (Fig. 7).

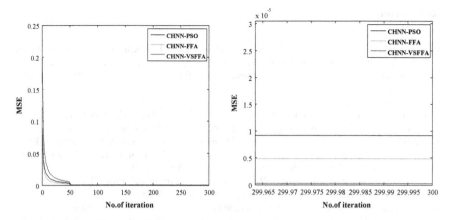

Fig. 6 Convergence curve of Chebyshev FLANN

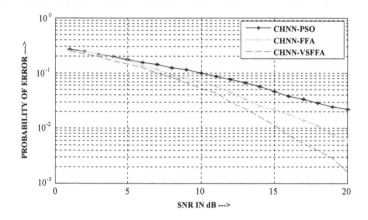

Fig. 7 BER plot for Chebyshev FLANN equalizer

8 Conclusion

In this work, a single-layer trigonometric expansion-based FLANN and Chebyshev expansion-based FLANN network have been projected for weight updation. Extensive simulation studies are carried out and performances of FLANN networks are compared on the basis of applied evolutionary optimization algorithms. From simulation results, it is proved that VSFFA-trained FLANN networks yields optimum outcomes than that of other two algorithms trained by FLANN networks.

References

1. Sandhya Yogi, K.R. Subhashini, J.K. Satapathy, "A PSO Functional link artificial neural network training algorithm for equalization of Digital Communication Channels", 5th International Conference on Industrial and Information Systems, ICIIS 2010, Jul 29–Aug 01, 2010.
2. Ravi Kumar Jatoth, M. S. B. Saithej Vaddadi, Saladi S. V. K. K. Anoop, "An Intelligent Functional Link Artificial Neural Network for Channel Equalization", Proceedings of the 8th WSEAS International Conference on Signal Processing, Robotics and Automation.
3. Jagdish C. Patra, and Alex C. Kot, "Nonlinear Dynamic System Identification Using Chebyshev Functional Link Artificial Neural Networks", IEEE Transactions on Systems, Man, and Cybernetics—Part B: Cybernetics, vol. 32, no. 4, August 2002.
4. J. Kennedy and R. C. Eberhart, "Particle swarm optimization," Proc. IEEE International Conference on Neural Networks, vol. 4, pp. 1942–1948, 1995.8.
5. X.S. Yang, Firefly algorithms for multimodal optimization. In: Proceedings of the 5th International Conference on Stochastic Algorithms Foundations and Applications, vol.5792. LNCS Springer, pp. 169–178, 2009.
6. Shuhao Yu, Shenglong Zhu, Yan Ma, Demei Mao, "A variable step size firefly algorithm for numerical optimization", Applied Mathematics and Computation, 263, pp. 214–220, 2015.

Real-Time FPGA-Based Fault Tolerant and Recoverable Technique for Arithmetic Design Using Functional Triple Modular Redundancy (FRTMR)

Shubham C. Anjankar, Ajinkya M. Pund, Rajesh Junghare and Jitendra Zalke

Abstract Single Event Upset (SEU) is a serious issue when considering the real-time process for critical time constraint applications. Scaling of the devices, in complex computing devices is sensitive to transient faults. Transient faults are not permanent but it causes the critical issues in real-time applications by flipping its bits. The proposed technique is an approach toward improving fault tolerance of the field-programmable gate array (FPGA) for Single Event Upset (SEU) soft error. The technique uses the Functional Fault tolerance and Recovery Technique using the Triple Modular Redundancy (TMR). Functionality and reliability are tested on ALTERA Cyclone III device by modeling 4-bit adder allow identifying and recovering soft error caused by Single Event Upset (SEU). The fault is injected and recovery time for the fault detection and restoring is less than 6 ns with the efficiency of 100% and 98% for single and multiple bit faults respectively.

Keywords Single event upset · Triple modular redundancy · SRAM FPGA

S. C. Anjankar (✉) · A. M. Pund · R. Junghare
Department of Electronics Engineering,
Ramdeobaba College of Engineering and Management, Nagpur, India
e-mail: shubhca90@gmail.com

A. M. Pund
e-mail: pundajinky@yahoo.com

R. Junghare
e-mail: rajesh.jhunghare@gmail.com

J. Zalke
Department of Electronics Design Technology,
Ramdeobaba College of Engineering and Management, Nagpur, India
e-mail: zalkej@rknec.edu

© Springer Nature Singapore Pte Ltd. 2018
V. Bhateja et al. (eds.), *Proceedings of the Second International Conference on Computational Intelligence and Informatics*, Advances in Intelligent Systems and Computing 712, https://doi.org/10.1007/978-981-10-8228-3_45

1 Introduction

Embedded systems with the multicore platform have emerged to be powerful and popular computing engines for much safety-critical application such as aviance, process control, and patient life supporting monitoring [1, 2–7]. The reliability has become a critical issue due to the increased complexity of the functions and architecture of electronics in space and avionics, with the rapid growth of science and technology. For designing embedded systems where the hard real-time operation, high reliability in the presence of transient faults, and low energy consumption are required which are offered by high-performance computing architectures, which have been employed for embedded applications offer new considerable opportunities for designing embedded systems [1]. Due to many factors, like operations in a harsh environment as encountered in space or nuclear applications, embedded systems with multicore platform especially those based on FPGA, are prostrate to transient faults. The reliability requirements are incompatible objectives for embedded system; it should be equipped with peculiar error detection and meets the timing constraint limiting factor [8, 9]. For many Very Large-Scale Integration (VLSI) systems, arithmetic operation such as addition is a most fundamental and required operation for Digital Signal Processors (DSPs) and microprocessors and microcontrollers. In a complex arithmetic circuit like multiplication, division, and address calculation, full adders are the nucleus of any system and used for a variety of operations [10, 11, 12].

For hardening circuits implemented and application on SRAM-based FPGAs, the most widely accepted and granted one is TMR. The feedback route of sequential logic circuits such as flip-flops (FFs) as well as the logic gates and circuits in combinational as well as sequential circuits, for digital circuits mapped on FPGAs, need to be hardened [8, 12, 13]. Redundancy, system redundancy, module redundancy, or logic element redundancy can apply TMR based on these different granularities. Finer the granularity higher the probability. For example, the original embedded system or process is replicated thrice and the output is gathered from the majority voter which votes the error consisting correct output is gathered from each system or process which is addressed as processor block [14, 13]. If the SEU causes the transient fault in one of the as processor block of the 4-bit adder, the TMR votes out the particular bit and mask it and propagates the corrected output as shown in the Fig. 1. TMR system having resistance against SEU, and can harden the embedded system or processor without affecting its proper normal operation provided with this method. However, TMR can resist and mask the single SEU at any particular time instant. For two faults also it can withstand but the output is not promising one. For this reason. logic element-level TMR could be applied. Allowing every logic element to flip or tolerate one failure for each logic element, which includes the logic gate and the Flip-Flop, by TMR hardening.

Fig. 1 TMR block system

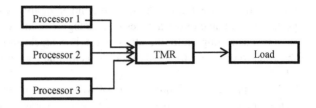

1.1 Single Event Upset (SEU)

A bit flip in the memory of a semiconductor device element is known as Single Event Upset (SEU). Having random behavior in these upsets, do not normally proximately cause damage to the device, and can be cleared by writing to the next memory location or by powering cycling device [13–15]. But the upsets for real-time and time constraint systems applications causes data corruption. Bombarding of the charged particle on the substrate of a silicon IC, leaves an ionization trail which is the main reason which causes a transient fault. For the researchers working in this field for online fault detection as well as correction techniques, it was challenging to detect these faults during offline testing. The collision of an atom in the substrate liberates a shower of energized charged particles, occurs due of a high-energy particle like neutron when emitted to the substrate, which then leaves ionization trail [2, 9].

1.2 Error-Correction Circuitry (ECC)

In the FPGA embedded block memory, Error-correction circuitry (ECC) can be added to memory devices for correction of single- and sometimes dual-bit errors. Triple-module redundancy technique is used for flip-flop of memory results in two good flip-flops overrule an upset flop [12]. Soft error or soft error rate (SER) physical phenomenon is referred to memory elements used for data storage only. It alters the functionality and behavior of the device if upset occurs in the configuration memory of an SRAM-based FPGA. Single-event function interrupt (SEFI) is referred if an upset occurs in the configuration of the logical cell or in a bit the controls routing [13].

2 Related Work

Commercial off-the-shelf (COTS) chips in spaceships, increases concerns about system reliability due to high probabilities of permanent faults in these chips as well as transient faults [12–16]. Data loss and device crash caused by injecting 50 MeV protons into three COTS devices, but no symptom was observed that implied permanent damage of the devices (e.g., a high current condition). Usually, this type of faults, as lasting up to one clock cycle, is characterized. The modified version of the voter to diagnose the faulty module technique was used and prepared in [1, 2–4, 11, 12].

Many prepared techniques use software boned method, while some hardware boned voter system technique is used for fault diagnose and voting which has some improper working, in the memory exceeded the predefined values as the number of transient faults stored is treated that transient fault as the permanent fault [3, 6].

Due to less space for the particle strike energy to dissipate, with the scaling of manufacturing technology, transients last longer, last up to two clock cycles or even longer [17, 18]. There is a high possibility that this duration will become longer, if scaling continues. The duration of intermittent fault that ranges from a few to billions of clock cycles [5, 7] caused by irregularities in chip voltage and temperature and various physical characteristics of chips. With the scaling of integration of many transistors into a chip and manufacturing technology becomes more and more susceptible to such physical characteristics for multicores and large caches.

3   Proposed System

3.1   One- and Multi-Bit Voter

The Configurable Fault Tolerant Processor (CFTP) needs to be able to detect an error by itself and is designed fault tolerant system by the software. The voter concept is generated to achieve CFTP [3]. Lashomb's thesis introduced the function of a 1-bit voter. After reviewing the concept of the 1-bit voter, in this section, we will start constructing the 3-bit voter, i.e., TMR assembly. Figure 2 shows that simple circuit of 1-bit voter consisting of only AND and OR gates. Table 1a shows the working functionality of 1-bit voter. It votes for the equal, i.e., majority bits as output of the system. For more than one incorrect bit, for example, 2 or 3 bits fault the voter can give the faulty output. System vitality fails for the ability to detect the fault and correct by masking two or more errors in a voter (e.g., the CFTP).

For the single error, the corrected output can be obtained, but only looking towards the output it cannot be specified about the error in the circuit. Hence, extra gates are added in the proposed system to report the occurrence of the faulty bits. Figure 2b shows a voter with error detection and Table 1b is its truth table. Error Detection (ERR) is high when un-identical input arrives in the inputs. The possibility of fault occurrence due to SEU within the voter in space application in CFTP is not ignored. Bit flip in voter itself can cause the faulty output of the system. For example, the bit flip in input C which changes the input 1101 instead of 1111 can change the output of the system to 1 instead of 0. Hence, it is not pleasant for the

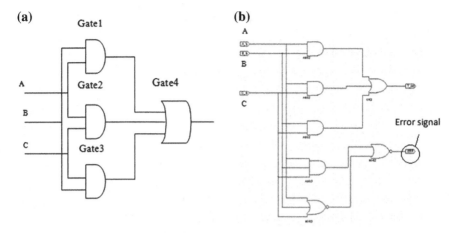

Fig. 2 **a** 1-bit voter. **b** 1-bit voter with error signal

Table 1 **a** 1-bit voter output. **b** 1-bit voter with ERR output

a

A	B	C	Y
0	0	0	0
0	0	1	0
0	1	0	0
0	1	1	1
1	0	0	0
1	0	1	1
1	1	0	1
1	1	1	1

b

A	B	C	Y	ERR
0	0	0	0	0
0	0	1	0	1
0	1	0	0	1
0	1	1	1	1
1	0	0	0	1
1	0	1	1	1
1	1	0	1	1
1	1	1	1	0

voter to catch an error itself, hence proposing the reliability of the voter itself. For the indication of which input is having an error, i.e., did not match with the two others, additional logic circuitry is added to the circuit. Figure 3 shows the schematic of this Voter circuit.

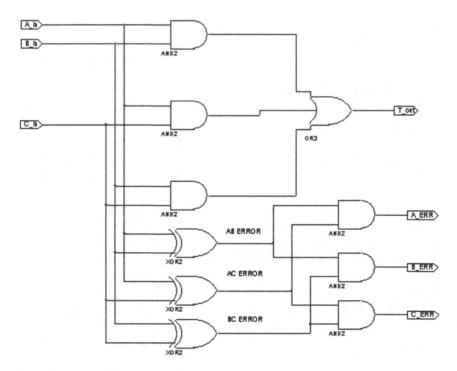

Fig. 3 Schematic of this voter circuit

Table 2 Majority voter output

A	B	C	V_ERR	Y	D_ERR	CID_0	CID_1
0	0	0	0	0	0	0	0
0	0	1	0	0	1	1	1
0	1	0	0	0	1	1	0
0	1	1	0	1	1	0	1
1	0	0	0	0	1	0	1
1	0	1	0	1	1	1	0
1	1	0	0	1	0	1	1
1	1	1	0	1	0	0	0

Table 2 shows the Majority Voter Output. One is the voted output representing the input in majority, and the other three represents which input contained the mismatch.

4 Simulations and Results

4.1 Compilation, Analysis, and Synthesis

Functional TMR (FTMR) for adder is compiled and simulated using Altera Quartus II 13.0 for Cyclone III FPGA using Total 25 logic elements having 44 input–output pins with 156 fan-out capability. The successful compilation was done for error detection and correction as well.

For Analysis and Synthesis Elapsed time of 0.5 s is required and average processor used is 1, it utilized virtual memory 197 MB. The filter takes 0.17 s and uses 1 processor, uses virtual memory 272 MB. Assembler takes 0.8 s and processor uses 1, utilizing virtual memory is 211 MB.

5 Discussion on the Proposed System and Results

Figure 4a describes the single error in the adder inputs with an error in A(2) bit of input A, Y = 0 hence voter is giving corrected output from the logic circuit Fig. 3 and D_RR, i.e., data error shows 4, i.e., "0100" means D_RR(2) bit is affected due to error which is corrected and the system is running without error. It also shows the error-correction output. A = B=15, i.e., "1111" and B = 11, i.e., "1011" showing bit flip in input B(2) which is detected and corrected using the FTMR showing the corrected output in t22 = t23 = t24 = 15 "1111" and these are the output pins.

(a)   (b)

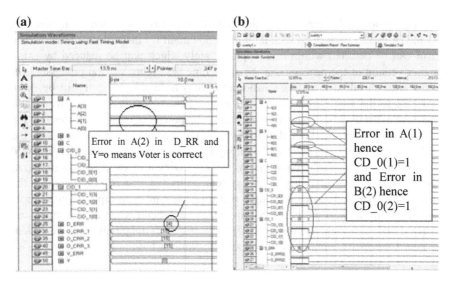

Fig. 4 **a** Single error output. **b** Multi (i.e., 2)-error output

Table 3 Comparing error detection and correction timing for two different FPGAs

All time is in "ns"	Cyclone III (DE0)			Cyclone II (DE2-70)		
No. of fault/s	1	2		1	2	
		Same	Different		Same	Different
Detection delay time	4.56	4.67	4.73	10.44	10.38	12.83
Corrected output time	5.18	5.08	5.15	10.79	10.73	10.91

Indicating the error-correction system is working successfully and giving the correct output within the time constraints. Figure 4b describes the multiple error detection and correction systems' output. Input A(1) and B(2) are affected due to SEU soft error and hence due to bit flip error arises. Error in the system is detected and masked by the error detection and correction module successfully keeping the time constraints.

Time constraint is the major issue in the real-time embedded systems. Timing analysis output for different errors in same input and multiple inputs is performed and found within the timing constraint limits. The minimum error detection is done in approximately 4 ns and corrected output is provided to the system in less than 1 ns time duration, having the overall error-correction and detection time duration is approximately 5 ns which is very less to provide the corrected output to the real-time embedded systems. Table 3 describes and compares the error detection and correction timing for two different FPGAs.

6 Conclusion

Error detection and correction for the multi-core embedded system using cyclone III FPGA can be performed with an arithmetic circuit for the real-time system. The error can be detected within 4 ns approximately which will be reduced by using higher version of the processor at the same time the higher version processor will come with shrink size and less nanometer (nm) technology which will be more sensitive to the single event upset soft errors. Permanent single-event upset fault removable can only be removed by replacing the processor which indicates that the removal of permanent fault due to SEU is not possible. The constraint in the real-time multi-core processor is the timing issue which is covered in this proposed scheme showing the faulty input in 4 ns and correcting it in less than 1 ns (0.52 ns) time increasing the reliability of the system. Proposed system is reliable for the real-time applications to detect single and dual error in same input or in multiple inputs any from A, B, C, but decision-making is difficult when all the inputs are affected due to single-event upset (SEU) bit latch as it is not possible for the system to detect the faulty input as all the inputs are same at the time.

References

1. Kashif Sagheer Siddiqui, Mirza Altamash Baig. "FRAM based TMR (Triple Modular Redundancy) for Fault Tolerance implementation". *Proceedings of The Sixth IEEE International Conference on Computer and Information Technology (CIT'06)*, 2005.
2. Ping-YehYin et al. "A Multi-Stage Fault-Tolerant Multiplier with Triple Module Redundancy (TMR) Technique". *4th International Conference on Intelligent Systems, Modelling 2013.*
3. C. W. Chiou. "Concurrent error detection in array multipliers for GF(2m) fields. Electron". Lett., vol. 38, no. 14, pp. 688–689, Jul. 2002.
4. M. Valinataj and S. Safari. "Fault tolerant arithmetic operations with multiple error detection and correction". in *Proc. IEEE Int. Symp. Defect and Fault-Tolerance in VLSI Syst.*, 2007, pp. 188–196.
5. D. Marienfeld, E. S. Sogomonyan, V. Ocheretnij, and M. Gossel. "A new self-checking multiplier by use of a code disjoint sum-bit duplicated adder". in *Proc. Ninth IEEE European Test Symp. (ETS'04)*, 2004, pp. 30–35.
6. B. K. Kumar and P. K. Lala. "On-line detection of faults in carry-select adders". in *Proc. Int'l Test Conf. 2003 (ITC'03)*, 2003, pp. 912–918.
7. R. Forsati, K. Faez, F. Moradi, and A. Rahbar. "A fault tolerant method for residue arithmetic circuits". in *Proc. IEEE Int. Conf. Information Managemen.*
8. Wei Chen, Rui Gong, Fang Liu, Kui Dai, Zhiying Wang. "Improving the Fault Tolerance of a Computer System with Space-Time Triple Modular Redundancy". *Proceedings International Conference on Dependable Systems and Networks*, pp. 389–98, 23–26 June 2006.
9. Chris Winstead, Yi Luo, Eduardo Monzon, and Abiezer Tejeda. "An error correction method for binary and multiple-valued logic". *41st IEEE International Symposium on Multiple-Valued Logic*, 2011.
10. Jun Yao, Ryoji Watanabe, et al. "An Efficient and Reliable 1.5-way Processor by Fusion of Space and Time Redundancies". *IEEE Transaction*, 2011.
11. Jakob Lechner. "Designing Robust GALS Circuits with Triple Modular Redundancy". *Ninth European Dependable Computing Conference*, 2012.
12. P. Akritidis, C. Cadar, C. Raiciu, et al. "Preventing Memory Error Exploits with WIT". in Proceedings of the IEEE Symposium on Security and Privacy, 2008.
13. Shubham C. Anjankar, M. T. Kolte "FPGA Based Multiple Fault Tolerant and Recoverable Technique using Triple Modular Redundancy (FRTMR)" Elsevier Procedia Computer Science 79 (2016) 827–834.
14. B. Pfarr, M. Calabrese, J. Kirkpatrick, and J. T. Malay. "Exploring the Possibilities: Earth and Space Science Missions in the Context of Exploration" in Proceedings of the Aerospace Conference, 2006.
15. S. K. Sahoo, M.-L. Li, P. Ramachandran, et al. "Using Likely Program Invariants to Detect Hardware Errors" in Proc. DSN, pp. 70–79, 2008.
16. Shubham C. Anjankar, M. T. Kolte. "Fault Tolerant and Correction System Using Triple Modular Redundancy (TMR). Paper published in International Journal of Emerging Engineering Research and Technology, IJEERT, volume 2, Issue 2, May 2014.
17. Mohammad Salehi, et al. "Two-Phase Low-Energy N-Modular Redundancy for Hard Real-Time Multi-Core Systems" IEEE Transactions on Parallel and Distributed Systems, vol. 27, no. 5, May 2016.
18. https://www.microsemi.Com/document–portal/doc_view/130765understanding-soft-and-firm-errors-in-semiconductor.

A New Approach of Grid Integrated Power System Stability and Analysis Through Fuzzy Inference PSO Algorithm for Handling of Islanding PDCS Mode Datasets

K. Harinadha Reddy and S. Govinda Raju

Abstract A network with large number computing nodes during the operation is a big challenge and complex task for system safe and secure operation. Data acquisition from large network has been well defined from the many methods in present scenario of world science and technology. Fuzzy-based inference system is elaborately presented with variable weighted random parameter of Particle Swarm Optimization (PSO), cognitive, and social terms. The distance between personal dataset and global set data points positions, at two previous instants, is taken for modified PSO. Main content this paper is that of data handling and concentrated in elaborating the modeling to complete this task for Peculiar Disturbance Conditional State (PDCS), operating mode of network. From application point of view, grid network with non-conventional energy systems like distribution generators and micro-grids is considered to manage and handle data in practical environment. Information from data availability is defined zone to be classified as accurately and precisely. This paper presents fuzzy inference-modified PSO algorithm for essential analysis and identification and differentiation of data on user-required satisfy level. When a network is extended by means of new small-scale energy systems and operating with main grid network, the possibility of re-existence of global best with PSO data is also reconsidered for acquisition of optimization.

Keywords Fuzzy inference system · Particle swarm optimization (PSO) Personal best datasets (PBDS) · Global best datasets (GBDS) Peculiar disturbance conditional state (PDCS) mode

K. Harinadha Reddy (✉)
Department of Electrical and Electronics Engineering, L B R College
of Engineering, Mylavaram, Krishna District, Andhra Pradesh, India
e-mail: kadapa.hari@gmail.com

S. Govinda Raju
Department of Electrical and Electronics Engineering, J N T University,
Kakinada, East Godavari District, Andhra Pradesh, India
e-mail: govinda4@gmail.com

© Springer Nature Singapore Pte Ltd. 2018
V. Bhateja et al. (eds.), *Proceedings of the Second International Conference on Computational Intelligence and Informatics*, Advances in Intelligent Systems and Computing 712, https://doi.org/10.1007/978-981-10-8228-3_46

1 Introduction

Data acquisition from large networks has been well defined from the many methods in present scenario of world science and technology. Data handling with complicate and minute grid network is now composed with conventional power stations and also extended with non-conventional energy systems like distribution generators and micro-grids. The information is considered as a data taken from the network concise with DGs and MGs at the small-scale level capacity energy systems. Data handling capacity depends on the convergence of problem in the specified region.

Many algorithms are come up for solving the different way and approach methods when large network operation difficulties. One of the other problems of these algorithms is the fact that they require complex computations for real-time communications involving rapidly altering network topologies such as the earlier-mentioned wireless unplanned networks [1]. Dijkstra [2] and Bellman–Ford [3] propose a deterministic algorithm, and they are normally used to solve the shortest path problem. Though these typical algorithms undergo from serious shortcomings, one of which is that they cannot be used for networks with negative weights of edges. There is a clear need for more efficient optimization algorithm for the shortest path problem. Newly, there has been a huge interest in the Particle Swarm Optimization (PSO) payable to its huge potential as evolutionary algorithm, which is based on the simple social deeds of flocks of birds and schools of fish [4]. The paper presents here, a novel report of the PSO, based on fuzzy logic inference engine, is proposed for computation of the shortest path problem like data handling with huge and complex variation in mode of network operation. For instance, in some communication networks, the weights can represent the transmission line capacity, and the negative weights depict the links with gain rather than loss. This can be of vast use in improving the steering in multi-jump communication networks at crucial mode of operation. However, the PSO itself is not perfect and it can fall into the local optimum catch and converges leisurely for obtaining global task-related positions. Using the PSO with fuzzy logic [5], the time taken for solutions of complexity in particular mode of problems can be easily handover to get solution.

2 An Overview of Particle Swarm Optimization

Particle Swarm Optimization (PSO) is a agent-based search engine to optimize the problem. PSO was introduced by James Kennedy and Russell C Eberhart in the year 1995. PSO is a stochastic, population-based EA search method, modeled after the behavior of bird flocks. The PSO algorithm maintains a swarm of individuals (called particles), where each particle represents a candidate solution. The reason why PSO has gained the popularity is because it has only a very few parameters that need to be adjusted. Even though PSO is at rest in its infancy, it has been used across a good range of applications. The second approach is to obtain the information on the actual operating point of the network, and these points are updated [6] according to the variation in environmental conditions. On the other hand, P&O method has simple structure and high reliability. The PSO method is well known for the optimization of complex problems with multivariable objective function. This method is also effective in the case of the attendance of multiple local maximum data point tracking.

The PSO is a swarm intelligence-based algorithm used to find the global optimal solutions. PSO techniques mostly rely on perturb and observe steps and use the hill-climbing concept in subsequent iterations. While doing so, these methods constantly compare present and previous data values, and when they reach the first local maximum, the algorithm stops progressing in the forward. The PSO algorithm is based on the cooperation of multiple agents that exchange in sequence obtained in their respective investigate process by personal best, P_i^t and global best of G_i^t at the instant of "t". Data from network is taken as particles used in PSO and updated velocity of particle is given by

$$v_i^{t+1} = w \; v_i^t + h_1 c_1 \left(P_i^t - x_i^t \right) + h_2 c_2 \left(G_i^t - x_i^t \right), \tag{1}$$

where v_i^t and v_i^{t+1} are the velocities of particle at instants t, t + 1 for i = 1, 2, 3..., n

n is the number particles used in modified PSO;
c_1 and c_2 are acceleration coefficients; and
h_1 and h_2 are random values between (0,1).

Particles follow a simple behavior; imitate the success of adjacent particles and its own achieved successes as shown in Fig. 1. The position of a particle is, therefore, influenced by the best particle in a neighborhood "P" best as well as the best solution found by all the particles in the entire population best. The effect of

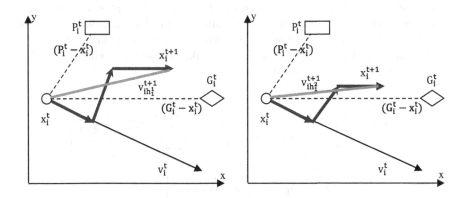

Fig. 1 Basic effect of random values h_1 and h_2 in particle movement in two steps

modifying position of personal best and global best essentially depends on the second and third terms of Eq. (1) as follows:

$$\left.\begin{array}{l} h_1 c_1 \left(P_i^t - x_i^t\right) \\ h_2 c_2 \left(G_i^t - x_i^t\right) \end{array}\right\} \begin{array}{l} \rightarrow \text{Congnetive Term} \\ \rightarrow \text{Social Term} \end{array}$$

Effect of Eq. (1) with the terms h_1 and h_2 is given by

The effect of particle movement in two steps at particular instant $t + 1$ is shown in Fig. 1. The second state of particular movement is observed as optimist first movement of particle. This conceptual phenomenon inspires to propose a modified PSO with fuzzy inference structure. To achieve these goals, various conventional single-stage and data tracking algorithms have been proposed and used to extract maximum possible data under different operating conditions [3]. The PSO algorithms are developed for two separate dataset points, and swarm region is obtained from the output of them. Random values h_1 and h_2 are sampled by membership functions instead of choosing constant random values.

3 Schematic Layout of Model

Large network data comprising with many number of data centers and it requires dead need of protection and secure constraints, especially in peculiar operating modes. One of the special operating modes is islanding state. This mode operation is represented as Peculiar Disturbance Conditional State (PDCS) mode throughout the presentation. In simple way, it is said to be islanding occurrence state of grid network. A simple layout of model for proposed method data handling and identification of peculiar operating mode, i.e., islanding, is shown in Fig. 2.

Many small-scale DGs are connected to main network and if any disturbance occurs on grid network side, the DG has to be disconnected. At this condition, data is read from main network and analyzed properly for estimation of islanding mode; otherwise, there is huge loss of human being or apparatus at main network side. PSO algorithm is modified with random values assigned for cognitive term and social term or Eq. (1). Fuzzy structure is elaborately given for tuning of h_1 and h_2 used in modified PSO algorithm. Hence, data taken network is optimized by proposed fuzzy inference PSO algorithm.

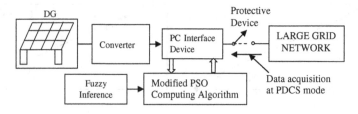

Fig. 2 Schematic layout diagram of model

4 Proposed Modified PSO for PDCS Mode of Network

Normal PSO model with acceleration term and cognitive and associative term is given by Eq. (1). By introducing modified random parameters, it is given by

$$v_i^{t+1} = w\,v_i^t + h_{mod\,1}c_1\left(P_i^t - x_i^t\right) + h_{mod\,2}c_2\left(G_i^t - x_i^t\right) \tag{2}$$

Modified values are determined by sampling the data points from PDCS operating mode of network. These modified values are auto-generated for next instant of PSO algorithm. These sampled random values are fuzzified-generated control vector value, u from fuzzy structure, and cognitive term of velocity of particle at instant $t + 1$ is

$$[\mu_{h1}(z).u].c_1\left(P_i^{t+1} - x_i^{t+1}\right) \tag{3}$$

Social term of velocity of particle fuzzified-generated control vector value, u at instant $t + 1$, is given by

$$[\mu_{h2}(z) \cdot u]c_2\left(G_i^{t+1} - x_i^{t+1}\right) \tag{4}$$

Datasets extracted from main network when operating at PDCS mode may be any nth order of mathematical equation. Function of nth-order polynomial is represented by

$$f(x) = a_n x^n + a_{n-1}x^{n-1} + a_{n-2}x^{n-2} + \cdots + b \tag{5}$$

The dataset from the main network is of any function with electrical quantity of grid side reference values i_{dref}, V_{dref}, or P_{ref}

$$x_i^t = fn^1\left(i_{dref1}^t, i_{dref2}^t, \cdots i_{drefm}^t\right); fn^2\left(v_{dref1}^t, v_{dref2}^t, \cdots v_{drefm}^t\right); fn^3\left(p_{dref1}^t, p_{dref2}^t, \cdots p_{drefm}^t\right) \tag{6}$$

where fn^1, fn^2, and fn^3 are considered as function variables of network at PDCS mode and "m" is referred to number of dataset points.

Step-by-step procedure with modified PSO is as follows:

$$\text{Choose initial population } x_i^0, i = 1, 2, 3 \ldots, n \tag{7}$$

$$\text{Set the initial velocities of each particles } v_i^0 = 0; i = 1, 2 \ldots, n \tag{8}$$

Evaluate objective function, $f(x)$ by Eq. (5).

Find personal best for each particle (for constraint)

$$P_{best, i}^{t+1} = x_i^{t+1} \text{ if } f_i^{t+1} < p_i^{t+1} \tag{9}$$

Find the global best $G_{best, i}^{t+1} = \min\{P_{best, i}^{t+1}\}$ (10)

Update particles velocities

$$v_i^{t+1} = w \, v_i^t + [\mu_{h1}(z).u]c_1 \left(P_i^t - x_i^t\right)c_1 + [\mu_{h2}(z).u]c_2 \left(G_i^t - x_i^t\right)c_2 \tag{11}$$

Find the new set of population $x_i^{t+1} = x_i^t + v_i^{t+1}$ (12)

Modified values in cognitive and social term majorly depend on distance of particle from person best and global best and are detailed in the following section.

5 Fuzzy Inference for Proposed Modified PSO for PDCS Mode of Network

Fuzzy inputs are taken from the cognitive and social term which are described by a distance of particle from person best, i.e., $(P_i^t - x_i^t)$ and distance of particle from global best, i.e., $(G_i^t - x_i^t)$. Output membership function is described by random parameters h_1 and h_2, respectively. Hence, output memberships are $\mu_{h1}(z), \mu_{h2}(z)$ for the respective distances of $(P_i^t - x_i^t)$ and $(G_i^t - x_i^t)$. Five linguistic terms are used for both inputs memberships, and they are denoted as Negative Big (NB), Negative Small (NS), Zero (ZE), Positive Big (PB), and Positive Small (PS). Five linguistic terms are used for output memberships and they are denoted as Very Low (VLOW), Low (LOW), Medium (MED), High (HIGH), and Very High (VHIGH). Here, output membership is not used the zero linguistic term because of only to avoid zero values of cognitive and social terms. Input membership of distances $(P_i^t - x_i^t)$ is given by at instant $t + 1$ that is taken at two previous instants $t, t - 1$, i.e., $d_t = (P_i^t - x_i^t)$ and $d_{t-1} = (P_i^{t-1} - x_i^{t-1})$. These input membership functions are used for fuzzified-generated control vector value, u (Fig. 3, Table 1).

Rules for inference fuzzy system are constructed with five conditional statements. Hence, the size of rule base table is 5×5. Fuzzy structure with 5×5 rule base is used.

Overall flow chart along with fuzzy inference for proposed modified PSO for PDCS operating mode of network is shown in Fig. 4. Similarly, for random parameter, h_2 can be obtained by fuzzy inference-modified PSO.

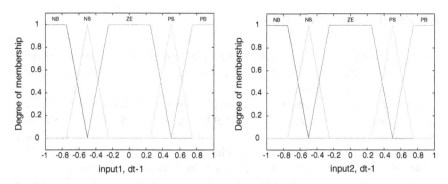

Fig. 3 Input-1, Input-2 membership functions, $d_t(z)$ and $d_{t-1}(z)$

Table 1 Few fuzzy inference rules for PDCS mode of network

Rule no.	Inputs		Output	Rule no.	Inputs		Output
	$d_t(z)$	$d_{t-1}(z)$	u		$d_t(z)$	$d_{t-1}(z)$	u
1	NB	NB	VHIGH	3	PB	PB	VLOW
2	NB	NS	HIGH	4	PB	NS	LOW

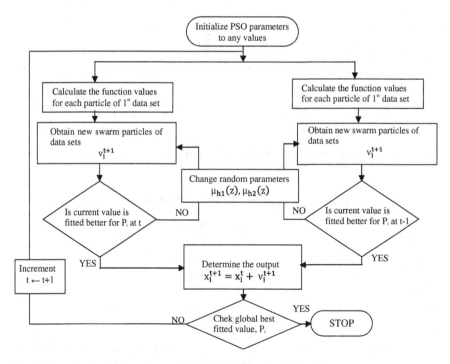

Fig. 4 Flow chart of fuzzy inference PSO for PDCS mode of network

6 Results and Discussion

Simulation is carried out by Matlab programming code for PSO algorithm. Distance between the particles at instant t, t − 1 instants is plotted in Figs. 5 and 7 for variation of random parameters.

For variation of h1 and h2, PSO results gave in negative region of swam datasets. But proposed fuzzy inference PSO gave optimist distance and shown in Figs. 5 and 7. Results shown in Figs. 5 and 6 are simulated for lower number dataset points, i.e., lower number particles in modified PSO. Particles of PDCS mode set quickly converge with a velocity term determined with Eq. 12.

The DG integrated grid network datasets are also plotted to determine the performance of proposed fuzzy inference PSO and shown in Fig. 7. Power variations are considered for datasets of PSO algorithm, and these datasets are tabulated.

Voltages and power at DG terminals are better with proposed MPPT-PSO method. As application considered, condition of grid network can be identified and

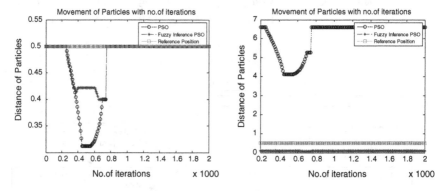

Fig. 5 Fuzzy inference PSO for PDCS mode of network for h1 = 0.2(right) and h2 = 0.3(left)

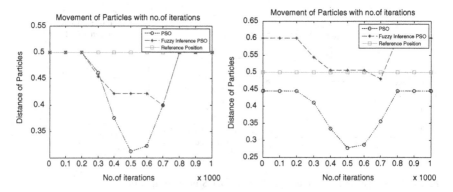

Fig. 6 PSO for h1 = 0.43(right) and h2 = 0.59(left) for lower number of particles

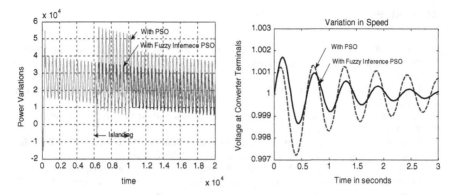

Fig. 7 PDCS mode of network and its power variations

Table 2 Few fuzzy inference rules for PDCS mode of network

No. of iterations at instant of islanding	Detection time	
	PSO (s)	Fuzzy inference PSO (s)
4100	0.214	0.189
4591	0.250	0.194

lead to protection and secure measures, especially in the requirement islanding mode of operation. Converge takes place in shortest distance, and hence the detection time is faster. Detection time and number of iterations reduced by the proposed method is shown in Table 2. Hence, the stability parameters like voltage, power, and speed variations tested in different operating conditions.

7 Conclusion

This paper presented a novel approach of modified PSO algorithm for large network systems like grid network to handle the dataset, especially in peculiar operating conditions. The approach proposed is tested at PDCS operating mode of grid network. Fuzzy-based inference system is elaborately presented with variable weighted random parameter of modified PSO. The distance between personal dataset and global dataset for both negative and positive regions of positions is given mathematical representation and fuzzy inference rules. Also, power and voltage variations are depicted during the test condition that is maintained to improve the profile margin with proposed fuzzy inference PSO method. The comparative results between PSO and fuzzy-based PSO optimized results proved that proposed method gives considerable satisfactory results. Hence, state of grid network can be known and identified in view of protection and secure actions.

References

1. Ammar W. Mohemmed, Nirod Chandra Sahoo, and Tan Kim Geok.: Solving Shortest Problem using Particle Swarm Optimization: Applied Soft Computing, 8(4), pp. 1643–1653 (2008)
2. E. W Dijkstra.: A note on two problems in connection with graphs. Numerische Mathematik, 1, pp. 269–271 (1959)
3. E. L. Lawler.: Combinatorial Optimization: Networks and Matroids. Holt, Rinehart, and Winston, New York, pp. 59–108 (1976)
4. J. Kennedy and R. C. Eberhart.: Particle swarm optimization:, Proceedings of the IEEE Int. Conf. on Neural Networks, Perth, Australia, pp. 1942–1948 (1995)
5. M. H. Noroozi Beyrami, and M. R. Meybodi.: Improving Particle Swarm Optimization using Fuzzy Logic: Proceedings of the Second Iranian Data Mining Conference, Amirkabir University of Technology, Tehran, Iran, pp. 108–120 (2008)
6. Yi-Hua Liu, Shun-Chung Wang.: Determining Membership Functions of FLC-based MPPT Algorithm Using the Particle Swarm Optimization Method: International Congress on Advanced Applied Informatics (2016)
7. Shyi-Ming Chen, and Chu-Han Chiou.: Multiattribute Decision Making Based on Interval-Valued Intuitionistic Fuzzy Sets, PSO Techniques and Evidential Reasoning Methodology: IEEE Transactions on Fuzzy Systems, vol. 23, no. 6, pp. 1905–1916 (2015)
8. D. F. Li.: TOPSIS-based nonlinear-programming methodology for multiattribute decision making with interval-valued intuitionistic fuzzy sets: IEEE Transactions on Fuzzy Systems, vol. 18, no. 2, pp. 299–311 (2010)
9. J. Q. Wang and H. Y. Zhang.: Multicriteria decision-making approach based on Atanassov's intuitionistic fuzzy sets with incomplete certain information on weights: IEEE Transactions on Fuzzy Systems, vol. 21, no. 3, pp. 510–515 (2013)
10. Bing Xue, Mengjie Zhang, and Will N. Browne.: Particle Swarm Optimization for Feature Selection in Classification: A Multi-Objective Approach: IEEE Transactions on Cybernetics, vol. 43, no. 6, pp. 1656–1671 (2013)

Predictive Methodology for Women Health Analysis Through Social Media

Ajmeera Kiran and D. Vasumathi

Abstract Health is an important factor that contributes to human well-being and economic growth. Women's health can be examined in terms of multiple indicators, which vary by geography, socioeconomic standing and culture. Currently, women face a multitude of health problems. To make health services more equitable and accessible for women and to adequately improve the health of women, multiple dimensions of well-being must be analysed in relation to global health average and also in comparison to men. Proposed system collects and analyses the information regarding the issue of women's health through social media regarding women's health to know about the diseases suffered by them. To accomplish this task, we track online health-related conversations about women from Twitter like maternal health, cancer, cardiovascular diseases, etc. These are analysed and outcomes are represented as graphs. This work helps in taking necessary preventive measures to control the diseases.

Keywords Hadoop · MapReduce · Hive · Social media · Health indicators

1 Introduction

The healthcare industry historically has generated a large amount of data, driven by the record-keeping compliance and regulatory requirements, and the patient care [1]. Social media provides an enormous amount of information regarding the health of women helping and taking necessary measures. Big data in health refer to electronic health dataset so large and complex that are difficult feeds (called tweets) related to women's health and analysing to know about the diseases suffered by them. Social media is that a means and environment for social interactions via the

A. Kiran (✉)
JNTUHCEH, Hyderabad, Telangana 500085, India
e-mail: Kiranphd.jntuh@gmail.com

D. Vasumathi
Department of CSE, JNTUHCEH, Hyderabad, Telangana 500085, India

© Springer Nature Singapore Pte Ltd. 2018   511
V. Bhateja et al. (eds.), *Proceedings of the Second International Conference on Computational Intelligence and Informatics*, Advances in Intelligent Systems and Computing 712, https://doi.org/10.1007/978-981-10-8228-3_47

internet [2]. Social media provides an enormous amount of information that can be taped to create measures that potentially serve as both substitutes and complements to traditional source of data from surveys and administrative records. It is important that social media is not merely viewed as a communication channel alone [3].

This study integrates social media into health communication campaigns, and activities allow health communicators to leverage social dynamics and networks to encourage participation, conversation and community, all of which can help spread key message and influence health decision-making. Social media also help people groups to achieve when, where and how they need to get well-being messages; it enhances the accessibility of substance and may impact fulfilment and trust in the well-being messages conveyed. Like, tapping into personal networks, presenting information in multiple formats, spaces and sources help to make messages more credible and effective.

Health is not just illness and treatment but is a state of well-being. It is also free from physical, mental and social stress. There are other social, cultural and economic and environment factors that affect a person's health. The women's health also depends on the above factors. There is an urgent need to have a broader understanding of health as the interaction of social economic, based on the primary healthcare approach. The health problems that are being faced by women are like maternal problems, breast cancer, HIV/AIDS mental health, obesity, pregnancy problems, etc. [4–6].

These are some of the problems and we are going to give clear view status regarding the women's health in India. So, in this paper, we are going to give a clear view of data on women's health in India through social. The health-related information can be extracted using social media and represented them as graphs and analysed them. The potential of online conversations to complement official statistics, by providing a qualitative picture demonstrates how people are feeling towards women's health. In addition to this, it will be helpful to the government to take preventable measures, in which part of India women and children are suffering more and with what health issue they are facing.

2   Motivation

Every year, millions of women's and children are dying from preventable causes. There are some problems which women face like mental stress, breast cancer, obesity, asthma, etc. Earlier, it was a problem with the survey but at present it is much easier to collect the data using social media, i.e. we can extract huge data which is generated by much number of people from social media and analysed them [7]. Various problems are faced by women of different geographical regions, social classes and indigenous and ethnic groups across the world. Currently, women are facing a multitude of health problems, which ultimately affect the aggregate economy's output.

3   Problem Statement

Health analysis is done by collecting data from all hospitals and social websites, and then processing and analysing them. This is a hectic task. Instead, data collection from different social networking sites can help to accomplish the task easily. Here arises a problem, that is, data is an unstructured format.

4   Existing System

In the existing system, the data collection was limited to certain hospitals only. Using social media, the data from all around the world can be collected. The traditional RDBMS had scalability problem. It cannot store unstructured data. It was difficult to handle large amount of data. The one way to collect data was by surveys which were time-consuming and expensive work to perform [8].

The key challenges include the following:

i. Preprocessing is required.
ii. High cost.
iii. Difficult to achieve scalability.
iv. Could store only structured data.

5   Proposed System

Big data use Hadoop Distributed File System (HDFS) which can store petabytes of data [9]. It can store both structured and unstructured data [10]. The proposed system collects the data from the social networking site (Twitter). The collected data may be an unstructured data. Using HDFS, the data is stored. The data is processed using MapReduce [11], a processing framework. It uses social media for survey. The proposed system is totally an automatic system. It uses the social media data to conduct a survey and generate the same result as it has done manually.

1. Extraction of data from social media (Twitter). Here, we use the data acquisition tool for extracting the data from the social media.
2. Based on the acquired data, we analyse the sentiments of the people for the data analysis.

3. Segregating the collected data and store it in proper tables to form databases, so that we can analyse the data clearly.
4. Analysing the data from database and generating analytical graphs based on the collected data.
5. The generated graphical outputs are stored and made a conclusion for each and every generated graph.

It has the following advantages:

   i. Data is available in real time and at high frequency.

  ii. Low-cost source of valuable information.

 iii. No wastage of manpower and time.

 iv. It can answer the questions that would have known through surveys in advance.

  v. Social media offers a distinctive window approach.

Projected Architecture

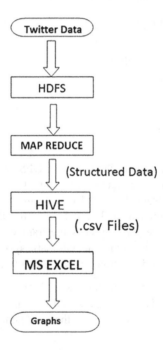

Proposed study begins with collection of women's health-related tweets from Twitter using flume-in with the support of Twitter Streaming API. These tweets are stored in HDFS. The collected tweets are in unstructured (Jason) format. These are converted to structured format using a MapReduce code which is fed as an input to hive. Using HQL queries, .csv files are produced which are converted to graphs using MS Excel.

6 Experimental Results

Figure 1 gives brief details about successful execution of MapReduce code.

Figure 2 gives brief details about count and categories of tweets on name node which is run on the MapReduce and it shows number of tweets on the particular diseases in different countries.

Figure 3 gives brief details about count and categories of tweets on the particular diseases like pregnancy obesity, breast cancer, infertility, etc. in different countries.

Figure 4 shows the number of steps in initialization of hive and creation of a table in hive environment. On top of the Hadoop, hive is a data warehouse framework. Using hive, we can write SQL like queries to process and analyse the data stored in Hadoop file system. The advantage of hive is scalability and performance, it works like normal SQL language and runs fast over big datasets.

Figure 5 gives brief details about count and categories of tweets on the particular diseases like pregnancy obesity, breast cancer, infertility, etc. in different countries which is produced using hive.

Figure 6 gives brief details about count and categories of tweets on the woman diseases like pregnancy obesity, breast cancer, infertility, etc. in different countries.

Figure 7 gives brief details about count and categories of tweets on the woman diseases like breast cancer in different countries like USA, UK, India, Australia and

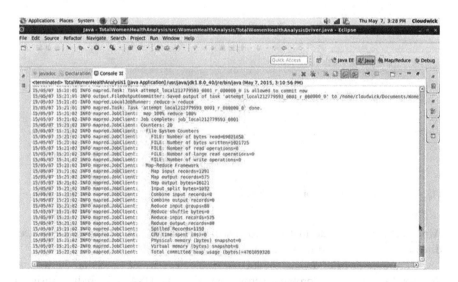

Fig. 1 Screenshot of successful execution of MapReduce code

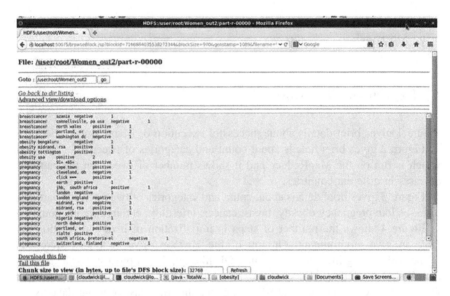

Fig. 2 Screenshot showing count and categories of tweets on name node

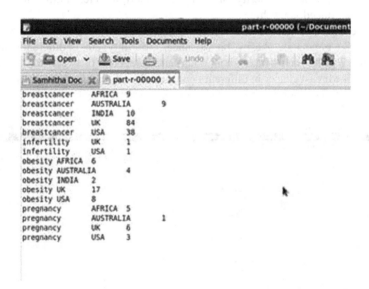

Fig. 3 Screenshot of output of MapReduce code

Africa. The results state that most of the people in UK are facing breast cancer disease when compared to other countries.

Figure 8 gives brief details about count and categories of tweets on the woman diseases like pregnancy problem across countries like USA, UK, India, Australia

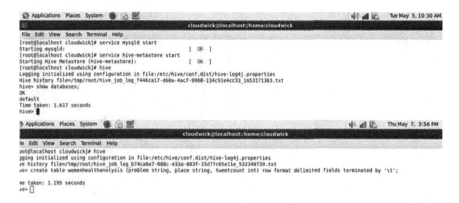

Fig. 4 Screenshot of initialization creating table of hive

Fig. 5 Screenshot of table produced by hive

File Edit View Search Terminal H			
Table default.wha stats: [num_part			
ze: 296, raw_data_size: 0]			
OK			
Time taken: 0.89 seconds			
hive> select * from wha;			
OK			
breastcancer	AFRICA	9	
breastcancer	AUSTRALIA		9
breastcancer	INDIA	10	
breastcancer	UK	84	
breastcancer	USA	38	
infertility	UK	1	
infertility	USA	1	
obesity	AFRICA	6	
obesity	AUSTRALIA		4
obesity	INDIA	2	
obesity	UK	17	
obesity	USA	8	
pregnancy	AFRICA	5	
pregnancy	AUSTRALIA		1
pregnancy	UK	6	
pregnancy	USA	3	
Time taken: 0.231 seconds			
hive>			

and Africa. The results state that most of the people in UK are facing breast cancer disease when compared to other countries.

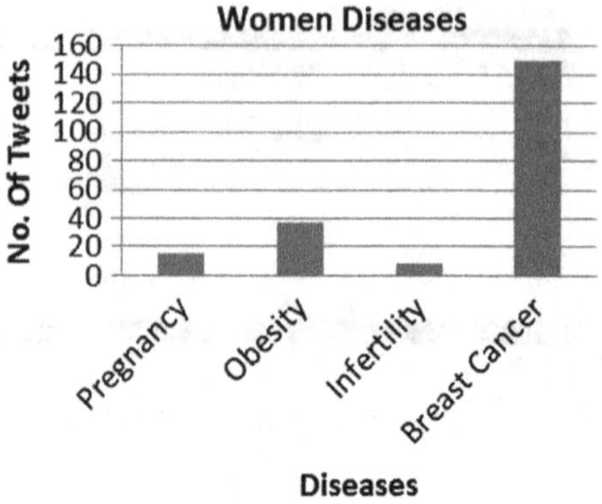

Fig. 6 Graph showing number of tweets regarding women diseases

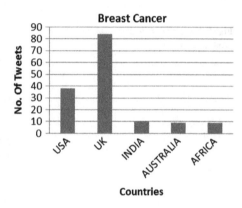

Fig. 7 Graph showing the number of tweets regarding breast cancer

Fig. 8 Graph showing number of tweets across countries regarding pregnancy problem

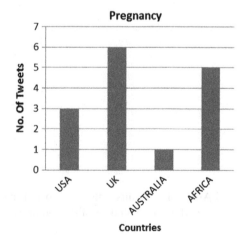

7 Conclusion

Improving and taking care of health-related issues are always given priority. To improve health, statistical data and information related to health are collected and a conclusion is drawn by processing it. Such health-related data is collected from social media; in this work, Twitter has generated a large amount of unstructured data related to health issues posted by various users from worldwide. This data is collected, processed and analysed using various software tools. Hadoop playing an Important role in Predicting and analysing Women Health Problems, because hadoop works on hetrogeneous data.

Also, the processed data is further analysed for generating the output in the form of graphs for the easy understandability; these statistics generated can be stored and help in improving the health conditions by creating the awareness in the society based on the statistics. The statistics also help to predict the future health conditions.

This paper helps the health organizations to have a clear view of diseases suffered by women belonging to different categories, take preventive measures and enhance women's health, and take necessary preventive measure to control the diseases.

8 Future Work

The work can be further extended by finding out the reason of illness through tweets. We can extract information from the articles shared on twitter for more accurate results. We can also continue the work by extending it to children or men or people as a whole.

References

1. H. C. Koh and G. Tan, "Data Mining application in Healthcare", Journal of Healthcare Information Management, vol. 19, no. 2, (2005).
2. Atallah, L., Lo, B., Yang, G.Z. Can pervasive sensing address current challenges in global healthcare? Journal of epidemiology and global health 2012; 2(1):1–13.
3. M. Silver, T. Sakara, H. C. Su, C. Herman, S. B. Dolins and M. J. O'shea, "Case study: how to apply Data mining techniques in a healthcare data warehouse", health. Inf. Manage, vol. 15, no. 2, (2001), pp. 155–164.
4. R. Kandwal, P. K. Garg and R. D. Garg, "Health GIS and HIV/AIDS studies: Perspective and retrospective", Journal of Biomedical Informatics, vol. 42, (2009), pp. 748–755.
5. Rafaqat Alam Khan "Classification and Regression Analysis of the Prognostic Breast Cancer using Generation Optimizing Algorithms" international Journal of Computer Applications (0975-8887) Volume 8, No. 25, April 2013.

6. 2nd International Symposium on Big Data and Cloud Computing (ISBCC '15), "Predictive Methodology for Diabetic Data Analysis in Big Data". Dr Saravana Kumar, Eswari, Sampath & Lavanya.
7. Alagugowri S, Christopher T. Enhanced Heart Disease Analysis and Prediction System [EHDAPS] Using Data Mining. International Journal of Emerging Trends in Science and Technology 2014; 1:1555–1560.
8. White Paper by SAS, How Government are Using the Power of High Performance Analytics, 2013.
9. Konstantin Shvachko, HairongKuang, Sanjay Radia, Robert Chansler, the Hadoop Distributed File System, IEEE, 2010.
10. The 4th International Workshop on Body Area Sensor Networks (BASNet-2015), "Stream processing of healthcare sensor data: studying user traces to identify challenges from a big data Perspective", Rudyar Cortes, Xavier Bonnaire, Olivier Marin, and Pierre Sens.
11. Skew Zhihong Liu, Qi Zhang, Reaz Ahmed, Raouf Boutaba, Yaping Liu, and Zhenghu Gong, "Dynamic Resource Allocation for MapReduce with Partitioning", IEEE TRANSACTIONS ON COMPUTERS, VOL. 65, NO. 11, NOVEMBER 2016

A Reversible Data Embedding Scheme for Grayscale Images to Augment the Visual Quality Using HVS Characteristics

T. Bhaskar and D. Vasumathi

Abstract Data embedding into multimedia is growing hastily due to the raising concern to handle the security issues. The reversible data embedding schemes have become popular due to its demand in various applications such as military communication, remote sensing, etc. There is a tradeoff between embedding capacity and visual quality while maintaining the reversibility. We propose a reversible scheme which focuses on ameliorating the visual quality by minimizing the usage of nonzero quantized DCT coefficients during the embedding process.

Keywords Data embedding · Reversible · DCT · Visual quality
HVS characteristics

1 Introduction

Data embedding into multimedia has become a popular method in addressing the security issues that arise during the use, storage, and transmission of multimedia data. Various applications from watermarking and steganography [1, 2] include authenticity (e.g., tracing copyright, banknote, etc.) and concealing any file into multimedia, etc. Embedding the secret data into multimedia data such as image, audio, or video with the compression technique is prominently increasing as it helps to reduce resource usage such as storage space or transmission capacity [2–4].

"we take the full responsibility of the ethical issue of the work. We have taken permission to use the information in our work. In case of any issue we are only responsible".

T. Bhaskar (✉)
JNTUHCEH, Hyderabad, India
e-mail: bhalu7cs@gmail.com

D. Vasumathi
Department of CSE, JNTUHCEH, Hyderabad 500085, Telangana, India

V. Bhateja et al. (eds.), *Proceedings of the Second International Conference on Computational Intelligence and Informatics*, Advances in Intelligent Systems and Computing 712, https://doi.org/10.1007/978-981-10-8228-3_48

The most promising frequency transformation mechanism in demand is Discrete Cosine Transform (DCT) [5]. The DCT is a technique for converting an image from spatial domain to frequency domain, which is widely used in image compression. The data embedding in transform domain helps in improving the visual quality by making use of the visual characteristics of frequency coefficients [6, 7]. The data embedding can be performed in either irreversible or in reversible manner. After embedding the data into the multimedia, the original form of the multimedia cannot be restored back in irreversible embedding, whereas restoring back of secret data along with the original form of multimedia can be achieved through reversible embedding [7–9].

A number of modifications are being carried out in the multimedia for achieving the reversibility [7, 10]. And because of these modifications, the visual quality of image or video degrades. To maintain the visual quality of an image or video, we must consider the effect of embedding capacity and other parameters that may affect the visual quality. Some measures such as PSNR can be used for evaluating the visual quality of degraded image or video. But PSNR is not a good measure for evaluating the visual quality of degraded image or video when data is embedded in frequency domain in which HVS characteristics are considered. Thus, to measure the visual quality, we must consider the metrics which take HVS characteristics into account. We propose a DCT-based reversible data embedding scheme for grayscale images which ameliorates visual quality using HVS characteristics. The HVS-based measures SSIM and SSIM_INDEX have been used to assess the visual quality of degraded image [11, 12].

2   Proposed Scheme

Embedding the data in middle-frequency coefficients is in practice to improve visual quality of the embedded image. In S. Gujjunoori et al. scheme [7, 13], the alteration of nonzero coefficients during the elimination of ambiguity process caused more visual distortions. So, we propose a reversible data embedding scheme by minimizing the use of nonzero coefficients to ameliorate the visual quality. The proposed scheme is referred as Ameliorate Visual Quality (AVQ) which aims to improve visual quality of embedded image. The middle-frequency quantized DCT coefficients of a grayscale image are considered for embedding the data as shown in Fig. 1. The embedding capacity of an image is calculated by considering the three consecutive zeroes. During the process of eliminating ambiguity, a few nonzero values are partially neglected which results in amelioration of the visual quality. We compare the results with S. Gujjunoori et al. scheme [7].

Let I be an image and S be the secret data to be embedded. The image I of size $M \times N$ is partitioned into 8 × 8 blocks. These 8 × 8 blocks are transformed into frequency domain using 2D DCT [4]. If considered image is a square, then size of image will be $N \times N$.

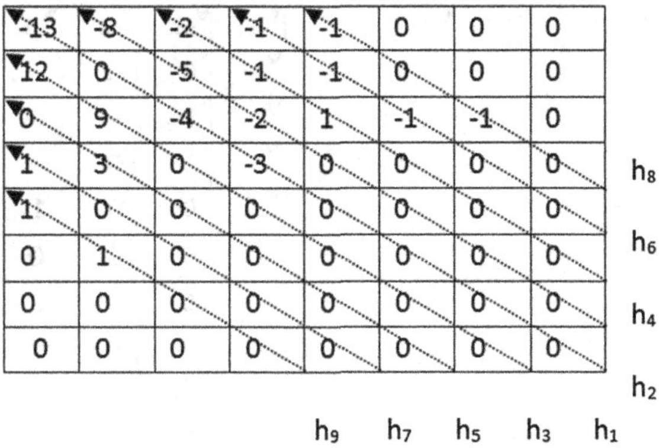

Fig. 1 An 8×8 DCT quantized block before embedding

-13	-8	-2	-1	-1	0	0	0	
12	0	-5	-1	-1	0	0	0	
0	9	-4	-2	1	-1	-1	0	
1	3	0	-3	0	0	0	0	h_8
1	0	0	0	0	0	0	0	
0	1	0	0	0	0	0	0	h_6
0	0	0	0	0	0	0	0	h_4
0	0	0	0	0	0	0	0	h_2

$$h_9 \quad h_7 \quad h_5 \quad h_3 \quad h_1$$

Consider a transformation of an original image x to transformed image X as in (1)

$$X(K_1, K_2) \leftarrow x(n_1, n_2), \tag{1}$$

where n_1 and n_2 are the coordinates in spatial domain, and K_1 and K_2 be the coordinates in transformed domain.

The forward 2D DCT transform of an image x of size $N \times N$ is as in Eq. (2)

$$X(K_1, K_2) = T \sum_{n_1 = 0}^{N-1} \sum_{n_2 = 0}^{N-1} x(n_1, n_2) AB, \tag{2}$$

where K_1, K_2, n_1, $n_2 = 0, 1, 2, \ldots, N-1$.
and $A = cos\left(\frac{\pi(2n_1 + 1)K_1}{2N}\right)$

$$B = cos\left(\frac{\pi(2n_2 + 1)K_2}{2N}\right)$$

$$T = \sqrt{\frac{4}{N^2}} Y(K_1) Y(K_2)$$

$$Y(K) = \begin{cases} \frac{1}{\sqrt{2}} & for K = 0 \\ 1 & otherwise \end{cases}$$

The corresponding inverse transform is given in (3)

$$x(n_1, n_2) = \sum_{K_1 = 0}^{N-1} \sum_{K_2 = 0}^{N-1} TX(K_1, K_2) AB \tag{3}$$

Let $I = \{I_1, I_2, I_3, \ldots I_m\}$, where I_j is the jth 8×8 block of I and $m = (N \times N)/64$. Let $C = \{C_1, C_2, C_3, \ldots C_m\}$ be the set of DCT transformed blocks. Let Q be the

Fig. 2 An 8×8 DCT
quantized block after
embedding

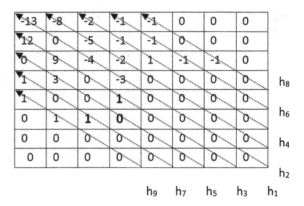

8×8 quantization block. Let $CQ = \{CQ_1, CQ_2, CQ_3, \ldots, CQ_m\}$ be the set of quantized DCT blocks of I. Let $E = \{E_1, E_2, E_3, \ldots E_m\}$ be the set of embedded blocks of image. Let $H = \{h_1, h_2, \ldots, h_9\}$ be the sets of middle-frequency coefficients in every CQ_i for embedding the data as shown in Fig. 1. We embed the data in mid of three ceaseless consecutive zeros present nearby either a nonzero coefficient or at the end of sets considered. In Fig. 1, consider h_7, h_5, h_3 sets having three consecutive zeroes from high frequency to low frequency. The secret data is embedded at the mid of three consecutive zeroes in sets h_7, h_5, h_3 as shown in Fig. 2. Let c_{h_i} be the number of ceaseless zeroes in the direction, high frequency to low frequency in the set h_i.

The proposed scheme is presented in Sects. 2.1 and 2.2 representing data embedding and data extraction procedures, respectively.

2.1 Data Embedding Procedure

The data is embedded in middle-frequency coefficients by considering the sets which satisfies the condition of having three consecutive zeroes from high frequency to low frequency (as shown in Fig. 1). During the process of eliminating ambiguity, the use of nonzero values is minimized, which results in improving the visual quality. The data embedding procedure uses the functions, *Amb* and *Emb*. The *Amb* is used to remove the ambiguity in advance that may result after embedding the data, if the care is not taken to eliminate the ambiguity before embedding. In (4), the tuple (a_h, a_m, a_l) denotes the continuous sequence of quantized DCT coefficients from high frequency to low frequency, nearby either a nonzero coefficient or at the end of sets considered as in Fig. 1. The function *Emb* is used to embed the data in the middle-frequency coefficients as in (5), where $s = \{0, 1\}^*$ denotes the random secret data, (a_h', a_m', a_l') denotes the tuple of sequence of quantized DCT coefficients from high frequency to low frequency after eliminating the ambiguity (as shown in Fig. 2), and (e_h, e_m, e_l) denotes the tuple containing the embedded data, where secret data s is embedded in the mid of three consecutive zeroes, as in (5). The data embedding algorithm is shown in Algorithm 1.

$$(a_h', a_m', a_l') = Amb(a_h, a_m, a_l) = \begin{cases} (a_h, a_m + 1, a_l) & if(a_h = 0, a_m = 1, a_l = 0) \\ (a_h, a_m, a_l) & otherwise \end{cases} \quad (4)$$

$$\begin{aligned} (e_h, e_m, e_l) &= Emb((a_h, a_m, a_l), s) \\ &= \begin{cases} (a_h, s, a_l) & if(a_h = 0, a_m = 0, a_l = 0) \\ (a_h', a_m', a_l') & otherwise \end{cases} \end{aligned} \quad (5)$$

where s denotes a secret bit 0 or 1.

Algorithm 1. Data embedding algorithm of AVQ

Step 1: Let I be the cover image and s = {0,1}*
Step 2: Partition the image I into 8 × 8 blocks,
   I = {I1, I2, I3, ..., Im}
Step 3: Apply 2D DCT on every 8 × 8 block, k
   for i = 1 to 8
      for j = 1 to 8
         $C_k(i, j) = DCT(I_k(i, j))$;
      end
   end
   C = {C1, C2, C3, ..., Cm} be the set of
   DCT transformed 8 × 8 blocks
Step 4: Apply quantization on every transformed
   8 × 8 block, k
   for i = 1 to 8
      for j = 1 to 8
         $CQ_k(i, j) = C_k(i, j)/Q(i, j);$
      end
   end
   CQ = {CQ1, CQ2, CQ3, ..., CQm} be the
   set of DCT quantized 8 × 8 blocks
Step 5: Consider $h_i(1 < i < 9)$ be the set of middle
   frequency coefficients to embed the data.
   Before embedding, eliminate the ambiguity in
   every 8 × 8 block using equation (4)
Step 6: foreach $I_j \in I$
Let be c_{h_i} be the number of ceaseless zeroes from
high frequency to low frequency in the set hi ,
$|h_i|$ denotes the size of hi.
   if $(c_{h_i} \geq \left\lceil \frac{|h_i|}{2} \right\rceil - 1)$
     Embed the data using the function Emb in (5)
   endif
   endfor
Step 7: Let E_j be an embedded block.
   Combine all the embedded blocks into
   E = {E1, E2, E3, ... Em} .

2.2   Data Extraction Procedure

Data extraction procedure is the reverse process of data embedding. The data extraction procedure uses the functions, *Ext* and *Rest. Ext;* as in (6), is used to extract the embedded secret data and *Rest,* as in (7) and is used to restore back the data to its original form after extracting the data. The data extraction procedure is shown in Algorithm 2. $ex = (ex_h, ex_m, ex_l) = Ext(e_h, e_m, e_l)$

$$= \begin{cases} 0 & if\,(e_h = 0, e_m = 0, e_l = 0) \\ 1 & if\,(e_h = 0, e_m = 1, e_l = 0) \end{cases} \tag{6}$$

$$Rest(ex_h, ex_m, ex_l) = \begin{cases} (ex_h, ex_m - 1, ex_l) & if\,(ex_h = 0, ex_m = 2, ex_l = 0) \\ (ex_h, ex_m, ex_l) & otherwise \end{cases} \tag{7}$$

Algorithm 2. *Data extracting algorithm of AVQ*
Step 1: Let E be the embedded image.
Step 2: Partition E into 8×8 blocks
Step 3: $foreach\ E_j \in E$
    Consider $h_i(1 < i < 9)$ be the set of middle
    frequency coefficients.
    $if\ c_{h_i} \geq \left\lceil \frac{|h_i|}{2} \right\rceil - 1$
        Extract the data using Ext as in (6).
    $endif$
    $endfor$
Step 4: After extraction, the restore the coefficients using
    $Rest$ as in equation (7)
Step 5: Let the extracted image be E'
Step 6: Apply de-quantization on every extracted 8×8
    block
    $for\ i = 1\ to\ 8$
        $for\ j\ =\ 1\ to\ 8$
            $R(i,j) = E'(i,j) \times Q(i,j)\,;$
        end
    end
Step 7: Apply $2D\ IDCT$ on every 8×8 block
    $for\ i = 1\ to\ 8$
        $for\ j\ =\ 1\ to\ 8$
             $R'(i,j) = IDCT(R(i,j))\,;$
        end
    end
Step 8: Let R'_i be the restored block and combine all R'_i into
    $R' = \{ R'_1, R'_2, R'_3, ..., R'_m \};$

3 Results and Discussions

The grayscale images of size 512 × 512 are used in the experiments which include images of Lena, Zelda, Elaine, etc. Some of the test images are shown in Figs. 3 and 4. The data-embedded images can be observed in Fig. 5. The visual quality of the degraded images is assessed using metrics such as PSNR (Peak signal-to-noise ratio) as in Eq. (8), SSIM (Structure Similarity) as in Eq. (9), and SSIM_INDEX as in Eq. (10) [12]. Here, the embedding capacity is defined as the total number of bits embedded in a 512 × 512 image.

$$\left\{ PSNR = 10 \log 10 \frac{255^2}{MSE} (1) \right. \tag{8}$$

$$\left\{ SSIM = \frac{(2*\bar{x}*\bar{y} + C1)(2*\sigma xy + C2}{(\sigma x^2 + \sigma y^2 + C2)(x^{-2} + y^{-2} + C1)} \right. \tag{9}$$

| (a) Lena | (b) Zelda | (c) Elaine |

Fig. 3 Original test images

| (a) Lena | (b) Zelda | (c) Elaine |

Fig. 4 Degraded images after embedding process

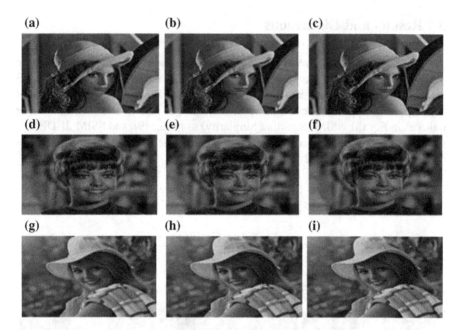

Fig. 5 **a, d, g** original images of Lena, Zelda, and Elaine, respectively. **b, e, h** are data-embedded images of Lena, Zelda, and Elaine, respectively using S. Gujjunoori et al. scheme. The images in **c, f, i** are data-embedded images of Lena, Zelda, and Elaine, respectively using the proposed scheme

$$\left\{ SSIM_{INDEX} = \frac{1}{M} \sum_{j=1}^{M} SSIM\left(x_j, y_j\right) \right. \tag{10}$$

From Table 1, we can observe that the visual quality in terms of PSNR, SSIM, and SSIM_INDEX has been improved. Figures 6, 7, 8, 9, and 10 show improved visual quality in terms of PSNR, SSIM, and SSIM_INDEX, respectively.

From Fig. 6, we can observe that proposed scheme (black) achieves good results for PSNR than existing scheme (blue color).

From Figs. 7 and 8, SSIM is improved from proposed scheme (black) when compared to the existing scheme (blue).

From Figs. 9 and 10, SSIM_INDEX is improved from proposed scheme (black) when compared to the existing scheme (blue).

Table 1 Visual quality measured in terms of PSNR, SSIM, and SSIM_INDEX

		Capacity	PSNR	MSSIM (SSIM)	MSSIM (SSIMJNDEX)
Barb512	S. Gujjunoori et al. scheme	33729	26.2109	0.904	0.6479
	Proposed scheme	28405	28.4345	0.9341	0.72
Airplane	S. Gujjunoori et al. scheme	36465	26.3099	0.7491	0.3822
	Proposed scheme	34984	28.7784	0.8267	0.4967
Baboon	S. Gujjunoori et al. scheme	28013	26.0568	0.9388	0.7745
	Proposed scheme	18906	26.903	0.954	0.7913
Boat	S. Gujjunoori et al. scheme	36154	26.3478	0.8893	0.5995
	Proposed scheme	31305	28.3597	0.9218	0.6683
Elaine	S. Gujjunoori et al. scheme	36698	26.3395	0.8683	0.513
	Proposed scheme	33141	28.0532	0.9024	0.5732
Crowd	S. Gujjunoori et al. scheme	36274	26.3236	0.9069	0.6188
	Proposed scheme	30195	28.8307	0.9366	0.7093
Building	S. Gujjunoori et al. scheme	35567	26.4371	0.8814	0.6358
	Proposed scheme	27379	28.3516	0.9161	0.6996
Couple512	S. Gujjunooriet al. scheme	36023	26.3696	0.8869	0.5926
	Proposed scheme	32233	28.4431	0.9202	0.6746
Goldhill	S. Gujjunooriet al. scheme	36336	26.3396	0.9007	0.6072
	Proposed scheme	31123	28.3936	0.9298	0.6814
Parrots	S. Gujjunoori et al. scheme	36109	26.7503	0.8373	0.5039
	Proposed scheme	33257	28.0405	0.8645	0.5489
Lena	S. Gujjunoori et al. scheme	34560	26.2535	0.8722	0.5685
	Proposed scheme	29225	29.16	0.9132	0.6686
Plane	S. Gujjunoori et al. scheme	36505	26.3004	0.8373	0.5144
	Proposed scheme	32408	28.8393	0.8895	0.6169
Zelda	S. Gujjunoori et al. scheme	36840	26.3678	0.8602	0.4807
	Proposed scheme	35690	28.7484	0.9055	0.5886

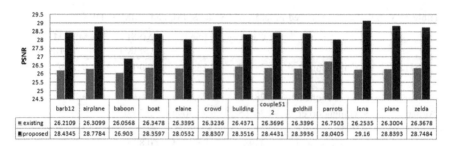

Fig. 6 Improved visual quality measured in terms of PSNR

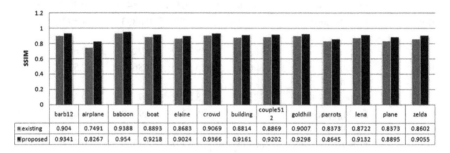

Fig. 7 Improved visual quality measured in terms of SSIM

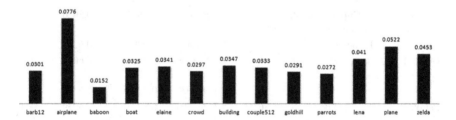

Fig. 8 Improved SSIM when compared to existing scheme

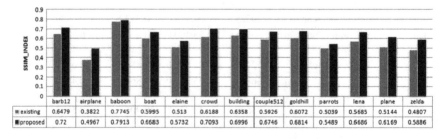

Fig. 9 Improved visual quality measured in terms of SSIM_INDEX

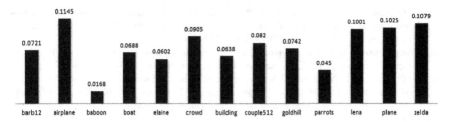

Fig. 10 Improved SSIM_INDEX when compared to existing scheme

4 Conclusion

The visual quality of an image is degraded more when nonzero coefficients are considered while eliminating the ambiguity that arises during embedding process. The proposed scheme AVQ embeds the data in middle-frequency coefficients by minimizing the use of nonzero coefficients while eliminating the ambiguity which ameliorates the visual quality. The visual quality of an image is improved in terms of PSNR, SSIM, and SSIM_INDEX. Thus, this scheme can be used for applications such as military communication, remote sensing, etc. which focuses on reversibility and visual quality.

References

1. Petitcolas, FAP; Anderson RJ; Kuhn MG, "Information Hiding: A survey," Proceedings of the IEEE (special issue), 1999, pp. 1062–1078.
2. Cox I, Miller M, Bloom J, Fridrich J, Kalker T, Morgan Kaufman, " Digital watermarking and steganography," 2008.
3. Furht B, "A survey of multimedia compression techniques and standards," Part I: JPEG standard. Real-time Imaging, Vol. 1, 1995, pp. 49–67.
4. Salomon D, "Data compression: the complete reference," Springer, 2007.
5. Wen-Hsiung Chen, C. Harrison Smith, And S. C. Fralick, "A Fast Computational Algorithm for the Discrete Cosine Transform," IEEE Transactions On Communications, Vol. 9, September 1977, pp. 1004–1009.
6. Chang CC, Lin CC, Tseng CS, Tai WL, "Reversible hiding in DCT-based compressed images," Information Sciences 2007, pp. 2768–2786.
7. S. Gujjunoori and B. B. Amberker, "A Reversible Data Embedding Scheme for MPEG-4 Video Using HVS Characteristics," Proceedings of the International Conference on Intelligent Systems and Signal Processing (ISSP2013), Gujarat, IEEE Explore, 2013, pp. 120–124.
8. Celik, M.U., Sharma, G.; Tekalp, A.M.; Saber, E., "Reversible data hiding," Proceedings of the IEEE International Conference on Image Processing, Vol. 2,2002, pp. 157–160.
9. S. Gujjunoori and B. B. Amberker, "Reversible Data Embedding for MPEG-4 Video Using Middle Frequency DCT Coefficients", Journal of Information Hiding and Multimedia Signal Processing, Vol. 5, No. 3, pp. 391–402, July 2014.
10. Lin CC, Shiu PF, "DCT-based reversible data hiding scheme," ICUIMC'09, Proceedings of the 3rd international conference on ubiquitous information management and communication, New York, NY, USA: ACM; 2009, pp. 327–335.

11. S. Gujjunoori and B. B. Amberker, "DCT based reversible data embedding for MPEG-4 video using HVS characteristics", Journal of Information Security and Applications, Elsevier, Volume 18, Issue 4, December 2013, Pages 157–166.
12. Wang Z, Bovik A, Sheikh H, Simoncelli E, "Image quality assessment: from error visibility to structural similarity," IEEE Transactions on Image Processing, 2004, pp. 600–612.
13. T. Bhaskar and D. Vasumathi," A reversible data embedding scheme for mpeg-4 video using non-zero ac coefficients of dct" https://doi.org/10.1109/iccic.2013.6724127, Conference: 2013 IEEE International Conference on Computational intelligence and Computing Research (ICCIC).

Optimal Feature Selection for Multivalued Attributes Using Transaction Weights as Utility Scale

K. Lnc Prakash and K. Anuradha

Abstract Attribute selection procedure is a key step in the process of Knowledge Discovery in Database (KDD). Majority of the earlier contributions of selection methods can handle easier attribute types. Such methods are not for multivalued attributes that comprise multiple values in simultaneously. Majority of the existing attribute selection methods can manage simple attribute types like the numerical and categorical. The methods cannot fit multivalued attributes, which are attributes that constitute multiple values simultaneously in the dataset for same instance. In this manuscript, a contemporary approach for selecting optimal values for features of multivalued attributes is proposed. In the proposed solution, the method is about adaptation of utility mining based pattern discovery approach. For evaluating the proposed approach, experiments are carried out with multivalued and multiclass datasets that are submitted to k-means clustering technique. The experiments show that the proposed method is optimal to assess the relevance of multivalued attributes toward mining models such as clustering.

Keywords Multivalued attributes · Utility scale · Transaction weight Data mining · Feature selection by frequency

1 Introduction

Clustering, which is an unsupervised learning method, is a predominant area in which many researches are taking place. Based on the chosen optimal feature sets, the process of clustering occurs. Feature selection is one of the popular models that is adapted for improving the clustering process, as clustering plays a vital role in the

K. Lnc Prakash (✉)
Department of Computer Science and Engineering, AITS, Rajampet, Andhra Pradesh, India
e-mail: klnc.prakash@gmail.com

K. Anuradha
Department of Computer Science and Engineering, GRIET, Hyderabad, Telangana, India
e-mail: kodali.anuradha@yahoo.com

© Springer Nature Singapore Pte Ltd. 2018
V. Bhateja et al. (eds.), *Proceedings of the Second International Conference on Computational Intelligence and Informatics*, Advances in Intelligent Systems and Computing 712, https://doi.org/10.1007/978-981-10-8228-3_49

data mining process and selection of subset features that represent significant information from the processed datasets, based on certain criteria [1].

In the cases of datasets with huge dimensionality, clustering process is widely recommended, as the data mining algorithms might need significant volume of computational effort upon using large number of attributes. Optimal feature selection process can ensure

- Enhanced performance of the mining process and eliminating unnecessary attributes which can affect the results.
- Supports in reducing the dimensionality of datasets.
- Reducing cost of computation and delivering more easily interpretable results.

In terms of conservative data mining, in which a sequential file or a single table denotes the dataset, the feature selection algorithms take into consideration only the simple attributes like the categorical and numerical. However, majority of the real-time datasets comprise multivalued attributes that characterized because of its ability to assume multiple values simultaneously.

For instance, the type of movies an individual likes to watch can be the attribute in a movie dataset, and one or more of the movie types like "drama," "action," and "thriller," etc., can be the value for the respective attribute. Such attributes often referred as multivalued attributes.

The multivalued attributes that are categorical with infinite range can be impacted using effective feature optimization process, as the dimensionality of the multiple values gets reduced by great extent. For instance if the sources in an attribute of network transactions comprising categorical values of an infinite dimension, for categorizing the given set of network transactions to normal or attack prone, the attribute sources with multiple values like the different combination of sources that are sparsely presented as values are discarded.

Nevertheless, the sources of the attribute are significant to label the network transactions as DDOS attack prone or not, as the DDOS attacks might get initiated from varied sources. In a different dimension, detecting of the concept drift over the streaming data might turnout inappropriate if the data comprise attributes having multiple values with the constraints as discussed above, as the combination of divergent values for a corresponding attribute reflects the change point in a concept. It might also lead to misclassification of the feature sets.

Many of the earlier contributions in the domain have focused on discovering optimal values pertaining to multivalued attribute, which is considered inappropriate as the selected optimal values might fail to correlate with other attribute's values. Considering such factors, it presumed that the optimal values of attributes comprising multiple values chosen under the context of the other single-valued attribute values.

Some of the literature in the domain has discussed the "multivalued attributes" with different connotation. In some of the studies, the attribute selection addressed in the instance of varied set of attributes having varied domain sizes [2], but not in the case of attributes that are observing multiple values simultaneously.

Relational mining is the research area that leads to developing methods for dealing with the databases structured as multiple tables [3]. It is imperative from the literature that numerous algorithms earlier proposed pertain to relational mining [4–8].

As a separate table for avoiding redundancy in the main table [9], research focusing on such type of attribute usually accomplishes the representation of multivalued attributes in a database with the object of present study emphasizing the context of multi-relational mining.

Though there are varied methods that handle the multivalued attributed direction, not much of solutions were contributed in the earlier studies, pertaining to address the importance of a multivalued attribute selection.

In [10], a solution proposed wherein the k possible values of a multivalued attribute in k binary attributed are used, which allows the application of traditional attribute selection model. Increasing dimensionality of the actual data is one of the key issues envisaged in the process and it might have a significant impact if k is large.

In order to address this gap, in this paper, a contemporary solution for feature selection from attributes comprising multiple values is discussed. The contributions of the proposal are utility scale, which are based on individual values of the attributes, and it is further used for selecting optimal values of attributes with multiple values.

The further sections of this paper are structured as follows: Sect. 2 comprises literature review of feature optimization techniques that comprise multivalued attribute optimization. In Sect. 3, the proposed model of multivalued attribute optimization using utility scale is discussed. Section 4 depicts discussions over the experiments and the evaluation of the proposed model. Section 5 depicts the conclusion of the proposed solution and scope for future work.

2 Related Work

Multivalued data mining is much complex than the mining of single-valued data. In addition, the multivalued data might reduce the performance of data mining algorithms, as the data obtained for mining might be more complicated. Discretization [11–15], a technique adapted for reducing the values in a chosen continuous attribute, is about dividing the range of attributes into finite set of adjacent intervals. The process of discretization is profoundly followed in the data mining algorithms that are highly sensitive for size of data, which is to handle the categorical attributes [16]. Empirical study of the process emphasize the potentiality of the technique in terms of speeding the learning process while improving the accuracy of the learning algorithms [17]. In [17], the authors have stated that the discretization approaches usually proposed along five lines as, the direct method of discretization versus incremental, static way of discretization versus dynamic, supervised method of discretization versus unsupervised, global method of discretization versus local and splitting method of discretization versus merging.

Supervised methods follow the method of discretizing attributes by considering class information, but the unsupervised methods ignore the class information [18–20]. Dynamic methods [21] discretize the continuous attributes while building a classifier in static methods, the discretization process completed even prior to learning the task. The global methods [20] uses all of the records in discretization scheme and it profoundly built with static methods.

In the other dimension, the local methods categorically associate with dynamic approaches in which only partial set of the records are used for discretization. Merging methods usually start the process with the complete list of all continuous values of attributes as cut points. It also eliminates few of them by merging intervals at every step. In the splitting methods, it starts with empty list of cut points and the new ones added at every step.

In the direct methods, human intervention is essential to decide the number of intervals and discretize the continuous attributes. However, in the case of incremental methods, it starts with the simple discretization scheme and gets through the refinement process, as some of them might need stopping criterion for terminating the discretization process.

Equal width and equal frequency [22] are two of the simple discretization algorithms that are unsupervised, static, splitting, and global and a direct method. The following are some of the prominent discretization algorithms in the lines of splitting versus merging.

Some of the splitting methods are equal width and equal frequency [22], Information Entropy Maximization (IEM), Maximum Entropy [23] Class-Attribute Interdependence Maximization (CAIM) [24], Class-Attribute Dependent Discretize algorithm (CADD) [25], Class-Attribute Contingency Coefficient (CACC) [26], and Fast Class-Attribute Interdependence Maximization (FCAIM) [27]. According to the experimental studies carried out [26], it is imperative that the CACC discretization algorithm is very efficient than the other splitting discretization algorithms.

A common characteristic of merging methods is to use the significant test for checking if two adjacent intervals merged. In [28], most typical merging algorithm was proposed, in which apart from the computational complexities, the other major issue envisaged is about the need for users to define various parameters during the application of algorithm. It includes the significance level and the maximal and minimal intervals too.

In [29], other algorithm model is proposed based on Chi Merge [28]. The model improved as it automatically estimates the value of significance level. However, even that algorithm needs the users to provide the inputs on inconsistency rate for stopping the merging procedure and it does not consider the freedom, which plays a vital role in the discretization schemes.

In [30], the previously mentioned problems of Chi2 addressed and it replace the parameter of inconsistency with quality of approximation as parameter after each iteration of discretization. Such mechanism makes necessary modification to a completely automated method for attaining better accuracy in prediction.

The model devised in [31] is an extension to Chi2 technique, which considers the labels of instances that often overlap in real-time environment. Extended Chi2 recognizes the predefined misclassification rate from data itself and it takes into account the effect of variance in two adjacent intervals. The modifications that incorporated can handle even an uncertain dataset. Experimental studies of the model emphasize that it outperformed the other bottom-up discretization algorithms as it discretization scheme, can attain higher levels of accuracy [31].

In a multivalued dataset, records might have varied number of values. Multi-valued classifier (MMC) [32] and the Multivalued Decision Tree Classifier (MMDT) [33] are two significant algorithms used for mining a multivalued dataset. Both MMC and MMDT are decision tree based approaches and MMDT is the extension of MMC. MMDT refines the goodness measurement of a splitting attribute proposed in MMC, in order to improve the classification accuracy. But the crux is that majority of the algorithms detailed above are not suitable for discretizing multivalued datasets, but only can discretize the single-valued data. Though some of the multivalued classifiers [32, 33] proposed, still there are no algorithms that can select features from multivalued datasets with global optimality.

3 The Utility Scale Design and Optimal Set of Values Selection for Multivalued Attribute

The theme of the optimal values selection for multivalued attributes proposed is the influence of utility-based mining that selects features based on their significance contribution than their frequency of occurrence. For instance, in market basket data, the profit gained against an item is more relevant than its frequency in transactions. Similarly, the values appeared for multivalued attributes must be selected based on their presence in significant transactions rather their frequency. This way of values selection for multivalued attributes reflects the relevance with values of the simple attributes. The detailed exploration of the model is proposed as follows:

Let DS be the dataset formed by set of records such that records contain one or more multivalued attributes. Let SA be the set of simple attributes (single-valued attributes) of the records in the said dataset DS. Let MVA be the set of multivalued attributes having in the records of dataset DS. The notations $v(sa_i)$ and $v(mva_i)$ denote the values assigned to simple attribute sa_i and multivalued attribute mva_i, respectively. Each entry in set $v(mva_i)$ is again a set representing a multivalue set of respective multivalued attribute in a record. However, each entry of set $v(sa_i)$ is a single value.

   The proposed utility-scale denotes the scope of the simple attributes involved in the records of the given dataset. Further, this utility scale enables to estimate the significance of the values of the multivalued attributes. In order to estimate the utility scale of the given dataset, initially estimates the coverage of each simple attributes value (see Eq. 1) as item utility, such that values of simple attributes as items. Then the transaction utility of a transaction, which is aggregate value of the item utility of all simple values those exists in the corresponding transaction assessed (see Eq. 2). Further assesses the corpus utility or dataset utility, which is the aggregate value of transaction utility of all transactions exists in the corresponding corpus or dataset (see Eq. 3). The product of the corpus utility and the utility threshold given is the utility scale (see Eq. 4). The utility scale threshold is usually greater than zero and less than one.

   Further, this utility scale is used to estimate the significance of each value assigned to the multivalued attribute. The overall process initially estimates the mean coverage of each value assigned to corresponding multivalued attribute (see Eq. 5). Afterward estimates transactional utility of each value of the multivalued attribute, which is the aggregate of the utilities obtain to transactions those contains the corresponding value of the multivalued attribute (see Eq. 6). The product of the mean coverage and transactional utility of a multivalued attribute's value is the value utility (see Eq. 7) of corresponding value of the multivalued attribute. If this value utility is greater than or equal to the utility scale, then the respective value of the multivalued attribute is optimal. The mathematical model of the description is as follows:

   The coverage of each value of simple attribute is referred as item utility, which is measured as follows:

$$\overset{|SA|}{\underset{i=1}{\forall}} \left\{ sa_i \exists sa_i \in SA \right\}$$

Begin

$w(sa_i) = ||v(sa_i)||$ //consider the values $v(sa_i)$ assigned to attribute sa_i as new set $u(sa_i)$, such that no duplicates exist

$$\overset{|w(sa_i)|}{\underset{j=1}{\forall}} \left\{ e_j \exists e_j \in w(sa_i) \right\}$$

Begin

$$u(e_j) = \sum_{k=1}^{|v(sa_i)|} \left\{ 1 \exists e_j \cong v_k \right\} \tag{1}$$

//v_k is the running value of the set $v(sa_i)$

   End

   End

The aggregate value of the utility of all items in a transaction referred as the transaction utility, and the aggregate of the utility of all transactions is referred as corpus or dataset utility that is measured as follows:

$u(DS) = 0; //$ dataset utility set zero initially

$$\overset{|DS|}{\underset{=1}{\forall}} \{t_i \exists t_i \in DS\}$$

Begin

$$u(t_i) = \frac{\sum_{j=1}^{|t_i|} \{u(v_j) \exists v_j \in \{t_i \wedge v(sa_j)\}\}}{|DS|} \qquad (2)$$

//aggregating the utility of all items existing in transaction t_i

$$u(DS) + = u(t_i) \qquad (3)$$

End

The product of the corpus utility $u(DS)$ and utility threshold τ (usually the threshold is in between 0 and 1) referred as utility scale usc:

$$usc = u(DS) \times \{\tau \exists 0 < \tau < 1\} \qquad (4)$$

Let the value sets of a multivalued attribute mva_i found in all transactions of data set DS be set MVS and then assess the utility of each entry in MVS as follows:
Let $UMVS$ be the set containing all unique values of MVS
Find the mean coverage of each value as

$$\overset{|UMVS|}{\underset{i=1}{\forall}} \{e_i \exists e_i \in UMVS\}$$

Begin

$$mc(e_i) = \frac{\sum_{j=1}^{|MVS|} \{1 \exists e_j \in MVS \wedge e_j \cong e_i\}}{|DS|} \qquad (5)$$

End

Find the transactions level utility, which is the aggregate of the utility of transactions those contains each entry of $UMVS$ as:

$$\overset{|UMVS|}{\underset{i=1}{\forall}} \{e_i \exists e_i \in UMVS\}$$

Begin

$$tu(e_i) = \sum_{j=1}^{|DS|} \{u(t_j) \exists t_j \in DS \wedge e_i \in t_j\} \tag{6}$$

End

Value utility of each entry of set *UMVS* is measured as the product of the mean coverage and aggregate utility of the transactions having that value, which is as follows:

$$\bigvee_{i=1}^{|UMVS|} \{e_i \exists e_i \in UMVS\}$$

Begin

$$vu(e_i) = tu(e_i) \times mc(e_i) \tag{7}$$

End

Then, select the entries of *UMVS* as optimal values of the respective multivalued attribute, such that the entries with value utility is greater than or equal to utility scale *usc*.

The optimal values selected for multivalued attribute through the proposed utility scale are significantly divergent from the values selected by their frequency.

The resultant values of multivalued attributes reflect the transaction level scope, contradicting to this, the values of multivalued attribute selected based on their transaction level frequency are less optimal.

4 Experimental Study and Performance Analysis

In this section of the study, the experimental setup, dataset, and the results are obtained from the experiments performed over the proposed model discussed. K-means algorithm [34] is used for clustering the unlabelled transactions of the dataset. Implementation of the proposed solution is carried out using Java 8 on a computer with 4 GB ram capacity, and i5 processor. Scripts are defined using R programming language [35] for assessing metrics on resultant cluster.

4.1 The Dataset

Experiments carried out on CORA [36] dataset comprise 2708 records and every record is a scientific publication from one of the seven chosen categories (Reinforcement learning, Case-Based Reasoning, Probabilistic Methods, Rule Learning, Neural Networks, Genetic Algorithm, Theory). Every record constitutes entries for subset of 1433 unique words that are simple attributes and set of values for two multivalued attributes termed as cited article and citing article.

Every record comprises subset of article ids for selected 5429 unique article ids as multivalue set for the multivalued attributes. Accuracy and performance of the proposed solution are explored using the cluster evaluation metrics, cluster purity, harmonic mean of clusters, inverse of purity, and harmonic mean of topics. For setting this, the recommended set of documents was chosen under the topic context as knowledge base and documents are clustered as corpus, that are grouped according to topic in order to facilitate cluster optimality evaluation as per the chosen metrics.

4.2 Assessment Metrics and Strategy

Metrics purity, inverse of purity, and harmonic mean play a vital role in the cluster evaluation process. The frequency of category in all the resultant clusters reflects as cluster purity [37]. Metric purity nullifies any kind of noise from the corresponding clusters, but does not have the ability to discover the relativeness of the documents, i.e., if each document is seen as one cluster, then it results in clusters having maximal purity. Hence, the metric called inverse of purity is significant in terms of identifying scope of documents of a cluster as same category. The metric inverse of purity is imperative in discovering the cluster with maximum recall value for every category.

Defining a cluster comprising all input documents results in maximum value for inverse of purity as the metric is not able to nullify the mix of documents from varied categories. It is imperative to consider the harmonic mean of clusters alongside the purity and inverse of purity. Harmonic mean of cluster is the inverse purity and combination of purity that are estimated by comparing every category with the cluster having higher combined precision and recall [38] termed as F-Measure.

4.3 Performance Analysis

Clusters formed from the records along with multivalued attributes are optimized to optimal extent with the usage of the proposed model, as their f-measure is high. The level of purity observed in every resultant cluster is with high accuracy. Statistics pertaining to the experimental study is depicted in Table 1.

To exhibit the significance of proposed model, K-means clustering technique is used on the records with multivalued attributes which optimized traditional frequency-based approach (In terms of selecting most frequent values of multivalued attributes). Cluster purity and f-measure observed with the clusters discovered from records optimized by the proposal are much significant compared to

Table 1 The statistics of the input data and results obtained

	Utility scale based multivalued set optimization	Frequency-based multivalued set optimization
Total number of records in CORA [36]	2708	2708
The maximum number of simple attributes	1433	1433
The maximum size of the multivalued set	5429	5429
The number of multivalued attributes	2	2
The number of classes (clusters)	7	7
The average of F-measure	0.94	0.81
Average cluster purity	0.97	0.85
Average clustering accuracy	0.95	0.77

Fig. 1 Cluster purity observed for divergent clusters

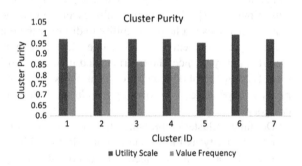

Fig. 2 F-Measure (harmonic Mean) observed for divergent clusters

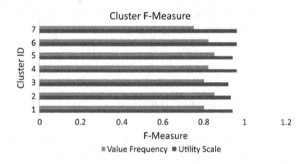

the results generated from clusters derived from records optimized using other traditional frequent value selection (as depicted in Figs. 1 and 2).

The cluster accuracy is observed for both models depicted in Fig. 3, which is the ratio between the resultant true documents count and actual true documents count of the respective cluster.

Fig. 3 Accuracy ratio of divergent clusters obtained from both models

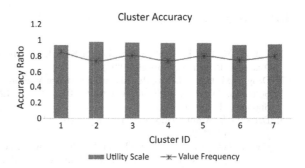

5 Conclusion

This work proposes a novel approach to select optimal value set for multivalued attributes, which facilitates to measure their importance for mining models. Unlike the general tendency of selecting values for multivalued attributes based on frequency, the proposed model selects the values based on the corresponding transaction weight. The devised model is influenced by the strategy of utility mining, where the values are selected by their significance rather their frequency. Similarly here, the values of multivalued attributes are selected based on their significance toward the weighted transactions rather their frequency. The significance of the values is estimated by the weights of the transactions in which those values exist. The transaction weights are estimated by the occurrence of the simple attribute values. The experimental study evinced that the proposed model is considerably best to select the optimal values for multivalued attributes compared to the value selection by frequency. In order to explore this, experiments are conducted on benchmark dataset called CORA [36], which is having a couple of multivalued attributes along with simple attributes. The optimality of the selected values is assessed by clustering the said dataset (with multivalued attributes having selected values) using K-means. The metrics such as cluster purity cluster harmonic mean (f-measure) and cluster accuracy are considered to notify the impact of the values selected by proposed model and value frequency on clustering process. The results obtained for the model proposed here in this manuscript are motivating our feature research in many directions, such as using the proposed utility scale as fitness function for evolutionary techniques to select optimal values for multivalued attributes; also, it is possible to device heuristic scales to select optimal values for multivalued attributes.

References

1. Liu, H., Motoda, H.: Feature Selection for Knowledge Discovery and Data Mining. Kluwer Academic Publishers, Norwell (1998).
2. Deng, H., Runger, G., Tuv, E.: Bias of Importance Measures for Multi-valued Attributes and Solutions. In: Honkela, T. (ed.) ICANN 2011, Part II. LNCS, vol. 6792, pp. 293–300. Springer, Heidelberg (2011).
3. Dzeroski, S.: Multi-relational data mining: an introduction. SIGKDD Explorations Newsletter 5(1), 1–16 (2003).
4. Garriga, G.C., Khardon, R., De Raedt, L.: On mining closed sets in multi-relational data. In: Proceedings of the 20th International Joint Conference on Artificial Intelligence, pp. 804–809. Morgan Kaufmann Publishers Inc., San Francisco (2007).
5. Goethals, B., Page, W., Mampaey, M.: Mining interesting sets and rules in relational databases. In: Proceedings of the ACM Symposium on Applied Computing, pp. 997–1001. ACM, New York (2010).
6. Nijssen, S., Jimenez, A., Guns, T.: Constraint-based pattern mining in multi relational databases. In: Proceedings of the 11th IEEE International Conference on Data Mining Workshops, pp. 1120–1127 (2011).
7. Siebes, A., Koopman, A.: Discovering relational item sets efficiently. In: Proceedings of the SIAM International Conference on Data Mining, pp. 108–119 (2008).
8. Spyropoulou, E., De Bie, T., Boley, M.: Interesting pattern mining in multi relational data. Data Mining and Knowledge Discovery 28(3), 808–849 (2014).
9. Elmasri, R., Navathe, S.B.: Fundamentals of Database System, 6th edn. Addison- Wesley, USA (2010).
10. Hall, M.A., Holmes, G.: Benchmarking attribute selection techniques for discrete class data mining. IEEE Transactions on Knowledge and Data Engineering 15(3), 1437–1447 (2003).
11. Cormen, Thomas H. Introduction to algorithms. MIT press, 2009.
12. Geiser, Jürgen. "Discretization methods with analytical solutions for a convection–reaction equation with higher-order discretization's." International Journal of Computer Mathematics 86.1 (2009): 163–183.
13. Mizianty, Marcin J., Lukasz A. Kurgan, and Marek R. Ogiela. "Discretization as the enabling technique for the Naive Bayes and semi-Naive Bayes-based classification." The Knowledge Engineering Review 25.04 (2010): 421–449.
14. Yang, Y., Webb, G.I., Wu, X. Discretization Methods. Data Mining and Knowledge Discovery Hand- book, 2nd ed. Springer, Berlin, pp. 101–116 (2010).
15. Zhou, Lu, and Birsen Yazici. "Discretization error analysis and adaptive meshing algorithms for fluorescence diffuse optical tomography in the presence of measurement noise." IEEE Transactions on Image Processing 20.4 (2011): 1094–1111.
16. Cios, Krzysztof J., and Lukasz A. Kurgan. "CLIP4: Hybrid inductive machine learning algorithm that generates inequality rules." Information Sciences 163.1 (2004): 37–83.
17. Liu, Huan, et al. "Discretization: An enabling technique." Data mining and knowledge discovery 6.4 (2002): 393–423.
18. Ferreira, Artur, and Mario Figueiredo. "Unsupervised joint feature discretization and selection." Iberian Conference on Pattern Recognition and Image Analysis. Springer Berlin Heidelberg, 2011.
19. Jiang, ShengYi, and Wen Yu. "A local density approach for unsupervised feature discretization." International Conference on Advanced Data Mining and Applications. Springer Berlin Heidelberg, 2009.
20. Zeng, an, Qi-Gang GAO, and Dan Pan. "A global unsupervised data discretization algorithm based on collective correlation coefficient." Modern Approaches in Applied Intelligence (2011): 146–155.

21. Wu, Qing Xiang, et al. "Improvement of decision accuracy using discretization of continuous attributes." International Conference on Fuzzy Systems and Knowledge Discovery. Springer Berlin Heidelberg, 2006.
22. Chiu, David KY, Andrew KC Wong, and Benny Cheung. "Information Discovery through Hierarchical Maximum Entropy Discretization and Synthesis." (1991): 125–140.
23. Wong, Andrew KC, and David KY Chiu. "Synthesizing statistical knowledge from incomplete mixed-mode data." IEEE Transactions on Pattern Analysis and Machine Intelligence 6 (1987).
24. Kurgan, Lukasz A., and Krzysztof J. Cios. "CAIM discretization algorithm." IEEE transactions on Knowledge and Data Engineering 16.2 (2004): 145–153.
25. Ching, John Y., Andrew K. C. Wong, and Keith C. C. Chan. "Class-dependent discretization for inductive learning from continuous and mixed-mode data." IEEE Transactions on Pattern Analysis and Machine Intelligence 17.7 (1995): 641–651.
26. Tsai, Cheng-Jung, Chien-I. Lee, and Wei-Pang Yang. "A discretization algorithm based on class-attribute contingency coefficient." Information Sciences 178.3 (2008): 714–731.
27. Kurgan, Lukasz A., and Krzysztof J. Cios. "Fast Class-Attribute Interdependence Maximization (CAIM) Discretization Algorithm." ICMLA. 2003.
28. Kerber, Randy. "Chi merge: Discretization of numeric attributes." Proceedings of the tenth national conference on Artificial intelligence. Aaai Press, 1992.
29. Liu, Huan, and Rudy Setiono. "Feature selection via discretization." IEEE Transactions on knowledge and Data Engineering 9.4 (1997): 642–645.
30. Tay, Francis EH, and Lixiang Shen. "A modified Chi2 algorithm for discretization." IEEE Transactions on knowledge and data engineering 14.3 (2002): 666–670.
31. Su, Chao-Ton, and Jyh-Hwa Hsu. "An extended chi2 algorithm for discretization of real value attributes." IEEE transactions on knowledge and data engineering 17.3 (2005): 437–441.
32. Chen, Yen-Liang, Chang-Ling Hsu, and Shih-Chieh Chou. "Constructing a multi-valued decision tree." Expert Systems with Applications 25.2 (2003): 199–209.
33. Chou, Shihchieh, and Chang-Ling Hsu. "MMDT: a multi-valued decision tree classifier for data mining." Expert Systems with Applications 28.4 (2005): 799–812.
34. Hartigan, John A., and Manchek A. Wong. "Algorithm AS 136: A k-means clustering algorithm." Journal of the Royal Statistical Society. Series C (Applied Statistics) 28.1 (1979): 100–108.
35. Ihaka R, Gentleman R. R: a language for data analysis and graphics. Journal of computational and graphical statistics. 1996 Sep 1; 5(3):299–314.
36. https://relational.fit.cvut.cz/dataset/CORA.
37. Zhao Y, Karypis G. Criterion functions for document clustering: Experiments and analysis.
38. Steinbach M, Karypis G, Kumar V. A comparison of document clustering techniques. In KDD workshop on text mining 2000 Aug 20 (Vol. 400, No. 1, pp. 525–526).

Android-Based Security and Tracking System for School Children

Kouthrapu Ravisankar and V. Vanitha

Abstract There has been a steep rise in crime on children. As a result, safety becomes the first priority of government. In this paper, a security system for school children is implemented by tracking children and school bus, which enables parents to track their children at runtime. They will be notified when the child is going out of the school and also when child is returning home to avoid any misplacement of the child. Security system for school children is implemented by using Global Positioning System (GPS). Mobile app is developed for parents and staff members to get location of children and school bus using GPS. The mobile application also contains contact details of bus in charge and principal so that they can be contacted by parents in case of emergency. Also, by using Radio Frequency Identification (RFID) technology, parents can find out the child's entry and exit status in the bus.

Keywords Security system · Tracking system · GPS · RFID
Android · ARM microcontroller

1 Introduction

Security arrangements in schools play an important role in the selection of school for admitting children by their parents. One of the problems that parents and schools are facing is that the school children are kidnapped when they are waiting to board the bus near the school premises. So, schools have to ensure that their security system is trustworthy and parents consider this as a positive aspect for admitting their children in a particular school.

K. Ravisankar · V. Vanitha (✉)
Department of Electrical and Electronics Engineering, Amrita School of Engineering,
Amrita Vishwa Vidyapeetham, Coimbatore, India
e-mail: v_vanitha@cb.amrita.edu

K. Ravisankar
e-mail: raviiit463@gmail.com

© Springer Nature Singapore Pte Ltd. 2018
V. Bhateja et al. (eds.), *Proceedings of the Second International Conference
on Computational Intelligence and Informatics*, Advances in Intelligent Systems
and Computing 712, https://doi.org/10.1007/978-981-10-8228-3_50

There are so many reasons to give a good security system for school children's. Several bitter incidents forced to develop a security system for providing safety to school children's. Missing of the student cases and kidnappings are increasing in advance. So, parents are not comfortable until the child resumes back to home. Solutions can be implemented on school-owned buses and school ID cards so that the child can be tracked continuously. Security arrangements in schools are nowadays improving because of improvement in technologies such as wireless networks, embedded controllers, etc. Normally, parents do not know any information about their children when they leave for school. By using the proposed security system, parents can know the status of their children. Through tracking school bus and children, parents can know the children's location and compare it with the bus location through mobile app. By doing so, they can know whether the child is on the bus or not. During the school timing, if the child goes out of the school, then parents will get a call and the parents can contact the child's caretaker immediately to know the reason. When the child is returning from school and the school bus reach near child's home location, parents get a call to pick up the child.

2 Related Work

Some authors implemented methods to give security for school children. These are the recent methods prepared for security. The usage of RFID in school security system reduces operating cost and it provides simplicity for storing and retrieval of data and security aspects are also maintained while using RFID. It manages student attendance system using a web design and stores the data in memory [1–3]. The existing GPS- and GSM-based system reduces the crime rate against school children by tracking the activities of children using Android-based database module [4–6]. For transmitting, this system uses ARM7 microcontroller, modules such as GSM, GPS, and voice recording and playback. Receiver module contains Android mobile and database monitoring station [7, 8]. Radhika N. et al. presented a smart grid test bed based on GSM technology. This is capable of fault detection and self-healing. The communication is achieved through GSM modules [9]. The security system is implemented for women by using RFID and GSM. This system stores the RFID information when women enter into any place and that information is sent to her friends through GSM. This system uses AT89C52 microcontroller and the location is tracked through GPS [10, 11].

Every system uses GSM to send information to parents continuously in a particular time period. It will be irritating to the parents for receiving texts from respective schools about their children status for every 5 min.

In this work, an Android app will be created for tracking the student and bus location whenever the user needs it. In addition to that parents can communicate with the drivers as the details of the driver will be updated regularly and using Google Maps, parents can navigate the direction of the bus.

3 Proposed System

The proposed system includes two separate modules, child module and bus module as shown in Fig. 1. In child module, both GPS and GSM are interfaced with microcontroller. There is a serial communication between microcontroller and GPS. The microcontroller used in this module is LPC2148. Every RFID tag has a unique number. The child module has a GPS based ID card which will be carried by the child while going to school. This will give updates about the status of the child to school principal and parents.

In bus module, RFID reader and GPS are interfaced with another microcontroller. RFID reader reads the RFID tag number of child according to which his/her entry and exit status are stored in database. GPS gives the location of bus to microcontroller every 1 min which is updated in database. To retrieve data from online database, one mobile application is created. This mobile application will have two login pages and location given by the GPS is traced by using Google Maps.

RFID is a technology where reader reads encoded digital data captured using radio waves. It is similar to bar coding, where data from a tag is captured by a reader and information is stored in any database. It belongs to a group which automatically identifies object and captures data. This group automatically identifies the tag, collects data about tags, and sends those data into any storage. RFID system includes RFID tag and RFID reader.

RFID tag consists of an antenna which is used to transfer digital data and the data are captured by RFID reader via radio waves. GPS is a satellite infrastructure based navigation system which orbits around earth with the use of network of

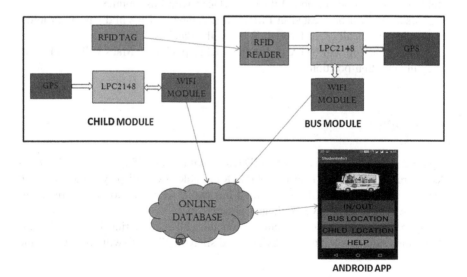

Fig. 1 Proposed security system

24 satellites and it contributes in tracking locations of users in all other conditions. Space segment, user segment, and control segment are present in GPS system.

Android is a software application built for mobile devices. Android app is written in Java programming language and uses Java core libraries. The Android market features both free and prices apps. Android applications constitute one or more application components. A different role is performed by each component in the overall application behavior and activation of each component can be done separately.

4 Methodology

4.1 Child Module

The proposed system is built on LPC2148 microcontroller board and to compute the position of child, a commercial GPS receiver is interfaced. A WiFi module is interfaced with microcontroller to store data in database.

4.2 Bus Module

After creating child module, bus module is built on LPC2148 microcontroller board and by using another commercial GPS, the bus positions will be tracked. RFID reader also interfaced with another UART port. RFID reader is to know whether the child entered into bus or not. WiFi is interfaced with bus module.

A database will be created in FIREBASE database. The child's position information is periodically sent through WIFI to the database.

An Android app will be created in the final step. Google Maps will be added to track the children position.

5 Results

Figure 2 shows the interfacing of RFID reader with LPC2148 using serial Universal Asynchronous Receiver/Transmitter (UART) cable. LCD displays entry and exit status of the child. The LPC2148 board has two UART pins for serial communication. A 5 V charger gives sufficient power.

Figure 3a shows the child status when the child is entering into school bus. When the child enters into bus, LCD will display "IN". LCD will also display the

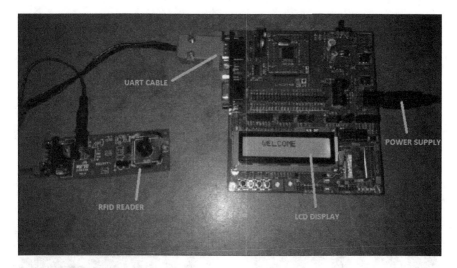

Fig. 2 Interfacing RFID reader and LPC2148

same when the card is read for first time. Figure 3b shows the child status when the child exits from the school bus. LCD will display "OUT" when the tag is read second time. If any other bus student tries to swipe the RFID card, it will display "TRY AGAIN".

GPS is interfaced with LPC 2148. This GPS gives GPGGA (Global Position Fix data) values which contains the longitude and latitude values. By using these two values, Google Maps gives the location of the child and school bus. Output of the GPS module is shown Fig. 4.

The main page of Android application is created, which contains two buttons for login as shown in Fig. 5a. First button is for staff members to log in. By clicking on the staff button, application goes to staff login page. By providing staff identification number and secured password, application goes to bus tracking page as shown in Fig. 5c. By clicking the buttons of particular bus location, it gives the location of the bus. Map button is to track the children location using Google Maps.

All bus in charge and principal contact details are added in help page. Figure 6a shows bus location and Fig. 6b shows help page.

As shown in Fig. 7c, parent login has four buttons. First button is to know whether the child is inside or outside the bus. Bus location and child location buttons give the location points of bus and child. Help button is for contact details of bus in charge and principal.

An account is created in FIREBASE database to store location points and children status as shown in Fig. 8.

(a)

(b)

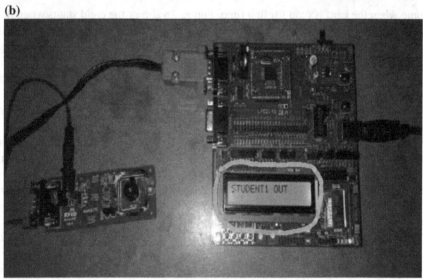

Fig. 3 **a** Interfacing RFID reader and LPC2148 **b** Student 1 exit status

Fig. 4 Getting the location of school bus from GPS

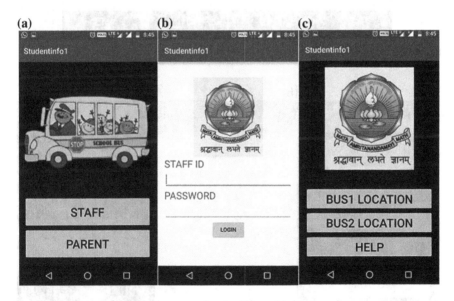

Fig. 5 **a** Main page; **b** staff login page; **c** bus tracking and help page

Fig. 6 a Bus position tracking; **b** help page

Fig. 7 a Main page; **b** parents login page; **c** bus and child tracking and help page

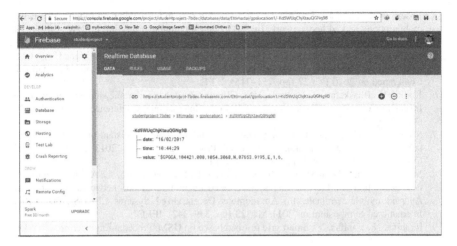

Fig. 8 Storing of the data in online database

6 Conclusion

The paper proposes design of a secure system for tracking school children, which is capable of giving updated information about the location of bus and children to their parents and school authorities at runtime. RFID reader and GPS in child module and school bus module are interfaced with LPC2148 controller which facilitates security system. Online database is created to store data given by microcontroller. An Android application is created to retrieve the data from the database. By using this application, child and the bus locations can be tracked with Google Maps. A wearable sensor device will be made available in future, which will enhance the security of children in an effective manner. A defined module which can even work in extreme conditions with enhanced cybersecurity issues will be considered in future.

References

1. Deenadayalan, C., Murali, M., Baanupriya, LR. Implementing prototype model for School Security System (SSS) using RFID. In: Computing Communication & Networking Technologies (ICCCNT), 2012 Third International Conference on 2012 Jul 26 (pp. 1–6). IEEE.
2. Hutabarat, DP., Hendry, H., Pranoto, JA., Kurniawan, A: Human tracking in certain indoor and outdoor area by combining the use of RFID and GPS. In: Wireless and Mobile (APWiMob), 2016 IEEE Asia Pacific Conference on 2016 Sep 13 (pp. 59–62). IEEE.
3. James, S., Verrinder, RA., Sabatta, D., Shahdi, A.: Localisation and Mapping in GPS-denied Environments using RFID Tags. In: Robotics and Mechatronics Conference of South Africa (ROBOMECH), 2012 5th 2012 Nov 26 (pp. 1–4). IEEE.

4. Ito, A., Ohta, T., Inoue, S., Kakuda, Y.: A security system for children on school route using mobile phone and ad hoc network. In: 3rd technical committee meeting on NW software of IEICE 2008 Feb.
5. Sunehra, D., Priya, PL., Bano, A.: Children Location Monitoring on Google Maps Using GPS and GSM Technologies. In: Advanced Computing (IACC), 2016 IEEE 6th International Conference on 2016 Feb 27 (pp. 711–715). IEEE.
6. Kodire, V., Bhaskaran, S., Vishwas, HN.: GPS and ZigBee based traffic signal preemption. In Inventive Computation Technologies (ICICT), In: International Conference on 2016 Aug 26 (Vol. 2, pp. 1–5). IEEE.
7. Saranya, J., Selvakumar, J.: Implementation of children tracking system on android mobile terminals. In: Communications and Signal Processing (ICCSP), 2013 In: International Conference on 2013 Apr 3 (pp. 961–965). IEEE.
8. Mori, Y., Kojima, H., Kohno, E., Inoue, S., Ohta, T., Kakuda, Y., Ito, A.: A self-configurable new generation children tracking system based on mobile ad hoc networks consisting of Android mobile terminals. In: Autonomous Decentralized Systems (ISADS), 2011 10th International Symposium on 2011 Mar 23 (pp. 339–342). IEEE.
9. Radhika, N., Vanitha, V.: Smart grid test bed based on GSM. Procedia Engineering. 2012 Jan 1;30:258–65.
10. Hussain, SM., Nizamuddin, SA., Asuncion, R., Ramaiah, C., Singh, AV.: Prototype of an intelligent system based on RFID and GPS technologies for women safety. In: Reliability, Infocom Technologies and Optimization (Trends and Future Directions) (ICRITO), 2016 5th International Conference on 2016 Sep 7 (pp. 387–390). IEEE.
11. Exauty, S., Gao, H., Wang, X., Yu, X.: Design of navigation system and transmitter based on GPS and GSM. In Logistics, Informatics and Service Sciences (LISS), 2016 International Conference on 2016 Jul 24 (pp. 1–4). IEEE.

Video Shot Boundary Detection and Key Frame Extraction for Video Retrieval

G. S. Naveen Kumar, V. S. K. Reddy and S. Srinivas Kumar

Abstract Shot boundary detection and key frame extraction are the primary steps for video indexing, summarization, and retrieval. We have proposed an advanced and sophisticated scale-invariant feature transform (SIFT) key point matching algorithm. This approach is based on capturing the changing statistics of different kinds of shot boundaries using SIFT, followed by extracting key frames from each segmented shot by means of image information entropy technique. We can enhance the performance of this algorithm by eliminating the redundant key frames by using the edge matching rate technique. The proposed algorithms have been applied to different categories of videos to detect shot boundaries and key frame extraction. The experimental results have been excellent and they show that these algorithms are effective and efficient in performance in terms of detection of shots and extraction of ultimate key frames.

Keywords SIFT · Shot boundary detection · Image information entropy
Key frame extraction · Edge matching rate · CBVR

1 Introduction

There has been the drastic increase in the size of video database due to the avail-ability of affordable video capturing devices, reduced prices of storage devices, and usage of abundant broadband connections. Hence, an advanced and sophisticated video database system is needed for browsing, searching, and retrieval [1, 2]. In the conventional video database system, we retrieve the videos basing on the text, manual video annotation, and adding of tags to videos. It consumes a lot of time

G. S. N. Kumar (✉) · V. S. K. Reddy
Department of ECE, Malla Reddy College of Engineering & Technology,
Hyderabad, India
e-mail: gsrinivasanaveen@gmail.com

S. Srinivas Kumar
Department of ECE, Jawaharlal Nehru Technological University, Kakinada, India

© Springer Nature Singapore Pte Ltd. 2018
V. Bhateja et al. (eds.), *Proceedings of the Second International Conference on Computational Intelligence and Informatics*, Advances in Intelligent Systems and Computing 712, https://doi.org/10.1007/978-981-10-8228-3_51

and involves more human efforts. To overcome this problem, an advanced video search engine called the content-based video retrieval system is needed [3]. The prerequisites for content-based video retrieval are shot boundary detection and key frame extraction [4]. We define a shot as the series of consecutive frames of a continuously captured video between the start and stop operation of a camera. A shot boundary or short transition separates two consecutive shots of a video. A shot transition in turn is classified into cut transition and gradual transition.

Shot boundary detection is an essential operation in indexing, classification, summarization, and retrieval of videos [5]. Hence, there has been a paradigm shift in research into shot boundary detection. The few algorithms that have been developed for this purpose are pixel-based method, block-based shot boundary detection, and histogram-based method. In pixel-based method [6], corresponding pixels in sequence of frames are compared. It is computationally simple and time-saving. Its bottle neck is a small motion in the camera or in the object, it makes two similar consecutive frames dissimilar. In addition, a change in the illumination of the object also affects the similarity detection. Block-based shot boundary detection [7]; the frames are divided into blocks. The corresponding blocks in consecutive frames are evaluated. When compared to the pixel-based method, this algorithm is not that sensitive. Its deficiency is that it is relatively slow due to difficulty in implementation. Moreover, this algorithm cannot recognize quick moving objects. Histogram-based shot boundary detection [8]: It uses the statistical representation of intensity values. It remains invariant to translation, rotation, and camera motion. Its drawback is that it is a failure in case of two dissimilar frames that have the same color distribution.

To overcome the limitations of existing algorithms, we proposed an innovative methodology for shot boundary detection based on scale-invariant feature transform (SIFT) in this paper. It is robust and invariant irrespective of image rotation, image scaling, noise, and variations in illumination [9]. In video summarization, key frame extraction follows shot boundary detection. Key frame extraction is selective or representative frame extraction. It contains significant information of the shot and it summarizes the characteristics of the rest of the frames. It eliminates the redundant information and reduces the amount of data for indexing and retrieval of videos. In the proposed algorithm, entropy values for all the frames in the shot are calculated. The frames that have different entropy values are considered as key frames. There is a possibility of identical key frames being that are repeated in different shots. These repeated key frames are called the redundant key frames that lead to an increase in the video representation data. Eventually, edge matching rate technique is used to remove the redundant key frames and in arriving at the ultimate key frames [10].

The structuring of the remaining sections is organized as follows: Shot boundary detection has been discussed in Sect. 2, key frame extraction is dealt with under Sect. 3, the framework of the proposed method is given in Sect. 4, the experimental results are covered in Sect. 5, and conclusion is given in Sect. 6.

2 Shot Boundary Detection

A shot is described as the series of consecutive frames of continuously recorded video with a single camera. A shot boundary or short transition is separated two consecutive shots of a video based on similarity and dissimilarity. A shot transition in turn is classified into cut transition and gradual transition [11]. A cut is a quick change from a shot to the subsequent shot. A cut exists only between the ending frame of the first shot and the starting frame of the second shot. A gradual transition takes place among the more number of frames. The gradual transition is classified into four types: fade in, fade out, dissolve, and wipe. A gradual transition between a scene and an image is known as "Fade in"; "Fade out" is a gradual transition between an image and a scene; A gradual transition between two scenes, viz., "Fade out" and "Fade in" is called "Dissolve"; "Wipe" is a gradual transition between two scenes separated by a partition line.

For detection of shots, feature extraction is required to verify similarity or dissimilarity between two frames. In this paper, we have proposed SIFT for the extraction of video features [12].

2.1 Scale Invariant Feature Transform

SIFT is used to extract and detect local feature descriptors of a frame. It is the best algorithm in an object matching from a large data base. It is a robust algorithm and remains invariant to changes in illumination, rotation and scaling of the image, and addition of noise. It consists of four stages: Detection of scale-space extreme, accurate key point localization, orientation assignment, and descriptor representation.

1. *Detection of Scale-Space Extreme*

In this stage, we detect interest points or isolated points in the scale space of an image. Isolated points are also called key points which become SIFT features. The key points are obtained by applying local extrema of difference of Gaussian (DoG) space to the frame. We can obtain DoG scale space from Eq. (1).

$$\begin{aligned} D(m, n, \sigma) &= (G(m, n, k\sigma) - G(m, n, \sigma)) * I(m, n) \\ &= L(m, n, k\sigma) - L(m, n, \sigma) \end{aligned} \tag{1}$$

$$L(m, n, \sigma) = G(m, n, \sigma) * I(m, n) \tag{2}$$

G(x, y, σ) is a variable-scale Gaussian kernel defined as

$$G(m, n, \sigma) = \frac{1}{2\pi\sigma^2} e^{-(m^2 + n^2)/2\sigma^2} \tag{3}$$

2. *Accurate Key Point Localization*

Candidate key point location and scale are determined in this stage. Key points are selected depending on the measure of stability. Accurate key points are extracted by removing low contrast and noise affected candidate key points.

3. *Orientation Assignment*

Here, we assign an orientation to each of the key points to construct the descriptor that is invariant to rotation. Orientation histogram of local gradients from the nearest smoothened image L(x, y, σ) is taken to compute the orientation of a key point. We compute the gradient magnitude and the orientation for each sample frame by using the pixel differences of the image.

$$M(m, n) = \sqrt{(L(m+1, n) - L(m-1, n))^2 + (L(m, n+1) - L(m, n-1))^2} \quad (4)$$

$$\theta(m, n) = \tan^{-1}\left(\frac{L(m, n+1) - L(m, n-1)}{L(m+1, n) - L(m-1, n)}\right) \quad (5)$$

The orientation histogram covering 360° of the major orientation with 36 bins.

4. *Descriptor Representation*

A descriptor is generated from gradients information of the local image for each key point. A 16 × 16 pixel region is used to calculate the feature vector descriptor with orientation histograms [13]. Each 4 × 4 region is projected as a histogram with 8 bins and 8 directions. This process forms a SIFT feature vector descriptor with 128 directions shown in Fig. 1.

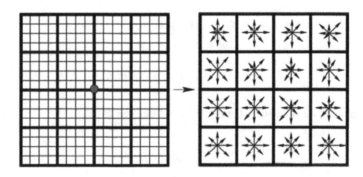

Fig. 1 Key point descriptor generation with 128 directions

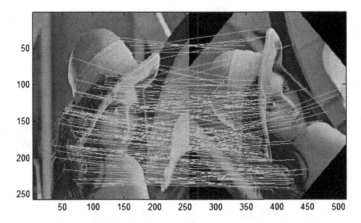

Fig. 2 Key point matching

2.2 Key Point Matching for Shot Boundary Detection

The key points in the current frame are compared with those of the next frame by using the k-nearest neighbor (kNN) search. If the key point has minimum Euclidean distance, it is called the nearest neighbor. kNN search algorithm is the simplest among the machine learning algorithms. kNN depends on the Euclidean distance to determine whether the key points of the current frame are matched with those of the next frame [14].

Euclidean distance is a technique that measures the distance between two feature vectors. Let **a** and **b** be two samples with **p** number of features.

$$d(a,b) = \sqrt{\sum_{i=1}^{p} (a_i - b_i)^2} \tag{6}$$

When the key points in two frames are matched with each other, lines are drawn between the matching key points. If the distance between any two key points is less than 0.6, the matching is accepted. An example key point matching is shown in Fig. 2.

If the number of matched key points between any two frames is greater than the threshold value, these two frames are related to the same region. Otherwise, shot boundary is detected.

3 Key Frame Extraction

Key frame extraction is an essential process in the video retrieval. A key frame is also called a selective or representative frame. It contains required information of the shot and summarizes the characteristics of the rest of the frames. It eliminates the unnecessary information and reduces the amount of data for indexing and retrieval of videos.

3.1 Image Information Entropy

The image information entropy is a technique that signifies the quantity of information contained in a frame [15]. For extracting key frames, information entropy of every frame in the shot is measured. The information entropy calculation for a frame is expressed in Eq. (7).

$$Entropy = - \sum_{i=1}^{L} p(x_i) \times \log(p(x_i)), \tag{7}$$

where

$p(x_i)$ is the probability of gray value x_i

$$p(x_i) = \frac{total\,(x_i)}{m \times n}, \; 0 \leq p(x_i) \leq 1 \; and \; \sum_{i=1}^{n} p(x_i) = 1$$

total (x_i) total number of pixels that have gray value x_i
L   number of gray levels in the frame
m   the height of the frame and
n   width of the frame.

The frames that have different entropy values are considered as key frames. In this process, the key frames are not similar in a given shot. When the other shots are considered, there may be similar key frames resulting in redundant key frames. Entropy depends on frame gray level distribution. The two similar frames may have different gray level distribution resulting in redundant key frames.

3.2 Extract Ultimate Key Frames Using Edge Matching Rate

Edge matching rate technique is used to eliminate the redundant key frames. Eventually, the ultimate key frames are derived. The redundant key frames keep varying based on the type of video. Redundancy is more in news videos, less in movie videos, and very less in the cartoon or animated videos. In edge matching rate, the edges of the candidate key frames are found out with Prewitt operator. Next, the edge of a candidate key frame is compared with that of the other candidate key frame. Edge matching rate is given a threshold value to determine similarity or dissimilarity. If the matching rate is greater than the threshold value, the key frames are similar. This implies that the key frames are redundant. By removing redundant key frames from candidate key frames, we got the ultimate key frames. The ultimate key frames keep varying with changes in threshold value. The edge matching algorithm is as follows:

Edge Matching Algorithm:

i. We apply Prewitt edge detection to the candidate key frames.
ii. Edge matching rate is computed between two candidate key frames using Euclidean distance.
iii. A threshold value is to be set.
iv. If edge matching rate is less than the threshold value, the candidate key frame is the ultimate key frame otherwise, it is a redundant key frame.
v. Step (iii) and Step (iv) are repeated for required results.
vi. All the redundant key frames are removed and the ultimate key frames are obtained.

4   Framework of the Proposed Method

The step-by-step procedure to implement the proposed method is shown in Fig. 3.

i. Input video is taken from video data set.
ii. All the frames are extracted from the video.
iii. SIFT key point feature vectors are extracted for every frame.
iv. Key points from the current frame are matched with those of the next frame for detection of the shot boundaries.
v. The image information entropy method is used to extract key frames from each of the shots.
vi. To remove the redundant key frames by using edge matching rate with different threshold values then we obtain ultimate key frames.

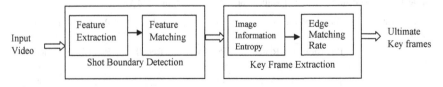

Fig. 3 Framework for the proposed method

5 Experimental Results

To test the efficacy of the algorithms: Shot boundary detection using SIFT, key frame extraction using image information entropy, and edge matching rate, five different categories of videos have been taken for video dataset viz. news, cartoon, entertainment movies, sports, and floral videos.

The performance of the shot boundary detection using SIFT algorithm is evaluated with respect to precision and recall are defined as follows:

$$Precision = \frac{correct\ detection}{correct\ detection + false\ detection} \times 100 \tag{8}$$

$$Recall = \frac{correct\ detection}{correct\ detection + missed\ detection} \times 100 \tag{9}$$

The experimental results show that the proposed method outperforms the existing methods as shown in Table 1.

The proposed shot boundary detection method has both high recall and high precision. Therefore, the performance of the proposed technique was assessed based on the comparison of the outcome of our method with regard to the other conventional methods as shown in Fig. 4.

The type of videos taken, the number of frames contained in every video, the number of shots detected, the number of candidate key frames extracted, and the number of ultimate key frames are mentioned in Table 2.

The candidate key frames that have been extracted from the flower video are shown in Fig. 5. The number of redundant key frames that have been eliminated by using the edge matching rate technique resulting in the ultimate key frames is shown in Fig. 6. Experimental results show that high accuracy and low redundancy have been achieved from the proposed system.

In this algorithm, the SIFT feature extraction method has been used for shot boundary detection. It is invariant to image scaling, rotation, change in illumination, and adding noise. Extraction of redundant key frames is the drawback in the other algorithms. Redundant key frames have been eliminated by using an edge matching

Table 1 The precision and recall values for different categories of videos

Type of query video	Block matching method		Histogram-based method		Proposed method	
	Precision (%)	Recall (%)	Precision (%)	Recall (%)	Precision (%)	Recall (%)
News	79.6	78.4	84.2	82.5	95.6	94.5
Cartoon	77.8	74.2	80.6	79.2	93.3	97.2
Movies	76.5	73.9	82.5	80.3	95.8	94.2
Sports	78.2	75.8	81.4	78.6	95.1	88.5
Flowers	79.2	76.3	83.9	81.4	95.4	93.8

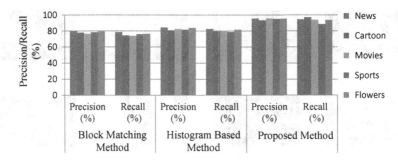

Fig. 4 Performance measurements of the proposed method with those of existing methods

Table 2 The extracted shot boundaries, key frames and ultimate key frames

S. no	The type of video	Number of frames	Shot boundaries	Candidate key frames	Ultimate key frames
1.	News	1250	11	53	21
2.	Cartoon	1580	25	81	54
3.	Movies	1780	21	65	51
4.	Sports	1150	8	37	28
5.	Flowers	850	16	41	34

Fig. 5 The candidate key frames extracted from the flower video

rate technique in the proposed algorithm. Correct shot boundaries and key frames have been extracted by making use of these techniques so this algorithm has shown high precision and high recall rate. Hence, it is robust and effective.

Fig. 6 The ultimate key frames extracted using edge matching rate

6 Conclusion

In this paper, shot boundary detection based on SIFT algorithm has been proposed. Next, we extract the key frames by calculating image entropy values. Later, the redundant key frames have been removed by using edge matching rate technique and the ultimate key frames are obtained. The experimental results and performance measurements show the strength and the accuracy of the proposed framework. Shot boundary detection and key frame extraction have effectively been achieved and this helps in proper functioning of video retrieval.

References

1. Hu, Weiming, et al. "A survey on visual content-based video indexing and retrieval." IEEE Transactions on Systems, Man, and Cybernetics, Part C (Applications and Reviews) 41.6 (2011): 797–819.
2. J. H. Yuan, H. Y. Wang, and B. Zhang, "A formal study of shot boundary detection", Journal of Transactions on Circuits and Systems for Video Technology, vol. 17, no. 2, pp. 168–186, February 2007.
3. Wu, Zhonglan, and Pin Xu. "Shot boundary detection in video retrieval." Electronics Information and Emergency Communication (ICEIEC), 2013 IEEE 4th International Conference on. IEEE, 2013.
4. Zhao Guang-sheng, A Novel Approach for Shot Boundary Detection and Key Frames Extraction, 2008 International Conference on Multimedia and Information Technology, IEEE.
5. Gygli, Michael, Helmut Grabner, and Luc Van Gool. "Video summarization by learning submodular mixtures of objectives." Proceedings of the IEEE Conference on Computer Vision and Pattern Recognition. 2015.

6. Upesh Patel, Pratik Shah, and Pradip Panchal, "Shot Detection using Pixel wise Difference with Adaptive Threshold and Color Histogram Method in Compressed and Uncompressed Video" International Journal of Computer Applications (0975-8887), Volume 64–No. 4, February 2013.
7. Yuan, Jinhui, et al. "A formal study of shot boundary detection." IEEE transactions on circuits and systems for video technology 17.2 (2007): 168–186.
8. Jordi Mas and Gabriel Fernandez, "Video Shot Boundary Detection based on Color Histogram", Digital Television Center (CeTVD), La Salle School of Engineering, Ramon Llull University, Barcelona, Spain.
9. D. G. Lowe, "Distinctive image features from scale-invariant keypoints," International Journal of Computer Vision, vol. 60, pp. 91–110, 2004.
10. Ren, Liping, et al. "Key frame extraction based on information entropy and edge matching rate." Future Computer and Communication (ICFCC), 2010 2nd International Conference on. Vol. 3. IEEE, 2010.
11. C. Cotsaces, N. Nikolaidis, and I. Pitas, "Video Shot Detection and Condensed Representation", IEEE Signal Processing Magazine, March, 2006, pp. 28–37, 2006.
12. Li, Jun, et al. "A divide-and-rule scheme for shot boundary detection based on SIFT." JDCTA 4.3 (2010): 202–214.
13. Hannane, Rachida, et al. "An efficient method for video shot boundary detection and keyframe extraction using SIFT-point distribution histogram." International Journal of Multimedia Information Retrieval 5.2 (2016): 89–104.
14. Liu, Gentao, et al. "Shot boundary detection and keyframe extraction based on scale invariant feature transform." Computer and Information Science, 2009. ICIS 2009. Eighth IEEE/ACIS International Conference on. IEEE, 2009.
15. Lina Sun and Yihua Zhou, "A key frame extraction method based on mutual information and image entropy," 2011 International Conference on Multimedia Technology, Hangzhou, 2011, pp. 35–38.

Performance Enhancement of Full Adder Circuit: Current Mode Operated Majority Function Based Design

Amit Krishna Dwivedi, Manisha Guduri and Aminul Islam

Abstract Paper reports a novel majority function based current mode operated compact sized robust design of 1-bit full adder (FA) circuit. The focus of this work is to reduce the power supply consumption required for performing arithmetic operations by introducing a novel and efficient way of computing 1-bit addition relying on current mode operation. Presented high speed FA design utilizes only 7Ts, per bit, to implement sum and carry functions. The majority function based proposed FA circuit requires lesser number of transistors as compared to the conventional 28Ts FA circuit. A current mirror (CM) circuit has been incorporated in the proposed design to act as a constant current reference to drive the whole circuitry. Further, to evince the uniqueness of the proposed FA design, comparisons have been drawn with various other standard FA designs in terms of different design metrics such as power consumption, supply voltage requirement, power delay product (PDP), energy delay product (EDP) and operating speed. Extensive simulations have been performed using Virtuoso Analog Design Environment of Cadence @ 90 nm technology to verify the proposed design.

Keywords Current mirror · Current mode operation · Full adder circuit
Majority function

A. K. Dwivedi (✉) · M. Guduri · A. Islam
Department of Electronics and Communication Engineering,
Birla Institute of Technology, Mesra, Ranchi 835215, Jharkhand, India
e-mail: a.k.dwivedi@ieee.org

M. Guduri
e-mail: manishaguduri@bitmesra.ac.in

A. Islam
e-mail: aminulislam@bitmesra.ac.in

© Springer Nature Singapore Pte Ltd. 2018   569
V. Bhateja et al. (eds.), *Proceedings of the Second International Conference
on Computational Intelligence and Informatics*, Advances in Intelligent Systems
and Computing 712, https://doi.org/10.1007/978-981-10-8228-3_52

1 Introduction

Evolution of nano-scaled technology demands compact sized electronic components with low power consumption [1]. In conventional integrated circuits, the conducting channel length determines the transit time i.e. gate delay as an important factor to limit the operating speed of the circuits [2]. Device scaling according to Moore's law is also reaching its limit. Apart from these issues, with increasing number of compact portable devices, low power consumption is required for the longer battery life. This is because the supply source or battery technology doesn't advent with the same rate to match the current technology trends which limits the availability of power to the components [1]. Moreover, reduction in power supply consumption also helps in increasing the durability of micro sized devices. This attracts researchers to develop reduced sized circuits with low supply consumption.

Recently, current mode circuits have emerged as an important class of circuits due to its various fertile properties as compared to its voltage mode counterparts. Utilizing current as an active parameter, in place of voltage, provides higher stability and sensitivity. Due to its advantageous properties, current mode circuit where current is used as the active variable in preference to voltage, have been introduced in a wide range of applications. Current mode approach provides superior performance as compared to voltage mode, even in cases where circuits have been synthesized from voltage-mode components due to the lack of suitable alternatives [3]. Current mode operation is explored in this paper to present a sensitive and low power consuming adder circuit.

Full adders play an important role in performing all the arithmetic operations such as addition, subtraction, multiplication and address generation. These FAs are further employed extensively as a building block of other complex logic circuits such as multipliers, convertors, comparators and so on. Thus, FA circuits determine the major sector of power consumption in a chip. Apart from this, a range of functional ICs require FAs in various application specific and microprocessor based logic circuits. Hence, computing addition acts as a critical path to determine the performance of the ICs. In general, XOR gates are integrated to act as FA [4]. However, performance of XOR based FA circuit is limited due to slower response of an XOR circuits employed. These significance demands to boost the performance FA which include power consumption, area efficient and high throughput with high operating speed.

In the view of above, this paper aims to present an efficient and potential way to compute 1-bit addition operation by employing current mode operation with minimum number of transistors.

The rest of the paper is organized as follows. Section 2 presents the elements of the proposed FA circuit. State-of-the-art of proposed FA circuit is mentioned in Sect. 3. Section 4 validates the proposed FA design with the simulation results. Finally, the concluding remarks are provided in Sect. 5.

2 Elements of the Proposed Full Adder

2.1 Current Mirror

Precise and efficient current mirror (CM) circuit has been incorporated as an active element, in the architecture of the proposed FA to generate constant current that can be used as a reference source. Further, to meet the low power consumption and to avoid mismatch of transistors in basic current mirror circuit, this paper utilizes a low voltage CMOS based Wilson current mirror that functions well at all current levels, ranging from weak inversion to strong inversion [5]. The basic structure of Wilson current mirror circuit is reported in Fig. 1a [6]. Mirrored current obtained from the current mirror circuit is utilized as main output variable, as can be seen in various high-speed operating circuits [7]. The output current (I_O) is the mirror current provided as the I_{REF} to the input terminal of the current mirror circuit.

The high-swing super-Wilson CM shown in Fig. 1a uses negative feedback mechanism to provide a high output resistance path. The CM formed by transistors MN5/MN6 mirrors the input reference (I_{REF}) as the output current (I_O). MN2/MN3 forms a basic CM circuit in which the gate terminals of MN2/MN3 are at same potential to provide $I_{REF} = I_O$. The output resistance of current mirror is directly proportional to the magnitude of the loop-gain of the feedback action formed by the output current I_O to the gate of the output transistor MN3. Considering transistors operating in saturation mode and large output resistance offered by current mirror circuit, the output resistance (r_{out}) of the high-swing super-Wilson CM is expressed as:

$$r_{out} \approx g_{m2} r_{o2} r_{o3} \tag{1}$$

where, g_{m2} is the transconductance of the MN2 and r_{o2} and r_{o3} are the output resistance of MN2 and MN3, respectively. The negative feedback loop-gain is

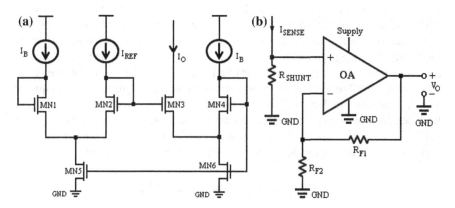

Fig. 1 **a** Low-supply consumption high voltage swing super-Wilson current mirror circuit [5]. **b** Current sensing scheme utilized in the proposed FA circuit to measure the output current (I_O)

credited to transistor MN2 which, in combination with current source load I_{REF}, forms a common-source amplifier which maintains the gate voltage of output transistor MN3 such that I_O traces the I_{REF}. Moreover, a diode-connected MN2 configuration is employed in the utilized CM design to avoid mismatch in the currents by keeping the drain of the mirror transistors MN5/MN6 at the same potential. Further, a cascade connect architecture is used in the design to enhance the low output resistance $(1/g_{m1})$ of the CM. Finally, MN5/MN6 forms a CM which drives the proposed FA design.

Constant current source (see Fig. 1a) utilized here operates at low voltage supply [6] which helps in driving proposed FA circuit with less power consumption. As the fan out of the CM is high enough to act as a constant current source to other circuit, a single CM can be utilized to drive different FA circuit with low supply consumption. This paper presents the FA design which operates using the constant current supplied by the mentioned current mirror circuit. The current mirror circuit acts as a driver to different circuits; hence a single current mirror circuit can be utilized to provide the constant current source supply to the various other FA circuits embedded in the same IC which reduces the area and power consumption.

2.2 Sensing Mechanism for the Proposed FA

The compelling reason for the current-mode operation is to reduce the power supply requirements of the digital circuits. In the reported design of FA circuit using current mode operation, the current at different nodes is analyzed to evaluate the sum and carry outputs. This requires a current sense amplifier (CSA) circuit to be employed in the proposed FA circuit to measure the amount of current flowing through the particular node. For the circuit level model of the proposed FA circuit shown in Fig. 2, the magnitude of current through the node1 and node2 determines the logic state of carry (C_{OUT}) and Sum as output of FA circuit, respectively. The CSA circuit utilized in this work is shown in Fig. 1b. CSA circuit provides output voltage proportional to the current flowing through the connected node.

For carry computation, the output voltage of CSA in the proposed design sets to logic state '1' if I_{SENSE} at node1 is greater than or equal to the voltage equivalent of $2 \times$ reference current (I_O); else it sets to logic '0'. Similarly, for the sum computation @ node2, if the voltage equivalent of the I_{SENSE} is greater than or equal to $3 \times I_O$, then output Sum of FA sets at logic '1', else it sets to logic '0'. The feedback resistance R_{F1} and R_{F2} connected to the sense amplifier helps in proper biasing of the sense amplifier. Further, to set the desired limit i.e. sense margin for logic '1' and sense margin for logic '0', values of feedback resistance are set as

$$V_{SENSE} = V_{SHUNT} \times \left(1 + \frac{R_{F1}}{R_{F2}}\right). \tag{2}$$

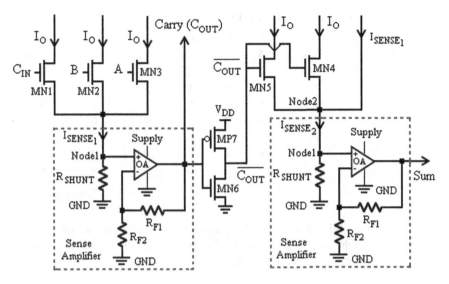

Fig. 2 Circuit level model of proposed FA circuit with current sensing scheme

3   State-of-the-Art of Proposed FA

3.1   Majority Function

The proposed FA circuit is implemented by means of the majority function (MF). Majority poll determines the logic state attained by a node/branch. Combinational logic circuit that determines the majority poll among the various odd numbers of input signals is termed as MF based circuit. Similar concept has been explored in the reported design which provides high logic state '1' when number of logic '1' are more than number of logic state '0' and vice versa. Here, 3/5 input signals are accepted to evaluate carry and sum, respectively, based upon the MF. Table 1 summarizes the MF based evaluation of the sum and carry bits by the proposed FA design.

From the table, it can be observed that majority poll of 3-inputs (A, B and C_{IN}) simply provides the carry (C_{OUT}) of the FA circuit. The NOT of MF ($\overline{C_{OUT}}$) is further used to provide Sum of the FA circuit by majority poll of 5-inputs (A, B, C_{IN}, $\overline{C_{OUT}}$, $\overline{C_{OUT}}$). Thus, carry (C_{OUT}) and sum can be expressed as

$$C_{OUT} = majoriy\ function\ (A,\ B,\ C_{IN}) \tag{3}$$

$$Sum = majoriy\ function\ (A,\ B,\ C_{IN},\ \overline{C_{OUT}},\ \overline{C_{OUT}}) \tag{4}$$

where, $\overline{C_{OUT}}$ is produced by inverting the 3-input MF (C_{OUT}) result bit. Equations (3) and (4) represent the carry and sum equations for the FA circuit. Figure 2

Table 1 Majority function for evaluation of sum and carry for 1-bit FA

3 I/P MF for C_{OUT}				5 I/P MF for sum					
A	B	C_{IN}	C_{OUT}	A	B	C_{IN}	$\overline{C_{OUT}}$	$\overline{C_{OUT}}$	Sum
0	0	0	0	0	0	0	1	1	0
0	0	1	0	0	0	1	1	1	1
0	1	0	0	0	1	0	1	1	1
0	1	1	1	0	1	1	0	0	0
1	0	0	0	1	0	0	1	1	1
1	0	1	1	1	0	1	0	0	0
1	1	0	1	1	1	0	0	0	0
1	1	1	1	1	1	1	0	0	1

clearly shows the MF output of the 3-input and 5-input functions as carry and sum respectively, for the proposed FA circuit. This paper bases (3) and (4) to implement the FA circuit with improved performance.

3.2   Implementing FA Using Majority Function

The majority function is realized by sensing the current flowing through the different branches. Magnitude of current flowing through a combination of 3-branches (3-inputs A, B and C_{IN}) and 5-branches (5-inputs A, B, C_{IN}, $\overline{C_{OUT}}$ and $\overline{C_{OUT}}$) are sensed to compute the carry and sum for the proposed MF based FA design. A constant current source i.e. CM is employed in the proposed design to generate reference current (I_O) through the different branches of the proposed FA circuit. Further, CSA senses current level at different nodes to finally evaluate the carry and sum bit. To realize the 3-inputs and 5-inputs MF with the current mode operation, following conditions are considered

(1) For 3-inputs MF, if sensed current (I_{SENSE}) $\geq 2 \times I_O$ (reference current) then carry (C_{OUT}) is set at logic '1' else C_{OUT} is set at logic '0'.
(2) For 5-inputs MF, if sensed current (I_{SENSE}) $\geq 3 \times I_O$ (reference current) then sum bit is set at logic '1' else sum is set at logic '0'.

Thus, the current mode operation is realized using CSA to perform the sum and carry computation for the proposed FA as expressed in (3) and (4). The circuit-level model of 7Ts design of FA realized using MF is shown in Fig. 2. Here, reference current (I_O) flowing through the different branches is sensed to find out the majority. For proper function of the proposed FA, the reference current (I_O) should be constant, hence CM is employed in the design. As per the logic states of the input signals (A, B, C_{IN}), provided at the gate terminals of the transistors MN1, MN2 and MN3 respectively, corresponding transistors turns ON to allow the flow of current. At node node1, if $I_{SENSE} \geq 2 \times I_O$ then C_{OUT} = '1', else C_{OUT} = '0'. Inverter composed of MN6/MP7 produces NOT of the 3-inputs MF i.e. $\overline{C_{OUT}}$, which is further used to evaluate the sum via 5-inputs MF. Based upon the 5-input signals (A, B, C_{IN}, $\overline{C_{OUT}}$, $\overline{C_{OUT}}$), logic value of sum is evaluated. Transistors MN1-MN5 are assisted based upon the logic states input at their gate terminals. Similarly, @ node2, if $I_{SENSE} \geq 3 \times I_O$ then Sum = '1', else Sum = '0'. Thus, only 7Ts are required to perform 1-bit addition by proposed MF based FA design.

4 Simulation Results and Discussion

The proposed adder circuit is realized using 90 nm technology node in Cadence platform. Further, various design parameters of the proposed FA are tabulated in Table 2. Parametric values are set as per the bias condition of the CSA such that the sensed output voltage is under acceptable range. Further, to draw proper comparison with the proposed FA circuit, various conventional FA circuits based upon different logic styles have been also studied. Although available FA circuits provides proper output but each type adder focuses on one or two aspects of the design metrics. Compared to other compact FA designs with fewer transistors such as pass transistor logic, the proposed FA design features reduced PDP and power consumption with higher operating speed. Also, pass transistor logic based FA suffers from threshold voltage loss problem [8].

Investigation of one of the authors of this work revealed that TG based FA (transmission gate full adder) [9] is the best in terms variability, and second best is the static CMOS topology. Therefore, we have included these two full adders for comparison in addition to a 14Ts full adder as mentioned below:

(1) Transmission gate CMOS (TG CMOS) [10] based FA design which utilizes 20Ts to implement the FA logic functions.
(2) Basic pull-up and pull-down approach based standard CMOS FA [11] design which utilizes 28Ts to implement sum and carry logic functions for 1-bit FA.
(3) For area performance comparison in terms of transistor count, a 14T [12] based compact design have also been studied.

Although all the FA circuits mentioned above perform the same task but the methodology utilized to generate the outputs differs, supply load and transistor count also varies significantly. For example, 14Ts and TG-CMOS generate XOR of

S. no.	Design parameters	Values
1.	R_{SHUNT}	1 KΩ
2.	R_{F1}	50 Ω
3.	R_{F2}	0.5 KΩ
4.	I_O	1 mA

Table 2 Design parameters of the proposed majority function based 1-bit FA

inputs $(A \oplus B)$ which is further utilized as a select signal to generate the final outputs using its complement and the signal itself. In contrast to this, CMOS adder generates the C_{OUT} and Sum using complimentary logic. Dual complementary function applied to this architecture produces switching loss at every stage of evaluation. Also, CMOS based design produces Sum output that depends upon the intermediate carry which causes one extra inverter delay. Apart from this, high transistor count consumes more power which degrades the FA performance.

In the view of above, the FA proposed in this work utilizes only 7Ts (MN1–MN7 in Fig. 2) to generate the Sum and C_{OUT}. The circuit-level model of the proposed FA with the CSA is shown in Fig. 2 in which the current through the branches having MN4 and MN5 is same and can be replaced with a single branch having twice the current flowing with either of the branch as both branches are assisted by the same gate signals (i.e. gate voltage of MN4/MN5 is same). Thus, only 6Ts are required to implement the proposed full adder circuit excluding CSA. This is very less number of transistors as compared to the design mentioned in [8–12]. CSA is not included because for comparison because it is a common circuitry for various other full adder cells, working in a same chip. Power consumption is also reduced in proportion with the number of transistors in the design. Other design metrics such as EDP and PDP are also reduced significantly due to reduced number of transistors in the proposed design. The current mode operation realized using CSA by analyzing magnitude of current at various nodes, adds novelty to the proposed design. However, including CSA in the proposed design increases the complexity of the chip, which needs to be addressed for further improvement in the design. The simulation results for the carry C_{OUT} and Sum are shown in Figs. 3

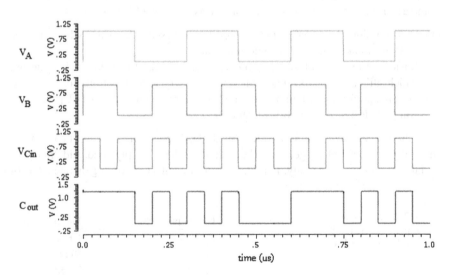

Fig. 3 Simulation waveform form for C_{OUT} calculation (3-I/Ps majority function)

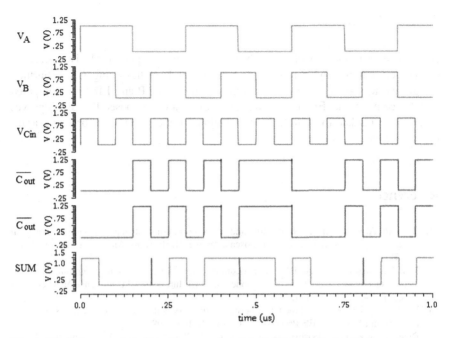

Fig. 4 Simulation waveform form for sum calculation (5-I/Ps majority function)

Table 3 Comparison between conventional CMOS and proposed 1-bit FA circuit @ 0.9 V supply

FA circuits	Delay × 10^{-12} (s)	Power × 10^{-7} (W)	EDP × 10^{-27}	PDP × 10^{-16} (J)	No. of transistors
Proposed FA	73.98	62.881 (excluding CSA)	844.11	4.6519	7-T
TG CMOS	98.76	87.302	911.47	8.6220	20-T
Standard CMOS	182.53	103.638	1018.70	18.917	28-T
14Ts	58.79	71.537	872.28	4.2056	14-T

and 4, respectively. Further, Table 3 summarizes the comparative results for the proposed FA circuit with the conventional FAs in terms of various other design metrics, implemented @ 90 nm technology node in Cadence environment.

5 Conclusion

A novel current mode operated MF based FA design is reported. MF is realized by sensing current flowing through odd number of branches using CSA. Further, design matrices such as power consumption, delay, PDP and EDP are compared with the conventional FA circuits. Results show that the proposed FA circuit proves deserving candidate to replace the conventional FA circuits used for larger arithmetic circuits.

References

1. Dwivedi, A. K., Islam, A.: Nonvolatile and robust design of content addressable memory cell using magnetic tunnel junction at nanoscale regime. IEEE Trans. Magn., 51(12), pp. 1–13. (2015)
2. Mal, A., Dwivedi, A. K., & Islam, A.: A comparative analysis of various programmable delay elements using predictive technology model. In: IEEE International Conference on MicroCom 2016, pp. 1–5. IEEE Press (2016)
3. Wilson B.: Recent developments in current conveyors and current-mode circuits. Circuits, Devices and Systems, IEE Proceedings G. 137(2). IET (1990)
4. Dwivedi, A. K., et al.: Performance evaluation of MCML-based XOR/XNOR circuit at 16-nm Technology node. In: Advanced Communication Control and Computing Technologies (ICACCCT), 2014 International Conference on, pp. 512–516. IEEE Press. (2014)
5. Minch B.: Low-Voltage Wilson Current Mirrors in CMOS. In: IEEE ISCAS, New Orleans, LA, USA, pp. 2220–2223. IEEE Press (2007)
6. Mazidi E.: Designing current mirror with Nano wire FET. In: IEEE Nanotechnology Materials and Devices Conference, pp. 339–342, IEEE Press (2010)
7. Toumazou C., Haigh D. G.: Design and Application of GaAs MESFET Current Mirror Circuits. Current mode analogue Circuits, IEE Colloquium on. IET (1989)
8. Lin, J. F., Hwang, Y. T., Sheu, M. H., & Ho, C. C.: A novel high-speed and energy efficient 10-transistor full adder design. IEEE Trans. on CAS I, 54(5), pp. 1050–1059. (2007)
9. Islam, A., Imran, A., & Hasan, M.: Robust subthreshold full adder design technique. In IEEE International Conference IMPACT, pp. 99–102, IEEE Press (2011)
10. Shams, A. M., Darwish, T. K., & Bayoumi, M. A.: Performance analysis of low-power 1-bit CMOS full adder cells. IEEE Trans. on VLSI Systems, 10(1), PP. 20–29. (2002)
11. Zimmermann, R., & Fichtner, W.: Low-power logic styles: CMOS versus pass-transistor logic. IEEE Journal of solid-state circuits, 32(7), pp. 1079–1090. (1997)
12. Abu-Shama, E., & Bayoumi, M.: A new cell for low power adders. In: IEEE International Symposium on Circuits and Systems, ISCAS '96., vol. 4, pp. 49–52. IEEE Press (1996)

A Tripartite Partite Key Generation and Management for Security of Cloud Data Classes

P. Dileep Kumar Reddy, C. Shoba Bindu and R. Praveen Sam

Abstract To enhance security at finer levels class-based cloud data security is studied. Data of high prioritized class is more confidential and frequently accessed and needed with high security parameters like strong encryption keys. In many of the class-based cloud approaches, a user needs to authenticate every class they want to access and because of this class, keys for authentication are to be generated in proportionate with the number of classes, i.e., more number of authentication keys are needed. As such the management of these huge keys is quite difficult. This paper presents class key generation and management for the user to traverse from class to class. Authenticated user being in his high prioritized class data may also access his low class data without additional key. A tripartite graph technique is used to mechanize the class key generation and management. The performance study of the approach is done by reducing the number of class authentication keys.

Keywords Cloud data · Prioritized data · Class keys · Tripartite
Lucas sequence

1 Introduction

Today's most trustable computing environment is cloud technology. Marked as the most secured platform handling user's data, the technology is still striving to strengthen the security parameters by studying its own loopholes. Enhancing data

P. Dileep Kumar Reddy (✉) · C. S. Bindu
CSE, JNTUA College of Engineering, Anantapur, India
e-mail: dileepreddy503@gmail.com

C. S. Bindu
e-mail: shobabindhu@gmail.com

R. Praveen Sam
G. Pulla Reddy Engineering College, Kurnool, India
e-mail: praveen_sam75@yahoo.com

V. Bhateja et al. (eds.), *Proceedings of the Second International Conference on Computational Intelligence and Informatics*, Advances in Intelligent Systems and Computing 712, https://doi.org/10.1007/978-981-10-8228-3_53

usability, data outsourcing has crept in reducing the burden of data management by many private enterprises. Data outsourcing has led to various cloud infrastructures: the private cloud and the public cloud. The private clouds are mostly maintained by the data owners where they limit the data access to few authenticated users. On the other hand, a service provider maintains the public cloud giving provision to store the personal data of various users. Data within the private cloud is totally out of owners control and a potentially threat to data security.

Addressing high storage and maintenance cost, class data partitioning is usually followed; where owner's data is partitioned to various classes. Literature has revealed the usual data classes are like: private, limited access, public; where private data can only be accessed by limited authenticated users. Under these types of class based data storages, managing the authentication keys of each class is a major issue. Literature has shown approaches where for N data classes N number of different keys are maintained. Maintaining these N keys for N classes of one user raises cost as well as increases difficulty in managing and maintaining them.

Literature has proposed with class-based partitioning approaches where a user authenticated to one class can access only that class with the generated key and cannot access his own other class data and also other users class data. But with today's data sharing, users even wanted to access other class data. With such class data accessing, key sharing becomes quite difficult.

In this paper, we propose a new class key generation method using tripartite cycles of a graph. We use a three-partitioned data where the user data is classified into three classes: Confidential (C1), less confidential (C2), and public data (C3). We used a tripartite graph connecting these three data partitions. Tripartite cycle that is generated from the graph is used to assign keys to various classes and authenticates these classes of users.

The rest of the paper is arranged as follows: Sect. 2 presents the literature survey. Section 3 presents the background and preliminaries. Section 4 presents tripartite class authentication approach. Section 5 presents the conclusion.

2 Literature Survey

Initially, many of the cloud storage technologies followed no classification method where the whole lot of user's data is encrypted and stored in the cloud. This type of whole data encryption has failed to increase data owners' trust, as he feared that data security is lost in case he wanted to view only a portion of the data which is important to him; where the mechanism restricts to decrypt the whole data even to view the small portion of his data. To raise this trust, data classification based on priorities is followed by clouds. The most reflected concern with data classification is:

1. How keys are generated to access various class data.
2. How these large number of class keys are managed.

A priority service scheme of cloud data is discussed in [1], where in the cloud services are classified on how frequently the data is requested. A cryptographic key generation method for multileveled data security is being discussed in [2]. The works focused on adding new classes even after the data levels are defined and the keys are generated. A hierarchical scheme is used to generate class keys and in using them.

A new key aggregate cryptosystem is discussed in [3]. Here, any set of secret keys is aggregated to be used as a compact single key. The ciphers generated using these several keys can be decrypted using the single aggregated key.

A hierarchical access control to various class data is discussed in [4]. The approach used a time-bound property to manage the keys. A light weight key management approach is proposed in [5]. The approach emphasizes on using a single key to access sensitive data of other users.

Works discussed in [6] presented a key management scheme for cloud data storage. Different key management schemes are used separately at client, server and at both sides. Works discussed in [7] projected on reducing the number of keys managed by each user.

3 Background and Preliminaries

Cloud computing is evolving as the most researches bread in giving a greater scope for varied research paths. As promising data storage technology cloud has inspired many private data owners toward it. Today as per the owners' requirements, many cloud started focusing on class based cloud data security where the owners' data is secured under various classes. The strength of class security is varied according to the priority of the class data. Various class keys are generated and used to provide class security. As the cloud classes vary the number of keys also varies, and the major problem faced is how to manage these huge set of class keys.

3.1 Preliminaries

This section discusses some preliminaries required to escalate the proposed work.

Many of the cloud infrastructures have taken data classes under various priorities like confidential(C), less prioritized (L), and public (P).

A tripartite graph is a structure to manage data under three partitions. The three partitions are the prioritized levels of the data (C, L, P)

Figure 1 shows a tripartite graph with 3 partitions, within each partition various data files like C1, C2, and C3.

The tripartite graphs usually have cycles of varied lengths. A 3-cycle is a tripartite cycle of length 3 called tricycle, with its 3 distinct nodes in each of the three classes like C1-L1-P1-C1, etc.

Fig. 1 Tripartite graph

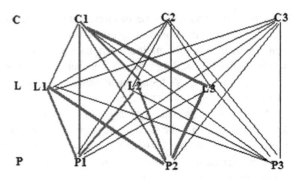

A restriction can be made that a tricycle can only start from high prioritized data of class C. Tripartite graphs with 3 files in each partition are represented as K(3, 3, 3) and there are 3 * (3 + 3) = 18, distinct 3-cycles. K(n, n, n) has n * (n + n) cycles of length n.

One category of number system is the Lucas numbers which are defined by

$$L_n = 2; \text{ if } n = 0;$$
$$= 1; \text{ if } n = 1;$$
$$= L_{n-1} + L_{n-2}; \text{ if } n > 1; \text{ for all } nEN$$

The primes from these Lucas numbers are collected as Lucas primes.

Based on the range of Lucas number, these can be partitioned into various Lucas partitions. We can define 3-Lucas partitions by dividing the whole range of Lucas Numbers to fall into 3-subranges of N:

[0, N/3) U [N/3, 2 N/3) U [2 N/3, N]. The 3-Lucas partitions are:

$$L_n^1 = L_{n-1} + L_{n-2}; \text{ n E } [0, N/3);$$
$$L_n^2 = L_{n-1} + L_{n-2}; \text{ n E } [N/3, 2N/3);$$
$$L_n^3 = L_{n-1} + L_{n-2}; \text{ n E } [2N/3, N];$$

Mathematically, any sequence has a generating function so as Lucas generating function.

For an RSA integer n with n = pq and a∈N, L (a) the Lucas sequence generated by a is $L_n (a) = L_{n-1}(a) + L_{n-2}(a)$; for which the generating function is $f(X) = X^2 + aX + 1$. Talking about the cryptographic strengths, the second-order equations are stronger than linear. This strengthens the proof that Lucas sequences are cryptically strong.

4   The Approach

Here, the proposed method takes the advantage of Lucas sequences and Lucas generating function to generate the random primes need for the class key pairs. We use a tripartite graph mechanism to manage the Key pairs to encrypt and decrypt the data. The proposed method also enhances that a user authenticated to access high prioritized class can also access data of other low prioritized classes. But a user authenticated to low prioritized class cannot access high prioritized classes.

4.1   Class Random Primes

The Lucas random primes to crypt the class C data are generated within the Lucas subrange n E [0, N/3); using the Lucas L^1 sequence: $L_n^1 = L_{n-1} + L_{n-2}$; to crypt class L data the primes are generate within the subrange n E [N/3, 2 N/3), using the Lucas sequence $L_n^2 = L_{n-1} + L_{n-2}$; and random for class P data are generated from $L_n^3 = L_{n-1} + L_{n-2}$; n E [2 N/3, N].

4.2   Generation of Key Pairs

Class C: let class C public parameters be (p^1, q^1, a); for Lucas primes p^1, q^1 belongs to L_n^1, n E [0, N/3), a belongs N. Choose a small integer e belongs to N; evaluate d such that gcd $((p^1)^2 - (q^1)^2, d) = 1$; computes $y = L_x(a)$, for a secret x belongs E [0, N/3); Now the key pair is public key(y, e) and private key(d, x).

Class L: let class L public parameters be (p^1, q^1, a); for Lucas primes p^1, q^1 belongs to L_n^1, n E[N/3, 2 N/3], a belongs N. Choose a small integer e belongs to N; evaluate d such that gcd $((p^1)^2 - (q^1)^2, d) = 1$; computes $y = L_x(a)$, for a secret x belongs E [N/3, 2 N/3]; Now the key pair is public key (y, e) and private key (d, x).

Class P: Since class P is public data, need not be encrypted.

4.3   Description of Class Data

The data of the confidential class C is again partitioned into three types. More frequently accessed (C1), less frequently accessed (C2), and very less frequently accessed (C3). Similarly, the Class L data is prioritized. The proposed method focuses on using the strong security parameters for more frequently accessed data.

   Encryptions: As each class data has three priorities, the encryption parameters vary in order to strengthen the mechanism.

Class C: Class C has three priority data classes C1, C2, C3.

C1 Encryption: C1 data is more frequently accessed hence our approach uses stronger security parameters for encryption of C1 data. A larger interval $[0, 2n]$ is used where from a secret k1 is choose; $0 < k1 < 2n$. A larger interval is analytically strong. Generate the cipher $CC1 = (CC_1^1, CC_1^2)$; where $CC_1^1 = L_{k1}(a)$; $CC_1^2 = K + L_e(C1)$ and $K = L_{k1}(y)$.

C1 Decryption: Compute $K = L_x(CC_1^1)$, $C1 = L_d(CC_1^2 - K)$ where (d, x) is private key.

The encryption and decryption mechanism are the same for L1 partitions where secrets from larger range $[0, 2n]$ are taken.

C2 Encryption: Data of C2 of C is less frequently accessed than C1 data; hence the cost of security of C2 reduced by defining less stronger parameters than C1. An interval $[0, n]$ is taken to choose a secret k2; $0 < k2 < n$. Generate the cipher $CC2 = (CC_2^1, CC_2^2)$; where $CC_2^1 = L_{k2}(a)$; $CC_2^2 = K + L_e(C2)$ and $K = L_{k2}(y)$.

C2 Decryption: When the user want to C2 data then the cloud forwards (CC_2^1, CC_2^2). User computes $K = L_x(CC_2^1)$, $C2 = L_d(CC_2^2 - K)$ where (d, x) is his private key.

The encryption and decryption mechanism are the same for L2 partitions where secrets from larger range $[0, n]$ are taken.

C3 Encryption: C3 data is less frequently accessed hence our approach uses bit less stronger security parameters for encryption of C3 data. A very smaller interval $[0, n/2]$ is used where from a secret k3 is chosen; $0 < k3 < n/2$. Generate the cipher $CC3 = (CC_3^1, CC_3^2)$; where $CC_3^1 = L_{k3}(a)$; $CC_3^2 = K + L_e(C3)$ and $K = L_{k3}(y)$.

C1 Decryption: Compute $K = L_x(CC_1^1)$, $C1 = L_d(CC_1^2 - K)$ where (d, x) is private key.

The encryption and decryption mechanism are the same for L3 partitions where secrets from smaller range $[0, n/2]$ are taken.

The encrypted data is thus stored in the cloud. Whenever the user requests particular prioritized class data, then the cloud has to send the needed cipher to the user. The user client module has to access the key hives in order to fetch the keys for decryption.

4.4 Tripartite Hives for Key Management

These key hives are maintained at the user module as well as at the cloud. The client key hives consist of secrets k1, k2, k3. The cloud key hives consist of cryptographic specifications. A tripartite graph is used to store these key hives. At the cloud at every vertex (priority class) of the tripartite graph, a key hive is maintained which stores the subintervals, random, and private keys used for encryption of that class. These hives are also prioritized using a partial ordering C < L<P. When an authenticated user requests class C data, which is a class of confidential data then it is an indication that the user can access low prioritized data also. The hives are arranged as ends of tricycle such that the secret k1 stored at hive of C of the user is

used to open the hives at L and the secret stored at hive of L, i.e., k2 is used to open the hive of C and the secret k3 stored at hive P is used to open the hive at C and is shown in Fig. 2.

When the user requests class C data the client module runs the tricycles with initial point as C; forwards secret k3 at hive P to the cloud, the hive at C can be opened to access the security specifications at the cloud and decrypt the data at level C using the security specifications stored at hive C the data is decrypted and is shown in Fig. 3.

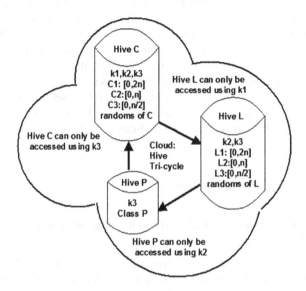

Fig. 2 Key hives at cloud

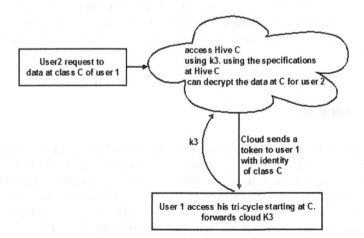

Fig. 3 Request for class C data

Table 1 No. of class key

No. of classes	No. of class priorities	No. of keys used	
		Tripartite approach	Usual class based
3	3	6	12
4	4	8	20
5	5	10	30

Being at C, a high prioritized class user is also authenticated to access low priority classes of L i.e., L1, L2, L3 and P, when the user is at level C the tricycle mechanism at user1 also passes the Keys k1, k2 to the cloud, using which hives of L, P can be authenticated and the user can access the data at L, P.

If the user is authenticated to level L data then he cannot access level C data. The mechanism defines a partial ordering between levels such that C > L > P, i.e., keys at hives of L cannot authenticate hives of C; and thus the user cannot access C data. If the user is accessing level L data and since L > P the secret key at hive L is populated to level P thereby data at level P can be accessed.

4.5 Performance comparisons

The performance comparison of the approach is discussed in a way to reduce the number of class keys used.

The Number of class keys: In the usual approach, the number of keys varies in proportion to number of classes and class priorities. If there are 3 classes and 3 class priorities in each, the usual mechanism takes 3 keys for each class accessing and 3 × 3 keys for each priority class accessing; hence a total of 3 + 9 = 12 keys. In our approach, we used the same secrets that are used for encryptions to authenticate the 3 hives. We used only three secret keys to encrypt the 3 class data; hence a total of 3 + 3 = 6. With any number of classes, we used reduced keys and is as shown in Table 1.

5 Conclusion

Finer leveled data security is achieved by class-based cloud security. Based on priorities, user data is stored under different classes within the cloud. Data of high prioritized class is more confidential and frequently accessed and needed with high security parameters like strong encryption keys. Authenticated user being in his high prioritized class data may also access his low class data. User authenticated to less prioritized data cannot access high prioritized data. In our approach, we used a three-partitioned data where the user data is classified into three classes:

Confidential (C1), less confidential (C2), and public data (C3). We used a tripartite graph connecting these three data partitions. Tripartite cycles that are generated from the graph are used to mechanize the class key generation and management. Key hives are stored at the ends of these tricycles. These key hives implement a key management mechanism to authenticate various class data. The theoretical description of the method performance is made by reducing the number of authentication keys used.

References

1. Xiaoming Nan, Yifeng He,Ling Guan "Optimal resource allocation for multimedia cloud in priority service scheme" Circuits and Systems (ISCAS), 2012 IEEE International Symposium on 20–23 May 2012, https://doi.org/10.1109/iscas.2012.6271425.
2. Lein Harn and Hung-Yu Lin, "A Cryptographic Key Generation Scheme for Multilevel Data Security", Elsevier Science Publishers Ltd., Vol. 9, No. 6, Pg No: 539–546.
3. Cheng-Kang Chu, Sherman S. M. Chow, Wen-Guey Tzeng, Jianying Zhou, and Robert H. Deng, Senior Member, IEEE "Key-Aggregate Cryptosystem for Scalable Data Sharing in Cloud Storage",. IEEE Transactions on Parallel and Distributed Systems. 25, (2), 468–477.
4. Tsu-Yang Wu, Chengxiang Zhou, Eric Ke Wang, Jeng-Shyang Pan, Chien-Ming Chen, "Towards Time-Bound Hierarchical Key Management in Cloud Computing" Intelligent Data analysis and its Applications, Volume I, PP 31–38, https://doi.org/10.1007/978-3-319-07776-5_4.
5. Zongmin Cui, Hong Zhu, Lianhua Ch" Light weight Key Management on Sensitive Data in the Cloud", security and communications networks, 2013; 00:1–9, https://doi.org/10.1002/sec.
6. Amar R.Buchade, Rajesh Ingle, "Key Management for Cloud Data Storage: Methods and Comparisons" 2014 Fourth International Conference on Advanced Computing & Communication Technologies, P. No: 263– 270, 978-1-4799-4910-6/14 $31.00 © 2014 IEEE. https://doi.org/10.1109/acct.2014.78.
7. Rui Zhang, and Ling Liu, "Security Models and Requirements for Healthcare Application Clouds", Cloud Computing (CLOUD), 2010 IEEE.

Transforming the Traditional Farming into Smart Farming Using Drones

R. P. Ram Kumar, P. Sanjeeva and B. Vijay Kumar

Abstract India bags second position into total arable land with the agricultural products including rice, wheat, sugarcane, onion, tomato, potato, beans, and mangoes. According to the statement "Agriculture is the backbone of our India," farming plays a significant role in India's economic growth. The objective of the paper is to improve the cultivation and cultivated products through Smart Farming by incorporating drones. The applications of drones, an unmanned aircraft, in the farming are addressed in this paper. The drones can be programmed and automated in the activities from monitoring the soil moisture to maintaining the livestock and concern logistics. The extensive study implies that incorporating drones transforms the traditional method of farming into Smart Farming.

Keywords Traditional farming · Smart farming · Internet of things
Drones · Global positioning system · Geographic information system

1 Introduction to Smart Farming

Due to the technological advancements in the Science and Technology, Internet of Things (IoT) makes its footprints into the agricultural domain too. In the Smart Farming (SF), farmers are informed with the following information through the help of reliable and higher end sensors. They are (a) Yielding of crops, (b) Expected rainfall, (c) Infection by pests, (d) Soil Nutrition, (e) Productivity, (f) Preventive Maintenance, (g) Preservation, and (h) Tracking the product lifecycle in real time. Rather than going to the traditional methods of cultivation, the SF incurs and

R. P. Ram Kumar (✉) · P. Sanjeeva · B. Vijay Kumar
Malla Reddy Engineering College (Autonomous), Secunderabad, India
e-mail: rprkvishnu@gmail.com

P. Sanjeeva
e-mail: psanjesus@gmail.com

B. Vijay Kumar
e-mail: vijaybru@gmail.com

© Springer Nature Singapore Pte Ltd. 2018
V. Bhateja et al. (eds.), *Proceedings of the Second International Conference on Computational Intelligence and Informatics*, Advances in Intelligent Systems and Computing 712, https://doi.org/10.1007/978-981-10-8228-3_54

589

involves intensive sensors and hardware for better yielding. Even though applying IoT in agriculture sounds good, the challenges need to address are summarized from [1].

Data Aggregation and Integration: The farmers need to know and track the machines and sensors to optimize the cultivation.

Design for Analytics factor deals with the identification of specific issues based on the data derived from the sensors.

Security concerns with blocking the malicious activities from unauthorized access and nonstandard protocols.

Manageability deals with conditioning of data (i.e., collecting, encrypting and determine what data to be sent, etc.,) in this phase.

Figure 1 shows the Dell's Agriculture IoT located at Dell Silicon Valley Solution Center, Santa Clara, CA [1]. Agriculture IoT Architecture functionalities are categorized into six phases. They are (1) Once the fundamental data is collected, then integration, correlation, and workflow orchestration process take place in the cloud environment, (2) Collected data is stored and analyzed, (3) Defining the alerts and action to be taken for peculiar scenarios, (4) Management of activities using sensors, gateways, inventory management, asset management through equivalent software, (5) Maintaining the data integrity using sensors and enterprise firewalls, and (6) Protection of collected data.

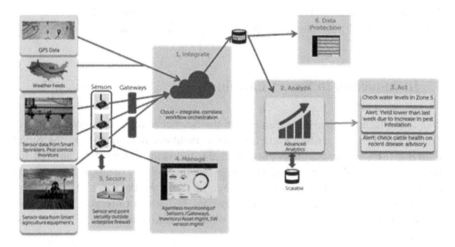

Fig. 1 Dell's agriculture internet of things lab process. (*Source* [1])

2 Need for Transforming the Traditional Farming

The term "IoT" is defined as the network of interconnected objects or things which can collect, transfer, and exchange data using sensors. The term "thing" or "object" refers to the quick communication among people-to-people, people-to-machine, and machine-to-machine. This discussion may be summarized as, "Anything can be connected or communicated to anywhere at any time." The applications of IoT may not be bounded by specific delimit. Applications of IoT put forth its footprints in almost all the domains are as follows [2–4]: (a) Smart City includes smart parking, smart roads to identify congestions, smart lighting, smart waste management, and intelligent buildings, (b) Smart Environment includes fire detection in forests, detection of the earthquake, monitoring snow level, controlling air pollution, and monitoring landscapes to prevent from avalanche actions, (c) Smart Watering deals with controlling the sea and river pollutions, monitoring the sewage and chemical intrusions in water bodies including sea, river, lake, and pond regions, (d) Smart Metering focuses on monitoring the grid system, water storage, transportation, and flow of oil and gas levels, (e) Security and Emergencies target monitoring the radiations and explosion of hazardous gas and maintaining at an affordable level(s), (f) Smart Retail Zones provides smart shopping, shipping and payment options with smart and intelligent supply chain management and control mechanism, (g) Smart Logistics has intelligent tracking of items with its conditions and exclusive storage and transportation for explosive materials, (h) Smart Industrial Control incorporates machine-to-machine communication and various monitoring features such as temperature, air quality, and ozone levels too, (i) Smart Agriculture focuses on greenhouses, monitoring soil moisture, irrigation zones, water resources, productions, forecasting climatic conditions, and weather reports, (j) Smart Animal Farming deals with monitoring and tracking the cattle (and concern animals) and their off springs, (k) Smart Home Annotation consists of intelligent intrusion/detection system, remote control appliances, metering the energy, and water consumptions, and (l) Smart eHealth Care arrives with ample of features which includes, intelligent surveillance for monitoring the patient's health, assistance for disabled and older adults, hospital drug tracking, ambulance telemetry, predictive and preventive maintenance, and Smart wearable's with enabled voice calls, text messages, and much more.

Gartner's report predicted that, by 2020, the number of connected things/objects would be around 20.6 million [5]. Table 1 summarizes the number of connected devices from the year 1990 and expected connected devices in the year 2025.

Table 1 Expected connected devices in the forthcoming years (*Source* [5])

Year	No. of connected devices
1990	0.3 million
1999	90.0 million
2010	5.0 billion
2013	9.0 billion
2025	1.0 trillion

Figure 2 shows the popularity and ranking of IoT applications summarized from [6]. The ranking is based on the three factors namely; (a) Most searched option on the Google, (b) Ultimate discussions on Linked-in, and (c) Most frequently discussed the topic on Twitter. Observation shows that the things/items required for day-to-day activities have the largest ranking due to its popularity. Table 2 shows the ranking order and popularity of IoT application according to IoT analysis from [6].

Observing Fig. 2 and Table 2 reveals that Smart Farming has the least ranking among top ten IoT Applications. Nevertheless, agriculture plays a vital role in the enhancement of overall economy of our nation. Agriculture, being the backbone of India, needs more refinement due to the following factors [7–10]: (a) Agriculture's

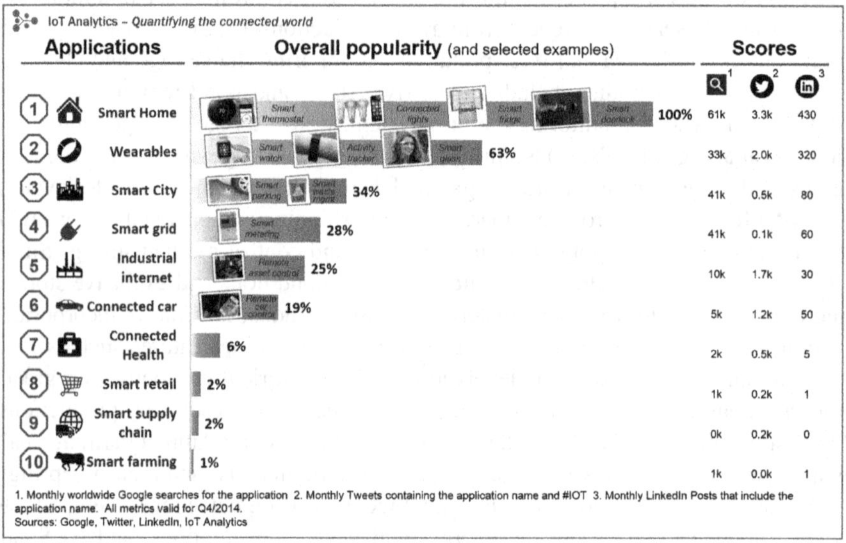

Fig. 2 Ranking of IoT applications. (*Source* [6])

Table 2 Ranking order of IoT applications

Applications	Popularity in percentages (%)
Smart home	100
Smart wearable	63
Smart city	34
Smart grid	28
Industrial internet	25
Connected car	19
Connected health	6
Smart retail	2
Smart supply chain	2
Smart farming	1

contribution to nation's income, (b) Contribution to the employment, (c) A significant rise in the Industrial Development, (d) Emphasis on National and International Trade, (e) Food source for Domestic consumption, (f) Million's occupation, and (g) Enhancement in the development of Agro-based Industries.

By considering these factors, the agricultural field needs breakthrough using applications of IoT. Henceforth, this paper focuses on transforming the traditional farming into Smart Farming by drones using IoT applications.

3 Existing Methods

The applications of drones in the agricultural domain are discussed in this section.

Paolo et al. [11] implemented a hybrid plowing technique using an RGB-D sensor which can be integrated with UAVs for better farming. In this approach, UAV was equipped with an RGB-D camera and Asus Xtion Pro to act as both fixed-wing drone and multicolor system. Fusing the GPS and Inertial Navigation System (INS) sensing modalities with drones, estimation of fidelity localization becomes higher. Further, the soil characteristics were analyzed based on a vision-based technique using drones at three different depths (open field, 25 and 50 cm). Experimental results revealed that drones could discriminate and determine the plowing depth of the fields which in turn improves the productivity.

Wireless Sensor Network (WSN) and integrated optical sensors are considered as a primary step toward Smart Agriculture. Takaharu and Hashimoto [12] focused on optical sensing methods for plants with WSN and color analysis for the images taken in agricultural fields through X-ray fluorescent (XRF) and Mid-Infrared (MIR) spectroscopy. XRF spectroscopic method quantified the nutrient element contents accurately while MIR spectroscopic method identified the organic component information especially saccharides and nitrate nitrogen. The tomato leaves were studied, and peak bands were analyzed for the presence of calcium. Further, the color changes were analyzed by transforming RGB color system to HSL color system. The color calibration of agricultural products helped in the surface color analysis. Finally, the authors concluded that with the advent of sensing technology through WSNs, agricultural products' quality might be improved thereby resulting in Smart Farming.

Susanne et al. [13] plotted a mobile app ecosystem called MyJohn Deere Mobile (MJDM), which is used for customer apps evaluation, feeds the concepts of architectural design, and also the reusable assets of the production environment. Data synchronization and data modeling were the ground mechanisms in the architectural design of MJDM. The data flow between mobile devices, agricultural machines, and the cloud even in offline is the major design in MJDM. The app was in the pilot study and developed as a high-quality prototype.

Tom stated that agricultural drones are not meant only for Crop Scouting [14]. The utility of drones integrated with software becomes active only if issues are identified accurately by the imagery followed by a timely intervention that would result in low input cost, high yield, and subsequently high profit. In the Smart Farming, identification of issues and the remedial steps by the drones puts the farmers in the safe zone.

Colin discussed the opportunities and challenges of UAS in Precision Agriculture [15]. The author analyzed and justified the different users of drones such as Producers/Growers, Researchers, Agronomists, Local equipment dealers, Crop Insurance companies, and Seed companies. The mere existence of drones cannot improve Smart Farming. The usage of hybrid technology, a combination of drones with ground-based sensors and software algorithms provide growers with necessary data for Smart Farming. The author concluded that drone service providers failed to provide the expertise to interpret the drone's data. Further, the study revealed that the correct interpretation of data within the stipulated time is essential to increase the yield from the fields.

Ioanna and Apostolos stated that the UAV is a major tool in precision agriculture [16]. The aerial images of the crops captured using UAV are enhanced for high resolution using specific indices such as Green-Red Ratio Vegetation Index (GRVI), Normalized Green-Red Difference Index (NGRDI), and Leaf Area Index (LAI). The author reviewed many case studies and proved that high-resolution images and index evaluation helped to estimate the crop growth and the current status of the soil. The authors concluded that UAVs which capture high precision images must be produced at an affordable cost to make UAV a key component in precision agriculture.

4   Case Studies: Applications of Drones in Agriculture

This section deals with the various case studies about the applications of drones in the agricultural field.

The case study deals with the usage of drones in agriculture to improve the yield efficiency [17]. EAZYPILOT controlled 400 mm Quadcopter with EAZFLY capabilities was used on a hectare size paddy farm to implement "Precision Agriculture". The aerial images in various stages of cropping cycles such as Beginning of cropping cycle, Irrigation mapping, Mid-cropping cycle assessment and Pre-harvest provides and educates the farmer more about the harvest cultivation. Incorporating drones in Smart Farming provides regular input at every stage of the crop cycle. The experimental study concluded that the farmer has to follow certain prerequisites (based on soil nature, water flow, and climatic conditions) for the productive yielding throughout the year.

Fera tied up with Strawson Ltd to estimate the potato yield using UAV [18]. The UAV of Fera has a twin sensor payload having RGB and a Near Infrared sensor which enables Fera to collect four bands multispectral imagery. The images obtained using Fera UAV were processed using Pix4D photogrammetry and ERDAS Imagine image processing software and analyzed using the eCognition software. Further, to identify potato plants individually, image segmentation and recognition methods were applied. Experimental analysis of the case study revealed that Strawson could extract the particular field area(s) or the entire region and able to estimate the corresponding yielding.

The study focused on determining the budding facts of in agriculture using drones to monitor the crops by three different case studies [19]. They are;

Case Study A: UAVs were used in Crete, Europe (where 1/4th of the total area was covered with olive trees) to detect Xyella fastidiosa (Olive Quick Decline Syndrome), a destructive disease. To test the method, DJI Phantom 3 was used in an olive grove. The UAV successfully identified 26 olive trees with dried yellowish foliage due to a fire accident.

Case Study B: DJI Phantom 3 Advanced UAV was deployed to detect any abnormality in the crown of palm trees at Afrathias, Tympaki. The bird's eye view of the palm tree was subjected to evaluation. Based on the analysis, the palm trees were categorized as (i) dead ones (ii) healthy ones and (iii) infected ones

Case Study C: The synergy of E-trap and UAVs in the field was used for target spraying. The DroneKit, Kit-sitl, and Mavproxy were used for simulation. Once the base station receives a signal about the increase in the threshold value of E-trap, a UAV is sent to that place to perform rectangular flight, and necessary steps were initiated. The study augments the possible uses of UAVs in the island of Crete, Greece for cultivation. Moreover, this technology increased crop protection and reduced crop monitoring.

5 Proposed Method

In Smart Farming, the automated machine does the profuse activities like weeding/ planting, applying fertilizers/pesticides, grain harvesting, monitoring and postharvesting activities. Further, these machines are equipped with Global Positioning System (GPS) to identify and decide the activities to be carried out, for example, to choose the required quantity of seeds, pesticides, and water during the particular climatic conditions. A joint venture of GPS and Geographic Information System (GIS) with the current scenario's data, the real-time decision(s) and its contour measures are initiated. An example of such situation is applying the pesticides. Based on the GIS and sensor's data, the nozzle's arm height and droplet size may be adjusted while applying the pesticides to the forms.

Figure 3 shows the proposed Smart Farming method. The various modules in the proposed Smart Farming include (a) Mapping of resources in the field using Sensors, (b) Monitoring the remote equipment, (c) Monitoring the crops, (d) Predicting the analysis for livestock, (e) Observing the climate and forecasting, (f) Tracking the livestock and Geo-farming, (g) Maintaining the statistics regarding the production and utilization of Feeding Activities, and (h) Monitoring the automated logistics and concerned warehousing activities.

The traditional farming method has some challenging activities, such as monitoring the crops throughout its lifetime, livestock, cattle feeding, and logistics. The

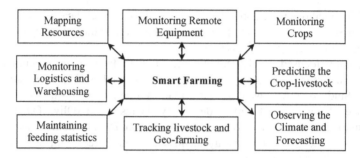

Fig. 3 Modules in the proposed smart farming approach

proposed method utilizes the functionalities of drones for transforming the traditional farming into Smart Farming. Drones, also called as Unmanned Aerial Vehicle (UAV) or Unmanned Aircraft System (UAS), are defined as a flying robot equipped with embedded systems, sensors, GPS, and GIS.

Employing the drones in the Smart Farming has significant advantages [20]. They are (a) Increase the timeline of farming, (b) Highest Return on Investment (RoI), (c) Even through the installation of UAV or UAS incurs complexity, the ease of use puts the farmers in the comfort zone, and (d) Integrating GPS with GIS improvises the security by establishing a virtual barrier over the geographical area; thereby the current scenario (real-time data) is conveyed to the farmer immediately, (e) Monitoring the health of the crop by imaging at regular (predefined) intervals, (f) Based on the cattle and their food, automating/maintaining the live-stocking requirement is carried out, (g) The drones put the farmers away from the unfeasible scenarios by monitoring and managing the logistics and warehousing activities, and (h) Identification and isolation of sickle and infected cattle from herd become ease. Thereby, the disinfected cattle's safety is ensured.

Figures 4 show some of the applications of drones in Smart Farming.

Fig. 4 Applications of drones **a** Monitoring the field **b** Monitoring the crop **c** Watering the field **d** Predicting the harvest by monitoring the climate **e** Monitoring the cattle

6 Discussions and Conclusion

The application of drones in Smart Farming is a boon to farmers. The peculiarities of drones make it pertinent for numerous applications. The essential characteristics of drones are summarized as follows: (a) Drones can be deployed at unique locations with the capability to withstand malleable payloads, (b) Drones can be reprogrammed and customized according to requirements, and (c) It can be used to measure anything, anywhere, and any time.

These unique characteristic features of drones are customized and programmed in such a way to suit for the agricultural purposes. Thus, the majority of farmer's role (including, monitoring of fields, monitoring crops, monitoring the physical equipment and resources, monitoring and maintaining the soil nutrition, watering the field and ensuring appropriate soil moisture content, observing the productivity, cattle feeding and supervision, livestock, logistics, and warehousing activities) are taken care by drones. Thereby, all the usual requirements of farmers are fulfilled immediately and efficiently by the applications of drones.

There is no doubt that the applications of drones in the agriculture transform the traditional method of farming into Smart Farming. Drones implications in agriculture not only improve the crop production but also ships the farmers to protected scenarios. Thus, employing the drones transforms the traditional farming into Smart Farming.

References

1. Statistica-Dell Community, http://en.community.dell.com/techcenter/information-management/statistica/?pi42011=3.
2. Top 50 Internet of Things Applications-Ranking, http://www.libelium.com/resources/top_50_iot_sensor_applications_ranking/.
3. Application Areas of the Internet of Things, http://www.ti.com/ww/en/internet_of_things/iot-applications.html.
4. Internet of Things Devices, Applications & Examples, http://www.businessinsider.com/internet-of-things-devices-applications-examples-2016-8?IR=T.
5. 10 Real World Applications of Internet of Things, https://www.analyticsvidhya.com/blog/2016/08/10-youtube-videos-explaining-the-real-world-applications-of-internet-of-things-iot/.
6. IoT Analytics-The 10 most popular Internet of Things applications right now, https://iot-anaytics.com/10-internet-of-things-applications/.
7. What is the importance of agriculture in Indian economy?, https://www.quora.com/What-is-the-importance-of-agriculture-in-Indian-economy.
8. Importance of Agriculture in the Indian Economy, http://www.yourarticlelibrary.com/agriculture/importance-of-agriculture-in-the-indian-economy/40227/.
9. Importance of Agriculture in Indian Economy, http://www.importantindia.com/4587/importance-of-agriculture-in-indian-economy/.
10. Importance of Agriculture in Indian Economy, http://www.economicsdiscussion.net/articles/importance-of-agriculture-in-indian-economy/2088.

11. Paolo, T., Massino, S., Giacomo, D., Emanuele, R., Carlo, A.A.: Towards Smart Farming and Sustainable Agriculture with Drones. In: International Conference on Intelligent Environments, pp. 140–143 (2015).
12. Takaharu, K., Atsushi, H.: Optical Sensing for Plant toward Science based Smart Farming with Wireless Sensor Network. SRII Global Conference, pp. 230–231 (2014).
13. Susanne, B., Ralf, C., Mattihas, N.: Piloting a Mobile-App Ecosystem for Smart Farming. pp. 9–14. IEEE Software (2016).
14. Agricultural Drones: What Farmers need to know, https://agribotix.com/whitepapers/farmers-need-know-agricultural-drones/.
15. Colin, S.: The Truth about Drones in Precision Agriculture. pp. 1–6, Skylogic Research (2016).
16. Ioanna, S., Apostolos, T.: The use of Unmanned Aerial Systems in Agriculture. In. 7th Intl. Conf. Information Communication Technologies Agriculture, Food Environment, pp. 730–736, (2015).
17. SSAI Case study, https://www.mygov.in/sites/default/files/user_comments/SSAI%20case%20study.pdf.
18. Using Fera's UAV to produce an Automated Potato Crop Count, http://fera.co.uk/agriculture-horticulture/environment-and-land-use/drone-services/documents/potato-counts.pdf.
19. Panagiota, P., Vasileios, S., Panagiotis, E., Ilyas, P.: Use of Unmanned Aerial Vehicles for Agricultural Application with emphasis on Crop Protection: Three Novel Case studies. Int. J. Agricultural Science Technology, 5, 30–39, (2017).
20. Benefits of Drones in Agriculture, http://www.agrotechnomarket.com/2016/09/benefits-of-drones-in-agriculture.html.

Implementation of Power Law Model for Assessing Software Reliability

S. Renu Deepti, B. Srivani and N. Sandhya

Abstract To have the unwavering quality of the product and equipment, we utilize nonhomogeneous Poisson process (NHPP). The power law process is the model used from repairable frameworks. In this paper, the proposed framework majorly concentrates on recognizing and assessing the disappointments inside of the product task or equipment venture. Here, to predict the methodology on the execution of parametric PLP model, the statistical regression approach is used. One of the imperative methodologies is regression methodology and PLP model utilizing mean time between failures (MTBF) capacity. The proposed framework evaluates the errors and failures in the process and demonstrates the analysis made from the proposed model.

Keywords The power law process model · Regression estimation approach
Maximum likelihood estimation · Time interval between failures
Mean time between failures

1 Introduction

The Power Law Process (PLP) is a famous vast NHPP model used to depict the dependability of repairable frameworks by taking into account the investigation of watched failure information, this model was taken from the equipment unwavering quality zone. Taking the traditional perception into consideration, a significant measure of writing on the PLP model is available. Duane [1] was the first to propose the PLP model. Additionally, he detailed a numerical relationship for

S. Renu Deepti (✉) · B. Srivani · N. Sandhya
CSE Department, VNRVJIET, Hyderabad, India
e-mail: renudeepti_s@vnrvjiet.in

B. Srivani
e-mail: srivani_b@vnrvjiet.in

N. Sandhya
e-mail: sandhyanadela@gmail.com

© Springer Nature Singapore Pte Ltd. 2018
V. Bhateja et al. (eds.), *Proceedings of the Second International Conference on Computational Intelligence and Informatics*, Advances in Intelligent Systems and Computing 712, https://doi.org/10.1007/978-981-10-8228-3_55

foreseeing and watching the dependability improvement as a component of combined failure time. He led some equipment practices wherein the pace of failure event was in a force law structure in working time. Amid the investigation with the failure information, the view of a linear line was shown on a two-dimensional log-log graph for the combined MTBF against total working time. Then, this representation exhaustively concentrated on by Crow [2], and he figured the relating model as a nonhomogeneous Poisson process (NHPP) with a force power law. Programming reliability needs to do with the likelihood of discovering failures in the operation of a given programming using a timeframe. Failures are a result of issues presented in the code. The need for software reliability models has expanded following the product part turned into a critical issue in numerous engineering ventures. A vital class of these models is the supposed development models. They utilize reports of past disappointments keeping in mind the end goal to fit the disappointment bend, or to foresee either the failure rate or the remaining failures in a given time interim. Some software reliability growth models taking into account nonhomogeneous Poisson related strategies have been proposed years prior. They are broadly utilized are still matter of study [3–6]. Different models consider failure relationship on the other hand bunches [7, 8].

2 Proposed Model

In this work, an examination of these models is made utilizing a few failure reports. The estimation of the parameters included in both models is another vital issue. It is surely understood that the maximum likelihood function is not generally accomplished in nonhomogeneous Poisson process models, see [1, 4, 9], and by and large, comparisons are minimal confounded to comprehend. Since there is a solitary parameter to gauge in a one of a kind circulation which does not rely on upon time, estimation procedure is less demanding in compound Poisson models. Some estimation strategies have been proposed in the writing. Nonetheless, a critical point must be commented here. Since the integrity of attack of a software reliability model is assessed by the quantity of remaining failures anticipated, and it is acquired as the normal number of failures from the dissemination capacity, then, any impartial estimator for the mean, gives the same forecast as the straightforward Poisson likelihood capacity, as it will appear. Though, one-sided estimators demonstrating a lessening in the measure of gatherings of failures give better results. Execution of Compound Poisson and NHPP models has been thought about in [2, 10, 11]. In this work, the comes about beforehand appeared in [2, 10] are expanded, a more definite investigation of the mode estimator displayed and the middle estimator is likewise presented for examination. The PLP model can be connected in programming constant quality also a few issues might emerge, it has been utilized as a part of numerous effective applications, especially in the safeguard business. To demonstrate the failure rate of repairable frameworks, the PLP model is received by the United States Army Materials System Analysis Activity and called the Crow-AMSAA model. In quality approaches, it is expected that the repeat rate is the same for all frameworks that are watched. Bain [10] broke down autonomous proportionate multi-framework by utilizing the PLP model. Much hypothetical work depicting the PLP model was performed (samples; Lee, L. also, Lee, K. [12], and Engelhardt and Bain [13]).

The PLP model has been broadly utilized as a part of unwavering quality development, in repairable frameworks, and programming dependability models (see; Calabria et al. [14], Yamada et al. [15], and Tian et al. [6]). Littlewood [16] altered this model to conquer an issue connected with it, the issue of limitless disappointment rate at time zero. Rigdon [1] demonstrated the misleading quality of the case that a direct Duane plot infers a PLP, and the change is false as well. The techniques proposed in this paper, when connected to genuine programming failure information, gives less mistake as far as all the estimation criteria contrasted with other well-known strategies from writing. Trial results demonstrate that the regression approach offers an extremely encouraging procedure in programming dependability development displaying and forecast.

Utilization of the power law model for good quality development test information for the most part accept the accompanying:

1. While the test is continuous, framework upgrades are presented that create nonstop changes in the rate of framework repair.
2. Over a sufficiently long timeframe the impact of these changes can be displayed satisfactorily by the consistent polynomial repair rate change model $\alpha t - \beta$. At the point when a change test closes, the MTBF stays consistent at its last accomplished quality.
3. At the point when the change test closes at test time T and no further change activities are progressing, the repair rate has been lessened to $\alpha T - \beta$. The repair rate stays consistent from that point on at this new (enhanced) level.

The force law process, frequently misleadingly called the Weibull procedure, is a valuable and basic model for depicting the disappointment times of repairable frameworks. We show basic properties of the force law procedure, for example, point estimation of obscure parameters, certainty interims for the parameters, and tests of speculations. It is demonstrated that a proper change of the disappointment times can prompt an integrity of-fit test. We likewise examine a portion of the phrasing utilized for repairable frameworks and balance this with wording utilized for non-repairable frameworks.

Constant quality development philosophy is a significant device to quantify item dependability change either through arranged, devoted testing or the progressive overhaul and change of the handled item. The approach is well considered with proper procedure numerical presumptions so when connected, it gives suitable and reasonable data and following of unwavering quality change. The standard Homogenous Poisson Process (HPP) with accepted steady failure rates and failure intensities is consistently expected and utilized as a part of all unwavering quality examination and particularly testing. The Nonhomogenous Poisson Process

(NHPP) sufficiently communicates step changes of failure rates resultant from item outline or procedures change, by fitting them with the constant power law curve. All unwavering quality development demonstrating are done in this way, for example, the take-off models Duane and AMSAA/CROW are legitimate and important. This paper proposes a few changes in the bookkeeping of the aggregate test times for the situations when the various units are seen in test or in the field which might yield more fitting determination of failure rates and their parameters. This paper calls attention to that the basic practice in unwavering quality development test information investigation with augmentations of test times of various test things at the seasons of individual failures might be improper for the situation where efficient failure plan or assembling hones blemishes are watched.

In this way, the paper proposes utilization of the first NHPP power law, which is the model, followed in deductions other current efficient quality development, the proposed expository technique will rectify the mistakes presented when various units are tried or watched, and will likewise give consistency of the information investigation. The mistakes do not occur when a solitary things' efficient quality development is observed. They get to be obvious just in the instances of uncommon things, and are relative to the quantity of watched things and the absence of synchronization—starting, consummation, and redesigns times [17]. To wipe out those mistakes, which could be somewhat vast when efficient quality development of handled items is taken after, this paper proposes taking after the standards of the first power law model in the majority of the RG information examination and counts of dependability results. It may be additionally fitting that the options are done in the distributed material to give this direction.

3 Related Work

A product is said to contain a flaw if, for info information, the yield result is inaccurate. A deficiency is dependably a current part in programming codes. In this way, the procedure of programming troubleshooting is a principal errand of the life cycle of a product framework. Amid this period, the product system is tried commonly with the goal of finding shortcomings contained. At the point when a disappointment is watched, the code is assessed to discover the issue which brought about the product disappointment. The deficiency is typically evacuated by adjusting the product codes. Subsequently, one anticipates that the product unwavering quality will increment amid the testing stage as more blames are uprooted. The unwavering quality change wonder is called dependability development. The size and the many-sided quality of the product bundles make it difficult to discover and redress every single existing issue. The best thing is to give programming a dependability necessity and to attempt to accomplish an objective by testing the product and redressing the identified issues. Be that as it may, acquiring the required programming dependability is not a simple assignment. Hence, high dependability is generally assessed by utilizing suitable models connected on disappointment information from the product disappointment history. A software unwavering quality model is a numerical portrayal of the troubleshooting and altering process worked in the accompanying three unique stages:

1. Model structure is chosen.
2. The free parameters in the model are tuned on the premise of the exploratory information.
3. A standard is given to utilize the assessed model for prescient purposes. A product unwavering quality model falls into two classifications that rely on upon the working space.

In this manner, the most famous models depend on time. Their principle highlight of unwavering quality measures, for example, the disappointment power which is determined as an element of time. The second sorts of programming unwavering quality models have an alternate methodology. This methodology is made by utilizing operational inputs as their primary components, which measure dependability as the proportion of fruitful races to aggregate runs. The second approach has a few issues, for example, numerous frameworks keep running of expansive lengths with yield measures that are contrary to the time-based measures. Because of these issues, the work of this thesis has been given to time-space models. The time-space model utilizes either the watched time between disappointments or the quantity of found disappointments per time period. In this manner, these two methods were created to assess the model parameters from either disappointment number information or time between disappointments.

In this way, programming unwavering quality displaying and estimation can be gathered into two classifications of general appropriateness:

1. Disappointment numbering portrayal (FCD).
2. Disappointment interim portrayal (FID).

Littlewood-Verrall Model (Lv):
Progressive times to disappointments are seen as arbitrary variables with exponential circulation where the force capacity is no more a diminishing deterministic capacity. The disappointment force is thought to be a stochastic diminishing irregular variable with a Gamma capacity of parameters α and $\theta(i)$. The generous contrast between the JM and LV models is that in the JM model, a fix dependably prompts a decrease of the disappointment power, which has dependably had the same quality. While in the LV model, the disappointment force takes after an arbitrary example so that the size of its varieties is not as a matter of course a steady. Moreover, the indication of the variety might change; that is, a fix may not bring about an unwavering quality change.

4   Results and Analysis

4.1   Existing System

In the current framework, the expectation of disappointments and there is a heaps of contrast between real time and unique time.

1. Failures are brought about by inadequacies in configuration, generation, and upkeep.
2. Failures are because of wear or some other vitality, parts related marvels, yet one can stretch out beyond time.
3. Preventive upkeep is accessible and makes the framework more dependable.
4. Dependability is time related. Failure rates might be diminishing, expanding or steady concerning working time.
5. Dependability is identified with natural conditions.

4.2   Proposed System

The proposed system has mainly based on the three parameters that are the bias, the Mean Squared Error (MSE), and MEOP.

The Bias: The addition of difference between curve to be estimated and actual data.

Mean Squared Error (MSE): For prediction, the most widely used method is mean squared error. The focus is made on the deflection between the expected values with the original values,

Mean Error Of Prediction (MEOP):
   It is defined as difference between the data that is real and actual values,
   Here a percentage of the upsides of proposed framework,

1. Disappointments are principally because of outline shortcomings in the product. Adjusting the outline can make it vigorous versus conditions that could trigger an inability to make the repairs.
2. No wear marvels in the product. Programming mistakes happen without past notice. Old codes can show an expanding disappointment rate as a component of mistakes prompted while making updates.
3. Disappointments happen when the rationale way that contains a mistake is executed. Unwavering quality development saw as mistakes in the product can be identified and redressed.
4. Outer natural conditions do not influence the product unwavering quality, while the interior ecological conditions influence the dependability, these inward conditions are lacking memory and unseemly clock speeds.

5. Information of configuration, use, and natural anxiety components are not elements in foreseeing the unwavering quality.
6. We can enhance dependability by troubleshooting and expanding the read access memory.
7. Replicating so as to unwavering quality can be enhanced the same blunder.

4.3 Results

This mechanization programming relapse estimation utilizing power law is the model used to discover the relapse estimation in the product improvement process. Chart demonstrates the execution of every procedures with relapse estimation. Here, we evaluated values for polygons for the product that draws 1, 2, 3, and 4 polygons What's more, the chart is produced as taken after the software has drawn the polygon. Power law model is defined for scheming the stress that is the errors in a software or hardware especially for software. In this process, we have chosen a software that draws polygons and that generates the graph between MTBF and cumulative time. The software that draws polygon is as follows:

Polygon 1

See Fig. 1.

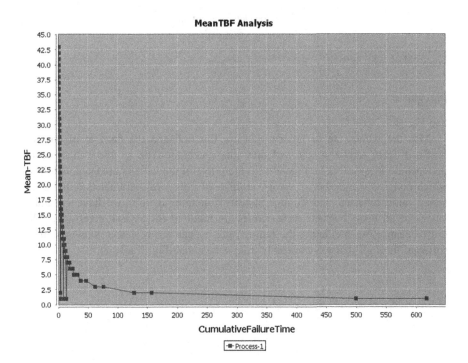

Fig. 1 Mean TBF analysis

Polygon 2

See Fig. 2.

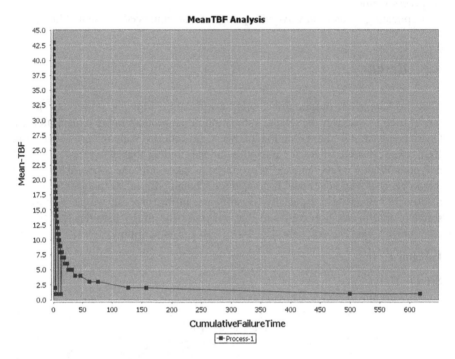

Fig. 2 Mean TBF analysis

Polygon 4

See Fig. 3.

Also the values are of failure intensity function and the number of errors is obtained from above graph and after the calculation of all the values of the parameter and mean value function for the polygon the table is obtained as follows which consists of number of errors, number of polygons, parameters a and b and mean value function and failure intensity function. In this, we have calculated values of all these because these values are required for assessing software reliability because by calculating the failure rate, we can easily detect efficiency of the software through that we can easily assess the reliability (Table 1).

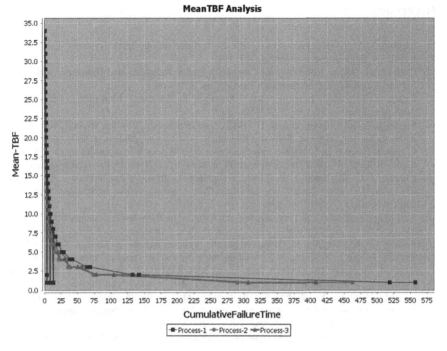

Fig. 3 Mean TBF analysis

Table 1 Software reliability assessment

S. no	Number of polygon	No errors	a	b	m(t)	ʎ(t)
1	1	40	0.253	0.315	1.79	0.00049
2	2	57	0.2307	0.450	3.94	0.00322
3	3	63	0.4365	0.222	3.96	0.0065
4	4	70	0.6360	0.175	4.02	0.00933

5   Conclusion

Regression estimation utilizing power law is characterized as failures principally because of outline issues in the product. In this paper, the proposed framework concentrates on predisposition, Mean Squared Error (MSE), Mean Blunder Of Expectation (MEOP). The software designed in this process is used to draw polygon which is used represent the stress in the form of graph that is drawn between MTBF and cumulative time. The stress is represented in the form of dots in the graph through that we calculated the number of errors since stress is nothing but errors. Through those errors, we have calculated failure intensity function finally, we generated a graph between number of errors and failure intensity function through which we assess the software reliability.

References

1. Rigdon, S. E. (2002). Properties of the Duane plot for repairable systems. Quality and Reliability Engineering International 18, 1–4.
2. Crow, L. H. (1974). Reliability for Complex Systems, Reliability and Biometry. Society for Industrial and Applied Mathematics (SIAM), pp. 379–410.
3. Duane, J. T. (1964). Learning Curve Approach to Reliability Monitoring, IEEE Trans. Aerospace Electron System, 2, 563–566.
4. Kyparisis, J. and Singpurwalla, N. D. (1985). Bayesian Inference for the Weibull Process with Applications to Assessing Software Reliability Growth and predicting Software Failures. Computer Science and Statistics: Proceedings of the Sixteenth Symposium on the Interface (ed. L. Billard), 57–64, Elsevier Science Publishers B. V. (North Holland), Amsterdam.
5. Rigdon, S. E. and Basu, A. P. (1990). Estimating the Intensity Function of a Power Law Process at the Current Time: Time Truncated Case. Communications in Statistics Simulation and Computation, 19, 1079–1104.
6. Guo-Liang Tian, Man-Lai Tang, and Jun-Wu Yu, (2011). Bayesian Estimation and Prediction for the Power Law Process with Left-Truncated Data. Journal of Data Science, 445–470.
7. Ascher, H. and Feingold, H. (1984). Repairable Systems Reliability, Decker, New York.
8. Ryan S. E, and L. S. Porth, A Tutorial on the Piecewise Regression Approach Applied to Bedload Transport Data, Gen. Tech. Rep. RMRS-GTP-189. Fort Collins, CO. U. S. Department of Agriculture, Forest Service, Rocky Mountain Research Station, 2007.
9. Xie M., Hong G. Y., Wohlin C.: A Practical Method for the Estimation of Software Reliability Growth in the Early Stage of Testing, Proc. of 7th Intl. Symposium on Software Reliability Engineering, Albuquerque, USA, 116–123 (1997).
10. Bain, L. J. (1978). Statistical Analysis of Reliability and Life Testing Models, Decker, New York.
11. Huang, Y. and Bier, V. (1998). A Natural Conjugate Prior for the Non-Homogeneous Poisson Process with a Power Law Intensity Function, Commun. Statist. 27(2): 525–551.
12. Lee, L. and Lee, K. (1978). Some Results on Inference for the Weibull Process, Technometrics, 20, 41–45.
13. Engelhardt, M. and Bain, L. J. (1986). On the Mean Time Between Failures for Repairable Systems. IEEE Transactions on Reliability, R-35, 419–422.
14. Calabria, R., Guida, M. and Pulcini, G. (1992). Power Bounds for a Test of Equality of Trends in k Independent Power Law Processes. Comm. Statist. Theory Methods, 21(11): 3275–3290.
15. Yamada S., Ohba M., Osaki S.: S-Shaped Software Reliability Growth Models and Their Applications, IEEE Trans. Reliability, 33, 289–292, (1984).
16. Littlewood, B. (1984). "Rationale for modified Duame model", IEEE Trans. Reliability, Vol. R-33 No. 2, pp. 157–9.
17. Crow, L. H. (1982). Confidence Interval Procedures for the Weibull Process with Applications to Reliability Growth. Technometrics, 24, 67–72.
18. Calabria, R. Guida, M. and Pulcini, G. (1988). Some Modified Maximum Likelihood Estimators for the Weibull Process. Reliability Engineering and System Safety, 23, 51–58.
19. Yu, J. W., Tian, G. L. and Tang, M. L. (2008). Statistical inference and prediction for the Weibull process with incomplete observations. Computational Statistics and Data Analysis 52, 1587–1603.
20. Higgins, J. J. and Tsokos, C. P. (1981). A Quasi-Bayes Estimate of the Failure Intensity of a Reliability Growth Model, IEEE Trans. on Reliability.

Energy Distribution Using Block-Based Shifting in a Smart Grid Tree Network

Boddu Rama Devi and K. Srujan Raju

Abstract The major challenge at the smart grid is to handle the load during peak hours. Using various techniques like peak clipping, shifting, valley filling, load shaping, energy conservation, etc., the demand side management balances the peak and nonpeak load in a day. This work mainly focuses on handling the peak and nonpeak load using the proposed adaptive block-based intelligent shifting (*ABIS*). The main objective of the proposed work is to distribute the energy to the users and maintain the balance between the available energy of the grid with the demand with high fairness. The simulation results show that the proposed shifting technique maintains load balance by maintaining a small gap between the load request and the available energy at the grid.

Keywords Adaptive block-based intelligent shifting · Demand side management
Load shifting · Peak clipping · Smart grid · Valley filling

1 Introduction

The advanced technology, latest appliances, modernized living, residual home, and industrial automation lead to huge demand of energy on the smart grid. To solve these challenges, various approaches at smart grid [1, 2] are applied like demand side management (*DSM*), utilizing secondary and renewable energy resources properly, integrating various micro-grids, adopting intelligent distribution with scheduling and shifting, etc.

B. R. Devi (✉)
Department of ECE, Kakatiya Institute of Technology & Science, Warangal,
Telangana, India
e-mail: ramadevikitsw@gmail.com

K. Srujan Raju
Department of CSE, CMR Technical Campus, Medichal, Hyderabad,
Telangana, India
e-mail: ksrujanraju@gmail.com

© Springer Nature Singapore Pte Ltd. 2018
V. Bhateja et al. (eds.), *Proceedings of the Second International Conference
on Computational Intelligence and Informatics*, Advances in Intelligent Systems
and Computing 712, https://doi.org/10.1007/978-981-10-8228-3_56

The key elements of a smart grid are: energy generation, transmission, distribution, and communication and control system [3]. To handle the demand side management, various techniques [4–11] are available in the literature for load management, integrating multiple micro-grids, residual user load based, and industrial applications.

In smart grid, we use various demand side management techniques. They are: (i) Load Shifting: shifting load to off-peak; (ii) Peak Clipping: clipping load in peak; (iii) Conservation: energy conservation using energy efficient techniques; (iv) Load Building: enhancing throughput by increasing the consumption; (v) Valley Filling: increasing load in off-peak; and (vi) Flexible Load Shape: special offers to control the demand. Various load scheduling approaches [12–17], valley filling [18], and load shifting [19–24] are available in the literature.

The energy generated using the primary resources is limited by the geographical resources available, financial status, and policies of the country. To meet the demand, at each level from the smart grid to users, everyone should concentrate on secondary source using renewable sources [25–27] like wind, solar, etc.

The energy distribution in smart grid (neglecting transmission and other losses) can be modeled as a distributed tree network [28]. Various energy allocation schemes in a distributed smart grid tree network (SGTN) and its fairness are evaluated in [29]. To handle the dynamic user load request with high fairness, a dynamic weight energy allocation with load is proposed in [30]. Later, to meet the future demands, renewable energy generation at the user with load request is introduced in [31]. Energy allocation and distribution to shiftable and nonshiftable type user loads by dynamic weight with load category is proposed in [31]. An adaptive intelligent technique is also proposed to handle the high demand urban scenario in [31]. Later, a thermometer approach based energy distribution with fine calibration is proposed in [32]. With this fine calibration, we achieve 100% distribution efficiency with above 98% fairness of the grid. Based on the above simulation results in [29–32], by considering few best findings, we propose a block-based shifting technique hereunder.

In this paper, we propose an adaptive block-based intelligent shifting energy distribution (ABISED) algorithm for different categories of users in a SGTN. The major objectives of the proposed work are mentioned here below.

- To find the demand on the grid based on the load request.
- To distribute the energy using adaptive block-based intelligent shifting.
- To maintain the balance between energy available at the grid and demand with high fairness.

The rest of the work is organized as follows. Section 2 explains the analysis of energy demand at grid based on a load request in SGTN. Section 3 describes the proposed adaptive block-based intelligent shifting energy distribution algorithm. Simulation results are findings are given in Sect. 4, and concluded in Sect. 5.

2 Energy Demand at Grid Based on Load Request in a SGTN

A three-level distributed *SGTN* with different category of users is shown in Fig. 1. The energy generated at grid G is transmitted to the primary substations P_S and then it is distributed to the secondary substations S_S. From S_S, the energy is distributed to the users of different categories. To make a smart grid system more and more efficient, the demand at the grid is calculated by the load request of the users. The energy allocation and distribution should be controlled via Communication and Control Centers at S_S, P_S, and G. Communication and Control Center is a computer-based intelligent control system integrated with the communication system.

For simplicity, we consider only two categories of load at the user. They are; (i) Nonshiftable (L_N) and (ii) Shiftable loads (L_S). All the users have different appliances. All the appliances of a user are classified into L_N or L_S. In view to meet future demand on the grid, all the users are self-sufficient of generating energy using renewable energy E_R to meet their requirements and reduce the burden at G. The load request in a three-level SGTN is calculated [31] as follows.

1. **Level 0: Load energy request at Grid G**

$$E_G^{t_i} = \sum_{p=1}^{P_S} \sum_{s=1}^{S_S} \sum_{u=1}^{U} E_{LR}^{t_i}(p, s, u) = \sum_{p=1}^{P_S} E_{LR}^{t_i}(p), \tag{1}$$

where

P_S Number of primary substations connected to G,
S_S Number of Secondary substations connected to a primary substation p,
U Number of users connected to the secondary substation s,

Fig. 1 Three-level distributed smart grid tree network

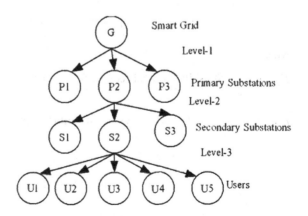

u User connected to s,

$E_{LR}^{t_i}$ Load request during time slot t_i.

2. *Level 1: Load energy request at Primary substation p*

$$E_{LR}^{t_i}(p) = \sum_{s=1}^{S_S} \sum_{u=1}^{U} E_{LR}^{t_i}(p, s, u) = \sum_{s=1}^{S_S} E_{LR}^{t_i}(p, s). \tag{2}$$

3. *Level 2: Load energy request at Secondary substation s*

$$E_{LR}^{t_i}(p, s) = \sum_{u=1}^{U} E_{LR}^{t_i}(p, s, u) - E_R^{t_i}(p, s), \tag{3}$$

where

$$\text{Total renewable energy at } s = E_R^{t_i}(p, s) = \sum_{u=1}^{U} E_R^{t_i}(p, s, u) \tag{4}$$

$E_R^{t_i}(p, s, u)$ Energy generated by the user using renewable energy sources.

The total renewable energy at secondary substation is considered during distribution to avoid fractional energy wastage at the user.

4. *Level 3: Load energy request at the user*:

Load energy request of the user is the load request from shiftable and non-shiftable loads.

$$E_{LR}^{t_i}(p, s, u) = L_N^{t_i}(p, s, u) + L_S^{t_i}(p, s, u), \tag{5}$$

where

$L_N^{t_i}(p, s, u)$ Nonshiftable load request from nonshiftable user appliances,

$L_S^{t_i}(p, s, u)$ Shiftable load request from shiftable appliances of the user at time slot t_i.

5. *Demand at the Grid*:

Let $E_{AG}^{t_i}$ is the available energy in the grid from primary and secondary energy generation. The demand varies based on the load and available energy on the grid. The energy demand E_D is given by

$$E_{DG}^{t_i} = \begin{cases} High, & E_{AG}^{t_i} < E_G^{t_i} \\ Low, & E_{AG}^{t_i} > E_G^{t_i} \end{cases}. \tag{6}$$

3 Adaptive Block-Based Intelligent Shifting Energy Distribution

Based on the demand, energy is allocated or distributed to various levels.

A. *Level 1: Energy allotted from Grid (G) to Primary substation (p)*:

$$E_A^{t_i}(p) = \begin{cases} E_{LR}^{t_i}(p), & when \ E_{DG}^{t_i} = Low \\ \lfloor \frac{E_{LR}^{t_i}(p)}{E_G^{t_i}} \rfloor E_{AG}^{t_i}, & when \ E_{DG}^{t_i} = High \ (for \ p \neq 1) \\ E_{AG}^{t_i} - \sum\limits_{p=2}^{P} E_A^{t_i}(p), & when \ E_{DG}^{t_i} = High \ (for \ p = 1) \end{cases}. \tag{7}$$

B. *Level 2: Energy available at secondary substation (s) with renewable energy, after the energy allocation from primary substations (p)*:

$$E_A^{t_i}(p, s) = \begin{cases} E_{LR}^{t_i}(p, s); & when \ E_{DG}^{t_i} = Low \\ \lfloor \frac{E_{LR}^{t_i}(p, s)}{E_{LR}^{t_i}(p)} \rfloor E_{AG}^{t_i} + E_R^{t_i}(p, s); & when \ E_{DG}^{t_i} = High \ (for \ p \neq 1) \\ E_{AG}^{t_i} - \sum\limits_{p=2}^{P} E_A^{t_i}(p, s) + E_R^{t_i}(p, s); & when \ E_{DG}^{t_i} = High \ (for \ p = 1) \end{cases}. \tag{8}$$

C. *Level 3: Energy distributed from secondary substation (s) to users (U)*

$$E_D^{t_i}(p, s, u) = \begin{cases} E_{LR}^{t_i}(p, s, u); & \text{when } E_{DG}^{t_i} = Low \\ \text{Use Adaptive Block based Intelligent Shifting}; & \text{when } E_{DG}^{t_i} = High \end{cases}$$

(9)

When demand is low, the user gets the requested energy. When demand is high, the energy is first distributed to the nonshiftable loads L_N, and then allotted to shiftable loads L_S. Once the shiftable loads get allocated they are treated as nonshiftable load as they require continuous slots to finish the work [31]. If they (L_N) do not get the energy due to high demand, then they are intelligently shifted to the other slots of low demand [31]. This, also leads to rise in the load demand in other slots. Block-based division idea is taken from [33]. To avoid the over demand due to shiftable loads, we propose an adaptive block-based intelligent shifting energy distribution (*ABISED*) algorithm. In *ABISED* algorithm, the remaining unallocated load at each secondary substation (s) is considered as a block. The steps in *ABISED* are as follows:

  i. Calculate the energy allocated to the secondary substations, primary substation using (7) and (8).
 ii. Calculate the energy unallocated at s, p [from (3), (4) and (7), (8)] in high demand time slots and energy left at grid at low demand time slots. The total unallocated load at s is considered as a block.
iii. Shift the unallocated blocks of (p, s) to various time slots with excess energy.
 iv. After shifting, distribute the energy based on the demand and available energy.

4 Simulation Results

A three-level smart grid tree network is considered for implementation using MATLAB [34]. Twelve time slots are considered with 4 slots of moderate, 4 slots of high, and 4 slots of low demand to analyze the energy distribution. The simulation parameters used are shown in Table 1. The energy is allocated based on (7), (8) and maintains high fairness index of 96–98% at G, p and s [30–32].

The load request at the primary substations and grid; available energy in the grid before ABISED during different time slots is shown in Fig. 2. The load is moderate during the time slots $t_i = 1$–4, high at $t_i = 5$–8 and low at $t_i = 9$–12. Available energy on the grid, load request at the grid before shifting; load request at the primary substations and grid after *ABISED* is shown in Fig. 3.

Table 1 Simulation parameters used in ABISED

Parameter	Value
Number of levels in SGTN	3
Number of primary substations	3
Number of secondary substations	5
Number of users connect to S_S	100
Category of users	2 (Shiftable, nonshiftable)
$E_{AG}^{t_i}$ (kWh)	850 ($t_i = 1$–4); 900 ($t_i = 5$–8); 800 ($t_i = 9$–12)
Unit energy	1 kWh
Total time slots considered	12 (4 slots with moderate request, 4 with high demand, 4 with low demand)

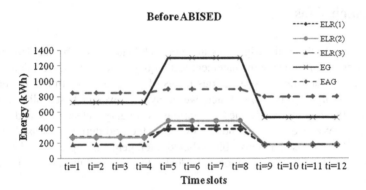

Fig. 2 The load request at the p and grid; available energy at G before ABISED

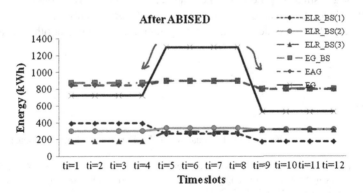

Fig. 3 Available energy at G, load request at G before shifting; load request at the primary substations and grid after ABISED

For the test case considered, the excess load of blocks $(1, 1)$, $(1, 2)$, $(1, 3)$, $(1, 4)$, $(1, 5)$ and $(1, 2)$ with 151 kWh at $t_i = 5$–8 are shifted to the time slots $t_i = 1$–4 as it has 125 kWh excess energy. Out of this, block $(1, 2)$ gets only 7 kWh in $t_i = 1$–4; so remaining 26 kWh in the block $(1, 2)$ are shifted to $t_i = 9$–12. The excess load of blocks $(2, 2)$, $(2, 3)$, $(2, 4)$, $(2, 5)$, $(1, 3)$, $(2, 3)$, $(3, 3)$, $(3, 4)$, $(3, 5)$ with 249 kWh and $(1, 2)$ partial of 26 kWh, i.e., a total load of 275 kWh is shifted to $t_i = 9$–12. At $t_i = 9$–12, only 270 kWh excess energy is available; hence 5 kWh is left unallocated and moves to consecutive slots where excess energy is available at G.

The peak load during $t_i = 5$–8 is shifted to $t_i = 1$–4 and $t_i = 9$–12 and makes the curve very smooth one, as shown in Fig. 3. From the simulation results, Fig. 3, it is clearly observed that the load request at the grid after *ABISED* is close to the energy available on the grid, which maintains a balance between the available energy of the grid and the load request.

5 Conclusions

The demand on the grid during peak and nonpeak loads should be handled properly using various demand side management techniques. In this paper, we calculated the load request from the users at various levels of the grid and estimate the demand during the various time slots. Later, we apply load weight based allocation and calculate the energy unallocated at various secondary substations and primary substations. We also calculate the energy left in the grid during various slots. Now we apply the proposed adaptive block-based intelligent shifting energy distribution algorithm and shift each unallocated blocks to various time slots with excess energy.

From the simulation results, it is observed that the load request at the grid after *ABISED* is close to the energy available in the grid. It maintains load balancing by maintaining a small gap between the load request and the available energy in the grid. It reduces the over queue or over demand due to shifting in any slot and handles the demand side management system functioning smoothly with high fairness.

References

1. Hassan Farhangi.: The Path of the Smart Grid. IEEE power & Energy Magazine. 18–28 (2010)
2. S Massoud Amin.: Smart Grid: Overview, Issues and Opportunities, Advances and Challenges in Sensing, Modeling, Simulation, Optimization and Control. European Journal of Control. 17 (5–60), 547–567 (2011)
3. V. K. Mehta, R. Mehta.: Principles of Power System: Including Generation, Transmission, Distribution. 24/e, S. Chand Publishing (2005)

4. Yee Wei Law, Tansu Alpcan, Vincent C.S. Lee, Anthony Lo, Slaven Marusic, Marimuthu Palaniswami.: Demand response architectures and load management algorithms for energy-efficient power grids: a survey. In: 2012 Seventh International Conference on Knowledge, Information and Creativity Support Systems, 134–141 (2012)

5. K. Nunna, S. Dolla.: Demand response in smart distribution system with multiple micro grids. IEEE Trans. Smart Grid. 3 (4), 1641–1649 (2012)

6. H. S. V. S. Kumar Nunna, Suryanarayana Doolla.: Energy management in micro grids using demand response and distributed storage a multi-agent approach. IEEE Trans. Power Deliv. 28 (2), 939–947 (2013)

7. H. S. V. S. Kumar Nunna, Suryanarayana Doolla.: Responsive end-user-based demand side management in multi micro grid environment. IEEE Trans. Ind. Inform. 10 (2), 1262–1272 (2014)

8. C. A. Babu, S. Ashok.: Peak Load Management in Electrolytic Process Industries. IEEE Transactions on Power Systems, 23 (2), 399–405 (2008)

9. Peng- Yong Kong.: Wireless Neighbourhood Area Network with QoS Support for Demand Response in Smart Grid. IEEE Transaction on Smart Grid. 7 (4), 1913–1923 (2015)

10. H. Chao, P. Hsiung.: A fair energy resource allocation strategy for micro grid. Journal of Microprocessors and Microsystems. 42, 235–244 (2016)

11. Ayan Mondal, Sudip Misra, Mohammad S. Obaidat.: Distributed Home Energy Management System with Storage in Smart Grid using Game Theory. IEEE Systems Journal. 99, 1–10 (2015)

12. Phani Chavali, Peng Yang, Arye Nehorai.: A Distributed Algorithm of Appliance Scheduling for Home Energy Management System. IEEE Transactions on Smart Grid. 5 (1), 282–290 (2014)

13. Samaresh Bera, Praveen Gupta, Sudip Misra.: D2S: Dynamic Demand Scheduling in Smart Grid using Optimal Portfolio Selection Strategy. IEEE Transactions on Smart Grid. 6 (3) (2015)

14. Feng Ye, Yi Qian, Rose Qingyang Hu.: Incentive Load Scheduling Schemes for PHEV Battery Exchange Stations in Smart Grid. IEEE Systems Journal. 99, 1–9, 23 (2017)

15. Mohammad Tasdighi, Hassan Ghasemi, Ashkan Rahimi-Kian.: Residential Microgrid Scheduling Based on Smart Meters Data and Temperature Dependent Thermal Load Modeling. IEEE Transactions on Smart Grid. 5 (1), 349–357 (2014)

16. Pengwei Du, Ning Lu.: Appliance Commitment for Household Load Scheduling. IEEE Transactions on Smart Grid. 2 (2), 411–419 (2011)

17. Alessandro Agnetis, Gianluca de Pascale, Paolo Detti, Antonio Vicino.: Load Scheduling for Household Energy Consumption Optimization. IEEE Transactions on Smart Grid. 4 (4), 2364–2373 (2013)

18. Zhenpo Wang, Shuo Wang.: Grid Power Peak Shaving and Valley Filling Using Vehicle-to-Grid Systems. IEEE Transactions on Power Delivery, 28 (3) (2013)

19. R. Lily Hu, Ryan Skorupski, Robert Entriken, Yinyu Ye.: A Mathematical Formulation for Optimal Load Shifting of Electricity Demand. IEEE Transactions on Big Data, PP (99), 1–1 (2016)

20. Nathanael Beeker, Paul Malisani, Nicolas Petit.: An optimization algorithm for load-shifting of large sets of electric hot water tanks. In: 29th International conference on ECOS, Portoroz, Slovenia (2016)

21. Lünsdorf, Ontje and Sonnenschein, Michael.: A pooling based A pooling based load shift strategy for household appliances. In: Integration of Environmental Information in Europe - 24th International Conference on Informatics for Environmental Protection, 734–743 (2010)

22. Shivaramu K.N, Roopashree N, Sneha K V.: Dynamic Quantum Shift Algorithm for Load Balancing in High Performance Clusters. International Journal of Applied Engineering Research. 11(7), 4954–4960 (2016)

23. Chaojie Li, Xinghuo Yu, Wenwu Yu, Guo Chen, Jianhui Wang.: Efficient Computation for Sparse Load Shifting in Demand Side Management. IEEE Transactions on Smart Grid. 8 (1), 250–261 (2017)

24. Chaimongkon Chokpanyasuwan, Tika Bunnang, Ratthasak Prommas.: Artificial Intelligence for Load Management Based On Load Shifting in the Textile Industry. International Journal of Engineering and Technology (IJET). 350–367 (2015)
25. V. Pilloni, A. Floris, A. Meloni, L. Atzor.: Smart home energy management including renewable sources: A QoE-driven approach. IEEE Transactions on Industrial Informatics. PP (99), 1–1 (2017)
26. M. J. Neely, A. S. Tehrani, A. G. Dimakis.: Efficient algorithms for renewable energy allocation to delay tolerant consumers. In: 1st IEEE Int. Conf. on Smart Grid Communications (2010)
27. Smart grids and renewable- a guide for effective deployment. International Renewable Energy Agency (IRENA) (2013)
28. Eli M. Gafni, Dimitri P. Bertsekas.: Dynamic Control of Session Input Rates in Communication Networks. IEEE Transactions on Automatic Control. AC-29 (11), (1984)
29. Boddu Rama Devi, Manjubala Bisi, Rashmi Ranjan Rout.: Analysis of Efficient Energy Allocation Schemes in a Tree Based Smart Grid. Pakistan Journal of Biotechnology (accepted)
30. Boddu Rama Devi.: Dynamic Weight based Energy Slots Allocation and Pricing with Load in a Distributive Smart Grid. Journal of Engineering and Applied Sciences (accepted)
31. Boddu Rama Devi.: Load Weight Based Energy Allocation and Energy Distribution Based on the Time Zone with Load Category in a Smart Grid. International Journal of Renewable Energy Research (accepted)
32. Boddu Rama Devi.: Energy Efficiency of Thermometer Approach based Energy Distribution with Fine Calibration in a Smart Grid Tree Network. (Submitted)
33. Wendong Yang, Yueming Cai.: A block based OFDM Decode-and-Forward Relaying Schemes for Cooperative Wireless Networks. In: ISCIT, pp. 938–942 (2009)
34. MATLAB, https://www.mathworks.com/products/matlab.html, dated: 31.05.2017

Properties and Applications of Hermite Matrix Exponential Polynomials

Subuhi Khan and Shahid Ahmad Wani

Abstract The Hermite matrix based exponential polynomials (HMEP) are introduced by combining Hermite matrix polynomials with exponential polynomials. Certain properties of the HMEP are also established. The operational representations providing connections between HMEP and some other special polynomials are also derived. The approach presented in this article is general.

Keywords Hermite matrix based exponential polynomials · Monomiality principle · Operational techniques

1 Introduction

Generalized forms of special functions are important from the applications point of view. This advancement provides tools to solve certain problems occurring in physics and engineering. The matrix forms of Hermite polynomials are considered in [1, 3, 4, 6, 7] for $\mathbb{C}^{N \times N}$ matrix, with eigen values lying in the right side of open-half plane.

Throughout the paper unless otherwise stated, M is considered to be a stable matrix (positive) in $\mathbb{C}^{N \times N}$, that is

$$Real\,(v) > 0, \forall\, v \in \xi(M). \tag{1}$$

Here, $\xi(M)$ denotes set of all the eigen values of matrix M.

Let E_0 denotes the cut part of the complex plane besides the left side of the real axis from the origin and $\log(\sigma)$ indicates the principal part of logarithm of σ, then $\sigma^{\frac{1}{2}}$

S. Khan (✉) · S. A. Wani
Department of Mathematics, Aligarh Muslim University, Aligarh, India
e-mail: subuhi2006@gmail.com

S. A. Wani
e-mail: shahidwani177@gmail.com

© Springer Nature Singapore Pte Ltd. 2018
V. Bhateja et al. (eds.), *Proceedings of the Second International Conference on Computational Intelligence and Informatics*, Advances in Intelligent Systems and Computing 712, https://doi.org/10.1007/978-981-10-8228-3_57

indicates $e^{\frac{1}{2}\log(z)}$. Let the matrix $\mathbb{C}^{N\times N}$ with $\xi(M) \subset E_0$, then $M^{1/2} = \sqrt{M}$ indicates the image by $\sigma^{\frac{1}{2}}$ of the matrix M.

The series form for Hermite matrix polynomials (HMP) $H_m(u, M)$ is given as [4]

$$H_m(u, M) = m! \sum_{k=0}^{[\frac{m}{2}]} \frac{(-1)^k (u\sqrt{2M})^{m-2k}}{(m-2k)!k!}, m \geq 0 \tag{2}$$

and it possess the following generating relation:

$$e^{(uw\sqrt{2M}-w^2 I)} = \sum_{m=0}^{\infty} H_m(u, M)\frac{w^m}{m!}. \tag{3}$$

Here, I is the identity matrix in $\mathbb{C}^{N\times N}$.

The two-variable Hermite matrix polynomials (2VHMP) $H_m(u, v, M)$ have the following series form [1]:

$$H_m(u, v, M) = m! \sum_{k=0}^{[\frac{m}{2}]} \frac{(-1)^k v^k (u\sqrt{2M})^{m-2k}}{(m-2k)!k!}, \quad m \geq 0 \tag{4}$$

and possess the following generating relation:

$$e^{(uw\sqrt{2M}-vw^2 I)} = \sum_{m=0}^{\infty} H_m(u, v, M)\frac{w^m}{m!}. \tag{5}$$

We note that 2VHMP have the following multiplicative and derivative operators:

$$\hat{A}_H := u\sqrt{2M} - 2v(\sqrt{2M})^{-1}\frac{\partial}{\partial u}, \tag{6}$$

$$\hat{B}_H := \frac{1}{\sqrt{2M}}\frac{\partial}{\partial u} \tag{7}$$

and thus can be considered as quasi-monomial.

Making use of the concept of quasi-monomiality, the isospectral problems [8] can be considered and new families can be introduced. For example, we use the following correspondence:

$$\begin{aligned} \hat{A} &\Leftrightarrow u, \\ \hat{B} &\Leftrightarrow \frac{\partial}{\partial u}, \\ p_m(u) &\Leftrightarrow u^m, \end{aligned} \tag{8}$$

to deal with such isospectral problems. The operational methods are required for investigating certain problems of physics.

From Eqs. (6) and (7), it follows that:

$$\hat{A}_H\{H_m(u, v; M)\} = H_{m+1}(u, v; M) \tag{9}$$

and also we have

$$\hat{B}_H\{H_m(u, v; M)\} = m \{H_{m-1}(u, v; M)\}. \tag{10}$$

Now from the expressions of the multiplicative, derivative operators $\hat{A}_H$, $\hat{B}_H$, we conclude that

$$\hat{A}_H \hat{B}_H\{H_m(u, v; M)\} = m \{H_m(u, v; M)\}, \tag{11}$$

which gives

$$\left(u\frac{\partial}{\partial u} - 2v(\sqrt{2M})^{-1}\frac{\partial^2}{\partial u^2} - m\right) H_m(u, v; M) = 0, \tag{12}$$

a second-order matrix differential equation. Assuming here and in the sequel $H_0(u, v; M) = 1$, then $H_m(u, v; M)$ can be explicitly constructed as:

$$H_m(u, v; M) = \hat{A}_H^m\{1\}, \tag{13}$$

from which it is easy to find the series definition (4) of $H_m(u, v; M)$.

The above equation implies that the expression for generating relation of $H_m(u, v; M)$ is given as:

$$\exp(w\hat{A}_H)\{1\} = \sum_{m=0}^{\infty} H_m(u, v; M)\frac{w^m}{m!}, \quad |w| < \infty, \tag{14}$$

which yields generating function (5).

The following commutation relation is satisfied for $\hat{A}_H$ and $\hat{B}_H$:

$$[\hat{B}_H, \hat{A}_H] = \hat{B}_H\hat{A}_H - \hat{A}_H\hat{B}_H = \hat{\mathbb{1}}, \tag{15}$$

which indicates the structure of Weyl group.

The exponential polynomial $\phi_m(u)$ [5] is a member of the associated Sheffer family and has the following generating relation:

$$e^{u(e^w - 1)} = \sum_{m=0}^{\infty} \phi_m(u)\frac{u^m}{m!} \tag{16}$$

and the binomial identity for $\phi_m(u)$ is as follows:

$$\phi_m(u + v) = \sum_{k=0}^{m} \binom{m}{k} \phi_k(u)\phi_{m-k}(u). \tag{17}$$

The exponential polynomials $\phi_m(u)$ also satisfy the following recurrence relations:

$$\phi_{m+1}(u) = u\{\phi_m(u) + \phi'_m(u)\} \tag{18}$$

and

$$\phi_{m+1}(u) = u\left\{ \sum_{k=0}^{m} \binom{m}{k} \phi_k(u) \right\}. \tag{19}$$

In this article, the Hermite matrix based exponential polynomials (HMEP) are introduced by combining Hermite matrix polynomials with exponential polynomials. In Sect. 2, certain properties of the HMEP are established. In Sect. 3, operational representations providing connection between HMEP and some other special polynomials are derived.

2 Hermite Matrix Based Exponential Polynomials

The two-variable Hermite matrix based exponential polynomials denoted by $_H\phi_m(u, v; M)$ are introduced by establishing the following result:

Result 1 The following generating relation for the HMEP $_H\phi_m(u, v; M)$ holds true:

$$\exp\left(u(e^w - 1)\sqrt{2M} - v(e^w - 1)^2 I \right) = \sum_{n=0}^{\infty} {}_H\phi_m(u, v; M)\frac{w^m}{m!} \tag{20}$$

Proof Replacement of u in of Eq. (16) by the operator $\hat{A}_H$ of two-variable Hermite matrix polynomial, it follows that:

$$\exp(\hat{A}_H(e^w - 1)) = \sum_{m=0}^{\infty} \phi_m(\hat{A}_H)\frac{w^m}{m!}. \tag{21}$$

Using the expression of $\hat{A}_H$ given in Eq. (6) and using the following notation for the HMEP in the right hand side by $_H\phi_m(u, v; M)$, that is

$$\phi_m(\hat{A}_H) = \phi_m\left(u\sqrt{2M} - 2v(\sqrt{2M})^{-1}\frac{\partial}{\partial u} \right) = {}_H\phi_m(u, v; M), \tag{22}$$

we find

$$\exp\left(\left(u\sqrt{2M} - 2v(\sqrt{2M})^{-1}\frac{\partial}{\partial u} \right)(e^w - 1) \right) = \sum_{m=0}^{\infty} {}_H\phi_m(u, v; M)\frac{w^m}{m!}, \tag{23}$$

that is

$$\exp\left(\left(u\sqrt{2M} - 2v\frac{\partial}{\partial(u\sqrt{2M})}\right)(e^w - 1)\right) = \sum_{m=0}^{\infty} {}_H\phi_m(u, v; M)\frac{w^m}{m!}. \qquad (24)$$

Now, disentangling the exponential in the left hand side of the above equation with the help of the Crofton-type identity [2]:

$$g\left(v + n\mu\frac{d^{n-1}}{dv^{n-1}}\right)\{1\} = \exp\left(\mu\frac{d^n}{dv^n}\right)\{g(v)\}, \qquad (25)$$

we have

$$\exp\left(-v\frac{\partial^2}{\partial(u\sqrt{2M})^2}\right)\exp(u\sqrt{2M}(e^w - 1)) = \sum_{m=0}^{\infty} {}_H\phi_m(u, v; M)\frac{w^m}{m!}. \qquad (26)$$

Expansion of the first exponential term in the left hand side of above equation and then its action on the second exponential term gives

$$\exp(-v(e^w - 1)^2)\exp(u\sqrt{2M}(e^w - 1)) = \sum_{m=0}^{\infty} {}_H\phi_m(u, v; M)\frac{w^m}{m!}, \qquad (27)$$

which on simplification yields assertion (20).

Result 2 The HMEP ${}_H\phi_m(u, v; M)$ are quasi-monomial with respect to the following raising and lowering operators:

$$\hat{A}_{He} := \left(u\sqrt{2M} - 2v(\sqrt{2M})^{-1}\frac{\partial}{\partial u}\right)(1 + D_U) \qquad (28)$$

and

$$\hat{B}_{He} := \ln(1 + D_U) \qquad (29)$$

respectively, where

$$\frac{D_u}{\sqrt{2M}} = D_U.$$

Proof Consider the following identity:

$$D_u \exp\left(u(e^w - 1)\sqrt{2M} - v(e^w - 1)^2 I\right)$$
$$= \sqrt{2M}(e^w - 1)\exp\left(u(e^w - 1)\sqrt{2M} - v(e^w - 1)^2 I\right). \qquad (30)$$

From the above equation, we have

$$\left(\frac{D_u}{\sqrt{2M}} + 1\right) \exp\left(u(e^w - 1)\sqrt{2M} - v(e^w - 1)^2 I\right)$$
$$= e^w \exp\left(u(e^w - 1)\sqrt{2M} - v(e^w - 1)^2 I\right). \tag{31}$$

Differentiating Eq. (21) partially with respect to w and using Eq. (22) in the right-hand side of the obtained equation, it follows that

$$\hat{A}_H \, e^w \exp(\hat{A}_H(e^w - 1)) = \sum_{m=0}^{\infty} {}_H\phi_{m+1}(u, v; M)\frac{w^m}{m!}, \tag{32}$$

which on using Eq. (14) with $w = (e^w - 1)$ in the right-hand side and in view of generating relation (5) becomes

$$\hat{A}_H \, e^w \exp\left(u(e^w - 1)\sqrt{2M} - v(e^w - 1)^2 I\right) = \sum_{m=0}^{\infty} {}_H\phi_{m+1}(u, v; M)\frac{w^m}{m!}. \tag{33}$$

Now using Eq. (20) in the right-hand side of Eq. (33) and in view of identity (31), it follows that:

$$\hat{A}_H \left(D_U + 1\right) \sum_{m=0}^{\infty} {}_H\phi_m(u, v; M)\frac{w^m}{m!} = \sum_{m=0}^{\infty} {}_H\phi_{m+1}(u, v; M)\frac{w^m}{m!}. \tag{34}$$

The coefficients of same powers of w in Eq. (34) are equated to give

$$\hat{A}_H \left(D_U + 1\right) {}_H\phi_m(u, v; M) = {}_H\phi_{m+1}(u, v; M), \tag{35}$$

which, from Eq. (9) allows us to conclude that the raising operator for ${}_H\phi_m(u, v; M)$ is given as:

$$\hat{A}_{He} = \hat{A}_H \left(D_U + 1\right). \tag{36}$$

Finally using Eq. (6) in the right-hand side of Eq. (36), assertion (28) follows.

Taking log on both sides of identity (31) and multiplying both sides of the resultant expression by $\sum_{m=0}^{\infty} {}_H\phi_m(u, v; M)\frac{w^m}{m!}$, we have

$$\ln(1 + D_U) \sum_{m=0}^{\infty} {}_H\phi_m(u, v; M)\frac{w^m}{m!} = w \sum_{m=0}^{\infty} {}_H\phi_m(u, v; M)\frac{w^m}{m!}, \tag{37}$$

which on simplification gives

$$\ln(1 + D_U) \sum_{m=0}^{\infty} {}_H\phi_m(u, v; M)\frac{w^m}{m!} = m \sum_{m=0}^{\infty} {}_H\phi_{m-1}(u, v; M)\frac{w^m}{m!}. \tag{38}$$

The coefficients of same powers of w in Eq. (38) are equated to give

$$\ln(1 + D_U)_H\phi_m(u, v; M) = m_H\phi_{m-1}(u, v; M), \tag{39}$$

which in view of Eq. (10) proves assertion (29).

Result 3 The HMEP ${}_H\phi_m(u, v; M)$ satisfy the following differential equation:

$$\left(\left(u\sqrt{2M} - 2v(\sqrt{2M})^{-1}\frac{\partial}{\partial u}\right)\left(1 + D_U\right)\ln\left(1 + D_U\right) - m\right){}_H\phi_m(u, v; M) = 0. \tag{40}$$

Proof Inserting Eqs. (28) and (29) in Eq. (11) with $H_m(u, v; M)$ replaced by ${}_H\phi_m(u, v; M)$, we get assertion (40).

In the forthcoming section, some operational connections of the HMEP ${}_H\phi_m(u, v; M)$ are derived.

3 Operational Representations

Result 4 The following operational connection between HMEP ${}_H\phi_m(u, v; M)$ and exponential polynomials $\phi_m(u)$ holds true:

$$_H\phi_m(u, v; M) = \exp\left(-v\frac{\partial^2}{\partial(u\sqrt{2M})^2}\right)\phi_m(\sqrt{2M}). \tag{41}$$

Proof In view of Eq. (22), we have

$$_H\phi_m(u, v; M) = \phi_m\left(u\sqrt{2M} - 2v\frac{\partial}{\partial(u\sqrt{2M})}\right), \tag{42}$$

which on using Crofton identity (25) in the right-hand side proves assertion (41).

Result 5 The following operational representation connecting the HMEP ${}_H\phi_m(u, v; M)$ with HMEP of one variable ${}_H\phi_m(u; M)$ holds true:

$$_H\phi_m(u, v; M) = \exp\left(-\frac{v}{2M}\frac{\partial^2}{\partial u^2}\right){}_H\phi_m(u; M), \tag{43}$$

where $_H\phi_m(u; M) = {}_H\phi_m(u, 0; M)$

Proof From Eq. (20), it follows that

$$\frac{\partial^2}{\partial u^2}\{_H\phi_m(u, v; M)\} = (-2M)\frac{\partial}{\partial v}\{_H\phi_m(u, v; M)\}. \tag{44}$$

Simplifying and integrating the above equation, we have

$$\ln\left(_H\phi_m(u, v; M)\right) = -\frac{v}{2M}\frac{\partial^2}{\partial u^2} + \ln(P). \tag{45}$$

Solving Eq. (44) with initial condition

$$P = {}_H\phi_m(u; M) := {}_H\phi_m(u, 0; M), \tag{46}$$

we get assertion (43).

In the present article, the matrix polynomials of special type are established by considering the use of identities of operational nature and concepts associated with the monomiality principle. The approach presented in the present paper is general in nature in view of its importance to obtain further other polynomial families of special type. The new polynomials and associated results can be used to solve many problems occurring in earth, biological and physical sciences, and also in emerging fields of engineering. In the field of engineering sciences, these polynomials play a significant and vital role in dealing with quantum, mechanical, and transport beam problems.

References

1. Batahan, R.S.: A new extension of Hermite matrix polynomials and its applications, Linear Algebra Appl. 419 (2006), no. 1, 82–92.
2. Dattoli, G., Ottaviani, P.L., Torre, A., Vázquez, L.: Evolution operator equations: integration with algebraic and finite-difference methods. Applications to physical problems in classical and quantum mechanics and quantum field theory, Riv. Nuovo Cimento Soc. Ital. Fis. 20(2) (1997), 1–133.
3. Defez, E., Jódar, L.: Some applications of the Hermite matrix polynomials series expansions, J. Comput. Appl. Math. 99 (1998), no. 1–2, 105–117.
4. Jódar, L., Company, R.: Hermite matrix polynomials and second order matrix differential equations, Approx. Theory Appl. (N.S.) 12 (1996), no. 2, 20–30.
5. Roman, S.: The Umbral Calculus, Academic Press, New York, 1984.
6. Sayyed, K.A.M., Metwally, M.S., Batahan, R.S.: On generalized Hermite matrix polynomials, Electron. J. Linear Algebra 10 (2003), 272–279.
7. Subuhi Khan, Raza, N.: 2-variable generalized Hermite matrix polynomials and Lie algebra representation, Rep. Math. Phys. 66 (2010), no. 2, 159–174.
8. Smirnov, Y., Turbiner, A.: Lie algebraic discretization of differential equations, Modern Phys. Lett. A 10(24) (1995) 1795–1802.

Economical Home Monitoring System Using IOT

Kunwardeep Singh Obheroi, Ankur Chaurasia,
Tanupriya Choudhury and Praveen Kumar

Abstract Home monitoring system is a device that is implemented using Internet of Things. The Internet of Things is the internetworking of physical devices, vehicles, buildings, and other items embedded with electronics, software, sensors, actuators, and network connectivity that enable these objects to collect and exchange data. The home monitoring system monitors doors and windows of your home and notifies you of any new access of your property via a data feed. Home monitoring system consists of sensors to detect intrusion and captures and sends multiple pictures of the intruder to the user anywhere through Internet. The system is implemented via the use of IOT, which is the interconnection of machines via Internet for advanced connectivity.

Keywords Home monitoring system (HMS) · Internet of Things (IOT)
Cloud networking · Arduino Uno

1 Introduction

1.1 Overview

Home monitoring system is an antitheft mechanism and is implemented using Internet of Things. The Internet of Things is the internetworking of physical devices, vehicles,

K. S. Obheroi (✉) · A. Chaurasia · T. Choudhury · P. Kumar
Amity University, Noida, Uttar Pradesh, India
e-mail: k.s.obheroi@gmail.com

A. Chaurasia
e-mail: ankurk069@gmail.com

T. Choudhury
e-mail: tchoudhury@amity.edu

P. Kumar
e-mail: pkumar3@amity.edu

© Springer Nature Singapore Pte Ltd. 2018
V. Bhateja et al. (eds.), *Proceedings of the Second International Conference on Computational Intelligence and Informatics*, Advances in Intelligent Systems and Computing 712, https://doi.org/10.1007/978-981-10-8228-3_58

buildings, and other items embedded with electronics, software, sensors, actuators, and network connectivity that enable these objects to collect and exchange data. The home monitoring system will be able to monitor doors and windows of your home and notify you of any unauthorized access of your property by capturing multiple pictures.

Home monitoring system is a system designed to monitor and keep a check on visitors. Home monitoring system consists of sensors to detect intrusion and record and provide a video of the visitor/intruder to the user anywhere through Internet. The system is implemented via the use of IoT, which is interconnection of machines via Internet for advanced connectivity.

1.2 Advantages of Home Monitoring System

In the recent years, devices working on Internet to provide various services have become more and more common leading advances in various fields. One such field is home security. Every house owner worries about the security of his property when he is away, and now there are various options available for him to monitor his home. But there are many advantages that the proposed system has over the existing one:

(1) System scalability and easy extension: The deployment of the proposed system is relatively easy and the system can be extended to add more cameras and sensors based on the needs. This is because of the minimum dedicated hardware requirement of the system.
(2) Integrity: The system provides on-site as well as off-site storage which prevents the data collected by the system from being compromised.
(3) Mobility: The system allows user to check the data collected by the system from anywhere in the world on a different device like his mobile phone.
(4) Economic: The proposed system is relatively cheaper to set up and maintain due to the methodology used in development of said system. The system can be set up using any existing computer system owned by the user.
(5) Low resources use: The system very low storage space and can work with a low-speed Internet connection which helps reduce the cost of maintaining the system further. This also allows the system to work in less technologically advanced countries where high-speed Internet connection is not available.

2 Related Work

2.1 Remote Home Supervising System by Gowthami T. and More

It is used for supervising homes items using a monitoring system. It uses a smart home for its security. The user can access the smart home using his mobile using ZigBee [1]. He can access and the security status of the home using multiple parameters and alter them according to users requirement. This application is mainly used when the user not at home and is present at some other location at that very moment.

2.2 Waste Electronics for Home Supervision System by Dingrong Yuan and More

As new electronics come into someone's home, old electronics go waste and are put aside. Not only these act as environmental pollutants if wasted and thrown into land as a garbage but their potential of being used in other applications is wasted [2]. This project focuses on creating a home supervision system using garbage electronics like these. It not only helps in security purposes but is also very cheap and saves a lost of user cost.

2.3 Home Intrusion System Using Raspberry Pi by Shivprasad Tavagad, and More

Home intrusion system uses Raspberry Pi for its development. See the importance of security systems, this project is developed to make a home intrusion system is a cheap rate [3]. This does not use RFID which are used by current technologies rather it uses other cheaper yet efficient means. It uses mobile phone to offer the security to the home and other items. The device is connected to an Internet providing dongle which offers it connectivity to the web.

2.4 Robot Monitored Home by Madhavi Shinde, and More

This project monitors the home using a robot which is connected to the Internet and it can be accessed by the user [4]. The robot provides a GUI of the home. It sends all the locations and activities that are happening in the house. If any change in any location is found, the robot gets an red alert and sends user information about it via Internet. If any intrusion happens then also it sends pictures of the thief to the user which could be used for future actions.

3 System Analysis

3.1 Problem Definition

Many such devices exist which provide similar feature of providing the power to monitor his home to the owner via Live Streaming video, alerts, etc. However, these devices either need a lot of cost to be set up, or a lot of resources to maintain, or both.

System's that streams videos need a lot of Internet bandwidth which may restrict the user's access to monitoring system if he lacks proper Internet connection.

In devices that record video and stores it on-site(home) of owner, there are chances of device being compromised in case of a break-in.

In devices that records and stores entire data on cloud, a lot of space on cloud and a very high-speed Internet connection and lots of storage space on cloud servers are necessary which increase the cost of setting up and maintaining the device.

3.2 Proposed System Feature

In the proposed system, the aim is to achieve a cheap alternate system that takes fewer resources to set up and maintain.

The setup consists of a smart board (Arduino Uno [5]), sensor, GSM module, camera, and a user interface module which allows us to monitor any activity. The camera records a picture/video whenever the sensor is triggered and stores it on cloud thus decreasing the storage space requirement at the cloud servers. This data can be accessed by user anywhere at relatively lower resources cost (even with poor Internet connection like 2G). This makes the system versatile and we can deploy it even when high-speed Internet is not available. The system uses open source software and plug-ins in the background to operate. It also uses free cloud storage (like GoogleDrive) to store the collected data which further reduces the capital required for maintaining the system.

4 System Design and Implementation

4.1 Proposed Home Monitoring System

The proposed model of the home automation system is shown in Fig. 1. The model consists of an Arduino Uno board which can be connected to various sensors. The model uses an ultrasonic sensor to detect any obstacle in its path and sends signal accordingly. The camera is also connected to the system which can click a stream of images whenever the sensor is triggered. The sensor transmits a constant stream of

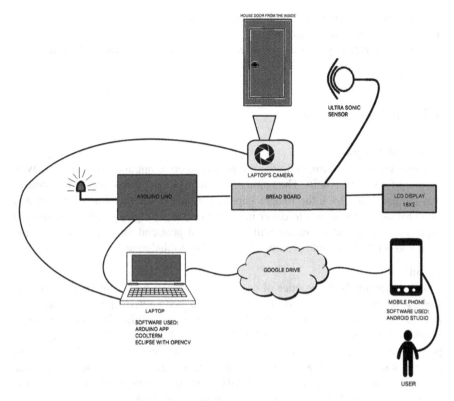

Fig. 1 Proposed system of home monitoring system

data which is read by a port reader and then with the help of java program, the system determines when to take screenshots. Once the conditions are met, the java program calls the camera which takes the pictures. The pictures are then stored in the computer system, the home monitoring system is running on as well as uploaded to the cloud server (Google Drive in this case). The user can then check the images clicked by the camera from anywhere in the world using his cell phone home monitoring application to access the cloud drive.

4.2 Proposed Home Monitoring System Functions

The proposed home monitoring system has various functions:

- Takes pictures of intruders
- Can have multiple sensors based on deployment scenario

- Can easily be extended based on user's need
- Stores pictures online and offline
- Has a dedicated application for user to access cloud from android device.

4.3 Software Design

Front-end Design:
The user interface is provided with the help of android application developed with the use of Android Studio. The application provides user quick access to cloud so that the user can check the images captured by the home monitoring system and also provides quick access to dialer in order to make emergency calls. The android application front end is created with the help of java and xml. Xml provides the front end while Java code provides functionality to the code.

Cloud:
Cloud computing is the practice of using remote servers on the Internet to manage, store, and process data instead of using a personal computer [6]. Cloud computing is a general term that is better divided into three categories: Infrastructure-as-a-Service, Platform-as-a-Service, and Software-as-a-Service. IaaS (or utility computing) follows a traditional utilities model, providing servers and storage on demand with the consumer paying accordingly. We are using Google Drive in our implementation, which provides us 15 GB of free data storage, which is more than what we need for the home monitoring system to function.

4.4 Implementation Setup

The home monitoring system works in the following steps (Fig. 2)

- Step 1—Movement is detected by the ultrasound sensor and Arduino generated a data log of distance object that moves. This also triggers the LED notification and LCD display displays a message.
- Step 2—A constant log is created by Arduino which is then recorded CoolTerm [7], which creates a text file of the recorded logs.
- Step 3—This log file is processed repeatedly by a JAVA program in search for a specific attribute. Once the condition is matched, the program triggers a camera to take pictures.
- Step 4—The pictures are then uploaded to Google Drive and a notification message is sent to user to check his Google Drive pictures log.
- Step 5—The user can remotely access his Google Drive in order to check the image of activity and decide the action that he wants to take.

Fig. 2 Sequence of activities
in HMS

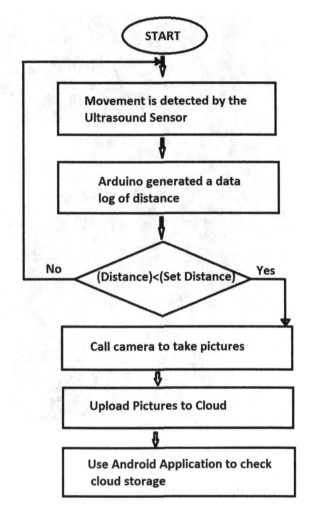

In Fig. 3, the cardboard acts as an open door. Whenever the door opens, the ultrasonic sensor changes the data stream it is generating and the change is detected by java code running in background and the pictures are then uploaded to cloud storage. This allows user to check the images from anywhere using the android application (Fig. 4) and take appropriate action.

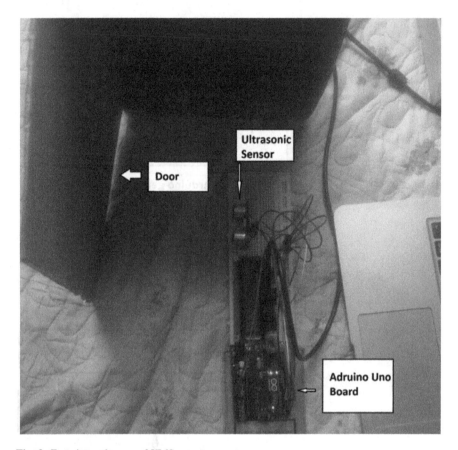

Fig. 3 Experimental setup of HMS

5 Results and Efficiency

After a successful connection to Internet, the HMS works as intended and uploads pictures taken by camera automatically to the cloud storage. The home monitoring system notifies and allows the user to monitor any access. In case of any access, the user gets notification on his mobile device to check his Google Drive. This is much more reliable and efficient as the device is not constantly generating video feed that leads to need of very high data rates and also, the off-site data storage on Google's cloud drive prevents the data from being compromised. The comparison of home monitoring system and CCTV camera (Current Technology) is shown in Fig. 5.

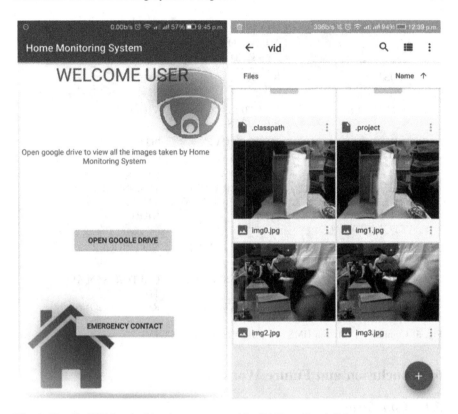

Fig. 4 User-End UI for checking images captured by HMS on GoogleDrive

EFFICIENCY—Home monitoring system's efficiency is calculated using the repeatability factor. The application was compiled and executed for multiple number of times and the ratio of pictures capturing the thief to the ratio of pictures not capturing the thief was calculated. The logic is illustrated below:

First Iteration: 2 out of 2 pictures capture the image of the thief: 100%
Second Iteration: 6 out of 7 pictures capture the image of the thief: 85%
Third Iteration: 3 out of 4 pictures capture the image of the thief: 75%
Fourth Iteration: 3 out of 3 pictures capture the image of the thief: 100%
Fifth Iteration: 5 out of 6 pictures capture the image of the thief: 83.3%
Sixth Iteration: 4 out of 5 pictures capture the image of the thief: 80%

According to the average or the mean of these ratios, the efficiency of the home monitoring system is (100 + 85 + 75 + 100 + 83.3 + 80)/6 = 87.2%.

Hence, the efficiency of the home monitoring system is **87.2%**.

ATTRIBUTE	HOME MONITORING SYSTEM	CCTV CAMERA
COST	Rs. 800	Rs. 5000
CAPTURES	MULTIPLE PICTURES	VIDEO
NETWORK REQUIREMENT	2G OR ABOVE	3G OR ABOVE
EFFICIENCY	QUICK IMAGE TRANSFER WITH 87.2% EFFICIENCY	QUICK VIDEO TRANSFER WITH 92% EFFICIENCY
STORAGE REQUIREMENT	LOW	HIGH
CAPTURE TRIGGER	ANY MOVEMENT IN DOOR	NO TRIGGER
REPEATABILITY FACTOR	CAPTURES IN ALL THE REPEATATIONS	CAPTURES IN ALL THE REPEATATIONS

Fig. 5 Comparison between HMS and current technology (CCTV)

6 Conclusion and Future Work

6.1 Conclusion

The HMS is a highly economical alternative to the existing technologies available in market and is able to work even with low-speed Internet connection. Home monitoring system senses any change in the door and thereby monitors the door. The camera is then alerted and it clicks multiple pictures. All the pictures are then uploaded to the Google Drive which are received by the user. The home monitoring system successfully monitors the home and guarantees security.

6.2 Future Work

The HMS has a wide application and can be installed in many areas and can even be modified to do different tasks based on requirements. In future, machine learning can be implemented to distinguish between visitors so that only unknown individual's image is uploaded to the cloud.

References

1. Gowthami.T and Dr. Adilinemacriga. G, "Smart Home Monitoring and Controlling System Using AndroidPhone", http://www.ijetae.com/files/Volume3Issue11/IJETAE_1113_71.pdf, 2013
2. Dingrong Yuan, Shenglong Fang and Yaqiong Liu, "The design of smart home monitoring system based on WiFi electronic trash", http://www.hades.in/BasePapers/Wireless/Journal/Iot/HWL%20(26).pdf, 2014
3. Shivprasad Tavagad, Shivani Bhosale, Ajit Prakash Singh and Deepak Kumar, "Survey Paper on Smart Surveillance System", https://www.irjet.net/archives/V3/i2/IRJET-V3I252.pdf, 2016
4. Madhavi Shinde, Rutuja Shinde, Neelam Thavare and Amol Baviskar, "security based home surveillance system using android application", http://citeseerx.ist.psu.edu/viewdoc/download?doi=10.1.1.684.6439&rep=rep1&type=pd, 2014
5. Arduino Systems Home, "Arduino System website", https://www.arduino.cc/
6. Cloud Computing Definition http://www.pctools.com/security-news/what-is-cloud-computing/
7. CoolTerm Open Source Port Monitoring Software http://freeware.the-meiers.org/

Hybrid Approach of Feature Extraction and Vector Quantization in Speech Recognition

Sarthak Manchanda and Divya Gupta

Abstract This paper examines the speech recognition process. Speech recognition has two phases: the front end which comprises of preprocessing of the speech waveform and the back end which comprises of feature extraction and feature matching. In this review, we discuss some feature extraction techniques like MFCC, LPC, LPCC, PLP, and RASTA-PLP. As these techniques have some demerits stated in the paper, we discuss a hybrid approach of feature extraction with some combinations of the above techniques. Feature matching helps in the recognition part of speech recognition. It is done by comparing the feature vectors of the current user to the feature vectors stored in the database. It can be optimized by vector quantization (VQ) in order to speed up the recognition process.

Keywords Feature extraction · Feature matching · Front end
Back end · Vector quantization

1 Introduction

Speech recognition is known as the ability of a machine to detect and identify the words or phrases uttered by the human being and to detect and convert this speech signal into machine read format. This concept of Artificial Intelligence is quite common these days and many advancements are being done toward it [1].

S. Manchanda (✉) · D. Gupta
Department of Computer Science and Engineering, Amity School of Engineering and Technology, Amity University, Noida, Uttar Pradesh, India
e-mail: sarthak120895@gmail.com

D. Gupta
e-mail: dgupta1@amity.edu

© Springer Nature Singapore Pte Ltd. 2018
V. Bhateja et al. (eds.), *Proceedings of the Second International Conference on Computational Intelligence and Informatics*, Advances in Intelligent Systems and Computing 712, https://doi.org/10.1007/978-981-10-8228-3_59

Speech recognition has front end and back end analysis.

Front End:

1. Preprocessing: It is used to increase efficiency for further processes. It helps in making the system more robust. It also helps in generating parametric representation of the speech signal.
2. Framing and Windowing: Most part of a speech signal processing depends on short time analysis done with the help of framing. The signal gets blocked into frames of samples N with a duration ranging between 10 and 30 ms.

Back End:

1. Feature Extraction: It is useful to extract useful information and remove the unwanted and redundant information. Certain features are extracted using this process which helps in speech recognizing. The goal of this technique is to compute the sequence savings of feature vectors which represent the computing signals. Different types of feature extraction techniques are: MFCC, LPCC, LPC, PLP, and RATSA-PLP [2].
2. Feature Matching: It is known as the process in which identification of the speaker is done by comparing the extracted features to the features which are stored in a database. It is basically done by first storing the input signal features in the database and then comparing the stored values to the input values of the unknown speaker. It gives the result as matched or not matched [2].

2   Type of Speech Uttered by Human Beings

There are many types of speech uttered by the human beings which differ in parameters [3]. These types are:

1. Isolated Speech:
   It requires single word at a time. This is one of the best speech types in which there is very less noise on both sides of the window.
2. Connected Speech:
   This type has minimum pause between the words uttered by the human.
3. Continuous Speech:
   This kind of speech has no pause between the words. It is basically what computer dictates and what human beings say the most.
4. Spontaneous Speech:
   This type of speech can be thought of as natural and without trying it before. An ASR system can handle spontaneous speech when it has every word with its meaning in the database.

3   Feature Extraction Techniques

It is the most important part in any ASR system. The meaning of extracting feature is to extract the useful information and remove unwanted information (Figs. 1 and 2).

1. Mel Frequency Cepstral Coefficients(MFCC):
   It is the most common and most popular feature extraction technique used for an ASR. Frequency bands in MFCC are placed in a logarithmical order so that it can approximate the human being system closely than any of the other system (Fig. 3).
2. Linear Prediction Coding(LPC):
   It is a good signal feature extraction method for linear prediction in speech recognition processes (Fig. 4). The basic idea behind LPC is that the speech signal can be approximated as a linear combination of past speech samples stored in the database [4].
3. Linear Prediction Coding Coefficient(LPCC):

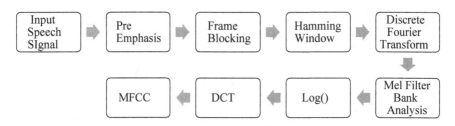

Fig. 1 Mel frequency cepstral coefficients(MFCC)

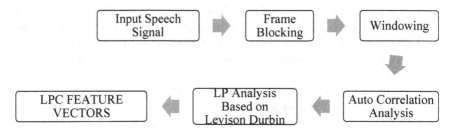

Fig. 2 Linear prediction coding(LPC)

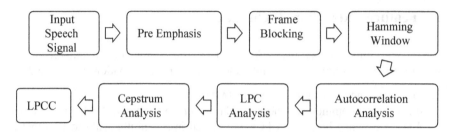

Fig. 3 Linear prediction coding coefficient(LPCC)

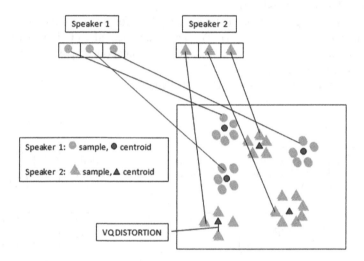

Fig. 4 Vector quantization

This technique works at a low bit by demonstrating the speech signal by a finite number of measure of the signals. It represents a mimic of speech after computing a smooth version of cepstral coefficient using autocorrelation method. In LPCC feature extraction, cepstrum analysis is done on the LPC analysis as seen below [4] (Fig. 3).

Other feature extraction techniques are PLP, RASTA-PLP etc.

4. Hybrid and Robust Technique:

In order to obtain a new and more effective feature extraction technique, certain hybrid algorithms make a distinction from the previous feature techniques. The previous techniques have some drawbacks which do not make the recognition process accurate and robust. The drawbacks are given in Table 1 [4].

Table 1 Demerits of feature extraction techniques

MFCC	LPC	LPCC	PLP	RASTA PLP
1. It is very much sensitive to mismatch of the channels between the training and testing phases 2. Performance of MFCC is effected by total number of filters used 3. Does not give accurate result in noisy environment	1. It is used in only a linear combination of speech signals	1. It is highly sensitive to Quantization Noise	1. It gives a higher error rate as compared to other techniques	1. It is better than PLP as it does the filtration but still error rate is still found in it

To overcome these drawbacks, a hybrid technique is used. Combination of previous features is taken to make the extraction more robust.

There is a generation of 13 parameter coefficients of these techniques

(MFCC, LPC, LPCC, PLP, and RASTA-PLP). In four experiments, different combinations of these techniques were used to give "39 coefficient parameters" in a single vector:

(a) 13(MFCC) + 13(LPCC) + 13(PLP)
(b) 13(MFCC) + 13(LPCC) + 13(RASTA PLP)
(c) 13(MFCC) + 13(PLP) + 13(RASTA PLP)
(d) 13(LPCC) + 13(PLP) + 13(RASTA-PLP).

4   Feature Matching

While the above feature extraction techniques were used in the front end to extract certain important characteristics, **vector quantization** is used in the back end to generate a correct decision while maintaining a codebook [5, 6].

It is a procedure to identify an unknown speaker by comparing the extracted features from the above approach with the data of the known speaker stored in the database [2].

Vector Quantization(VQ):

VQ is nothing more than "rounding off to the nearest integer." It is also known as the centroid model. It was introduced in the year 1980 and it has its roots from data compression. It has many advantages:

- speed up the recognition process
- for lightweight practical use
- ease of implementation.

Apart from these advantages, it can lead to loss of potential data.

In VQ, the signal vectors from a very large vector scale can be mapped on a finite region of space (Fig. 4). Every region of the space is known as "cluster" and is shown by a center called "codeword". The group of these code words is known as a "codebook" [7].

In the figure are two different speakers on a two-dimensional space. Triangle refers to the vector from "speaker 1" while the circle refers to the vector from "speaker 2". There are two phases in VQ. In the first phase which is also known as the training phase, a codebook is generated which is speaker specific for every single known speaker taking the training vectors in the database. The result of these code words is represented as the centroid of every triangle or circle as shown in the figure by Green Circles for "speaker 1" and Green Triangles for "speaker 2".

In the second phase which is also known as the recognition phase, speech input signal is "vector quantized" using a trained codebook and the total VQ-distortion is computed. The speaker having the codebook with the least distortion is discovered.

5 Conclusion and Future Work

The main objective of this research is to analyze why the hybrid approach is better to use for feature extraction and how can vector quantization help in fast and accurate feature matching. This paper gives the detail about various feature extraction techniques and how hybrid approach is better than these techniques. It shows some of the advantages of the hybrid approach in the table.

On the basis of the review, it is analyzed that the combination of techniques can help in achieving good results in a noisy environment also as variety of filters is used in hybrid approach. This hybrid approach is used for a flexible kind of environment.

For feature matching, vector quantization is used in which optimization is done which speeds up the recognition process. The future work is to develop an ASR System with high accuracy.

References

1. Karpagavalli, S., and Chandra, E "A Review on Automatic Speech Recognition Architecture and Approaches" International Journal of Signal Processing, Image Processing and Pattern Recognition Vol. 9, No. 4, (2016) pp. 393–404.
2. Geeta Nijhawan 1, Dr. M.K Soni 2 "Speaker Recognition Using MFCC and Vector Quantisation" ACEEE Int. J. on Recent Trends in Engineering and Technology, Vol. 11, No. 1, July 2014.

3. Akansha Madan, Divya Gupta "Speech Feature Extraction and Classification: A Comparative Review" International Journal of Computer Applications (0975–8887) Volume 90 – No 9, March 2014.
4. Veton Z. Këpuska, Hussien A. Elharati "Robust Speech Recognition System Using Conventional and Hybrid Features of MFCC, LPCC, PLP, RASTA-PLP and Hidden Markov Model Classifier in Noisy Conditions" Journal of Computer and Communications, 2015, 3, 1–9 Published Online June 2015.
5. Hemlata Eknath Kamale, Dr.R. S. Kawitkar "Vector Quantization Approach for Speaker Recognition" International Journal of Computer Technology and Electronics Engineering (IJCTEE) Volume 3, Special Issue, March-April 2013, An ISO 9001: 2008 Certified Journal.
6. Preeti Saini, Parneet Kaur "Automatic Speech Recognition: A Review" "FLEXIBLE FEATURE EXTRACTION AND HMM DESIGN FOR A HYBRID DISTRIBUTED SPEECH RECOGNITION SYSTEM IN NOISY ENVIRONMENTS" International Journal of Engineering Trends and Technology-Volume 4 Issue 2– 2013.
7. Dipmoy Gupta, Radha Mounima C. Navya Manjunath, Manoj PB "Isolated Word Speech Recognition Using Vector Quantization (VQ)" International Journal of Advanced Research in Computer Science and Software Engineering Volume 2, Issue 5, May 2012 ISSN: 2277 128X.
8. Veton Z. Këpuska, Hussien A Elharati "Performance Evaluation of Conventional and Hybrid Feature Extractions Using Multivariate HMM Classifier" Int. Journal of Engineering Research and Applications www.ijera.com ISSN: 2248-9622, Vol. 5, Issue 4, (Part -1) April 2015, pp. 96–101.
9. Pratik K. Kurzekar 1, Ratnadeep R. Deshmukh 2, Vishal B. Waghmare 2, Pukhraj P. Shrishrimal 2 "A Comparative Study of Feature Extraction Techniques for Speech Recognition System" International Journal of Innovative Research in Science, Engineering and Technology (An ISO 3297: 2007 Certified Organization) Vol. 3, Issue 12, December 2014.
10. Pawan Kumar, Astik Biswas, A. N. Mishra and Mahesh Chandra "Spoken Language Identification Using Hybrid Feature Extraction Methods" JOURNAL OF TELECOMMUNICATIONS, VOLUME 1, ISSUE 2, MARCH 2010.
11. H. B. Kekre, Tanuja K. Sarode "Speech Data Compression using Vector Quantization" International Journal of Computer, Electrical, Automation, Control and Information Engineering Vol: 2, No: 3, 2008.

Security in Home Automation Systems

A Brief Study of Security Aspects of Different Types of Home Automation Systems

Dinesh Bhasin and Shruti Gupta

Abstract This paper aims to discuss the problems regarding the security of home automation systems and highlight all the shortcomings in implementing such systems. Home automation systems are the future of context-aware and energy-efficient residential systems. The technology has been around for a while now, but is still not adapted by the masses because of the high costs and low reliability. The aim of this research was to find out the reasons why such systems are not available in the common markets right now.

Keywords Wireless networks · Smart homes · Internet of Things

1 Introduction

Home Automation is a very interesting application of Internet of Things, which consists of making a home "smart". This means that all the appliances can be controlled and monitored remotely and sometimes even programmed to work on their own. While it sounds really easy, implementing such a system is not an easy task. Various devices are used in a specific manner just for implementing a simple task such as switching on the lights. This may sound like the system is complicated, but the end product is user friendly and very easy to use with just a touch on the smartphone is required to accomplish any task. Not only this, but certain conditions or situations can be programmed into the system so that it functions even without the consent of the user. Such systems should be made energy efficient and should even help save energy from our regular appliances, as stated by the brilliant idea by Archana N. Shewale, and Jyoti P. Bari in their paper: [1]. This, along with the

D. Bhasin (✉) · S. Gupta
Department of Computer Science, Amity School of Engineering and Technology,
Amity University, Noida, Uttar Pradesh, India
e-mail: dinesh.bhasin@student.amity.edu

S. Gupta
e-mail: sgupta65@amity.edu

© Springer Nature Singapore Pte Ltd. 2018
V. Bhateja et al. (eds.), *Proceedings of the Second International Conference
on Computational Intelligence and Informatics*, Advances in Intelligent Systems
and Computing 712, https://doi.org/10.1007/978-981-10-8228-3_60

always-online nature of the system are the two biggest vulnerabilities in the field of home automation. Some of these vulnerabilities are given by Michelle L. Mazurek in his article [2].

2 Literature Review

2.1 Summary of Papers

The following research papers were studied over the research period for better understanding of the topic:

"Renewable Energy Based Home Automation System Using ZigBee"
In this paper, Archana N. Shewale and Jyoti P. Bari proposed a new approach to the ZigBee architecture by using renewable energy sources to power up an automation system. This was quite innovative as renewable energy sources are the need of the hour, and combining that with 24/7 energy monitoring, it could be the most efficient system.

"Access Control for Home Data Sharing: Attitudes, Needs and Practices"
In this paper, strategies were explained on how to keep our data center of the home, safe and measures to ensure total privacy. Michelle L. Mazurek describes their results of the data collected from their analysis of the different practices that people follow during the execution of the access control systems.

"Smart Home Automation Security: A Literature Review"
In this paper, Arun Cyril Jose and Reza Malekian present a detailed study of all the security flaws in different types of home automation systems and their own analysis of the definition of the words "Security" and "Intruder", with respect to the home automation systems. For this, they have used various literary sources and compared them wonderfully in their work.

3 History of Home Automation

Home Automation dates back to mid-1960s when the first smart device: ECHO was introduced. This device was capable of turning appliances on/off, computing shopping lists, and also controlling thermostats. Then came in the kitchen computer, which also stores recipes; but both of them were never commercially sold [3].

In 1990s a term, Gerontechnology was introduced which was a combination of Gerontology (Study of old age) and Technology, which led to several monitoring systems for the elder members of the society. This concept led to development of a similar system, for the babies, which we today see in baby-monitoring systems.

A similar research was done by S.P. Pande and Prof. Pravin Sen in their article on an automation system for the disabled: [4, 5].

In late-90s, the Silicon Valley was emerging at an exciting pace, which led to a decrease in the prices of silicon-based chips and microcontrollers, and thus, the home automation systems were becoming more affordable.

4 Home Automation Systems Today

Today's smart homes are based on two main focus points: Security and Eco-Friendliness. These systems are often sustainable and help us keep a check on our power consumption too. For security, there are monitoring systems that lock/ unlock the home when a certain breach is detected, like a wrong passkey entered or face detection fails to recognize the face.

On the other hand, eco-friendly systems monitor our consumption at all times, and can even be programmed to switch the lights off when nobody is home or when everybody sleeps. Not only this, backup generators connected with solar panels can be automatically switched when the load is not high. These practices save us a lot of money and also help the environment at an elementary level.

4.1 Wireless Home Automation Networks (WHAN)

To allow interaction of different smart objects along with advanced functionality, a new aspect of technology called the wireless sensor network (WSN) must be integrated along the so-called Internet of Things (IoT). A new gateway has opened the ways for emerging technology of wireless network sensors because of the recent advances in radio technology and microcontroller design. These sensors act as smart agents that come with high-performance microcontroller which consumes low power and has a very large memory space, and such devices are now maturing a lot of faces if companies that work in the field of smart devices and advanced surveillance systems. Even though today most of the technology in this area is still wired based, but replacing the cable will save time as well as make space for more flexible technologies for the end user and thus, a wireless home automation network or WHAN will act as the key technology in order to make our homes intelligent and aware of its environment and content.

A WHAN consists of a number of actuators and sensors such as motion sensor, light bulbs, accelerometer, and safety sensors that communicate with each other using an appropriate wireless architecture and thus smartly sharing data with each other. The same can be graphically described as Fig. 1. In a WHAN, an access point (AP) is used which starts the network, assigning all the resources to the end devices (ED) and checking the quality of the links. End devices are then spread over the area in which they want to communicate or the area of interest and it is the job

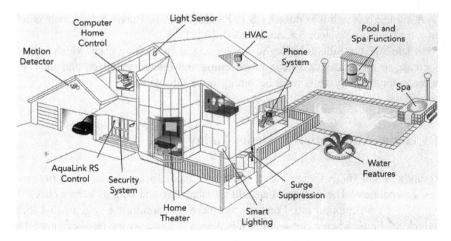

Fig. 1 Typical WHAN components

of the AP coordinator to ensure the communication takes place for which range extenders maybe used. Various architectures of WHANs are deeply studied and intuitively published as an article by Carles Gomez and Josep Paradells [6].

The main aim of WHAN design architecture is to provide cost-effective interactions amongst the end devices and in order to do so provide efficient software–hardware subsidiary infrastructure and middleware platforms.

4.2 Types of WHANs

Context-Aware Systems:
Such systems are designed in such a way that they are flexible to any type of connectivity that the user might prefer at any point of time. This makes the user's privacy to be at the edge. Similar concern freaks out people on knowing about government surveillance on private lives of people, as the automation system also knows about all the private and intimate details of the house such as when someone is taking a shower or when the appliances are switched ON.

Central Controller based Systems:
Such systems use a central server to process data for multiple rooms or even multiple buildings. This makes it much more cost-effective and alarming at the same time. The main concern is knowing how much data is actually accessible to the system at that given time. Because it contains sensitive data for multiple entities, thus leaving all of them vulnerable from just a single point.

Decentralized Systems:
This can be implemented by installing actuators into the WSN of the system, thus distributing the control. This reduces the risk of attack, but surely does not eradicate it. Such a system is very complex to manage and expensive to maintain and an attacker with prior knowledge of the WSN of the home can easily exploit the system.

4.3 Issues with Current Systems

With the growth in technology and specially after the emergence of Internet of Things (IoT) and embedded systems, even the most basic devices such as tube lights and smoke detectors are beginning to be connected to our smartphones but every coin has two sides. At one point where it makes our lives easier, it also has some pitfalls which a person must know about before they can invest in them and must be handled by every developer as well. Some of them are given below:

- Does not necessarily makes things easier: The most basic definition of technology is anything that makes things easier in terms of effort required, but it is not always the case when we talk about these home automation technologies. It is often noticed that instead of making things easier they end up complicating the process even further than it earlier was. For instance, lets talk about smart light bulbs, normally it would take just a second to switch on the analog light bulb but when it comes to smart bulbs the process becomes kind of cumbersome, finding one's phone then unlocking it, finding the correct app, and then navigating to the process that one wants to start. It certainly leaves the user more anxious than relaxed which basically defeats the sole purpose of smart homes. Given by Ahmed, M., Sharif, L. Kabir, M. Al-Maimani, M [7].
- Bad Integrating Applications: Any hardware is as good as the software that helps to integrate it. One is incomplete without the other, however many manufacturers often do not realize the importance of good software. The apps tend to have bad interface, lack the basic features and crash way to often, and even if they work properly, it is almost impossible for the user to handle the clutter of applications that they have installed. They have a separate app for thermostat, one for the light bulbs, another one for the media devices, and so on. But with time, the manufacturers have realized this and have opened their gates for third-party apps.
- Security Issues: Automated devices surely make one's life easier where a person can control their devices from practically anywhere in the world but it also leaves their devices vulnerable to hackers who just wait for an opportunity to cause havoc and the reports of such issues are increasing exponentially. Hence, the user need to be very diligent and must make sure that they do install all the updates that are being provided by the manufacturer. One such survey was done by B. Fouladi, S. Ghanoun [8].

- Cost: While everyone agrees that these smart home technologies help their users, cost is something that put these devices out of reach of most of the people. Even if people are prepared to spend some extra cash they will also think twice when they come to know that a smart device costs ten times the money than a normal device. Although, the prices are expected to drop over the years, the sky-high costs right now prevent the users to invest in smart devices.

4.4 Types of Home Automation Systems Based on WHAN

Different users prefer different types of connection systems for their smart homes, some might like to be connected to their home every time through the Internet and monitor every move, while others may be comfortable just receiving important texts about the events in the home. Thus, there are different types of network architectures available for home automation systems given as follows

Internet-Based Automation Systems:
Such systems are the most popular type of systems, as connecting to the Internet provides the ready architecture for hardware as well as network, as given by A. ElShafee, K. A. Hamed [9]. Not only this, but the cost involved in communication through the Internet is minimum, and the devices can connect/disconnect from the network very easily, shown in Fig. 2. The user interface (UI) is generally in the form of mobile applications or general Web pages which is intuitive and very easy to use and understand. Thus, this type of system provides peace of mind to the end user but increases the complexity for the developer of such systems.

Such systems are always online so they need to be secured accordingly. Thus, heavy encryption protocols and backup storage for the database is needed. So, a centralized approach is applied and all the devices in the house are channelized through a central server. This leaves a vulnerability to the server as it remains as the only point of attack for the intruder.

Bluetooth-Based Automation Systems:
Using Bluetooth-based systems can be tricky because of the short range of communication, so generally, a Bluetooth-based controller is connected to the user via the Internet or GSM. In the works of N. Srikanthan [10] and H. Kanma [11], (Also refer the works by R. Piyare, M. Tazil, [12] D. Naresh, B. Chakradhar, S. Krishnaveni [13]), the systems use different communication protocols over the basic Bluetooth protocol, in order to communicate with the microcontrollers that run the devices in the house, as described in Fig. 3.

The main issue with these systems is the low proximity. Bluetooth is also low on security, thus raising a concern for the safety of such systems. Also, power consumption of Bluetooth is high, thus eliminating the Green Computational approach.

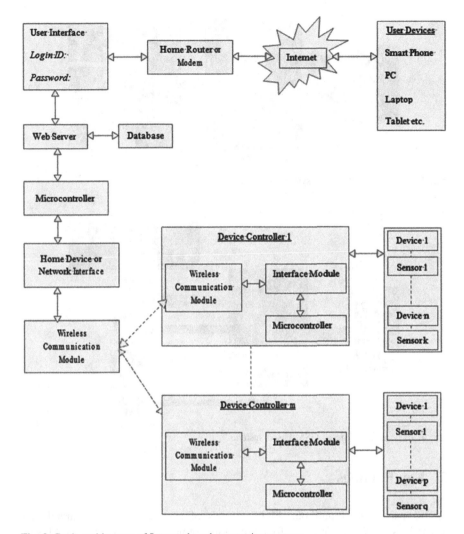

Fig. 2 Basic architecture of Internet-based automation systems

GSM-Based Automation Systems:
These systems are used in the oldest form of home automation systems available. In the study given by Mahesh N. Jivani [14], a GSM device gives instructions to the microcontroller, based on the input it receives in the form of text messages or calls, as described in Fig. 4.

Main concern while using these systems is the lack of information and low security. Anyone with the knowledge of the GSM SIM number of the house can easily access the mainframe. Also, the info received is generally in the form of basic texts, with little information about the house functionalities.

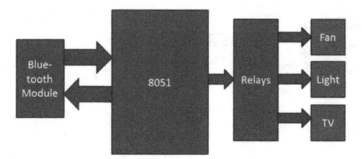

Fig. 3 Basic Bluetooth-based system architecture

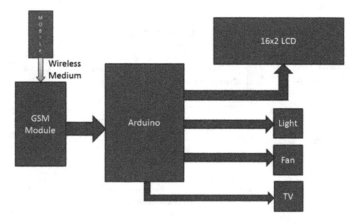

Fig. 4 Basic architecture of GSM-based systems

5 Key Findings

In this paper, we found out different types of systems that are available for researchers to explore. We also found that this segment, which is a beautiful fusion of Internet of Things and basic consumer electricals, is not explored much, or rather invested in much. This is the reason why there is no stability in these systems at the moment. We also found out the different connectivity technologies and their pros and cons. To prevent data misuse and tampering, all systems have measures that ensure the safety of the user's sensitive data, but up to a certain limit. None of the systems are completely secure. In order to ensure maximum safety, we found out that the best solution is to use an optimal combination of different systems, which will not only be more secure, but also much more versatile and easy to use for the end user.

6 Conclusion

The systems have their own pros and cons, but what makes their common is the ultimate goal of making the home act "smarter" and interactive for the user, while keeping a check on the power consumption, and applying any methods necessary for saving the energy usage. But each of the systems is vulnerable to regular usage. This is why such systems are not gaining common interest. This paper listed the different architectures and their pros and cons. Our focus was on the security aspect of these systems and how each system is vulnerable to intruders.

Security is important for the proper implementation and development of these systems for regular use. Moreover, it provides a sense of security to a home's inhabitants and puts their minds at ease.

References

1. Archana N. Shewale and Jyoti P. Bari, "Renewable Energy Based Home Automation System Using ZigBee" International Journal of Computer Technology and Electronics Engineering (IJCTEE) Volume 5, Issue 3, June 2015 pp. 6–9. Article.
2. Michelle L. Mazurek, "Access Control for Home Data Sharing: Attitudes, Needs and Practices," in CHI'10 Proceedings of the SIGCHI Conference on Human Factors in Computing Systems, pp. 645–654, 2010.
3. Drew Hendricks, "History of Smart Homes", iotevolutionworld.com, Article, April 2014.
4. S.P. Pande, Prof. Pravin Sen," Review On: Home Automation System For Disabled People Using BCI", International Conference on Advances in Engineering & Technology – 2014.
5. Yeh-Liang-Hsu, "Gerontechnology", Journal of Gerontechnology, Vol 16, Issue 1.
6. Carles Gomez and Josep Paradells, "Wireless Home Automation Net-works: A Survey of Architectures and Technologies" IEEE Communications Magazine • June 2010 pp. 92–101. Article.
7. Ahmed, M., Sharif, L. Kabir, M. Al-Maimani, M, "Human Errors in Information Security," International Journal, 1(3), 2012.
8. B. Fouladi, S. Ghanoun, "Security Evaluation of the Z-Wave Wireless Protocol," Black hat USA, Aug. 2013.
9. A. ElShafee, K. A. Hamed, "Design and Implementation of a WiFi Based home automation System," World Academy of Science, Engineering and Technology, vol. 6, 2012.
10. N. Sriskanthan, F. Tan, A. Karande, "Bluetooth based home automation system," Microprocessors and Microsystems, Elsevier, vol. 26, pp. 281–289, 2002. Article.
11. H. Kanma, N. Wakabayashi, R. Kanazawa, H. Ito, "Home Appliance Control System over Bluetooth with a Cellular Phone," IEEE Transactions on Consumer Electronics, vol. 49, no. 4, pp. 1049–1053, Nov. 2003. Article.
12. R. Piyare, M. Tazil "Bluetooth Based Home Automation System Using Cell Phone", IEEE 15th International Symposium on Consumer Electronics, 2011.
13. D. Naresh, B. Chakradhar, S. Krishnaveni, "Bluetooth Based Home Automation and Security System Using ARM9", International Journal of Engineering Trends and Technology (IJETT)– Volume 4 Issue 9 Sep 2013.
14. Mahesh N. Jivani, "GSM Based Home Automation System Using App-Inventor for Android Mobile Phone", Vol. 3, Issue 9, September 2014.

A Modified Technique for Li-Fi Range Extension

Sidhartha Singh, Tanupriya Choudhury, Parth Joshi
and Roshan Lal Chhokar

Abstract Li-Fi is a wireless communication technology that utilizes a duplex communication system. It is a kind of visible light communication. Li-Fi employees LEDs which turn on and off at very high rates, thus, transmitting data to a remote receiver. Li-Fi is thought of as an alternative solution to the RF bandwidth limitation problem existing in the Wi-Fi. Although Li-Fi provides and unmatchable data transfer speed over other prevailing technologies, there are few limitations and disadvantages of using it which further attribute to cost, maintenance, range, etc. The paper provides a theoretical method to implement range expansion of Li-Fi devices.

Keywords LED · Li-Fi · OWC · UV · VLC · Wi-Fi

1 Introduction

Li-Fi is a high-speed wireless communication technology similar to Wi-Fi. Li-Fi stands for light fidelity. It is a type of visible light communication and a subset of optical wireless communication. The term was framed by Harald Haas of University of Edinburg in 2011. The technology can be a complement to the RF communication techniques like Wi-Fi or even a replacement in contexts of data broadcasting.

S. Singh (✉) · T. Choudhury · P. Joshi · R. L. Chhokar
Department of Computer Science Engineering, Amity University, Noida,
Uttar Pradesh, India
e-mail: sidharthasingh2@gmail.com

T. Choudhury
e-mail: tchoudhury@amity.edu

P. Joshi
e-mail: parthjoshi0207@gmail.com

R. L. Chhokar
e-mail: rchhokar@amity.edu

© Springer Nature Singapore Pte Ltd. 2018
V. Bhateja et al. (eds.), *Proceedings of the Second International Conference
on Computational Intelligence and Informatics*, Advances in Intelligent Systems
and Computing 712, https://doi.org/10.1007/978-981-10-8228-3_61

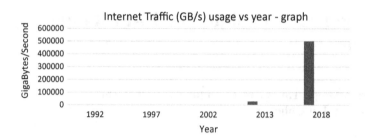

Fig. 1 Graph representing the year wise Internet traffic, globally since last two decades

It is wire and UV visible light communication or infrared and near-ultraviolet instead of radio frequency spectrum, part of wireless communications technology, which carries much more information and has been proposed as a solution to the RF bandwidth limitations. Since the Li-Fi does not limit its bandwidth, several problems existing in the today's world can be solved which demand higher bandwidth, more than what existing technology can provide.

2   Existing Technology

The current scenario consists of methods like Wi-Fi for wireless internets access or wired transmissions like cabled data transmission. Few statistics of the current scenario show the following things:

- The overall population, globally, of people using the Internet has grown rapidly since 2011. It has grown 60% from 2 billion users to more than 3.4 billion users till 2016 [1, 2].
- By the end of 2016, 90% of the world's data was created in last 2 years.
- Every day, users create 2.5 quintillion (10^{18}) bytes of data across the globe.
- 100 GB/day was the average Internet traffic in 1992, 100 GB/h in 1997, 100 GB/s in 2002, 28,875 GB/s in 2013 and is estimated to reach 50,000 GB/s till 2018 which is 4.32×10^7 times the state of its initial launch. In the graph represented below, the Internet traffic value in the year 1992 and 1997 is too small that it seems nothing in respect to other years [3] (Fig. 1).

3   Problems with Existing Technology

With the current scenario in view several predictions have been made about the coming years of the Internet. The problems may affect the storage and speed of the Internet.

- Bandwidth limitation for the Internet traffic.
- Lack of data centers
- With deployment and development platforms growing rapidly in number and versatility, security issues concerning data security is increasing.

Technologies which make the processes faster and compact will save the day when it comes to the problems related with the storage and speed. Some of them are:

- Better data compression algorithms
- Faster bandwidth of data transfer
- Faster data centers.

4 How Li-Fi Helps

Li-Fi has been tested to reach a speed of over 224 Gbit/s or 28 Gbyte/s when tested in a controlled environment. In general environment, Li-Fi is speculated to attain a speed of above 1 Gb/s [4].

When the technology is applied in a day to day transfer of data, the data transfer becomes faster. This sets an example for better bandwidth and can be seen as a substitution for a technology of better bandwidth. Few of the advantages of this technology are:

1. **Speed**: Light travels at a greater speed than the sound. The rounded off speed is around 3×10^8 m/s whereas the speed of sound on an average is only 3.4×10^2 m/s, thus giving a faster performance to the light-based technology.
2. **Efficiency**: Working on visible light technology, Li-Fi is very efficient when compared to others. Homes and offices already have LED bulbs for lighting purposes, from which, the source of light can be used to transmit data. Thus, it is very efficient in terms of costs as well as energy. Light must be on to transmit data, so when there is no need for light, it can be reduced to a point where it appears off to human eye, but is actually still on and working.
3. **Availability**: Every light source present anywhere has the ability to be an Internet source. Light bulbs are present everywhere—in homes, offices, shops, malls, and even planes, meaning that high-speed data transmission could be available everywhere.
4. **Security**: The biggest advantage that Li-Fi offers is security. Light does not have the ability to pass through opaque structures. Internet provided by Li-Fi is available to users within a room and cannot be breached or interfered by others users in other rooms or buildings. Thus, providing enough physical security to the network. This advantage can also prove to be a disadvantage sometimes, as the Li-Fi source need to be in direct communication with the Li-Fi receiver for proper data transfer.

5. **Cost**: Li-Fi is 10 times cheaper than the Wi-Fi and thus, is easily affordable and easy to set up in local environment.

5 Li-Fi—Working

Standard LED light bulbs use a constant current, which emits a constant stream of photons perceived by us as visible light. Li-Fi is different because the current it uses varies, meaning that the output intensity of the light fluctuates [5, 6].

LEDs are semiconductors so the current and the output can be modulated at high speeds, which is picked up via a photodetector device (the equivalent to a Wi-Fi networking card in PC). The optical output is then converted back into an electrical current, which is processed and sent to your device as data. The varying light intensity is invisible to the naked eye, making it about as noticeable as Wi-Fi signals.

The light emitted from various LEDs is used by the OWC technology (Optical wireless communication) as a medium to deliver networked, mobile, high-speed communication in a similar manner to Wi-Fi (Fig. 2).

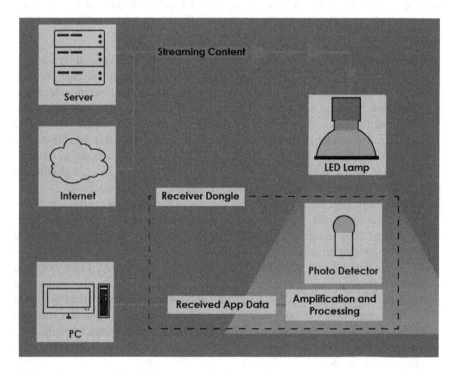

Fig. 2 Diagram showing the functioning of Li-Fi system

6 Limitations of Li-Fi

The Li-Fi technology possesses few limitations in term of functionality due to the properties of light. Few of them are:

- Light, as a photon, goes in a straight line. It does not really bounce very well. Think of a laser. Laser is essentially invisible except at each endpoint. For this reason, Li-Fi is a "line of sight" technology. Until and unless the receiver is in line of sight of the transmitter, the connection is not successful. Due to this, Wi-Fi has the biggest advantage over Li-Fi as radio waves can penetrate through substances and can be accessed without being in the line of sight of the transmitter [7].
- Without the presence of a light source, Internet cannot be used. This poses as a limitation to the locations and situations in which one can use Li-Fi [8].
- Interference with signals from other light sources is possible. One of the biggest potential drawbacks is the interception of signals outdoors. Sunlight will interfere with the signals, resulting in interrupted Internet.
- A whole infrastructure would be required to be built for Li-Fi.

7 Developing a Technique for Range Extension

Since the current prevailing scenario requires an upgradation in terms of technology stack for wireless communication, there exists a need to develop new techniques that increase the range of Li-Fi and possibly overcome the line of sight problem.

From the limitations of Li-Fi, it is clear that the biggest problem that exists in a Li-Fi system is that of finding a method that removes the system dependency on the "line of sight" of the receiver and the transmitter.

Due to the above-stated fact, Li-Fi can only be used inside the walls of our home as the light will not be able to penetrate the walls. Few techniques which could be implemented to extend the range of the transmitter are:

- Reflecting devices on doors.
- Reflecting floors.
- Multidirectional lighting system.
- Receiver–Transmitter module.
- Li-Fi—Wi-Fi hybrid system.

8 Reflecting Devices on Doors

This method can be utilized to extend the range of a Li-Fi system of a single room to various other adjacent rooms which are accessible through the doors [9].

1. **Method**: This method consists of fixing two concave mirrors at the two corners of the door. This is done in such a way that the center of the reflective surface of one of the mirrors is in line of sight of the Li-Fi source and the center of the reflective surface of the other mirror is in line of sight of both, the first mirror and the other room. This method provides a cheap alternative to Li-Fi range extension using just mirrors as the means (Fig. 3).
2. **Advantages**: Since the method involves non-circuitry device except the Li-Fi source, it has following properties:

   - Low installation cost.
   - Good range extensibility.
   - Low maintenance cost.

3. **Disadvantages**: The proposed method can have several disadvantages as well. The method is based on assumption that the environment is noise free and totally isolated. Several disadvantages that might befall such method implementation are:

   - Incomplete range coverage: The extensibility depends on the mechanical properties of the mirrors. If they incur any kind of fault in setup, texture, or alignment, possible data loss and connection breaks are possible.

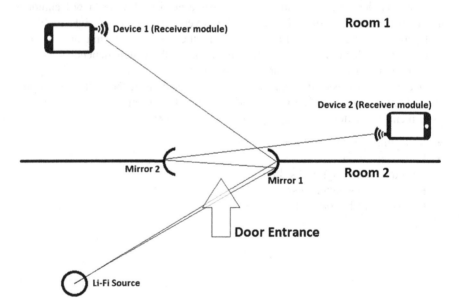

Fig. 3 Range extension technique using reflecting devices or door sides

- Noise accumulation: It is possible that the receiver might gather inputs from other light sources present in the room which give out infrared rays but are not Li-Fi devices. This kind of received data with is an error case. Further, it is possible that multiple reflections are taking place in the same room which causes error entries to the Li-Fi receiver.

9 Reflecting Floors

The method can be utilized to extend the range of Li-Fi source within a single room. This method increases the range of Li-Fi source in such a way that it can reach in indirect places which are not direct accessible from the source (Fig. 4).

1. **Methods**: This method involves laying the flooring of the particular room in which Li-Fi is to be used using a reflective substance. This would help in increasing the range of Li-Fi source that was not directly accessible to the receiver. This method is a bit expensive than the first one.
2. **Advantage**: Since this technique of Li-Fi range extension involves a large reflective surface for increasing the range of the source, the efficiency of the technique is better than the previously described technique. The technique has following advantages:

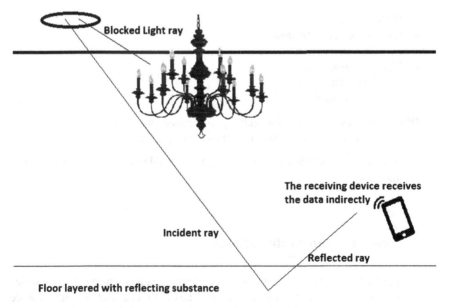

Fig. 4 Range extension technique utilizing the reflecting floors

- Better range extensibility.
- Low maintenance cost.
- Better suited for office environments.

3. **Disadvantages**: The techniques have few disadvantages:

- More wear and tear.
- Higher installation cost.
- Higher noise accumulation.

10   Multidirectional Lighting

This method incorporates a new ceiling system which organizes the LED bulbs in a three-dimensional pattern. This way we have the sources of Li-Fi light sending signals in several directions [1, 10].

1. **Method**: This method involves setting the ceiling design in a manner, different from usual. It involves setting the face of the Li-Fi bulbs in various directions. This way, the Li-Fi signals can be spread throughout the room evenly. This gives an experience just like Wi-Fi as the signals are spread evenly throughout the space.
2. **Advantages**: Since, the signals are spread evenly throughout the room, the system gives an experience just like Wi-Fi.

- Better range extensibility.
- Low noise accumulation.
- Low data corruption.
- Direct connectivity.
- Low-cost installation.
- Easy maintenance.

3. **Disadvantages**: Although the techniques increase the range of the Li-Fi source evenly, there are few disadvantages this technique has. They are:

- Due to higher number of LED bulbs, it is possible that connection mismatch may occur.

11   Receiver–Transmitter Module

In this method, the general Li-Fi system is accompanied by an additional device that helps in extending the Li-Fi source range. The device is a kind of normal device which has an extra inbuilt Li-Fi receiver and transmitter which helps in increasing the range of the system. The device takes data from the source and utilizes it. It

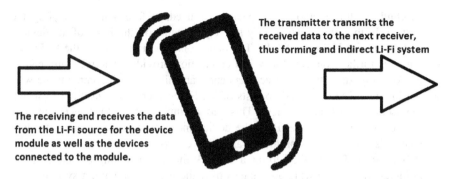

Fig. 5 A receiver–transmitter module functioning

passes on the data requested by the other devices connected to it. This system works like a Wi-Fi hotspot, the difference being, instead of Wi-Fi, Li-Fi is involved and instead of wired input to the module, wireless input takes place (Fig. 5).

1. **Method**: In this method, a third-party device is fixed with an extra pair of receiver and transmitter. This set is used to extend the connection on the receiving end of the device, thus forming a linear chain of connection. This linear chain can accommodate a long number of devices with strong network connectivity.

2. **Advantages**: This method is the one method in which the source of the Li-Fi source has no specific requirement. The advantages of this method are:

   - It can accommodate any number of devices.
   - Low data loss might occur.
   - Solves the "Line of sight" problem
   - Easy setup
   - Linear setup reduces noise.

3. **Disadvantages**: The disadvantages of this method are:

   - Data traffic across a branch may increase and affect the performance.
   - Relatively expensive setup.

12 Li-Fi—Wi-Fi Hybrid System

This method combines the pros of both of the systems that are, Wi-Fi and Li-Fi. The combination helps in increasing the data traffic and data traffic processing speed. Since this method involves two technologies in a combination, it can be called as a hybrid system. Li-Fi—Wi-Fi hybrid system is practically easy to design as both the technologies use IEEE 802.11 based protocols.

1. **Method**: In this method, we specify our needs before actually creating the system. In this phase, we specify the requirements on the basis of the devices that would use Li-Fi and Wi-Fi. The main connection provider is the Li-Fi unit which is further connected to Wi-Fi connection provider. The receiving end of the Wi-Fi module has Li-Fi receivers and transmitter, which communicate with the Internet. The other end consists of Wi-Fi receiver and transmitter to which the users connect their devices. This way a number of Wi-Fi devices are connected to a single source. Both the device modules would use same standard protocols. This reduces protocol translation efforts.

2. **Advantages**: This method forms a network mesh constituting of various technologies. It has following advantages over the general prevailing system:

   - It can easily support high network traffic.
   - It would work with all the devices having Wi-Fi antenna.
   - It solves the "line of sight issue" of the Li-Fi.
   - Easy installation.
   - Better networking capabilities for local networks.
   - No protocol translation, thus giving good speed to all the connected devices.
   - Suitable for office use.

3. **Disadvantages**: Since this method involves use of various technology stack in a single system, it is possible that conflicts occur in protocols. Therefore, the system has several disadvantages, which are:

   - Costly to set up such a hybrid system.
   - Not suitable for home use.

13   Discussion and Future Scope

The Li-Fi technology has been seen as a future replacement for the existing ways of wireless communication. Its high speed is the key feature that makes users to choose it over any other mode of wireless communication.

There is still a lot to be done in the Li-Fi technology. The paper proposes few techniques which can be used with a more practical approach to extend the range of Li-Fi sources. This would lay down the foundation of future wireless communication. Few modifications in the mentioned techniques can be done to intensify the outcomes.

14 Conclusion

The Li-Fi is a technology which is still in its development phase. Developed in 2011, it is taking the market with a boom. It is seen as a replacement for the prevalent and market dominant Wi-Fi technology.

Since, it is a new technology, there exist several issues with its compatibility and use in the local environment. The paper tries to focus on the limitations possessed by the Li-Fi system. The paper also proposes five techniques to overcome few of the limitations. The techniques are theoretical and may have less practical approach. It also develops a method of developing new technique for Li-Fi range extension, i.e., if all the proposed technologies are combined, the practicality of range extension increases.

References

1. Harald Haas. "Harald Haas: Wireless data from every light bulb". http://www.ted.com.
2. Nidhi Soni, Mayank Mohta, Tanupriya Choudhury, "The looming visible light communication Li-Fi: An edge over Wi-Fi" 5th Fifth International Conference on System Modeling & Advancement in Research Trends, 25th–27th November, 2016, College Of Computing Sciences & Information Technology, Teerthanker Mahaveer University, Moradabad, India.
3. Prof Amit Mishra and Mr. Lalit: "Li-Fi wireless communication Media" in International journal of innovative research in electrical electronics instrumentation and control engineering, Vol. 4, Issue 2, February 2016.
4. S. Dimitrov and H. Hass: "Principles of LED Light Communications: Towards Networked Li-Fi". Cambridge, U.K.: Cambridge Univ. Press, Mar. 2015.
5. Prof Harald Hass with Prof Svilen and Prof Thilo and Prof Irina: "Li-Fi modulation and networked Li-Fi attocell concept" 2012.
6. P.Kuppusamy, IT Research Lab, Department of CSE, Gnanamani College of Technology: "Survey and Challenges of Li-Fi with comparison of Wi-Fi", IEEE WIPSNET 2016 Conference.
7. "Li-Fi technology: Data transmission through visible light": Anurag Sarkar and Prof. Shalabh Agarwal. In IJARCSMS. June, 2015.
8. Harald Hass, Member IEEE, Liang Yin, Student member IEEE: 'What is Li-Fi?' published in Journal of Light wave technology, Volume: 34, Issue: 6, March15, 15 2016.
9. Dr. Harald Haas, 'Wireless data from every light bulb', TED Global, Edinburgh, July 2011.
10. www.arubanetworks.com/assets/wp/WP_WiFiCoreIntegration.pdf.

Accelerating EGENMR Database Operations Using GPU Processing

Shweta Malhotra, Mohammad Najmud Doja, Bashir Alam
and Mansaf Alam

Abstract Data is growing at faster speed. In cloud, fast processing has become essential need to process hefty data. The earlier EGENMR system was developed for process large amount of data present in cloud repositories with the help of MapReduce functions we concluded that the system is better than the latest techniques for processing large amount of data. In this paper, we are enhancing EGENMR by further enhancing the speed of database query operation by using GPU as a co-processor. A theoretical comparison is made in terms of time taken and complexity for hybrid query processing using GPU and EGENMR.

Keywords MapReduce · Query processing · GPU co-processing

1 Introduction

Data is growing at a faster speed. In cloud, it has become very essential to process data at a faster speed. So, there is a need to develop an efficient technique which can manage and process such massive amount of data at faster speed. The goal of the parallel programming techniques is to improve the speed of the computation. Earlier in [1], author implemented a model EGENMR (Enhanced Generalized query processing using MapReduce) which is used to process hefty amount of data presents at cloud repositories with the help of MapReduce codes [2]. Users send

S. Malhotra (✉) · M. N. Doja · B. Alam · M. Alam
Jamia Millia Islamia, New Delhi, India
e-mail: Shweta.mongia@yahoo.com

M. N. Doja
e-mail: mndoja@gmail.com

B. Alam
e-mail: babashiralam@gmail.com

M. Alam
e-mail: Mansaf_alam2002@yahoo.com

© Springer Nature Singapore Pte Ltd. 2018 669
V. Bhateja et al. (eds.), *Proceedings of the Second International Conference on Computational Intelligence and Informatics*, Advances in Intelligent Systems and Computing 712, https://doi.org/10.1007/978-981-10-8228-3_62

their queries in any of the RDBMS query languages and their queries get converted into MapReduce form. MapReduce [2] is a programming paradigm that is used to process such hefty amount of data in parallel fashion. Queries are processed with the help of these MapReduce functions and users get their results back. It has been concluded from the number of experiments that EGENMR proved to be a better technique than the latest techniques used to process large amount of data such as HIVE, PIG, HadoopDB, etc. EGENMR is implemented in .net and the author achieved significant speedup by using MapReduce functions. To further improve the model, EGENMR in terms of execution times of queries, in this paper, a prototype model is proposed which with the help of low-cost GPU can achieve significant processing speed.

CPU is used for control-intensive task and GPU is used for data-parallel computation-intensive task. If the problem exhibits massive data parallelism, GPU is a good choice as it provides large number of programmable cores, massive multithreading, larger peak bandwidth as compared to CPU. CPU supports heavy weight threads whereas GPU supports lightweight threads. CPU with quadcore processors and hyperthreading can run only 16 threads concurrently or 32. GPU supports 1536 threads per multiprocessors and GPU with 16 multiprocessors can support up to 24000 active threads. Context switch is slow and expensive in case of CPU where as in case of GPU if thousands of threads are waiting for execution in the queue it begins to start another set of threads. CPU cores are used to minimize latency whereas GPU cores are used to maximize throughput. CUDA is a programming language designed by NVIDIA to provide joint CPU and GPU execution of an application.

For many data-intensive applications, nowadays graphical processing unit (GPU) are proved to be an emergent and prevailing co-processor. GPUs are mainly used for high-performance computing (HPC) applications such as graphical 3D gaming problems, simulations, etc. HPC generally refers to get the high throughput with the help of multiple processors to do certain complex task. Databases are considered to handle huge amount of data. Now, researchers are attracted toward the use of GPU for accelerating database operations without or little change in the source code. But, there is one limitation associated with GPU for accelerating database operations which is input–output cost overhead incorporated, while using GPU as a coprocessor as it takes time to transfer data from CPU main memory (MM) to GPU device memory (DM) or vice versa. Which overpower the advantages of using GPU.

Recent research is concentrating on how to overcome this limitation of GPU. Researchers suggested [3] that GPU acceleration can be achieved by using data resides in the main memory instead of disk. For database operations such as Join [4], sort [5] GPU is used as an accelerator. A hybrid query processing approach [6], a combination of CPU and GPU processing is introduced in which the first step is to parse the query and then split into smaller parts, next step is to dispatch small parts of queries into CPU or GPU depending upon type of operator, type of data, and data size. In all of the above techniques, a lot of time is consumed in making the decision that which operator should dispatch to which processor depending upon the cost model. As earlier research [1] was based on the concept that every query is converted into MapReduce hence each query can run on GPU. Overhead in terms of time can be saved by using MapReduce codes. In this paper, a model is being proposed which can be used to process large amount of data at faster speed with the help of using GPU as a co-processor.

Main contributions of this work are as follows:

- Proposing a model which with the help of GPU as co-processor enhances the performance of EGENMR.
- Model takes care of the complex queries such as Join, Orderby.

The remainder of this paper is structured as follows: Sect. 2 presented the related literature work that has been done in this area, Sect. 3 presents the proposed model, and we concluded this work in Sect. 4.

2 Related Work

Based on the various research papers, we grouped the study into following categories: GPU architecture, GPU memory hierarchy, and GPU accelerator for database operations.

2.1 GPU Architecture

GPU is vastly used in many high process computing applications such as 3D gaming, physics simulations, image processing, networking, storage, and security. GPU is used to accelerate the execution of data with the help of vast parallelism. It is a common example of the hardware accelerator. NVIDIA GPU computing platform is enabled for the product families like Tegra (used for devices like mobile, tablets, and phones), GeForce (for consumer graphics), Quadra (for professional visualization), and Tesla (for data center parallel processing) [7]. Out of these two important GPU architectures, Fermi and Kepler of Tesla family are famous. GPU latest architecture is Tesla Maxwell introduced in 2015.

Fermi was released in 2010 and is used for accelerating application areas such as weather and climate modeling, seismic processing, computer-aided engineering. Kepler was released in 2012 that offers new features to optimize and increase parallel workload execution on GPU high-performance computing. There are two important features that describe GPU capabilities, first is number of CUDA cores (448 in Fermi and 2 * 1536 in Kepler) and other is memory size (6 GB in Fermi and 8 GB in Kepler) and two performance parameters of GPU are peak computational performance (1.03 Tflops trillion floating point calculations per s in Fermi and 4.58 Tflops in Kepler) and memory bandwidth (144 GB/s in Fermi and 320 GB/s in Kepler).

There are multiple Streaming Multiprocessors (SM) and each SM support concurrent execution of thousands of threads. Processing is based on Single Instruction Multiple Thread architecture. It manages and executes threads into a group of 32 called warps. It is a basic unit of execution in SM. Threads in thread block are further partitioned into warps. GPU thread hierarchy is decomposed into two levels: threads of block and grid of blocks. All threads in a grid share the same global memory. Different block threads cannot cooperate with each other.

Kepler K20 contains 15 streaming multiprocessors and six 64-bit memory controller. Three enhanced features of Kepler's are Enhanced SM, Dynamic Parallelism, and Hyper-Q. Each SM unit consists of 192 single precision CUDA cores, 64 double precision units, 32 special function units and 32 load/store unit, 4 warp schedulers and 8 instruction dispatchers enabling 4 warps to be executed concurrently on a single SM. It can schedule 64 warps per SM for a total of (64 * 32 = 2048) threads resident in a single SM at a time. Register size is 64 KB and it allows more partitions of on-chip memory between shared and L1 cache. It offers more than 1 tflop of peak double precision computing power. Figure 1 shows the basic architecture of GPU. Dynamic parallelism allows GPU to dynamically launch new grids. It allows any kernel to launch any other kernel to have inter-kernel dependencies needed to perform an additional task. It eliminates the need to communicate with CPU every time. Hyper Q allows adding more simultaneous

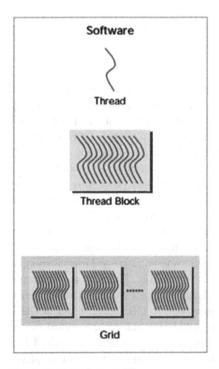 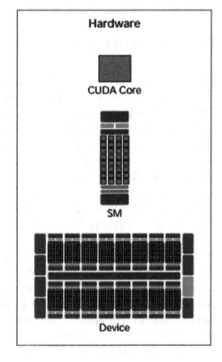

Fig. 1 GPU architecture [7]

connection between CPU and GPU, which leads to CPU to do more task on GPU. Which results in increased GPU utilization and reduced CPU idle time. It provides 32 hardware queues which enable more concurrency, maximize GPU utilization, and hence increases overall performance.

2.2 GPU Memory Hierarchy

Following are the memory hierarchies provided by GPU register, shared memory, local memory, constant memory, texture memory, and global memory. Register memory is the fastest memory of GPU. Register variable shares their lifetime with the kernel, once kernel completes execution variables in a register cannot be accessed again. Kepler architecture provides 255 registers per threads.

Variables that are eligible for registers but cannot fit into register space can be placed into local memory. It provides high latency and low bandwidth. Shared memory is on chip having high bandwidth and lower latency than local and global memory.

It performs best when all threads in warps read from the same memory address. Texture memory resides in device memory and is cached in SM as read-only cache. It is a type of global memory, and it is accessed through read-only cache. It supports hardware filtering, floating point interpolation. Global memory is largest, having highest latency, and most commonly used the memory of GPU.

2.3 GPU Accelerator for Database Operations

Due to GPUs parallel computing, it is used to accelerate applications and it is energy efficient too. Many researchers are using GPU or FPGA to accelerate database operations. The limitation with FPGA is that it is complicated and difficult to debug thus it has not taken the popularity in this field. Hence, NVIDIA GPU with CUDA (Compute Unified Device Architecture) using C programming language is used for the acceleration of database operations.

The author in [7] implemented Virginian Database using GPU and it works with SQLite and some subset of commands which directly executed on GPU. It can compute a large amount of data larger than GPU memory size as it uses efficient caching and GPU mapped memory. A Hybrid Query Processing Engine (HyPE) [8–10] has been developed which performs the task such as operators cost estimation in terms of data locality, hybrid query optimization and selects a query plan according to the optimization criteria. Optimization is based on three main components: self-tuning execution time estimator, algorithm selector, and hybrid query optimizer. CoGaDB is column oriented and GPU-accelerated database management system [9], in which instead of loading complete data on GPU it loads only a list of tuple identifier and hence it minimizes the data transportation between CPU and GPU. For accelerating database operation cross-join, a virtual machine model was used in [11]. The author implemented join queries with the help of GPUs parallel approach and got 2–5 times speedup. The authors in [4, 12] implemented primitive functions directly on the CUDA kernel such as map, reduce, split, and with the help of these primitives, the author evaluated various join algorithms such as nested loop join with and without indexing, hash join, sort-merge join. But in [11], a virtual machine model is used instead of primitive functions for implementing SQL statements. In [13], the author implemented the hybrid query processor model by automatically parsing query statements and splitting it into the parts rather than using any virtual machine model or primitives. Further query operations are dispatched to the suitable processor which lowers cost as well as reduce workloads. Mi-Galactica [14] an SQL like query accelerator which used GPU by offload geospatial computation. An efficient co-processing utilization based on cost-based optimization is proposed in [15] and they implemented heterogeneous scheduling strategy which distributes database operations on CPU or GPU based on minimal response time and cost models. GPU is used in accelerating MapReduce functions in [16].

3 Proposed Model

Researchers are continuously searching for techniques on how to take advantage of GPU as a co-processor in the field of database.

In this section, first we explained the model hybrid query processing which is taking advantage of using GPU as a co-processor for accelerating database operations, then we discussed and explained prototype model E-GENMR using GPU, also a comparison is made in terms of time complexity and time taken by both the techniques.

3.1 *Hybrid Query Processing Using GPU Co-processor*

To exploit GPU as a co-processor in database applications, hybrid query processing [6, 10] techniques have been implemented which shows that for some operations such as 'aggregation' GPU gives a better result than CPU and operation like 'selection' executes faster on CPU than GPU. In hybrid query processing through GPU, it takes time to split the query as well as to decide which query operator will go to which processor. The selection of processor is based on mainly three components which are types of operation (Sequential, Parallel, Sequential and parallel both), column data type (numeric or non-numeric), and data size (large or small) [13].

A key component of GPU CUDA programming is kernel execution. CUDA processing flow is as follows: the first step is to copy data from CPU memory to GPU memory. Second, kernel operation is executed on the data present in GPU memory then lastly is copy data back from GPU memory to CPU memory.

3.1.1 Hybrid Query Processing Works on the Following Algorithm

Input: Queries—q_1, $q_2...q_n$, Operations on queries—op_1, op_2 ...op_m

```
1    For each query qᵢ .....o(n)
2        Do parsing....qᵢ is parsed , split
3            For each operation opⱼ in qᵢ
4                if (opⱼ==simple)
5                    opⱼ will be executed on CPU or opⱼ will be executed using GPU with
     CPU
6                else if (opⱼ==complex)
7                    opⱼ will be executed on CPU or opⱼ will be executed using GPU with
     CPU
8                end if
9            end For
10   end For
```

First two lines show that query is first parsed and split. The query is broken into smaller operations, which takes $O(n)$ time. For simple operations (line 4), if the data is nonnumeric and operation is simple and size of the data is small then query operation will be executed on CPU with $O(n) + O(n) + O(n)$ times results are taken from [13]. But if the query is executed with the help of CPU and GPU, it takes $O(n) + O(1) + O(n)$ times. If the operation is complex and executed on CPU it takes $O(n) + O(n) \cdot O(n) + O(n)$ times, which is approximately $O(n^2)$. If it is executed on CPU with the help of GPU then it takes $O(n) + O(n) \cdot O(1) + O(n)$ times.

3.1.2 Total Time for Executing the Operations for Hybrid Query Processor on GPU Includes

$$T_{total} = \left(T_{split} + T_{Decision} + T_{op\,(CPU)} + T_{HM-DM} + T_{op(GPU)} + T_{DM-HM} \right)$$

- T_{split} is the time taken to split each query,
- $T_{Decision}$ which is a time to take decision for very split operation that whether it should be dispatched to CPU or GPU processor, $-O(n)$
- $T_{op(CPU)}$ is the time if the operation is executed on CPU $-O(n^2)$
- $T_{HM\text{-}DM}$, $T_{DM\text{-}HM}$ is the data transfer time between device memory and main memory, which is equal to $T_0 + data/Band$, where T_0 is the initial time to prepare data for transfer between device and main memory and data/band is the time for transferring data at bandwidth from main memory to device memory or vice versa.

- $T_{op(GPU)}$—is the time to execute an operation on GPU co-processor which includes $T_{memory} + T_{computation}$, where T_{memory} is the memory access time and $T_{computation}$ is the time taken for computing the operation which depends upon number of rows.

3.2 E-GENMR Model Using GPU Processing

In the previous work [1], the author achieved significant speedup with the help of parallel processing on data in the form of MapReduce functions. EGENMR system's compiler converts user queries into MapReduce codes as MapReduce codes provides parallel processing of large amount of data. Here, further enhancement of EGENMR is proposed with the help of GPU processing. Figure 2 shows the enhanced architecture model of EGENMR using GPU.

- Following is the working model of EGENMR using GPU processing
- Client data is horizontally partitioned and stored at the Datanode's of Racks present at cloud data centers
- For processing data user send query in any database languages (Oracle, SQL, DB2)

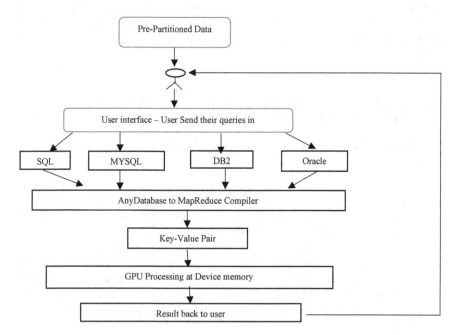

Fig. 2 Proposed model

- Queries get converted into MapReduce (Key-value pair) with the help of ENHANCED GENMR compiler.
- Instead of running codes of CPU GPU co-processor is being used for processing queries.

3.2.1 EGENMR Query Processing Works on the Following Algorithm

Input: queries q_1, q_2, ...q_n

1. For each query qi.
2.   Qi convert into map reduce
3.   All the operations will be executed on GPU.
4. End For

First line shows that query is first parsed which takes $O(n)$ time. For both simple or complex operations, it simply takes $O(n)$ time to execute operations on GPU.

3.2.2 Our Proposed Model Takes Total Time to Execute Query on GPU Includes

$$T_{total} = T_{con(map-reduce)} + T_{HM-DM} + T_{op(GPU)} + T_{DM-HM}$$

- $T_{HM-DM} + T_{op(GPU)} + T_{DM-HM}$ is the same as that of previous hybrid query processing technique.
- $T_{con(map-reduce)}$—is the time taken by the model to convert queries into MapReduce.

3.3 Comparison of Time and Complexity for Hybrid Query Processing and EGENMR Using GPU

For both the models, it can be seen that six time factors components ($T_{split} + T_{Decision} + T_{op (CPU)} + T_{HM-DM} + T_{op (GPU)} + T_{DM-HM}$) affecting the hybrid query processing and four time factors components ($T_{con (map-reduce)} + T_{HM-DM} + T_{op(GPU)} + T_{DM-HM}$) affecting EGENMR using GPU in which $T_{HM-DM} + T_{op(GPU)} + T_{DM-HM}$ is the same for both the models. Three time components $T_{split} + T_{Decision} + T_{op (CPU)}$ takes more time than one component $T_{con(map-reduce)}$ of EGENMR using GPU. With the help of algorithms for both the techniques, it can also be seen that worst-case complexity, in case of hybrid query processing is $O(n^2)$ and $O(n)$ in case of enhanced GENMR.

4 Conclusion

In this paper, EGENMR using GPU model is proposed. A theoretical comparison is made in terms of time taken and complexity for hybrid query processing using GPU and EGENMR. It can be seen that there are six time factor components affecting hybrid query processing which takes mode time as EGENMR using GPU processing is affected by four time components. With the help of algorithms for both the techniques, it can also be seen that worst-case complexity, in case of hybrid query processing is $O(n^2)$ and $O(n)$ in case of enhanced GENMR.

References

1. S. Malhotra, M.N. Doja, B. Alam and M. Alam, "E-GENMR: Enhanced Generalized Query Processing using Double hashing technique through Map Reduce in cloud Database Management System," *Journal of Computer Science,* 2017.
2. J. Dean and S. Ghemawat, "MapReduce: simplified data processing on large clusters," *Communications of the ACM,* vol. 51, no. 1, pp. 107–113, 2008.
3. P. Ghodsnia, "An In-GPU-Memory Column-Oriented Database for Processing Analytical Workloads," in *In proceedings of VLDB PhD Workshop,* 2012.
4. B. He, K. Yang, R. Fang, M. Lu, N. Govindaraju, Q. Luo and P. Sander, "Relational joins on graphics processors," *Proceedings of the ACM SIGMOD international conference on Management of data,* pp. 511–524, 2008.
5. N. K. Govindaraju, J. Gray, R. Kumar and D. Manocha, "GPUTeraSort: high performance graphics co-processor sorting for large database management," *Proceedings of the 2006 ACM SIGMOD international conference on Management of data,* pp. 325–336, 2006.
6. J. Cheng, M. Grossman and T. McKercher, Professional CUDA C Programming, Indianapolis, Indiana: John Wiley & Sons, 2014.
7. P. Bakkum and K. Skadron, "Accelerating SQL database operations on a GPU with CUDA," in *Proceedings of the 3rd Workshop on General-Purpose Computation on Graphics Processing Units,* Pittsburgh, Pennsylvania, 2010.
8. S. Breß, E. Schallehn and I. Geist, "Towards Optimization of Hybrid CPU/GPU Query Plans in Database Systems," *New Trends in Databases and Information Systems,* vol. 185, pp. 27–35, 2013.
9. S. Breß and G. Saake, "Why it is time for a HyPE: A hybrid query processing engine for efficient GPU coprocessing in DBMS," in *Proceedings of the VLDB Endowment,* Trento, Italy, 2013.
10. S. Breß, N. Siegmund, M. Heimel, M. Saecker, T. Lauer, L. Bellatreche and G. Saake, "Load-aware inter-co-processor parallelism in database query processing," *Data & Knowledge Engineering,* vol. 93, no. C, pp. 60–79, 2014.
11. K. Angstadt and E. Harcourt, "A virtual machine model for accelerating relational database joins using a general purpose GPU," in *Proceedings of the Symposium on High Performance Computing,* Alexandria, Virginia, 2015.
12. B. He, M. Lu, K. Yang, R. Fang, N. K. Govindaraju, Q. Luo and P. V. Sander, "Relational query coprocessing on graphics processors," *ACM Transactions on Database Systems,* vol. 34, no. 4, pp. 1–39, 2009.
13. E. Shehab, A. Algergawy and A. Sarhan, "Accelerating relational database operations using both CPU and GPU co-processor," *Computers & Electrical Engineering,* vol. 57, no. C, pp. 69–80, 2017.

14. H. H. O. Keh Kok Yong and V. V. Yap, "GPU SQL Query Accelerator," *International Journal of Information Technology,* vol. 22, no. 1, pp. 1–18, 2016.

15. S. Breß, F. Beier, H. Rauhe, K.-U. Sattler, E. Schallehn and G. Saake, "Efficient co-processor utilization in database query processing," *Information Systems,* vol. 38, no. 8, pp. 1084–1096, 2013.

16. Y. Chen, Z. Qiao, S. Davis, H. Jiang and K.-C. Li, "Pipelined Multi-GPU MapReduce for Big-Data Processing," in *Computer and Information Science. Studies in Computational Intelligence, vol 493*, Heidelberg, 2013.

Hardware Implementation of IoT-Based Image Processing Filters

Ajay Rupani, Pawan Whig, Gajendra Sujediya and Piyush Vyas

Abstract The image processing based on interfacing of FPGA and Raspberry Pi using the Internet of Things which is a recently introduced technique is a hot topic in the present scenario. The unimaginable interconnection of smart devices, smart cities, smart vehicles, and smart people throughout the globe is made possible by Internet of things. The hardware implementation of various filters used in image processing using Internet of things on an FPGA platform is presented in this dissertation. The Raspberry Pi and FPGA interfacing-based implementation of image processing filters using Internet of Things have gotten huge consideration from the exploration group in a previous couple of years. In this paper, we highlight that how one can access the design resources based on FPGA from any place. The primary point of this research is to highlight how the clients can get the FPGA-based outline resources from anyplace. In this manner, we exhibit an idea that abbreviates the utilization of immediately unused resources for executing different assignments automatically.

Keywords FPGA · Internet of things · Image processing

A. Rupani (✉) · G. Sujediya
RIET, Jaipur, India
e-mail: ajayrupani1991@gmail.com

G. Sujediya
e-mail: gajendrafromsawar@gmail.com

P. Whig
VIPS, Delhi, India
e-mail: pawanwhig@gmail.com

P. Vyas
JIET, Jodhpur, India
e-mail: piyush.vyas@jietjodhpur.ac.in

© Springer Nature Singapore Pte Ltd. 2018
V. Bhateja et al. (eds.), *Proceedings of the Second International Conference on Computational Intelligence and Informatics*, Advances in Intelligent Systems and Computing 712, https://doi.org/10.1007/978-981-10-8228-3_63

1 Introduction

Some factors introduce brightness variation in an image when there is no detail available for that image. There often occurs random variation and no particular pattern of variation is present [1, 2]. The quality of an image is reduced in many cases due to the variation.

1.1 Purpose of Image Processing

(a) Visualization: Observe the objects that are not visible.
(b) Image Sharpening and Restoration: To create a better image.
(c) Image Recognition: Distinguish the objects in an image.
(d) Image Retrieval: Seek for the image of interest.
(e) Measurement of Pattern: Measure various objects in an image.

1.2 Objectives of This Research

Not all image processing filtering methods are appropriate for a particular application. An objective and quantitative dimension on the appropriateness of each method for a specified application is of great consequence and usefulness. The following are the goals that explain the work in brief:

- Designing an IoT-based architecture having various image processing filters, where the user can select the type of filter operation just by clicking on the appropriate option.
- To implement the different image processing filters using the interfacing of Raspberry Pi and ZedBoard Zynq 7000 APSoC.
- To obtain the output of various image processing filters, viz., Gaussian filter, average filter, and sharpen filter for the Lena's image having a size 256 × 256 along with the on-chip power components report.

2 Related Work

In this section, a review of the literature has been carried out. Research papers, books, and articles have been used in preparing this research work. Various researches have been carried out in the field of image processing; some of them are described below.

D. M. Wu et al. (1995) displayed the outline and usage of an IBM-PC-based circulated image processing framework. An optional technique is accommodated ongoing image processing which is accessible only for more costly frameworks [3].

Zhu Bochun et al. (1996) presented a constant image processing framework in view of two-chip DSP TMS320C40 of Texas Instruments. There was a frame store available that could be able to configure again whenever required. This consolidated with the adaptable interface and powerful correspondence capacity of TMS320C40, makes the framework to be effectively designed as MIMD, SIMD pipeline, and parallel structures. The distortion present in the image processed by few components smartly maintains a strategic distance from without extra circuits [4].

In 2000, Jie Jiang et al. presented the execution and design of ongoing image processing system for Electronic Endoscope. The real-time image processing has been utilized as a part of Electronic endoscope effectively. The picture quality and the precision rate of finding are made strides. FPGA is embraced as the center of the framework for its alterable and reconfigurable element, so the framework can be effortlessly adjusted, in addition, the execution of the framework can be enhanced in the future, without changing the equipment. With a lot of logic gates in a solitary gadget, FPGA gives the framework high combination and magnificent execution [5].

In 2003, T. Sugimura et al. proposed a reconfigurable parallel image processing using FPGA for the superior real-time image processing framework. Parallel setup of logic units and interconnection network has effectively diminished arrangement information. A test chip was manufactured for this image processing using FPGA utilizing 0.35 pm CMOS technology [6].

In 2005, Pei-Yung Hsiao et al. proposed edge detection method for image processing system by means of an FPGA architecture design. There are two edge detection algorithms available which are appropriate for realization on hardware and not sensitive to noise. These two algorithms can create distinctive yields appropriate for various applications. The altered LGT algorithm presented in this paper preserves the original edge detection performance as well as extraordinarily diminishes the utilization of hardware asset [7].

In 2008, Jun Wang et al. presented the DSP- and FPGA-based image processing framework having all features, viz., extendibility, modularization [8].

In 2010, Chunfu Gao et al. presented the SOPC-based image processing system. For the algorithm, an image was collected by this system and that image was processed using the embedded Nios II soft-core processor, which has the elements of effortlessness and incredible information amount and parallelism. The implementation of this system was done using FPGA. Along these lines, the processing time of image was diminished that enhanced the continuous execution of framework adequately and understands the optimizing of cutting direction progressively. The system has various advantages, viz., small volume, reconfiguration of software and hardware, flexibility of programming, portability, strong generality, etc., [9].

In 2011, Hao Wei proposed multiple parallel processing of multiple video image data using the combination of both DSP and FPGA. Configuration takes the full preferred standpoint of programmable gadgets, for example, FPGA and FPGA equipment assets accessible inside the enhanced framework reconciliation and unwavering quality, additionally streamlines the framework equipment outline. The framework has been connected to a pseudo-review instrument venture, which have gotten attractive outcomes. After the genuine venture approval, the framework configuration is effective [10].

In 2012, Hayato Hagiwara proposed the FPGA implementation-based image processing system that can be used for the distant vision framework. A moving object is easily identified and tracked by the robot vision system. Additionally, this robot vision framework distinguishes the passageway limits amongst all the dividers and floors. The hardware–software integration for image processing framework was appropriate to different goals [11].

In 2014, Sunmin Song et al. presented a Gaussian filtering technique, i.e., window based. Both low-pass filter and average filter are used for blurring an image. As the weight average value provided by low-pass filter, a few pixels are more critical. Rather, the significance of other pixels is relinquished. The Gaussian filter having a place with an LPF (low-pass filter) has been an enthusiasm for PC vision field as well as digital image processing. The Gaussian filter is for the most part utilized as a part of the preprocessing of edge recognition [12].

In 2015, Jinalkumar Doshi highlighted how clients can get to FPGA plan assets from any place in combination with an FPGA lab and that lab can be easily accessible from a distance by users. Various languages like PHP, Python, etc., are used to create an environment, i.e., remotely accessible on an open-source operating system Ubuntu [13].

In 2015, Zalak Dave et al. proposed an image processing framework based on NoC. A Torus topology having 6 nodes is used to design the architecture of NoC [14]. This paper introduced display interface, a memory interface, various IP algorithms, and computer interface (utilizing Raspberry Pi). A versatile plot for remote applications has been provided by the Raspberry Pi.

As indicated by Shivank Dhote (2016), the system architecture which comprises of numerous logic blocks used for the execution of various point operations on the given pixel information. The given information can be recovered by means of Block Random Access Memory. The Universal Asynchronous Receiver/ Transmitter (UART) Buffer register is being used by the Block RAM for getting its input. The client interface transfers the input image on the File Transfer Protocol server over Raspberry Pi. By utilizing the Universal Asynchronous Receiver/ Transmitter protocol, the Spartan 6 FPGA development kit receives the image bitmap from the Raspberry Pi at a given point. The decoder is used to select appropriate filter type which is connected to a received image. By the utilization of Universal Asynchronous Receiver/Transmitter protocol, the output image is first stored in Random Access Memory and then transferred to the Raspberry Pi, when asked [15].

3 Proposed Methodology

The system architecture which comprises of numerous logic blocks used for the execution of various point operations on the given pixel information. The given information can be recovered by means of Block Random Access Memory. Then the input image is loaded to ZedBoard Zynq 7000 APSoC as.coe file. The Decoder is used to select appropriate filter type, which is to be implemented on the acknowledged image from Raspberry Pi to ZedBoard Zynq 7000 APSoC. The output image is first stored in the Block Random Access Memory and then fed to VGA monitor. According to the user's choice, the appropriate filter is selected using decoder circuit within the FPGA. The required logic block can be activated by generating a signal using the decoder. Raspberry Pi 3 MODEL B is an upgraded version of the enhanced network with Bluetooth Low Energy (BLE) and BCM43143 Wi-Fi on board [16]. ZedBoard Zynq 7000 FPGA development kit is an easy advancement board for APSoC (All Programmable SoC) [17]. The VGA driver is used for displaying the outputs of various filters by means of FPGA. Data from Random Access Memory is sent to the VGA driver as input. The obtained output of various image processing filters has been displayed on 640 × 480 VGA display. On horizontal section, the image of size 256 × 256 has been displayed from pixel number 200 to pixel number 456, while on the vertical section the image is displayed from pixel number 110 to pixel number 366 pixels on VGA Screen (Fig. 1).

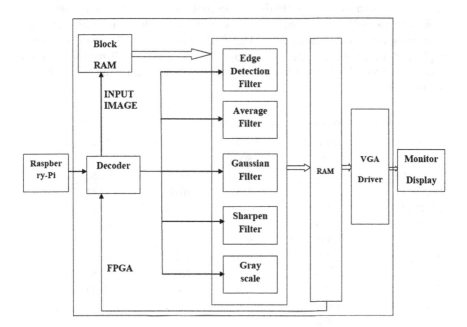

Fig. 1 Proposed system architecture

Fig. 2 Hardware implementation of the proposed image processing system

The hardware implementation of proposed IoT-based image processing system, i.e., the interfacing of Raspberry Pi and FPGA1 is shown in Fig. 2.

4 Experimental Results

The code written in Verilog Hardware Description Language is simulated using Xilinx Vivado software to obtain the output of various filters, viz., Gaussian filter, sharpening filter, and average filter. The ZedBoard Zynq 7000 APSoC is used to implement the design, i.e., used in this paper. The total power utilized for the whole system is 40.580 W. The tabular representation of power utilization for the system is shown in Table 1.

Table 1 Power utilization report of on-chip components

On-chip	Power (W)	Used	Available	Utilization (%)
Slice logic	18.119	38069	–	–
LUT as logic	15.292	9879	53200	18.56
Register	2.593	18557	106400	17.44
CARRY4	0.225	118	13300	0.88
BUFG	0.009	2	32	6.25
Signals	12.541	19526	–	–
Block RAM	7.935	72	140	51.42
DSPs	0.410	3	220	1.36
I/O	0.486	22	200	11.00
Static power	1.089			
Total	40.580			

Fig. 3 Input image (Lena)

Fig. 4 Output of Gaussian filter

The input image and output images captured from VGA monitor are shown below.

Figure 3 shows the input image on which various filtering operations have been performed. From Fig. 4, it is clear that the Gaussian filter gave gentler smoothing and the edges have been preserved superior to anything a comparably sized mean filter. Figure 5 displays the output of sharpening operation and Fig. 6 shows the output after grayscale operation. From Fig. 7, it is clear that the effect of the edge detection indicates how easily or unexpectedly the values of image pixel change at each point in the whole image and in this way we can say how likely the segment of

Fig. 5 Output of sharpening

Fig. 6 Output of grayscale

Fig. 7 Edge detection output

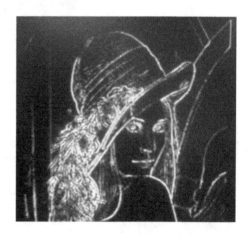

Fig. 8 Average filter output

Table 2 SNR for filtered output images

Image parameter	PSNR (dB)
Filter type	
Average filter	21.6416
Gaussian filter	18.2617
Sharpening filter	16.9686
Grayscale	15.2847
Edge detection	8.5470

the picture denotes an edge. We have observed that the harder edges have appeared as brighter (higher values) in the results. Figure 8 shows the effect of applying a 3 × 3 mean (average) filter. From the Figure, it is clear that the noise is less apparent, but the image has been softened. We observed that the pixels will be nearer to each other in value; accordingly, hard edge has been eliminated in this situation.

Also, we have calculated the SNR values for images obtained from various image processing filters that are shown in following table (Table 2).

5 Conclusion

The various best suitable image processing filters are available to enhance a noisy image. The decision of applying which specific filter strongly depends upon various levels of noise present at various pixel locations. In this paper, all-inclusive filter architecture used for image processing has been presented. This gives the solution for multi-practical image processing for various applications. The design and implementation of various image processing filters have been presented. All of the modules have been designed and implemented on a ZedBoard Zynq 7000 APSoC.

Hence, image processing and IoT can together have been produced the adequate results. We have observed that the average-filtered output image is having maximum SNR, whereas edge detection-filtered output image has the lowest SNR.

6 Future Scope

To make the system architecture more comprehensive, we can add more filters to our system architecture presented in this paper. We can apply this technique to the videos for getting the better motion of an object or the motion of an object can be enhanced. In future, using this technique we can focus on an image object by eliminating the image background or by changing the background with respect to an object.

References

1. Trimberger S. M, "Field – Programmable Gate Array Technology", Kluwer Academic Publishers, 1995.
2. R. C. Gonzalez and Richard E. Woods, Digital Image Processing – Second Edition, Addison-Pearson Education Publishing Company, 2005.
3. D. M. Wu, L. Guan, G. Lau and D. Rahija "Design And Implementation of a Distributed Real-Time Image Processing System", *Engineering of Complex Computer Systems, 1995 held jointly with 5th CSESAW, 3rd IEEE RTAW and 20th IFAC/IFIP WRTP*, Proceedings, pp- 266–269, 1995.
4. Zhu Bochun, Guo Chengming, Wang Zhaohua "A Parallel Real-time Image Processing System Based on TMS320C40" *3rd International Conference on Signal Processing* Proceedings, Vol. 1, pp- 653–656, 1996.
5. Jie Jiang, Daoyin Yu, Zheng Sun, "Real-time Image Processing System Based on FPGA for Electronic Endoscope", *IEEE APCCAS- IEEE Asia-Pacific Conference on Circuits and Systems proceedings*, pp- 682–685, 2000.
6. T. Sugimura, Jeoung Chill Shim, H. Kurino, M. Koyanagi, "Parallel Image Processing Field Programmable Gate Array for Real Time Image Processing System", *IEEE International Conference on Field-Programmable Technology (FPT) Proceedings*, pp- 372–374, 2003.
7. Pei-Yung Hsiao, Le-Tien Li, Chia-Hsiung Chen, Szi-Wen Chen, Sao-Jie Chen, "An FPGA Architecture Design of Parameter- Adaptive Real-Time Image Processing System for Edge Detection" *Conference on Emerging Information Technology*, pp- 1–3, 2005.
8. Jun Wang, Lei Peng, Li Wei, "Design and Implementation of a Real-Time Image Processing System with Modularization and Extendibility", *International Conference on Audio, Language and Image Processing*, pp- 798–802, 2008.
9. Chunfu Gao, Zhiyong Luo, Xinsheng He, "Study on The Real-time Image Processing System of Cutting Robot Based on SOPC", *International Conference on E-Product E-Service and E-Entertainment*, pp- 1–4, 2010.
10. Hao Wei, "Designing and Development of Multi-DSP real-time image processing system based on FPGA" *Proceedings of 2011 International Conference on Computer Science and Network Technology*, vol. 2, pp- 1263–1265, 2011.

11. Hayato Hagiwara, Kenichi Asami, Mochimitsu Komori, "Real-time Image Processing System by Using FPGA for Service Robots", *The 1st IEEE Global Conference on Consumer Electronics 2012*, pp- 720–723, 2012.
12. Sunmin Song, SangJun Lee, Jae Pil Ko, Jae Wook Jeon, "A Hardware Architecture Design for Real-time Gaussian Filter", *IEEE International Conference on Industrial Technology (ICIT)* pp- 626–629, 2014.
13. Jinalkumar Doshi, Pratiksha Patil, Zalak Dave, Ganesh Gore, Jonathan Joshi, Reena Sonkusare, Surendra Rathod, "Implementing a Cloud Based Xilinx ISE FPGA design Platform for Integrated Remote Labs", *International Conference on Advances in Computing, Communications and Informatics (ICACCI)*, pp- 533–537, 2015.
14. Zalak Dave, Shivank Dhote, Jonathan Joshi, Abhay Tambe, Sachin Gengaje, "Network on Chip Based Multi-function Image Processing System using FPGA", *International Conference on Advances in Computing, Communications and Informatics (ICACCI)*, pp- 488–492, 2015.
15. M. Shivank Dhote, Pranav Charjan, Aditya Phansekar, Aniket Hegde, Sangeeta Joshi, Jonathan Joshi, "Using FPGA-SoC Interface for Low Cost IoT Based Image Processing", *International Conference on Advances in Computing, Communications and Informatics (ICACCI)*, pp- 1963–1968, 2016.
16. https://www.raspberrypi.org/products/raspberry-pi-3-model-b/.
17. http://zedboard.org/product/zedboard.

An Extensive Study on IoRT and Its Connectivity Coverage Limit

Ayushi Gupta, Tanupriya Choudhury, Suresh Chandra Satapathy
and Dev Kumar Chaudhary

Abstract IoT is the network of physical object devices embedded with electronic/sensors—that enables these devices to connect/share data through standard network protocols. IoRT is an evolution of IoT introduced by ABI research. Unlike IoT, it provides an active sensorization. This paper will propose time-effective connectivity of user with all the devices through a medium interface, and motion control strategy for good connectivity between robot and devices.

Keywords Internet of robotic things—IoRT · Internet of things—IoT
Sensors · Connectivity limit

1 Introduction

In the twenty-first century, the rapid development in technology has brought forward a demand for hyper-connectivity. Internet connectivity is necessary for implementation. And in the present scenario, it is plays a pivotal role in every sector. Nevertheless, these devices are still foremost things on the Internet that require more human intercommunication and surveil. We want to be connected with anything, anytime and anywhere. Some worldwide technologies like Ubiquitous Computing, Ambient Intelligence, Sensors, Actuators content and most of our prerequisite of smart world are not tightly coupled with the Internet. The IoT

A. Gupta (✉) · T. Choudhury · D. K. Chaudhary
Amity University, Noida, Uttar Pradesh, India
e-mail: gupta.ashi111@gmail.com

T. Choudhury
e-mail: tchoudhury@amity.edu

D. K. Chaudhary
e-mail: dkchaudhary@amity.edu

S. C. Satapathy
P.V.P. Siddhartha Institute of Technology, Kanuru, Vijayawada, India
e-mail: sureshsatapathy@gmail.com

© Springer Nature Singapore Pte Ltd. 2018
V. Bhateja et al. (eds.), *Proceedings of the Second International Conference
on Computational Intelligence and Informatics*, Advances in Intelligent Systems
and Computing 712, https://doi.org/10.1007/978-981-10-8228-3_64

(IoT) is a core component, which is ideally emerging to influence the Internet and hyper-connectivity. It is a convergence of technologies—Ubiquitous Computing, Sensors, Actuators, Internet Technologies, Ambient Intelligence, Communications Technologies, Embedded systems, etc. IoT have wide range of applications ranging from smart home, wearables, smart city, smart grids, industrial Internet, connected car, connected health (digital health/telehealth/telemedicine), smart retail, smart supply chain, smart farming to many others. The IoT is an up-to-date epoch of intelligence computing and it provides benefits to communicate all over the world.

But sensors constituting IoT prototype are acquiescent hitherto, summing a dynamic role for sensors will be required, in furtherance to ameliorate the systems where they are ad hoc. ABI Research introduced [1] IoRT for such need in 2014. IoRT is a smart device that can track events, put together data from diverse sources, use local and distributed intelligence to figure out smartly the most suitable procedure, and then respond according to control or manipulate objects in the physical world. Various IoRT applications are smart agriculture, environment monitoring, exploration, disaster rescue, etc. The objective of IoRT is Everything, Everyone, Everytime, Everyplace, and Every service. Figure 1 describes the seaming of C's and E's. This acknowledge that humans and devices can stay connected everytime, everyplace, with everyone, ideally by using artificial intelligence and network.

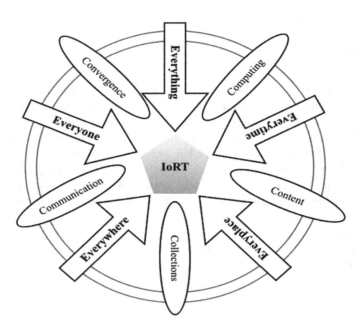

Fig. 1 Objectives of IoRT

2   Literature Survey

S. K. Anithaa et al. [2] did a survey and stated "The Internet of Things is a up to date epoch where the day-to-day objects can be mannered with sensing, networking and processing capabilities that will allow them to get across with one another round the internet to attain divers intent."

K. Swarnalatha et al. [3] determined that "Architecture to enable the users to control and monitor smart devices through internet. It creates an interface between users and smart home by using GSM and internet technologies, or it simply creates GSM based wireless communication from the web server into the smart home."

Cristanel Razafimandimby et al. [4] presented "A neural network and IoT-based scheme for performance assessment in IoRT. It proposed IoRT-based and a neural network control scheme to efficiently maintain the global connectivity among multiple mobile robots to a desired quality of service (QoS) level. The IoT-based approach was based on the computation of the algebraic connectivity and the use of virtual force algorithm. The neural network controller, in turn, is completely distributed and mimics perfectly the IoT-based approach."

Ethan Stump, Ali Jadbabaie et al. [5] developed a framework for improving a communication bridge between a stationary and the user [5]. Measured the quality of the connectivity and provides information about how to move the robots to improve the connectivity in "Making networked robots connectivity-aware".

Maria Carmela De Gennaro and Ali Jadbabaie [6] proposed a decentralized algorithm to control connectivity for many agent systems.

Devices interact and communicate with each other with the help of internet and we call it as Internet of Things (IoT). The advancement in the IoT affecting human lifestyle in a positive manner in the field of healthcare, automation, transportation, and emergency response to manmade and natural disasters where it is hard for the human to make decisions, as well as in a negative manner as it makes humans more dependable on devices and technology. The Internet will be heretofore assumed as the network of computers.

Nevertheless, it will be convoluted with the billions of smart devices accompanied with the embedded systems. As a result, Internet of Things (IoT) will enormously boost its size and scope, contributing an advanced way of opportunities, as well as objectives.

Devices are still dependent on the User. IoT can interact and monitor devices but it's maintenance and physical work has to be done by the user. Using IoT user gives the command to any medium interface like smartphone using GSM and other internet technology. And various robots can work together to complete a task.

3   Use of an Interface

In this era, there are around 20 billion devices ready to be connected by the Internet and it is expected to grow to anywhere from 50 to 212 billion by 2020. These devices are present everywhere. From Industrial districts to Government properties to Home Automation and it will be extremely difficult for the user to connect with all the devices and give commands to it. So here is the place where there is a need to build an interface which will act as an interface between the user and devices (Fig. 2).

To be able to connect with everything and still have a clean, easy to use and manageable interface is very important. Building an interface or an automaton is required. That automaton or simply a robot will become an interface between the user and devices. Where it's job will be to accept the commands from the user and projecting it to the devices. And look after the maintenance of devices, which user has to monitor. This will have a tremendous impact on user-device intercommunication as it will be time effective for the user. Now, the user will give all the commands to the robot instead of having one-to-one intercommunication with all the devices. And IoRT is a new era of active sensorization and intelligent computing. An interface will be a Robot based on IoRT (Fig. 3).

Fig. 2   User and devices

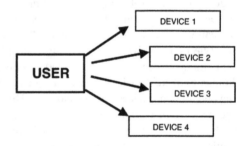

Fig. 3   Use of an IoRT robot

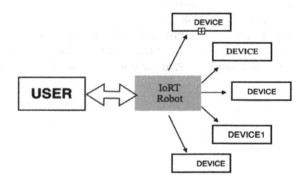

4 System Description

I. The user uses the Internet for sending commands and receiving confirmation results from the Robot.

II. Robot use built-in hardware and software to send and receive data from the user via various communication protocols. The robot can hop on the net via WiFi signals. Robot run mostly via the command of the user or their own embedded software or firmware, telling them what to do. This sending and processing of the sensor data are often nearly, allowing the interface to react in real time.

III. To carry out cooperative tasks, Robot needs to connect to other devices via a wireless link like Bluetooth and other local communication methods. They may also communicate with other smart devices in the vicinity. Smart devices can work in conjunction with tagging technology including RFID tags, QR codes, barcodes.

IV. Robot *Instructs* the devices. They can *gather* various sorts of readings and gather data, like city road systems that signal smartphones to help monitor traffic conditions.

The user needs to input all the device. Selection of devices can be application dependent and might involve ambiguity. So, to prevent it from ambiguity we involve multiple criteria such as the maintenance level, closest device, etc.

In some situation, ambiguity might arise. And robot calculates the best move towards the target device that keeps the robot in connection with other devices. If such a move is not found then robot select another target device and again calculate the best move towards the target device that keeps the robot in connection with other devices.

After finding such a move it moves to the target device, a complete task for the respective device and updates user. This loop of selecting a device, calculating the best move, updating the user repeats till the task is done for all the devices.

The following algorithm summarizes our approach.

Algorithm 1:

Input : A set of devices

Output : Next move to the device, without getting disconnected to other devices.

```
1   begin
2       while taskdone() do
3               target ← the device and calculate the best
                move towards the target device that keep
                the robot in connection with other devices.
4           if such a move was found then
5               move to the position of device;
6           else
7               repeat target;
8           end
9           if has updates then update the user
10      end
11  end
```

5 Application

IoT has already made his mark in the world. Every advanced city in the world is trying to incorporate this technology. Nowadays, Even the term "smart" means that the device it is referring to is ready to be manipulated by the internet. Smart city, smart homes, smart devices, smart grids and now smart hospitals are emerging into existence using the IoT. Not only this but it also gave birth to the IoRT. Robots are supposed to take over the human jobs by taking charge of simple yet time-consuming jobs. IoRT can be used to manage a large number of devices and by managing I mean getting the work efficiency of the device, reporting the newly installed part to the device, i.e., what kind of enhancements and improvements the device made after getting upgraded.

IoRT also has a ton of applications that mainly lies in the area which consumes a lot of human-time. Robots are supposed to decrease the time consumed by the management process and providing more accurate results.

6 Case Study

If a company owns 100 Air Conditioners of different brands. All of them have different functions. And those 100 AC need maintenance of filters, coils, and fins regularly. Now, the user has to interact and monitor all of them, which can be time consuming for the user. IoT can reduce the human intercommunication with AC but not monitoring, as the user has to do all the physical work like maintenance of filters, coils, and fins. IoRT Robot can interact as well as can monitor those AC. They use built-in hardware and software to send and receive data from the user and command the AC via a wireless link like Bluetooth and other local communication methods. They can monitor them. They look after its maintenance of filters, coils, and fins regularly. Besides the maintenance, if there is any technical issue, they can inform the user and place a complaint to its technician.

7 Connectivity Coverage

Maintaining a good communication and connection between Robot and Devices is a pivotal affair. Many approaches have been drafted to guard the connectivity of multi-robot. It is important to develop an approach for maintaining the connectivity of Robot and devices. There are two types of connectivity:

1. Local connectivity
2. Global connectivity

Global connectivity is preferred, as local connectivity maintenance has some limitations. Over and above of connectivity, discovering the coverage limits is an important issue.

8 Robot–Device Connectivity (RDC)

Robot–Device Connectivity (RDC) is model by an undirected graph $G(V, E)$ where, V is set of devices in a walled environment and E is set of edges and it is defined as the distance between Vn^{th} device and IoRT robot. It can be defined as

$$E_{RobotVn} = \{(a - b)|Vn^{th} \neq R \wedge (a - b) \leq C\}, \tag{1}$$

where

a_n is the position of Vn^{th} device and the collection of all the devices position is $a = [a_1, ..., a_n]$

b   is the position of R robot, and

Vn^{th} be the device which trade information with IoRT robot R. Vn^{th} is defined as

$$Vn^{th} = \{a \in V | (a - b) \leq C\}. \tag{2}$$

Now, E determines the connectivity coverage.

$$E \propto D^{-1} \tag{3}$$

The distance between Vn^{th} device and IoRT robot is inversely proportional to connectivity. Connectivity will increase as distance decrease and the robot moves closer to the devices.

The IoRT Robot is inbuilt with a protocol that allows connectivity with all the devices. This connectivity is range dependent, with a quality that varies between 0 and 1. The quality of connectivity between Robot and Device **Vn** is set to be

$$Fij = \begin{cases} 0 & \|Da - Db\| > p \\ 1 & \|Da - Db\| < p \\ e^{\frac{-5(\|Da - Db\| - p)}{R - p}} & p \leq \|Da - Db\| \leq R \end{cases} \tag{4}$$

where

R   is connectivity limit,

p defines a saturation distance, i.e., the connectivity remains constant when Robot gets closer to Device,

D_a is position of device which remains constant, and

D_b is the position of Robot.

At the distance $\| r_{ab} \| = R$, there exist some breach in connectivity (Fig. 4).

Fig. 4 F_{ij} is a function matrix of the distance between device and IoRT robot

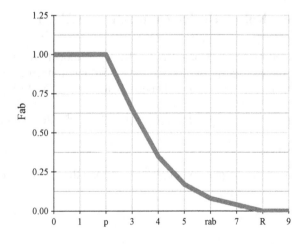

Connectivity coverage remains constant till a distance and starts to decrease as distance increase. At a point, connectivity coverage reduces to zero and after that point, it gets disconnect to the device.

So, we implemented a motion control strategy algorithm. All the algorithms in this paper were implemented in several test cases.

9   Motion Control Strategy

After a point, our robot gets disconnect to the devices. So, we need to solve this problem to maintain a good connectivity between robot and device all the time while it is in motion. And build a motion control strategy which allows the robot to move within the connectivity coverage limit.

An IoRT robot with tremendous computational proficiency, compute and surveil the connectivity with all devices. The position of all the devices will remain constant. We presume that each IoRT robot is aware of its own position as well as the device's position through GPS sensors or alternative localization system. IoRT robot needs to apply an algorithm to restrain its movement. This algorithm will be established on the requirements of maintenance and technical issue.

Now, in furtherance of keeping the requisite distance and hence the desires connectivity quality with V^{th} device, R robot should move within the limits of connectivity coverage.

Connectivity coverage will have limits: $E < p$.
And if $E > p$, it should move close.
Where Emax is the maximum desired distance between an IoRT and robot.

Now, if R robot is moving to a particular device it needs to maintain distance with other devices simultaneously.

$$W = \sqrt{\left[(a_0 - b_0)^2 + (a_1 - b_1)^2 + \cdots + (a_n - b_n)^2\right]} \tag{5}$$

W is the distance of robot with respect to to all the other devices.
The following phases summarize our approach:

Phases of Execution:

Phase (1) *Selected target device.*

Phase (2) *Determine the position (j) of the Device.*

Phase (3) *Determine position with other devices.*

(To maintain distance with other devices)

Phase (4) *Calculate the distance to device.*

Compute E using Formula (1)

Phase (5) *Calculate the distance with other devices and*

Compute W using Formula (4)

Phase (6) *Check $D < D_{max}$ for each device.*

if $E < p$ for each device, then Phase (6)

else, Repeats Phase 1–6

Phase (7) *R robot moves to the device.*

The motion control strategy is a protocol to help Robot to maintain connectivity with all the device and prevents disconnectivity. Robot needs to select the target device without arising any ambiguity. Then determine the position of all the devices. The robot moves to the target device, maintaining the desired distance with other devices to prevent disconnectivity.

10 Interpretation of the Results

An architecture model of the proposed system is described with simulation parameters. It describes how our IoRT Robot maintains connectivity with devices. The importance of taking account the desired distance is important to maintain a good connectivity is also highlighted.

10.1 Architecture Model

The architecture of working of IoRT Robot between the user and devices proposed in this paper is divided into four-tier, which constitutes:

- Bottom Tier: It is responsible for all the devices. It has data of all the devices and shares all the data with the user. It collects the commands from the user. It is known as **Data Tier**.
- Intermediate Tier 1: This tier is also know as **Communication Tier**. It is responsible for the communications between the user and devices through sensors.
- Intermediate Tier 2: It is responsible for data management and starts executing the command of the user. All the calculation and management is done in this layer that is why it is known as **Execution Tier**.
- Top Tier: It is also known as **Application Tier**. It is responsible for the application of user's command and results are generated.

Level	Tier	Responsibility
Bottom	Data tier	IoRT sources, data generations, and collections
Intermediate	Communication tier	Communication between sensors, relays, base stations, the Internet, etc.
Intermediate	Execution tier	Data management and processing
Top	Application tier	Application and usage of the data analysis and action generated

10.2 Simulation Parameters

Our first Simulation parameter is physical. Physical Simulation parameter includes determination of radio wave propagation between IoRT Robot and the devices which can be determined through 2 Ray Ground Reflection model which predicts the lost path when the signal receives multipath component formed by a single ground reflected wave.

IoRT Robot is movable and needs to compute new position with respect to to the device every time (Table 1).

10.3 Simulation Results

In the following results, it is assumed that the topology is totally connected at the beginning of the simulation. IoRT Robot contains the data of all the devices. And has an active connection with the user through its characteristics of active sensorization, through which the robot receives the user's command and status of all the devices. Every device has it's own connectivity limit (p), now when the IoRT Robot is moving, it has to maintain connectivity with all the devices simultaneously. Since IoRT Robot has all the data about the devices and has active communication with the user every time, it can execute the command anytime user want. When IoRT robot executes the user's command, it calculates the best move

Table 1 Simulation parameters: physical, mobility, and topology

Physical	
Reflection model	Two-ray ground
Error model	Real
Communication range	250 m
Mobility	
Computation of the new position	Two-ray ground
p	212 m
Topology	
Topology width	3000 m
Topology height	3000 m

towards the needed device while maintaining connectivity with all the other devices simultaneously. Connectivity limit of the devices are different and it tends to reach connectivity limit as the distance increase. IoRT robot moves to the devices after calculating the best move. And finally executes the user's command.

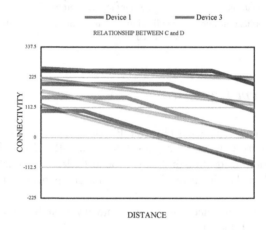

The graph depicts the relationship of connectivity with distance for IoRT Robot and different devices. It is the most crucial part for maintaining a good connectivity between devices and IoRT Robot.

11 Conclusion

This paper proposed time-effective IoRT-based interface between a user and devices, and Motion Control Strategy to maintain good connectivity between IoRT Robot and the devices by calculating its connectivity limit. An algorithm is proposed to sum up our approach, with phases of execution to describe all the phase thoroughly, in order to select a virtuous device to get connected and follow the algorithm proposed. An architecture model is put forward to cleave the responsibilities into various tiers and sub-tiers. It was scrutinized by various simulation parameters. And the relationships were derived, analyzing simulation parameters.

References

1. ABI Research. Internet of robotic things. https://www.abiresearch.com/market-research/product/1019712-the-internet-of-robotic-things. Accessed November 2, 2015.
2. S. K. Anithaa, S. Arunaa, M. Dheepthika, S. Kalaivani, M. Nagammai, Mrs. M. Aasha, Dr. S. Sivakumari stated "The Internet of Things - A survey".

3. K. Swarnalatha, Amreen Saba, Asfiya Muhammadi Basheer, P. S. Ramya proposed "Web Based Architecture for Internet of Things Using GSM for Implementation of Smart Home Applications", EISSN 2392–2192.
4. Cristanel Razafimandimby, Valeria Loscri, Anna Maria Vegni presented "A neural network and IoT-based scheme for performance assessment in Internet of Robotic Things", I4T - 1st International Workshop on Interoperability, Integration, and Interconnection of Internet of Things Systems, Apr 2016, Berlin, Germany.
5. Ethan Stump, Ali Jadbabaie, and Vijay Kumar. Connectivity management in mobile robot teams. In *Robotics and Automation, 2008. ICRA 2008. IEEE International Conference on*, pages 1525–1530. IEEE, 2008.
6. Maria Carmela De Gennaro and Ali Jadbabaie. Decentralized control of connectivity for multi-agent systems. In *Decision and Control, 2006 45th IEEE Conference on*, pages 3628–3633. Citeseer, 2006.

Statistical Analysis Using Data Mining: A District-Level Analysis of Malnutrition

Reyhan Gupta, Gaurav Raj and Tanupriya Choudhury

Abstract Children in India suffer from the highest level of undernourishment in the world, having serious effects on health. Bihar, in turn, has the highest incidence in India. This study was taken up to ascertain the effect of clusterization on a large dataset with respect to analyzes of the association of multitude of variables with malnutrition indices. Raw data were obtained using secondary data sources, especially government reports pertaining to malnutrition indices, demographic, social, nutritional, economic and medical factors causing malnutrition. In stage one the variables from un-clustered data were correlated with malnutrition indices (Stunting, wasting and underweight). Subsequently, the data was split using RapidMiner Studio (version-7.2.003) into three clusters. This segregation was done by the software using k-means and hierarchical agglomerative clustering. In the second phase, each of these clusters was again analyzed using the software and the correlation results were compared. Significant variation was observed in most of the correlations in the clustered and un-clustered datasets. This indicates the importance of clusterization in reaching the truth when a statistical analysis is carried out, as clusterization excludes/segregates the outliers/extremes of values. This has significant implications in policy making for malnutrition control, through identifying the most relevant variables/factors responsible.

Keywords Clustering · Malnutrition analysis · Data analysis
Data mining

R. Gupta (✉)
Amity School of Engineering and Technology, Noida, India
e-mail: reyhangupta@gmail.com

G. Raj · T. Choudhury
Amity University, Noida, Uttar Pradesh, India
e-mail: graj@amity.edu

T. Choudhury
e-mail: tchoudhury@amity.edu

© Springer Nature Singapore Pte Ltd. 2018
V. Bhateja et al. (eds.), *Proceedings of the Second International Conference on Computational Intelligence and Informatics*, Advances in Intelligent Systems and Computing 712, https://doi.org/10.1007/978-981-10-8228-3_65

1 Introduction

As per the World Bank Report, 2015, India fares very poorly in Global Hunger Index. India scores 29 points and its malnutrition condition is stated as 'serious', as it is ranked even below Bangladesh, Sri Lanka and sub-Saharan Africa. Bihar lies at the bottom of malnutrition table in India, with child undernourishment rates to be as high as 80% [1]. Though several policies have been initiated to address malnutrition in the state, it seems either the schemes presently running in the state are not reaching the beneficiaries or they are flawed. It is also possible that the policies for malnutrition control are not based on sound presumptions, which are meant to be dependent and, are as good as, relevant data generation, analysis, interpretation and application [2].

The latest NFHS 4 data released by the Government of India in 2017 gives the district-level details of many determinants of Malnutrition in Bihar [3]. A maze of factors, ranging from poverty, economic status of family, medical condition of child, infections, nutritional status of women/mothers, demographic and social factors (literacy, female literacy), political will, poor implementation of food policies qualify as determinants of malnutrition. The relative importance of these determinants is critical in selecting them to be targeted for effective malnutrition control. Here lies the importance of efficient data mining which could delineate the most important undernutrition determinants in a given district (or set of districts with similar social/demographic/medical attributes). Once such a set/cluster of similarly placed districts is segregated, it is easy to analyze it further with respect to the most critical determinants prevalent as a cause of undernutrition in that district/cluster. Since there are infinite permutations and combinations possible within these clusters, it is not humanly possible to pick the right ones up and process it. The data miner enables us to perform this cluster analysis so as to focus on the most important determinants and to prioritize them for selective policy-making and intervention. The present study is undertaken to appreciate the utilization of RapidMiner Studio in data analysis with special emphasis on the importance of clusterization in the correct interpretation of results in a large set of data.

2 Literature Survey

2.1 Data Mining Methods

Data mining is a field which helps us to find unknown patterns in data which may otherwise seem incoherent. The data mining methods that have been used are clustering, regression, association, classification, followed by statistical analysis [4].

2.2 Clustering

Clustering is a technique to group different parameters based on the similarities that exist between them and their interrelationship with each other. It enables us to understand how data is distributed and to understand the similar and dissimilar objects in the data. There are many ways to implement this technique, the important ones are

1. Centroid-Based Clustering	3. Distributed Clustering
2. Hierarchical Method	4. Density-Based Clustering

These help us to find clusters in our data following different approaches. The algorithms which are popularly used are the k-means algorithm, fuzzy C means method, hierarchical clustering and others alike, which are all used widely for achieving similar goals [5]. In this paper. the k-means clustering algorithm and hierarchical agglomerative clustering were implemented which are partition and hierarchical algorithms for clustering, respectively. According to the k-means algorithm, if we have a dataset having n objects, we can partition it into k different partitions (or clusters) such that k $\leq$ n. We need to make sure that each data point gets associated with a cluster [6]. The optimal value of k can be determined by the Davies–Bouldin index.

According to this index, the most suitable value of k is decided by the smallest Davies–Bouldin index which tells about the closeness of values within the cluster, i.e. lesser the DB index, more optimal the value of k to be selected for further analysis [7].

The k-means algorithm generates clusters based on the following pseudocode:

Procedure k-means
  Set $a_1...a_k$ be distinct randomly   selected
  inputs from x_i ... x_k
    repeat
    for i=1...n do

  $y_{ij} = 1$ *if* $j = argmin_j \left\| xi(j) - cj \right\|^2$
  $y_{ij} = 0$ *otherwise*
    end for
    for j=1...k do
  $n_j = \sum_{i=1}^{n} yij$
  $a_j = \dfrac{1}{N} \sum_{i=1}^{n} yij\ xi$
    end for
    until convergence
  end procedure
  Return a_1 ... a_k

The following expression is used during analysis of k means:

$$J = \sum_{j=1}^{k} \sum_{i=1}^{n} \left\| xi(j) - cj \right\|^2$$

J = Objective function,
k = number of clusters,
n = number of cases,
c_j = centroid for cluster j,
x = case number

The hierarchical agglomerative clustering algorithm was also applied to the dataset to compare and contrast the results obtained with the k-means algorithm. This algorithm uses a bottom-up clusterization approach, which leads to the creation of subclusters and repeatedly creates more of these. Clusters are generated using the following steps:

(1) Each object is assigned to a separate cluster	(4) The clusters with the shortest distance between them are selected and removed from the matrix
(2) Respective distances between various clusters formed is evaluated	(5) These are then merged into a single cluster
(3) A distance matrix is then created using the values obtained	(6) The above process is repeated till the matrix is left with a single element and further comparisons are not possible

2.3 Davies–Bouldin Index

The k-means algorithm can be implemented for varying degrees of clusters expressed by k. The optimal value of K can be found by considering the lowest Davies–Bouldin Index which can be found using the expression shown hereunder:

$$DB = \frac{1}{N} \sum_{i=1}^{N} Di$$

Here, the count of clusters is shown by N, Di is the cluster distance ration such that average distance between each point and its corresponding centroid in that cluster are considered, for all the clusters in the dataset.

2.4 Pearson's Correlation

The r value can be found using the Pearson's correlation formula: $r = \dfrac{\sum XY - \frac{\sum X \sum Y}{N}}{\left\{ \sum X^2 - \sqrt{\sum Y^2 - \frac{(\sum Y)^2}{N}} \right\}^{1/2}}$	Correlation can be divided into four categories: $0 < r < 0.25$: Poor $0.25 < r < 0.50$: Fair $0.50 < r < 0.75$: Good $r > 0.75$: Excellent where r is the Pearson's correlation coefficient.

3 Methodology

(1) **Scope of Study**: The present work is an in-depth district-level exploratory analysis where information has been gathered applying the 'Secondary data collection' approach. The data pertained to all the 38 districts of Bihar. Twenty-seven potential determinants of malnutrition of economic, health, medical, social and demographic categories were shortlisted and studied. Correlations were studied for all variables and the significant ones were analyzed in depth.

(2) **Period of Study**: The present study was undertaken in late 2016, and results were analyzed and interpreted in 2017.

(3) **Data Sources**: All requisite variables were not available at one place. Relevant social, medical and malnutrition-related data were accessed primarily from NFHS4. Demographic, social, medical and economic data were obtained from Indian Census, RSOC and Economic Surveys.

(4) **Data Extraction**: The data with reference to Bihar was analyzed and the pertinent variables were listed in an Excel sheet and further imported into the RapidMiner Studio.

(5) **Statistical Analysis and Interpretation**: Data was analyzed using RapidMiner Studio-7.2.003(rev: a8c41d-WIN64). This is an open-source tool which provides an environment for data mining, business analytics and other applications alike. An in-depth analysis was undertaken using clustering algorithms, correlation, regression, plotting, etc., to understand critical determinants of malnutrition.

3.1 Steps Involved

(1) **Obtaining the Data**: The primary source for social and medical data was NFHS4 of 2017, Government of India (GoI). The demographic data was derived from Census of India, 2011 and the economic data from Economic Survey of GoI [3].

(2) **Data Cleaning**: Data was cleaned and prepared before analysis. The few missing entries were replaced with respective averages. Columns with no significance to the study were not considered. All values were brought to percentage form for uniformity and ease of analysis. Certain data which are traditionally depicted as 'per 1000' (e.g. Sex Ratio) were also converted into percent for the sake of uniformity in the analysis.

(3) **Importing Data**: Data was imported in .xlsx format (Excel file) and saved in the Local Repository of the RapidMiner Studio, ready for analysis.

(4) **K-means Clustering Algorithm**: To proceed with the analysis, the data is divided into clusters by using the k-means clustering algorithm.

(5) **Hierarchical Agglomerative Algorithm**: The same data was run through the hierarchical agglomerative algorithm to compare the results with that of k-means, and selecting the most suitable approach.

(6) **Davies–Bouldin Index**: The k-means algorithm can be implemented for different counts of clusters denoted by k. The optimal value of k was found by considering the smallest Davies–Bouldin index value obtained after calculating it for different k values. The optimal k value was then selected.

(7) **Correlation Analysis**: Correlation is found out by using the Pearson's correlation coefficient (embedded in the software) to determine how strongly the values are correlated with dependent variables (malnutrition) [8].

A flowchart has been shown alongside which shows the steps involved in the process (Fig. 1).

3.2 Experimental Setup

The dependent variables for the study were Malnutrition indicators, i.e. Stunting, Wasting, Severe Wasting and Underweight. These were analyzed against the independent variables, i.e. Demographic, Education, Mother/Child Health and

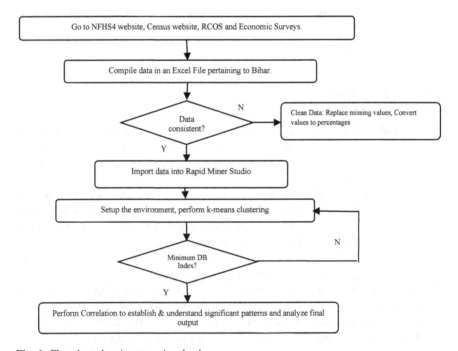

Fig. 1 Flowchart showing steps involved

Table 1 Attributes of dataset (The determinants of malnutrition have been highlighted in bold, and we try to find out how the other independent factors—shown in white—affect them)

Parameter (%)	Definition	Parameter (%)	Definition
Electricity	Percentage of villages with electricity	Exclusive breastfed	Infants fed with only breast milk till 6 months of age
Water source	Percentage of villages with clean water source	Full ANC visits	Women who visited doctor 4 times for check-up and took TT, iron tablets during pregnancy
Sanitation	Percentage of villages with toilets at home	Full immunization	Children receiving full course of vaccination
Clean fuel	Percentage of villages with clean fuel	Women BMI < 18.5%	Women with BMI < 18.5, (malnourished)
Female literacy	Women who could read/write(Census)	Women anaemic	Women with Haemoglobin < 10 g%
Anaemia children	Children with haemoglobin < 10 g%	Women married before 18%	Women aged 19–45 years of age, married before age of 18 years
Women 10th class	Women who had passed Class 10th or more	Adequate diet	Infants getting adequate diet
Colostrum	Newborns given the first breast milk	Sex Ratio	Number of women per 1000 men in the district
Underweight	**Children falling below standards for weight for age criteria**	**Severely underweight**	**Children falling below standards for weight for age criteria (<3SD)**
Wasted	**Children falling below standards for weight for height (<2SD)**	**Severely wasted**	**Children falling below standards for weight for height criteria (<3SD)**
Stunted	**Children falling below standards for height for age (<2SD)**	**Severely stunted**	**Children falling below standards for height for age criteria(<3SD)**

Health Services-related factors. The important parameters have been described in Table 1.

3.3 Model Structure

RapidMiner Studio was used for analyzing the data, running clustering and correlation algorithms and to generate results. The Excel data sheet was imported and saved in the local repository. It was then passed through a replace missing value attribute, where the data fields with no values were fixed by replacing them with average values.

Table 2 Optimal Davies–Bouldin index (shown in bold)

K = 3	K = 4	K = 5	K = 6	K = 10
DB	DB	DB	DB	DB
Index = −0.303	Index = −0.399	Index = −0.457	Index = −0.350	Index = −0.387

This data was then plugged into a nominal to numerical filter. This was carried out as the dataset comprised of alphanumeric and numeric entries and the algorithm was not able to work on such datasets.

After cleaning and preparing the data, the k-means clustering algorithm was applied to the dataset using the k-means clustering operator. The results generated were fed into a cluster distance performance operator to analyze the cluster with respect to centroid distance, Davies–Bouldin index, etc., to get a clearer understanding of the clustering carried out by the operators [9].

The same dataset was passed through a hierarchical agglomerative algorithm. The flattening of clusters was used to limit the number of clusters to the most appropriate value and a complete link mode was implemented, subsequently. To determine the optimal value of k, the clustering algorithm was run for various different values of k. Davies–Bouldin index was noted afterwards.

The optimal value of k was selected for the point at which the Davies–Bouldin index was minimum. As seen in Table 2, the optimal k value was 3 and using this value further results were obtained and analyzed.

4 Result and Analysis

4.1 Clusterization

K-means clustering was implemented on the dataset with k = 3 giving optimal Davies–Bouldin index. The hierarchical agglomerative algorithm was also used, with the flattening value set at 3 for most accurate results. It was observed that this approach yielded the same results as the k-means algorithm. Hence, we could conclude that the clusterization was carried out in an accurate manner, and the results would be closest to the truth. However, in this paper, the k-means algorithm was used for further analysis because it has a few advantages over the other, which include more efficient performance, reduced time for execution, consistent performance results even when dataset increases, which are not seen in the hierarchical agglomerative algorithm [10].

Since the dataset contains a multitude of variables, with widely varying figures over a large magnitude, the conventional analysis of this dataset as a whole, may give fallacious results. For example, certain extreme values (very low or very high) which are present as outliers do not give the correct interpretation of analysis. It is, therefore, vital to segregate the dataset into clusters with relatively similar attributes, for more meaningful analysis [11]. RapidMiner was used to make clusters,

Table 3 Clusters and corresponding number of objects

Cluster number	Objects: (Districts)
Cluster 1	1 District: (Patna)
Cluster 2	5 Districts: (Begusarai, Bhagalpur, Munger, Muzzafarpur, Rohtas)
Cluster 3	30 Districts: (Araria, Arwal, Aurangabad, Banka, Bhojpur, Buxer, Darbhanga, Gaya, Gopalganj, Jamui, Jehanabad, Kaimur, Kathiar, Khagaria, Kishanganj, Lahisarai, Madhenpura, Madhubani, Nalanda, Nawada Pash, Champaran, Purb, Champaran, Purnia, Saharsa, Samastipur, Saran, Sheikhpura, Sheohar, Siwan, Sitamarhi)
	Total = 36 districts

applying k-means algorithm and the hierarchal agglomerative algorithm. The clusters obtained in this dataset are as shown in Table 3.

To observe the relevance of results obtained after clusterization, a baseline analysis was carried out on original data prior to clusterization as well. The results have been tabulated in Table 4.

4.2 Correlations

Correlations were worked out between the dependent variables (i.e. the malnutrition indicators, namely stunting, wasting, severe wasting and underweight with the independent variables namely demographic, education, mother/child health and health services-related factors. The important correlations are depicted in tables below.

The results of statistical analysis, using RapidMiner Studio are summarized in the subsequent tables. Table 4 depicts results before clusterization, from the original data involving all 36 districts of Bihar

The data is segregated by RapidMiner Studio into clusters with similar attributes for more efficient analysis with the objective of finding meaningful trends in the dataset, and the results are depicted in Tables 5 and 6.

Cluster 1: District Patna, the most urban, economically and socially developed district of Bihar was found to be an outlier with significantly differing parameters as compared to other districts. Hence, it was segregated as just one object in that cluster, so no correlations were possible and no tables could be drawn for it and was, therefore, could not be analyzed further.

Cluster 2: This cluster contained 5 districts. Attributes are summarized in Table 5.

Cluster 3: This cluster contained 30 districts/objects. Attributes are summarized in Table 6.

Table 4 Correlations before clusterization (36 districts): different variables and malnutrition

Capability deprivation indicator	Stunting (r value)	Wasting (r value)	Severe wasting (r value)	Underweight (r value)
Demographic				
• Sex ratio	−0.34	−0.20	−0.10	−0.37
• Women married < 18 years	0.48	0.06	−0.04	0.32
Education				
• Female literacy	−0.31	0.17	0.15	−0.12
• Women studied till 10th	−0.28	−0.23	0.22	−0.09
Mother's health				
• Malnourishment (BMI < 18.5)	−0.49	0.21	0.90	0.51
• Anaemia	0.09	0.24	0.12	0.23
Child health				
• Episode of diarrhoea	−0.14	−0.53	0.42	−0.44
• Episode of acute respiratory infections	−0.31	−0.37	−0.35	−0.41
• Anaemia	0.07	−0.11	0.08	0.01
Availability of health services				
• AWW	0.15	−0.28	−0.20	−0.07
• ASHA	0.17	−0.16	−0.03	0.08
Utilization of health services				
• ANC	−0.41	0.11	0.13	−0.25
• 4 ANC visits	0.26	0.25	0.31	0.21
• Immunization				
Child feeding practices				
• Breastfed in 1 h	0.33	0.16	0.034	0.30
• Exclusive breastfeeding	−0.30	−0.28	−0.17	−0.39
• Colostrum fed	0.07	0.14	−0.01	0.23
• Adequate diet to infant	0.03	−0.15	−0.17	0.01

5   Discussion

If the entire data collected is analyzed as a whole, as one dataset containing many variables, with numbers and figures varying over large magnitude, it may end up with erroneous results. For example, certain extreme values (very low or very high) which are present as outliers do not give the correct interpretation of analysis. It is, therefore, very important to segregate the dataset into clusters with relatively similar attributes, for reaching closer to truth through meaningful analysis.

Inter-cluster Analysis

The data analysis has been oriented to appreciate this effect. It was, therefore, decided to compare the correlations obtained in the original information set and the three clusters generated, based on common attributes. This inter-cluster analysis

Table 5 Primary: Basic malnutrition outcome/dependent variables

Capability deprivation indicator	Stunting (r value)	Wasting (r value)	Severe wasting (r value)	Underweight (r value)
Demographic				
• Sex ratio	−0.54	−0.55	−0.11	−0.86
Economic				
• Per capita income	−0.55	−0.30	−0.3	−0.13
• Ruralization	0.53	−0.64	0.02	0.04
Mother's health				
• Anaemia	−0.19	0.74	0.24	0.30
Child health				
• Acute respiratory infections	0.35	−0.59	−0.52	0.05
• Anaemia	−0.36	0.85	0.83	−0.37
Availability of health services				
• AWW	−0.32	−0.26	0.06	−0.75
• ASHA	0.25	−0.74	−0.13	0.08
Utilization of health services				
• ANC	−0.34	0.08	0.25	−0.15
Child feeding practices				
• Breastfed in 1 h	−0.30	0.05	−0.26	−0.39
• Colostrum fed	0.89	−0.43	−0.30	0.82
• Adequate diet to infant	−0.30	0.26	−0.31	−0.16
Health facilities				
• Water	0.18	−0.53	0.23	−0.26
• Sanitation	−0.90	0.31	−0.15	−0.64

helps the researcher to get a more realistic effect of different independent variables on the malnutrition indicators.

The effect of inter-cluster analysis is observed that for many attributes there are wide variations in the correlations obtained for the original (un-clustered) dataset and the two other clusters. For example, the correlation between sanitation and severe wasting varies starkly for un-clustered data (minimal at r = −0.19) and for clusters 2 and 3 at r = −0.90 and r = −0.05, respectively. This indicates that, had we been interpreting this data merely using the traditional system of analyzing the un-clustered original data, we would have found poor correlation between sanitation and malnutrition as the Pearson's correlation coefficient is found to only at −0.19. This indicates that there is a negligible association between sanitation and malnutrition in children.

But when the data is segregated into three clusters, the results are starkly different. The first cluster of one district (Patna), which had extreme values has been left out of analysis as it will vitiate the results and give fallacious interpretations. The other two clusters have been defined by the software using certain common

Table 6 Basic malnutrition outcome (Cluster 3)

Capability deprivation indicator	Stunting (r value)	Wasting (r value)	Severe wasting (r value)	Underweight (r value)
Demographic				
• Sex ratio	−0.32	−0.32	−0.21	0.28
• Female literacy	−0.22	0.14	0.17	−0.17
• Women 10th pass	−0.22	0.10	0.18	−0.13
• Woman married before 18	0.54	0.12	0.01	0.15
Economic				
• Per capita income	−0.28	−0.01	−0.04	−0.06
• Below poverty line	0.09	0.35	0.28	−0.32
Mother's health				
• Anaemia	0.09	0.26	0.17	−0.16
• Diarrhoea	−0.24	−0.54	−0.41	0.30
• Women BMI < 18.5	0.46	0.36	0.17	−0.10
Child health				
• Anaemia	−0.01	−0.04	0.21	0.12
Availability of health services				
• AWW	0.16	−0.28	−0.24	0.20
• ASHA	0.02	0.03	0.20	−0.24
Utilization of health services				
• ANC	−0.48	−0.33	−0.23	−0.01
• Four ANC Visits	−0.42	−0.09	−0.04	−0.04
Child feeding practices				
• Exclusive breastfeeding	−0.44	−0.28	−0.15	−0.04
Health facilities				
• Water	0.07	−0.45	−0.27	0.14
• Clean fuel	−0.32	−0.22	−0.20	0.20

attributes. Thus the remaining 35 districts (excluding Patna) have been segregated into two clusters of 5 and 30 districts. Here the results are totally different. The second cluster of five districts gives a very good Pearson's correlation coefficient of r = −0.90. This indicates a very strong association between sanitation and malnutrition.

The third cluster, however, shows a result with correlation at r = −0.05. This indicates a negligible correlation for the same—between sanitation and malnutrition.

These varying correlation values have major implications for the prevention and control of malnutrition. In case if the correlations are found to be 'strong', it indicates that measures are required to be taken against that independent variable (sanitation), and if there is poor/no correlation then the relative importance of that determinant is negligible and it does not really need a modification, in order to control malnutrition. To illustrate the present example, there was a minimal correlation seen between sanitation and malnutrition when the entire (un-clustered) data was analyzed (r = −0.19). This indicates that since sanitation has minimal effect on malnutrition, no measures are required to be taken to improve malnutrition in the state of Bihar.

However, when Cluster 2 was analyzed independently, it indicated a correlation coefficient of −0.90, which depicts a very strong association, thus highlighting the undisputed importance of sanitation in the improvement of malnutrition in this cluster. Hence it is imperative that all measures must be taken to improve sanitation in this cluster of districts, in order to improve malnutrition.

On the other hand in Cluster 3, where the r value was found to be minimal, probably that much emphasis on sanitation is not needed.

It is inferred from the above discussion that clusterization not only orients the data analysis in the correct direction, but also helps in the most prudent data interpretation, as it is seen in the instant case that the correct data interpretation would lead to correct policymaking, for malnutrition control (Fig. 2).

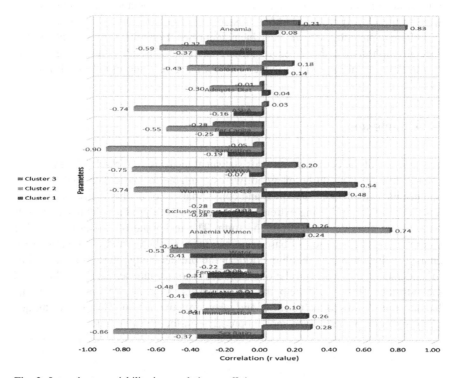

Fig. 2 Inter-cluster variability in correlation coefficients

Table 7 Parameters to target for curbing malnutrition in Bihar

Cluster with districts	Parameters to be targeted
Cluster 1 (Patna)	Patna, the most urban, economically and socially developed district of Bihar was found to be an outlier with no specific parameters to be targeted
Cluster 2 (Begusarai, Bhagalpur, Munger, Muzzafarpur, Rohtas)	Improving sanitation and water source, adequate diet for infant, acute respiratory problems, maintaining a balance of sex ratio, preventing anaemia in women and children, increasing Anganwadi workers' presence
Cluster 3 (Araria, Arwal, Aurangabad, Banka, Bhojpur, Buxer, Darbhanga, Gaya, Gopalganj, Jamui, Jehanabad, Kaimur, Kathiar, Khagaria, Kishanganj Lahisarai, Madhenpura, Madhubani, Nalanda, Nawada Pash, Champaran, Purb, Champaran, Purnia, Saharsa, Samastipur, Saran, Shikhpura, Sheohar, Siwan, Sitamarhi)	Improving clean fuel availability, water facilities, breastfeeding to be continued for at least the first 6 months of infant's birth for better health, curbing diarrhoea, providing adequate diet for women and awareness should be spread for women to get married only after 18

Table 7 mentions the clusters, the districts in them and the parameters to target for curbing malnutrition in Bihar. This was the result obtained after the clusters were formed, dividing Bihar into 3 clusters based on similarity in attributes.

Different parameters have thus been identified from the analysis above, and special emphasis can be given to these specifically instead of targeting all and giving them equal importance, which has not been very successful up till now.

6   Conclusion

The purpose of data collection and its statistical analysis is to reach (closest to) the truth. While there might be many softwares available for basic data analysis, each one has its own attributes and pros and cons. Analysis of the big dataset as a whole has the limitation of having extremes in the data, which produces fallacious results on analysis, thus deviating from the truth. The technique of clustering of a big dataset into smaller groups based on common attributes and their subsequent analysis helps overcome this limitation. In the present study, clustering enabled us to understand the real correlations between various independent variables and malnutrition. This would help the policymakers' shortlist the most appropriate factors contributing to malnutrition and take corrective action. Thus this technique of data mining is unique and most appropriate in the given setting.

References

1. Gupta A, Gupta SK, Baridylne n. Integrated Child Development Scheme (ICDS): A journey of 37 years. Indian J Community Health. Vol 25 (1) (2013)
2. Report of the Comptroller Auditor General on Performance audit of ICDS. Report No. 22. New Delhi (2013)
3. National Family Health Survey 4, Government of India, http://rchiips.org/nfhs/nfhs4.shtml
4. Anwesha Mal, Prof (Dr)Bebo White "Analysis and Clustering of PingER Network Data" IEEE Xplore Digital Library, https://doi.org/10.1109/confluence.2016.7508127 (2016)
5. Wagstaff, Kiri, et al. "Constrained k-means clustering with background knowledge." ICML. Vol. 1. (2001)
6. Jyoti Yadav, Monika Sharma, "A Review of K-means Algorithm", International Journal of Engineering Trends and Technology, Volume 4 (2013)
7. Jian Hua Yeh, Fei Jie Joung, Jia Chi Lin, "CDV Index: A Validity Index for Better Clustering Quality Measurement", Journal of Computer and Communications, 2014, 2, 163–171 (2014)
8. Chandan JS. Statistics for business and economics. Vikas Publishing House Pvt Ltd. New Delhi. (2003)
9. McCallum, Andrew, Kamal Nigam, and Lyle H. Ungar. "Efficient clustering of high-dimensional data sets with application to reference matching." Proceedings of the sixth ACM SIGKDD international conference on Knowledge discovery and data mining. ACM, (2000)
10. Manpreet Kaur, Usvir Kaur, "Comparison between K-mean and Hierarchical Algorithm Using Query Redirection", Volume 3, Issue 7, July (2013)
11. Ngai, Eric WT, Li Xiu, and Dorothy CK Chau. "Application of data mining techniques in customer relationship management: A literature review and classification." Expert systems with applications 36.2 2592–2602 (2009)

Traversal-Based Ordering of Arc–Node Topological Dataset of Sewer Network

Ankita Gupta, Vikram Singh and Gaurav Raj

Abstract Stream ordering and analysis of tree networks are necessary for understanding graph data structures. The idea of stream ordering finds its base in hydrology and open networks. Large dataset representations of such networks are difficult to process for analysis. In this paper, the stream ordering concept has been extended to a general class of structures characterized by open network semantics. A new algorithm has been proposed for analyzing tree networks and hierarchal subdivision of network segments through traversal, with minimal and generic computation over Arc–Node topological data sets. The algorithm has been implemented as a case study for understanding the complexity and calculation of manhole invert levels of sewer network of wastewater utility in the capital city of Delhi.

Keywords Arc–node · Algorithm · GIS · Stream ordering
Traversal

1 Introduction

GIS literature is vast due to the broad variety of domains that utilize geographic data. This data is collected through primary and secondary geographic data capture techniques [1]. Arc–Node topology is a common model for analyzing geospatial data for a system of networks. The captured data is voluminous and unintelligible without mapping and analytical processing.

A. Gupta (✉) · G. Raj
Amity School of Engineering and Technology, Noida, India
e-mail: ankita.gupta9@student.amity.edu

G. Raj
e-mail: graj@amity.edu

V. Singh
Indian Institute of Technology, Delhi, India
e-mail: vs_djb@yahoo.com

© Springer Nature Singapore Pte Ltd. 2018
V. Bhateja et al. (eds.), *Proceedings of the Second International Conference on Computational Intelligence and Informatics*, Advances in Intelligent Systems and Computing 712, https://doi.org/10.1007/978-981-10-8228-3_66

GIS applications are quite large; apparently, most work is done in the field of river basin hydrology and natural hydrology. Stream ordering approaches are used to build, analyze and understand the complexity of such networks by stratification into hierarchy of classes or orders [2]. Various strategies of ordering drainage systems like Horton [3], Scheideggar, Shreve (1967) [4], and Strahler [5] have been devised for this classification [6]. This paper deals with an approach inclined towards developing an algorithm and analytical understanding of data sets dealing with topographical networks by extending the principles of natural hydrology to analyze general class of networks such as stormwater systems, wastewater networks, social networks, L-Systems. This generalized strategy may be applied for a critical understanding of complexity and fault detection in any data set pertaining to similar networks.

A case study data set is used to illustrate the network analysis using the developed algorithm. A small abstract dataset pertaining to wastewater network from the city of New Delhi is collected through primary geographic data capture techniques and is modeled using ArcGIS software. The algorithm helps define to a corollary between the hydrology streaming principles and manmade wastewater networks. Computation of node-based values (e.g., invert levels of manholes) becomes feasible through digital calculations and traversal methods.

2   Literature Review

Various implementations of stream ordering have been devised. Some common methods include breadth-first search of tree structures and naming streams in post order (implemented in QGIS, ArcGIS software [7]). Other methods include pruning of networks to the parent nodes and assigning orders at each level of pruning. Another approach devised by Gido Langen [2] that uses structured query language for ordering prerequisites upstream/downstream data, thus limiting the application to hydrology domain. While these approaches are suitable for small networks, highly branched and sophisticated networks can reduce the efficiency of computation in terms of space and time complexity [8].

Algorithms that store tree structures using linked lists/adjacency matrices require extrinsic data structures for computation which is unfeasible for large networks [9]. Spiral traversal-based system [10] implemented by Mohan P. Pradhan et al. focuses on raster scanning of skeleton structure of the network. Thus arc–node topology generalize the data for varied applications, hence a generalized algorithm for complexity analysis can have varied uses too.

3 Methodology

3.1 Prerequisites

Any network data serving as input data for this following algorithm implementation should be fully structured and modeled in terms of topology. Each dataset entry should be uniquely identified with Arc_ID, from_node, to_node along with other related attributes. The network can be divergent whereas circular networks are not permissible.

3.2 Dataset

Dataset must be composed of Arc Table (attributes: Arc_ID, from_node, to_node) and Node Table (Node_Number) respectively. Dataset used here consists of wastewater network comprising of pipelines and manholes in Excel format. Every pipeline is denoted by an Arc and a unique Arc_ID, while every manhole is represented by a node identified by unique Node Number that connects one arc to another (Fig. 1).

Other attributes corresponding to pipeline data set are diameter of every pipeline (in millimeters) and length of every pipeline (in meters). In manhole dataset (Node_Table) ground level, the manhole (calculated from mean sea level in meters) is taken as one of the node attributes. Invert level of manholes will be computed using the algorithm. The network is highly branched and spread over the entire geographical area. The dataset can be manually collected and adjusted as per the topological requirements. Here the data is exported from ArcGIS tool under the prescribed format.

3.3 Data Structure Employed

The algorithm can be executed on Excel file itself with each cell serving a numerical value for the computation. Apart from this, the entire data can be stored in arrays for faster computation. No extrinsic data structures are used for this method. The sample network is represented in two tables: **Arc** and **Node** (Table 1).

The stream order number is stored with incremental flag values of traversal for different orders of the stream.

ARC_ID	From_Node	To_Node		Node_Number

Fig. 1 Arc table and node table attributes respectively

Table 1 **a** Arc table and **b** node table of dataset respectively

a

Arc_ID	From_Node	To_Node	Length	Diameter
1	1	2	14.75	150
2	3	4	16.72	150
3	5	3	15.11	150
4	2	6	18.92	150
5	7	8	14.78	230
6	9	10	15.11	230
7	11	9	17.56	150
8	6	12	18.0	150
9	13	7	15.76	230
10	14	5	15.90	230
11	15	11	16.0	150
12	12	15	17.01	150
13	16	16	18.0	150
14	17	19	14.78	230
15	18	20	15.25	230

b

Node number	Node count	Ground level	Invert level
1	1	217	0.0
2	1	217	0.0
3	1	217	0.0
4	2	217	0.0
5	2	218	0.0
6	2	217	0.0
7	1	218	0.0
8	1	217	0.0
10	2	217	0.0
11	2	217	0.0
12	2	218	0.0

3.4 Conceptual Modeling

This algorithm uses iterative cycles of traversal from most branched endings of network toward the parent source node. A parent source node is given as input to the algorithm along with the data set. Each traversal cycle initiates from the dendritic end of the network and traces the upward path of stream to the closest encountered junction in the network. The path so traversed is classified as lowest order of stream. After traversal, the traversed paths can be neglected from the network and new dendritic ends of the network are exposed for the next iteration of procedure. With every iteration, the ordering of the traversed stream is incremented as per the stream ordering system employed. The process terminates when the source parent node is encountered.

3.5 Automation Rules

The Arc and Node relationship defines the entire network and the location of particular Arc and it semantics in respect to other segments depends on the location of Node's geographical coordinates. Hence, these semantics of Arc's can be translated onto the nodes that connect these arcs through an index defined as Frequency/Count.

Frequency

The frequency of a particular node can be defined as the number of arcs that connect to the given node. Theoretically, it can be calculated by counting the segments that associate with a node by either originating or terminating at a particular node.

Any node can serve either of the two profiles to the network:

End Node—End nodes initiate an arc but no arc terminates at these nodes, in other words, these are free ends of the network.

Junction Node—Junction nodes initiate a higher order arc as well as terminates a lower order arc, in other words, these nodes are the bounded nodes of the network.

The frequency of each node can be associated with this profile by the following rules.

Rule 1—For any iteration i, every corresponding End Node has frequency 1 and for any iteration, the corresponding Junction node has frequency > 1.

If N = set of all nodes of node table

Ni = node with frequency i

$$\text{for iteration i, } \forall \text{ End Node} = n_1; \ \forall \, n \, \varepsilon \, N$$

$$\text{for iteration i, } \forall \text{ Junction Node} = n_k; \ k > 1, \forall \, n \, \varepsilon \, N$$

The frequency of a node can be experimentally calculated by finding the number of occurrences of the node under from-node AND to-node columns. A separate entry for the frequency is appended in the Node Table.

Rule 2—Every path defined starting from node with frequency 1 and terminating at frequency > 2 forms a traversal stream.

Rule 2 is implemented iteratively where all the paths satisfying the given conditions are traversed in a particular traversal. Hence, all these traversed paths have same stream orders greater than the orders assigned in previous iterations. Also, this traversal strategy allows calculation of node attributes that are cumulative and dependent on previous node values (e.g., invert levels of a node in this case).

Fig. 2 Frequency of end node

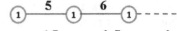

Frequency =1 Frequency =2 Frequency =2

Fig. 3 Frequency of junction node

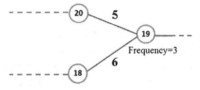

Rule 3—After every traversal, frequency counts of visited nodes is decremented to 0; while the terminating nodes are decremented by 1.

$$\text{for iteration i, } N_k \rightarrow N_0; \; \forall \, (N_k \, \varepsilon \, N) \cap (i < \, = 2)$$
$$\text{for iteration i, } N_k \rightarrow N_{k-1}; \forall \, (N_k \, \varepsilon \, N) \cap (i > 2)$$

This rule helps to eliminate the traversed network from consideration in further traversals (Figs. 2, 3 and 4).

Rule 4—The number of iterations gives the order of network and the number of Iterations terminates when a traversal stream encounters the source node.

$$\text{Order of Network} = \text{number of iterations} = \text{i}$$

This rule terminates the process of iterations and is also useful in the determination of faulty networks and hanging structures within the network.

Identification of Faulty Networks and Loops

Certain parts of the network can be disjoint from the majority part in which case a traversal stream that initiates from the node with frequency 1 again encounters a node with frequency 1. Such a structure can be identified with necessary highlighting in the network. The algorithm does not traverse loops, hence loops remain part of un-traversed network at the end of execution of the algorithm.

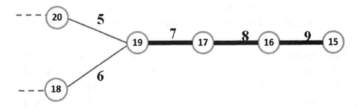

Fig. 4 A traversal stream from node 16 to node 19

4 Algorithm

The algorithm reads through to-node and from-node data values and the corresponding frequency values from the node table to traverse the entire network.

A node N with frequency i is denoted as Ni. Ns is the parent source node of network. Index[N] denotes the frequency of node N. The initial stream order for all segments is 0. Every iteration generates a new set of Index denoted by New_Index that serves for the next iteration. The pseudocode for the algorithm is given as follows:

Order(Arc Table, Node Table, Ns)

```
i=1
Order: do
{
  do{
      Search for Node N1
      { N=N1
        New_Index[N]=0
        do {
            Search    node  N  in  Arc
Table where stream order=0
            if(N ∈to-Node)
              {e=read      corresponding
from - Node Value}
            else if(N ∈from-Node)
              {e=read  corresponding  to-
Node Value}
            Set stream order for N=i

            N=e;
            if ( Index [e] ==1 && e!=Ns )
              { Network fault;  highlight  N,e;}
            else if ( Index [e] ==2 )
              { New_Index [e] =0
            else if (Index[e] > 2)
              { break;}
            else
              { break Order; }
            }
        while ( Index [e] ==2 )
            New_Index [e] =Index [e]-1
        }
      while( Node N1 exists)
        i++;
        Index=New_Index;
}
```

The algorithm does not account for closed loops in the networks as these loops do not exist in tree structures. This traversal-cum-ordering approach focuses primarily on the classification of the network instead of ordering. Since entire network is traversed in any case, all ordering strategies (Strahler, Horton, and Shreve) can be implemented using this algorithm irrespective of logic of stream ordering.

Complexity analysis

The above algorithm is certainly more efficient in terms of space complexity than other algorithms involving breadth-first search, since it requires a declaration of adjacency matrix as input parameter having space complexity equal to $O(n^2)$ [11]. For larger datasets, this amount of space is not feasible for computation. Further, it is an enhancement over the method proposed by Gido Langen [2] as it does not require flow direction for stream ordering.

Since the input is in the form of number of nodes n, the complexity can be analyzed in terms of n. Since tree networks are minimal graphs, for n number of nodes there exists (n − 1) arcs in Arc Table. The searching algorithms employed in traversal can be linear/binary search. The complexity of algorithm can be reduced by using binary search for finding N_i item in Node table with n Nodes. Complexity of Binary search for n inputs is given by

$$Time\ complexity\ of\ Binary\ Search = O(\log_2 n). \tag{1}$$

For searching items in Arc_Table with n − 1 items, we use linear search whose complexity is given by

$$Time\ complexity\ of\ Linear\ Search = O(n - 1). \tag{2}$$

Since for traversal each node is searched and accessed once in the Arc_Table and once in the node table. The search in the Arc_Table is linear search since conditions checks all occurrences of a particular node in the table, while searching the respective node in Node_Table is implemented by binary search since node values are recorded in ascending order. For network with n nodes, the time complexity can be given by

$$Time\ complexity = O(n(\log n + (n - 1)))$$
$$= O(n \cdot \log n + n^2 - n)$$
$$= O(n^2).$$

5   Implementation

The data set comprises of about 1500 record sets in Arc Table relating to sewer network of wastewater utility of capital city of Delhi.

5.1   Problem Statement

The efficiency of municipal sewer network can be ascertained by developing hydraulic model of the network. The exercise becomes more complex in case of evaluation for as-build networks since it requires mapping the pipe layout and invert levels of the manholes.

The GIS tools are effectively utilized for the layout mappings. However, the measurements of inverts levels by field surveys are not always feasible as the opening of large numbers of manholes is a cumbersome exercise. There is a need to identify other feasible approaches to determine invert levels by using graph computation theories. The sewer networks based on gravity flow are essentially open networks comprising of tree structures. The data required for analysis and modeling of the network can easily be generated by use of network traversal.

Each pipeline segment is laid at a particular slope with the horizontal ground that depends on the diameter of the pipe segment. For instance, for a pipe segment of diameter 250 mm the ratio of depth to length is 1:190. Using these calculated depth values for each segment during traversal the invert levels of each manhole can be calculated.

5.2 Implementation Process

The algorithm here is used for implementing Strahler Ordering of the network. Implementation requires following steps (Fig. 5).

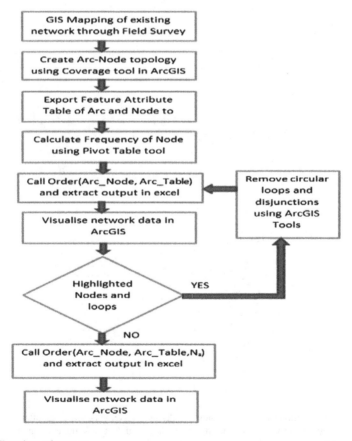

Fig. 5 Flowchart of process

5.3   *Observation*

See Figs. 6 and 7.

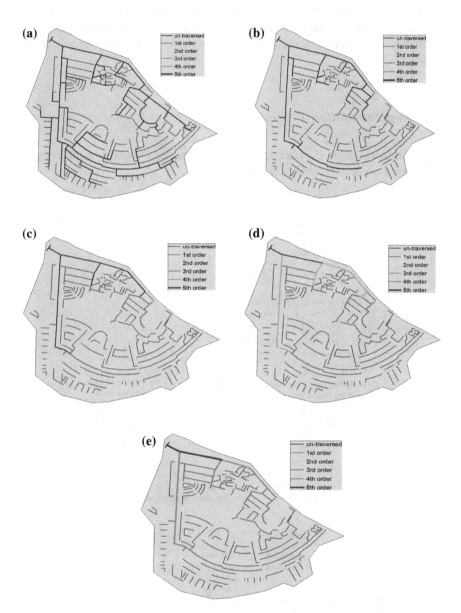

Fig. 6 a Network traversal after first-level ordering. **b** Network traversal after second-level ordering. **c** Network traversal after third-level ordering. **d** Network traversal after fourth-level ordering. **e** Network traversal after fifth-level ordering

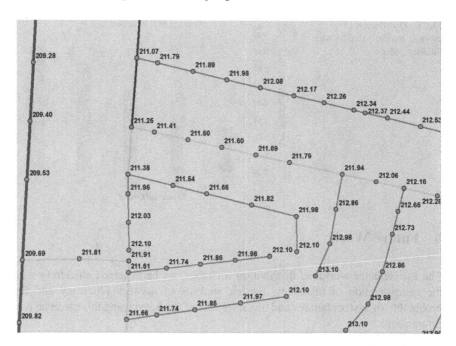

Fig. 7 Invert levels calculated for nodes. The invert levels of manholes are computed as below

5.4 Result

The network is analyzed for its complexity using the proposed algorithm and invert levels are calculated as per the relationships between length and diameters of pipeline segments. The subdivision of segments obtained is as follows (Table 2).

Figure 8 shows the distribution of manhole depths with calculated Strahler orders. The graph shows a linear increase in the depths with increasing orders.

Table 2 Distribution of segments according to order

Order	Number of segments	Average depth (in meter)
1	972	1.320
2	355	2.326
3	121	2.377
4	84	3.193
5	18	3.685

Fig. 8 Distribution of
average manhole depths with
Strahler order

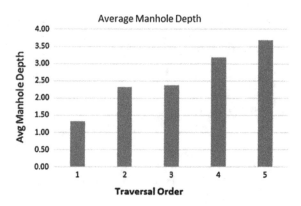

6 Future Work

The maintenance workload distribution for labor can be managed efficiently with the categorization of manholes. Further analysis of network efficiency and the probability of system failures and blockage can be predicted using this traversal as a future scope.

References

1. Michael F. Goodchild and Robert P. Haining; "GIS And Spatial Data Analysis: Converging Perspectives", Papers in Regional Science, October 2003, Volume 83, Issue 1, pp. 363–385.
2. Griffith, Jerry A. "An alternative method for automatic coding of stream order on digital stream network data." (2012).
3. Horton, R.E., "Erosional development of streams and their drainage basins: hydrophysical approach to quantitative morphology", Bulletin of the Geological Society of America 56, 275–370.
4. Shreve, R., 1967. "Infinite topologically random channel networks". Journal of Geology 75, 178–186.
5. Strahler, A. N., "Quantitative analysis of watershed geomorphology." Transactions American Geophysical Union, 38, No. 6, 913–920., 957.
6. I Yekutieli and B B Mandelbrot; "Statistical Law of Stream Numbers".
7. JarosŁaw Jasiewicz, Markus Metz; "A new GRASS GIS toolkit for Hortonian analysis of drainage networks", Computers and Geosciences, Volume 37, Issue 8, August 2011, pp. 1162–1173.
8. Alper SEN, Turkay GOKGOZ, "Clustering Approaches For Hydrographic Generalization", GIS Ostrava 2012—Surface models for geosciences.
9. Fatih Gülgen, "A stream ordering approach based on network analysis operations", Journal of Geocarto International, Volume 32, Issue 3, 2017.

10. Mohan P. Pradhan, M.K. Ghose, Yash R. Kharka; "Automatic Association of Strahler's Order and Attributes with the Drainage System", International Journal of Advanced Computer Science and Applications, Vol. 3, No. 8, 2012.

11. Gleyzer, A., M. Denisyuk, A. Rimmer, and Y. Salingar. 2004. "A fast recursive GIS algorithm for computing Strahler stream order in braided and nonbraided networks." Journal of the American Water Resources Association 40(4), 937–946.

Author Index

© Springer Nature Singapore Pte Ltd. 2018
V. Bhateja et al. (eds.), *Proceedings of the Second International Conference
on Computational Intelligence and Informatics*, Advances in Intelligent Systems
and Computing 712, https://doi.org/10.1007/978-981-10-8228-3